PHYSICS RESEARCH AND TECHNOLOGY

QUANTUM DOTS AND QUANTUM CELLULAR AUTOMATA

RECENT TRENDS AND APPLICATIONS

Physics Research and Technology

Additional books in this series can be found on Nova's website under the Series tab.

Additional e-books in this series can be found on Nova's website under the e-book tab.

Nanotechnology Science and Technology

Additional books in this series can be found on Nova's website under the Series tab.

Additional e-books in this series can be found on Nova's website under the e-book tab.

PHYSICS RESEARCH AND TECHNOLOGY

QUANTUM DOTS AND QUANTUM CELLULAR AUTOMATA

RECENT TRENDS AND APPLICATIONS

DEBASHIS DE
SITANSHU BHATTACHARAYA
AND
K. P. GHATAK
EDITORS

nova publishers
New York

For permission to use material from this book please contact us:
Telephone 631-231-7269; Fax 631-231-8175
Web Site: http://www.novapublishers.com

Library of Congress Cataloging-in-Publication Data

Quantum dots and quantum cellular automata : recent trends and applications / [edited by] Debashis De, Sitanshu Bhattacharaya and K. P. Ghatak (Department of Computer Science and Engineering, West Bengal University of Technology, Salt Lake City, Kolkata, India, and others).
pages cm
Includes bibliographical references and index.
ISBN: 978-1-62257-920-4 (hardcover)
1. Quantum dots. 2. Cellular automata. I. De, Debashis, editor of compilation. II. Ghatak, Kamakhya Prasad, editor of compilation.
QA267.5.C45Q364 2013
621.3815'2--dc23
2012034754

Published by Nova Science Publishers, Inc. † New York

This Edited book is dedicated to Professors D. Bimberg, University of Berlin, Germany, H. L. Hartnagel, Technische Universitat Darmstadt, Germany, A. K. Roy, Brunel University, U.K., H. Sharda and J. Williams, University of Western Australia, for numerous academic discussions. We are grateful to the members of our research team for confining us in the infinitely deep quantum wells of Ramanujan and Rabindranath.

CONTENTS

PREFACE

The impact of Quantum Dots (QDs) and Quantum Cellular Automata (QCA) on modern nano-science and technology is well-known since their inception and there has been considerable study in the recent literature on these two modern topics. In this edited book an attempt is made by the Nova Science Publishers to present the theoretical investigations of different aspects of QDs and QCA for researchers, post graduate students, professionals in the fields of Solid State Science, Nano-science and technology in addition to Materials Science.

In the first chapter, an attempt is made to study the Einstein's Photoemission(EP) from QDs of non-linear optical, III-V, II-IV, n-GaP, n-Ge, Te, Graphite, $PtSb_2$, zero gap, II-V, GaSb, stressed materials,Bi, IV-IV, Lead germanium telluride, Zinc and Cadmium diphosphides, Bi_2Te_3 and Antimony on the basis of respective carrier energy spectrum. We have also investigated the EP from III-V,II-VI,IV-VI and HgTe/CdTe Quantum Dot Superlattices(QDSL) with graded interfaces together with the QD effective mass superlattices of the afore mentioned materials. It has been found taking QDs of n-Cd Ge As_2, n-GaAs, CdS, GaP, Ge, Te, $PtSb_2$, HgTe, $Pb_{1-x}Ga_xTe$, GaSb, stressed InSb, Bi, PbTe, CdSb, CdP_2, ZnP_2 Bi_2Te_3 and Sb as example that the EP increases with increasing carrier degeneracy and decreasing film thickness exhibiting spikes for specified values of the said variables and also increases with increasing photo energy in a step like manner exhibiting signature of the 3D quantization of the wave vector space of the carrier in each case. The EP from HgTe/CdTe,HgTe/$Hg_{1-x}Cd_xTe$, CdS/ZnSe and PbSe/PbTe QDSL with graded interfaces and QD effective mass SLs of the said compounds increases with increasing electron concentration, shows oscillatory dependence with increasing film thickness and increases with increasing photon energy in quantum step fashion. The appearance of the humps of the respective curves is due to the redistribution of the electrons among the quantized energy levels when the quantum numbers corresponding to the highest occupied level changes from one fixed value to the others. Although the EP varies in various manners with all the variables as evident from all the curves, the rates of variations totally depend on the specific dispersion relation of the particular 3D structure. Under certain limiting conditions, the results as derived in this chapter get transformed in to well known formulas of the EP and the electron statistics and thus confirming the compatibility test. The contents of this chapter find two applications in the field of quantized structures.

The chapter 2 investigates the photoemission from QDs and QDSLs of Opto-Electronic Materials on the basis of newly formulated electron dispersion laws in the presence of microwaves within the framework of k.p formalism. It appears taking QDs of $Hg_{1-x}Cd_xTe$ and $In_{1-x}Ga_xAs_yP_{1-y}$ lattice matched to InP as examples that the photoemission from QDs of optoelectronic materials decreases with increasing film thickness exhibiting prominent trapezoidal variation for relatively large thickness bandwidth. These types of variations are the special signatures of 3D confinement in QDs of optoelectronic materials respectively in the presence of microwaves. The normalized photo-emitted current density from the QD structures increases with increasing carrier degeneracy and for relatively large values of the same variable; it exhibits quantum jumps for all types of band models when the size quantum number changes from one fixed value to another. It appears that the photoemission from $HgTe/Hg_{1-x}Cd_xTe$ and $In_xGa_{1-x}As/InP$ QDSLs with graded interfaces increases with decreasing film thickness with larger oscillations again exhibiting the quantization effect. The photoemission from the said systems increases with decreasing intensity and decreasing wavelength respectively in various manners. The content of this chapter finds two applications in the field of semiconductor science.

In chapter 3, the Seebeck coefficient under large magnetic field (S) has been investigated in QDs of heavily doped nonlinear optical, III-V, II-VI, GaP, Ge, Te, $PtSb_2$, II-V, GaSb, stressed materials, IV-VI, Lead Germanium Telluride, Zinc and Cadmium diphosphides and Bi_2Te_3 on the basis of newly formulated carrier dispersion laws respectively. We have also studied the S in heavily doped III-V,II-VI,IV-VI, HgTe/CdTe and strained layer QDSL with graded interfaces together with the effective mass superlattices of the afore mentioned materials by formulating new carrier energy spectra in each case. It has been found that the S for the said heavily doped QDs and QDSL oscillate with increasing thickness and changes with increasing electron concentration in various manners for all types of heavily doped systems with two entirely different signatures of quantization as appropriate in respective cases of the aforementioned quantized structures.

Transistor Laser is a Hetero junction Bipolar Transistor (HBT) the base of which usually incorporates a Quantum Well (QW). The stimulated recombination in the QW gives rise to laser action. It can thus work both as a transistor and a laser. The chapter 4 considers the structures in which a layer of QD is inserted in the QW (Dots-in-Well) in the base and examines theoretically if there is improvement in the performance over the usual QW structure. To put the study and comparison in proper perspectives, the usual TL structure with QW in its base (QW-TL) and the basic operation of the device are discussed first. The analytical models for terminal currents, light power output and small signal modulation characteristics for QW-TL are then presented. Similar theoretical models for TL with Dots-in-Well (QD-TL) are then developed. The model for QW-TL is tested against experimental results and good agreement obtained signifies the validity of the model. The theoretical value for threshold base current for QD-TL is found to be one order lower than the value for QW-TL. The small signal modulation bandwidth for QD-TL cannot however exceed the corresponding values for QW-TL. The analysis for QD-TL is based on excitonic models with and without excited states in the QD and a comparison is made between the results obtained from the two excitonic models. The role of separate electron-hole dynamics is also examined. The experimental values for QD-TL are not available in the recent literature.

In this chapter 7 an attempt is made to introduce the logic synthesis with tile nanostructure in QCA. Tile nanostructure supposed to be most empirical and robust in nature. Logic design with QCA tile structure is considered here. There are two possible type of QCA tile have been discussed. Depending on the population of all QCA cell, the QCA tile is classified into Fully populated and Non-Fully populated type. The studies on fully populated and non-fully populated QCA tile have been discussed in this chapter. Each type of QCA tile structures are also classified into active and passive type. The cascading effects on QCA tile structure have been represented here for functional development. The coupled majority – minority voter (CMmV) is most promising QCA tile structure which is actually used for designing n- variable symmetric functions. The Defect in QCA tile nanostructure is also discussed. There are several possible defects that may occurs during fabrication are considered here. The extra dot defect in QCA cell does not produces an error but only increases numerical complexity. The extra cell deposition and Missing cell deposition are discussed for different QCA tile structure.

The chapter 8 investigates Reversible Logic which is a promising computing paradigm with applications in low-power quantum computing and Nanotechnology. Quantum Circuit representation of the Basic Reversible Logic Gates is also described in this chapter. Different Reversible logic circuit design using the basic Reversible logic gates and their implementation using QCA is the major part of this chapter. Different fault analysis technique like Parity Preserving Technique, Majority Multiplexing Technique and the Hardware Redundancy Technique are described elaborately in this chapter.

In chapter 9 an attempt is made to study the Magnetic Quantum Cellular Automata which uses the phenomenon of magnetism. Nanometer-scale magnets are designed for digital computation in magnetic quantum-dot cellular automata (MQCA) systems. We have also studied the fabrication of MQCA on silicon wafer by electron beam lithographic technique. The clocking of MQCA using Spin Transfer Torque (STT) and corresponding storage property is also explained. The operations of MQCA devices like binary wires and majority voter operations are also briefly demonstrated .The timing analysis of Majority Voter is demonstrated by suitable simulations results using OOMMF simulator. Low power consumption by magnetic QCA using spin torque transition current fed input circuit and multilayer structures are explained. Low power MQCA voter operation using NiFe polycrystalline nanomagnets are demonstrated. The basic principal of spintronics, its application and drawbacks are also demonstrated in this chapter. Digital Logic using atomic scaling of MQCA is also explained.

In the era of advancement in Nano-technology, the devices dimension has been reduced rapidly. According to Moores' law the numbers of transistors in an integrated circuit is to be doubled in every eighteen months, hence the device complexity, power dissipation increases and other short channel effects greatly influenced on circuit performance. Therefore the reliability analysis of a nano device circuit is very important aspect in the field of Nano-technology. Due to these difficulties of nano-device performance we have discussed in chapter 10 about the probability analysis of switching activity of a Electrostatic Quantum-dot Cellular Automata (EQCA) cell. Besides, this chapter will also help to derive the Boolean difference of the nano-scale circuitry. Reliability analysis is very much needful for every nano scale device performance or operation to get a reliable or error free outcome.

Since it is very much difficult to find out the reliability of a device in nano-scale in this chapter the algebraic method is used to find the reliability using Probability Transfer Matrix. Boolean Difference and Signal Activity is introduced into the proposed work.

The success of an edited book significantly depends on the contributors of the different chapters. The job of the editors is to present it in a systematic fashion and it may be noted that the errorless first edition of any book from every point of view enjoys permanently the domain of "Impossibility Theorems" which is very true for this edited book also. Last but not the least we do hope that our humble effort will fire the desire of anyone engaged in nano-science and technology either in academics or in industries to delve deeper into these two potential topics of research.

ACKNOWLEDGMENTS

K P Ghatak is grateful to his father Late Dr. A P Ghatak for the priceless inspiration. S. Bhattacharya and D. De together with K.P.Ghatak are grateful to DST, India for sanctioning the project and the fellowship under ''SERC Fast Track Proposal of Young Scientist'' Scheme SR/FTP/ETA-37/08 and SERC/ET-0213/2011 respectively. Besides, K.P. Ghatak and D.De are grateful to UGC for sanctioning the research project having the No-F 40-469/2011(SR) under which this paper has been completed. They are grateful to Prof. P.K.Bose, Director, National Institute of Technology, Agartala, Tripura for his continuous help without which the completion of this edited book would be a mere dream. Last but not the least they wish to thank Mrs. Swati De, Mr S. Ghosh, Md.T Pir and Mr R. Basak for placing their combined effort towards the development of this book in the DO-LOOP of a computer.

Kolkata D.De
Bangalore S.Bhattacharya
Kolkata K.P. Ghatak

In: Quantum Dots and Quantum Cellular Automata
Editors: D. De, S. Bhattacharaya and K.P. Ghatak
ISBN: 978-1-62257-920-4

Chapter 1

ON THE EINSTEIN'S PHOTOEMISSION FROM QUANTUM DOTS AND QUANTUM DOT SUPER LATTICES OF NON PARABOLIC SEMICONDUCTORS: SIMPLIFIED THEORY AND RELATIVE COMPARISON

M. Mitra[1], S. Bhattacharya[2], D. De[3,4], N. Pradhan[3], P. K. Sarkar[5], S. Banerjee[6], B. Nag[1], M.Mitra[7], S. Ghosh[7], N. Paitya[8] S. Pahari [9] S. Ghatak [10], S. Mukherjee [5] S. K. Biswas[11] and K. P.Ghatak[5*]

[1]Department of Applied Physics, University of Calcutta, Kolkata, India
[2] Nanoscale Device Research Laboratory, Center for Electronics Design and Technology, Indian Institute Science, Bangalore, India
[3] Department of Computer Science and Engineering, West Bengal University of Technology, Kolkata, India
[4] School of Physics, University of Western Australia, The University of Western Australia, Crawley, Perth, Western Australia
[5] Department of Electronic Science, University of Calcutta, Kolkata, India
[6]Department of Electronics and Communication Engineering, Academy of Technology,Adisaptagram, West Bengal, India,
[7]Department of Electronics and Telecommunication Engineering, Bengal Engineering and Science University, Howrah , India
[8]Department of Applied Electronics and Instrumentation Engineering, Sikkim Manipal Institute of Technology, Majitar, Rangpo, East Sikkim, India
[9] Administration Department, Jadavpur University, Kolkata, India
[10]Department of Computer Science and Engineering, Sikkim Manipal Institute of Technology, Majitar, Rangpo, East Sikkim, India
[11] Department of Physics, University of Calcutta, Kolkata, India

ABSTRACT

In this chapter an attempt is made to study the Einstein's Photoemission(EP) from Quantum Dots(QDs) of non-linear optical, III-V, II-IV, n-GaP, n-Ge, Te, Graphite, $PtSb_2$, zero gap, II-V, GaSb, stressed materials,Bi, IV-IV, Lead germanium telluride, Zinc and Cadmium diphosphides, Bi_2Te_3 and Antimony on the basis of respective carrier energy spectrum. We have also investigated the EP from III-V,II-VI,IV-VI and HgTe/CdTe Quantum Dot Superlattices(QDSL) with graded interfaces together with the QD effective mass superlattices of the afore mentioned materials. It has been found taking QDs of n-Cd Ge As_2, n-GaAs, CdS, GaP, Ge, Te, $PtSb_2$, HgTe, $Pb_{1-x}Ga_xTe$, GaSb, stressed InSb, Bi, PbTe, CdSb, CdP_2, ZnP_2 Bi_2Te_3 and Sb as example that the EP increases with increasing carrier degeneracy and decreasing film thickness exhibiting spikes for specified values of the said variables and also increases with increasing photo energy in a step like manner exhibiting signature of the 3D quantization of the wave vector space of the carrier in each case. The EP from HgTe/CdTe,HgTe/$Hg_{1-x}Cd_xTe$, CdS/ZnSe and PbSe/PbTe QDSL with graded interfaces and QD effective mass SLs of the said compounds increases with increasing electron concentration, shows oscillatory dependence with increasing film thickness and increases with increasing photon energy in quantum step fashion. The appearance of the humps of the respective curves is due to the redistribution of the electrons among the quantized energy levels when the quantum numbers corresponding to the highest occupied level changes from one fixed value to the others. Although the EP varies in various manners with all the variables as evident from all the curves, the rates of variations totally depend on the specific dispersion relation of the particular 3D structure. Under certain limiting conditions, the results as derived in this chapter get transformed in to well known formulas of the EP and the electron statistics and thus confirming the compatibility test. The contents of this chapter find two applications in the field of quantized structures.

1.1. INTRODUCTION

With the advent of fine line lithography [1], molecular beam epitaxy [2], organometallic vapor-phase epitaxy [3] and other experimental techniques, low-dimensional structures [4-6] having quantum confinement of the charge carriers in one, two, and three dimensions (such as ultrathin films (UFs), nipi structures, inversion and accumulation layers, quantum well superlattices, carbon nanotubes, quantum wires (QWs), quantum wire superlattices, quantum dots (QDs), magneto inversion and accumulation layers, quantum dot superlattices, etc.) have, in the last few years, created tremendous passion among the interdisciplinary researchers not only for the potential of these quantized structures in uncovering new phenomena in low dimensional science but also for their new diverse technological applications. As the dimension of the UFs increases from 1D to 3D, the degree of freedom of the free carriers decreases drastically and the density-of-states function changes from the Heaviside step function in UFs to the Dirac's delta function in QDs [7].

The QDs can be used for visualizing and tracking molecular processes in cells using standard fluorescence microscopy [8-10]. They display minimal photo-bleaching [11], thus allowing molecular tracking over prolonged periods and consequently, single molecule can be tracked by using optical fluorescence microscopy [12]. The salient features of quantum dot lasers [13] include low threshold currents, higher power, and great stability as compared with the conventional one and the QDs find extensive applications in nano-robotics [14], neural networks [15] and high density memory or storage media [16]. The QDs are also used in nano-photonics [17] because of their theoretically high quantum yield and have been suggested as implementations of qubits for quantum information processing [18]. The QDs also find applications in diode lasers [19], amplifiers [20], and optical sensors [21]. High-quality QDs are well suited for optical encoding [22] because of their broad excitation profiles and narrow emission spectra. The new generations of QDs have far-reaching potential for the accurate investigations of intracellular processes at the single-molecule level, high-resolution cellular imaging, long-term in vivo observation of cell trafficking, tumor targeting, and diagnostics [23]. The QD nanotechnology is one of the most promising candidates for use in solid-state quantum computation [24]. It may also be noted that the QDs are being used in single electron transistors [25], photovoltaic devices [26], photoelectrics [27], ultrafast all-optical switches and logic gates [28], organic dyes [29] and in other types of nano devices.In this chapter we shall study the EP from QDs and quantum dot super lattices of non-parabolic semiconductors on the basis of the respective dispersion relation of the carriers.

It is well known that the Einstein's photoelectric effect occupies a singular It is well known that the Einstein's photoelectric effect occupies a singular position in the whole arena of materials science and related disciplines in general together with the fact that the photoemission from the electronic materials is also a vital physical phenomenon from the viewpoint of modern optoelectronics and photoemission spectroscopy [30].

The classical equation of the photo-emitted current density is [31] $J = \left[4\pi e m^* g_v (k_B T)^2 / h^3\right] \boldsymbol{exp}\left[(h\nu - \phi)/(k_B T)\right]$, where e, m^*, g_v, k_B, T, h, $h\upsilon$ and are the electron charge, effective electron mass at the edge of the conduction band, valley degeneracy, the Boltzmann constant, temperature, the Planck constant, incident photon energy along z-axis and work function respectively. The afore-mentioned equation is valid for both the charge carriers and in this conventional form it appears that, the photoemission changes with the effective mass, temperature, work function and the incident photon energy respectively. This relation holds only under the condition of carrier non-degeneracy.

The EP has different values for different materials and varies with doping and with external fields which creates quantization of the wave-vector space of the carriers leading to various types of quantized structures. The nature of these variations has been studied in [31-63] and some of the significant features are as follow:

1. The EP from bulk materials increases with the increase in doping.
2. The EP exhibits oscillatory dependence with inverse quantizing magnetic field because of the Shubnikov de Haas (SdH) effect.
3. The EP changes significantly with the magnitude of the externally applied quantizing electric field in electronic materials.
4. The EP from quantum confined Bismuth, nonlinear optical, III-V, II-VI and IV-VI materials oscillate with nano-thickness in various manners which are totally band structure dependent.
5. The nature of variations is significantly influenced by the energy band constants of various materials having different band structures.
6. The EP has significantly different values in quantum confined semiconductor superlattices and various other quantized structures.

The Section 1.2.1 of this chapter investigates the EP in QDs of non-linear optical materials (taking n-$CdGeAs_2$ as an example) which find applications in non-linear optics and light emitting diodes [63]. The quasi-cubic model can be used to investigate the symmetric properties of both the bands at the zone center of wave vector space of the same compound [64]. Including the anisotropic crystal potential in the Hamiltonian, and special features of the non-linear optical compounds, Kildal [65] formulated the electron dispersion law under the assumptions of isotropic momentum matrix and the isotropic spin orbit splitting constant, respectively, although the anisotropies in the two aforementioned band constants are the significant physical features of the said materials [66].

In this context, it may be noted that the III-V compounds find potential applications in infrared detectors [67], quantum dot light emitting diodes [68], quantum cascade lasers [69], quantum well wires [70], optoelectronic sensors [71], high electron mobility transistors [72] etc. The III-V, ternary and quaternary materials are called the Kane type compounds since their electron energy spectra are being defined by the three-band model of Kane [73]. In Section 1.2.2, the EP from QDs of III-V materials has been studied and the simplified results for two-band model of Kane and that of wide gap materials have further been demonstrated as special cases. Besides Kane, the conduction electrons of III-V materials also obey another six different types of electron dispersion laws as given in the literature. The EP has also been investigated for all the cases for the purpose of complete presentation and relative assessment among the energy band models of III-V compounds.

The II-VI compounds are being extensively used in nanoribbons, blue green diode lasers, photosensitive thin films, infrared detectors, ultra high-speed bipolar transistors, fiber optic communications, microwave devices, photovoltaic and solar cells, semiconductor gamma-ray detector arrays, semiconductor detector gamma camera and allow for a greater density of data storage on optically addressed compact discs [74-76]. The carrier energy spectra in II-VI materials are defined by the Hopfield model [77] where the splitting of the two-spin states by the spin-orbit coupling and the crystalline field has been taken into account. The Section 1.2.3 contains the investigation of the EP in QDs of II-VI compounds, taking p-CdS as an example.

The n-Gallium Phosphide (n-GaP) is being used in quantum dots light emitting diode [78], high efficiency yellow solid state lamps, light sources and high peak current pulse for high gain tubes. The green and yellow light emitting diodes made of nitrogen-doped n-GaP possess a longer device life at high drive currents [79]. In Section 1.2.4, the EP in QDs of n-GaP is studied. The importance of Germanium is already well-known since the inception of transistor technology and in recent years, memory circuits, single photon detectors, single photon avalanche diode, ultrafast all-optical switch, THz lasers and THz spectrometers [80-81] are made of Ge. In Section 1.2.5, the EP has been studied in QDs of Ge.

Tellurium (Te) is also an elemental semiconductor which has been used as the semiconductor layer in thin-film transistors (TFT) [82]. Te also finds extensive applications in CO_2 laser detectors [83], electronic imaging, strain sensitive devices [84] and multichannel Bragg cell [85]. The Section 1.2.6 contains the investigation of EP in QDs of Tellurium. The importance of graphite is already well-known in the whole spectrum of materials science and the low dimensional graphite is used instead of carbon wire in many practical applications. Graphite intercalation compounds are often used as suitable model for investigation of low dimensional systems and in particular for investigation of phase transition in such systems [86]. In Section 1.2.7, the EP in QDs of graphite has been explored. Platinum Antimonide ($PtSb_2$) finds applications in device miniaturization, colloidal nanoparticle synthesis, sensors and detector materials and thermo-photovoltaic devices [87]. The EP in QDs of p-$PtSb_2$ has been investigated in Section 1.2.8.

Zerogap compounds are used in optical wave-guide switch or modulators that can be fabricated by using the electro-optic and thermo-optic effects for facilitating optical communications and signal processing. The gapless materials also find extensive applications in infra-red detectors and night vision cameras [88]. The section 1.2.9 contains the study of EP in QDs of the same taking p-HgTe as an example.

The II-V materials are used in photovoltaic cells constructed of single crystal materials in contact with electrolyte solutions. Cadmium selenide shows an open-circuit voltage of 0.8V and power conservation coefficients near 6 percent for 720-nm light [89]. They are also used in ultrasonic amplification [90]. The thin film transistor using cadmium selenide as the semiconductor has been developed [91-92]. In Section 1.2.10, the EP in QDs of II-V materials has been studied taking CdSb as an example. Gallium antimonide (GaSb) finds applications in the fiber optic transmission window, heterojunctions and quantum wells. A complementary hetero-junction field effect transistor (CHFET) in which the channels for the p-FET device and the n-FET device forming the complementary FET are formed from GaSb. The band gap energy of GaSb makes it suitable for low power operation [93]. In Section 1.2.11, the EP from QDs of GaSb has been studied.

It may be noted that the stressed materials are being widely investigated for strained silicon transistors, quantum cascade lasers, semiconductor strain gages, thermal detectors and strained-layer structures [94]. The EP in QDs of stressed materials (taking stressed n-InSb as an example) has been investigated in Section 1.2.12. In recent years, Bismuth (Bi) nano-lines are fabricated and Bi also finds use in array of antennas which leads to the interaction of electromagnetic waves with such Bi-nano-wires [95]. Several dispersion relations of the carriers have been proposed for Bi. Shoenberg [96] experimentally verified that the de Haas-Van Alphen and cyclotron resonance experiments supported the ellipsoidal parabolic model of Bi, although, the magnetic field dependence of many physical properties of Bi supports the two-band model [97]. The experimental investigations on the magneto-optical [98] and the ultrasonic quantum oscillations [99] support the Lax ellipsoidal non-parabolic model [97]. Kao [100], Dinger and Lawson [101] and Koch and Jensen [102] demonstrated that the Cohen model [103] is in conformity with the experimental results in a better way. Besides, the Hybrid model of bismuth, as developed by Takoka et. al. also finds use in the literature [104]. McClure and Choi [105] devised a new model of Bi and they showed that it can explain the data for a large number of magneto-oscillatory and resonance experiments. In section 1.2.13, we have formulated the EP in QDs of Bi in accordance with the aforementioned energy band models for the purpose of relative assessment.

Lead chalcogenides (PbTe, PbSe, and PbS) are IV-VI compounds whose studies over several decades have been motivated by their importance in infrared IR detectors, lasers, light-emitting devices, photovoltaics and high temperature thermoelectric [106]. PbTe, in particular, is the end compound of several ternary and quaternary compoumds [107]. It has been used not only as bulk but also as films [108], quantum wells [109], superlattices [110], nanowires [111], and colloidal and embedded nanocrystals [112]. PbTe films doped with various impurities have also been investigated [113]. These studies revealed some of the interesting features that has been observed in bulk PbTe, such as Fermi level pinning and, in the case of superconductivity [114]. In Sections 1.2.14 and 1.2.15, the EP in QDs of IV-VI materials $Pb_{1-x}Ge_xTe$ has been studied.

The diphosphides find prominent role in biochemistry where the folding and structural stabilization of many important extra-cellular peptide and protein molecules, including hormones, enzymes, growth factors, toxins, and immunoglobulin are in concerned [115]. Besides, artificial introduction of extra diphosphides into peptides or proteins can improve biological activity [116] or confer thermo-stability [117]. The asymmetric diphosphide bond formation in peptides containing a free thiol group takes place over a wide pH range in aqueous buffers and can be crucially monitored by spectro-photometric titration of the released 3-nitro-2-pyridinethiol [118-119]. In Section 1.2.16, the EP in QDs of zinc and cadmium diphosphides has been investigated.

Bismuth telluride (Bi_2Te_3) was first identified as a material for thermoelectric refrigeration in 1954 [120] and its physical properties were later improved by the addition of bismuth selenide and antimony telluride to form solid solutions [121]. The alloys of Bi_2Te_3 are very important compounds for the thermoelectric industry and have extensively been investigated in the literature [121]. In Section 1.2.17, the TPSM in QDs of Bi_2Te_3 has been considered. In recent years, antimony has emerged to be very promising since glasses made from antimony are being extensively used in near infrared spectral range for third or second order non linear processes. The chalcogenide glasses are in general associated with high non linear properties for their Infrared transmission from 0.5-1μm to 12 – 18 μm [122]. Alloys of Sb are used as ultra high frequency indicators and in thin film thermocouple [122]. In Section 1.2.18, the EP in QDs of Sb has been studied.

In this context it may be noted that the said modern fabrication techniques have also generated altogether a new dimension in the arena of quantum effect devices through the experimental realization of an important artificial structure known as semiconductor superlattice (SL) by growing two similar but different semiconducting compounds in alternate layers with finite thicknesses. The materials forming the alternate layers have the same kind of band structure but different energy gaps. The concept of SL was developed for the first time by Keldysh [123] and was successfully fabricated by Esaki and Tsu [124]. The SLs are being extensively used in thermal sensors [125], quantum cascade lasers [126], photodetectors [127], light emitting diodes [128], multiplication [129], frequency multiplication [130], photocathodes [131], thin film transistor [132], solar cells [133], infrared imaging [134], thermal imaging [135], infrared sensing [136] and also in other micro-electronic devices.

The most extensively studied III-V SL is the one consisting of alternate layers of GaAs and $Ga_{1-x}Al_xAs$ owing to the relative easiness of fabrication. The GaAs and $Ga_{1-x}Al_xAs$ layers form the quantum wells and the potential barriers respectively. The III-V SL's are attractive for the realization of high speed electronic and optoelectronic devices [137]. In addition to SLs with usual structure, other types of SLs such as II-VI [138], IV-VI [17] and HgTe/CdTe [139] SL's have also been investigated in the literature. The IV-VI SLs exhibit quite different properties as compared to the III-V SL due to the specific band structure of the constituent materials [140]. The epitaxial growth of II-VI SL is a relatively recent development and the primary motivation for studying the mentioned SLs made of materials with the large band gap is in their potential for optoelectronic operation in the blue [141]. HgTe/CdTe SL's have raised a great deal of attention since 1979, when as a promising new materials for long wavelength infrared detectors and other electro-optical applications [142]. Interest in Hg-based SL's has been further increased as new properties with potential device applications were revealed [142,143]. These features arise from the unique zero band gap material HgTe

[144] and the direct band gap semiconductor CdTe which can be described by the three band mode of Kane [145]. The combination of the aforementioned materials with specified dispersion relation makes HgTe/CdTe SL very attractive, especially because of the tailoring of the material properties for various applications by varying the energy band constants of the SLs.

We note that all the aforementioned SLs have been proposed with the assumption that the interfaces between the layers are sharply defined, of zero thickness, i.e., devoid of any interface effects. The SL potential distribution may be then considered as a one dimensional array of rectangular potential wells. The aforementioned advanced experimental techniques may produce SLs with physical interfaces between the two materials crystallographically abrupt; adjoining their interface will change at least on an atomic scale. As the potential form changes from a well (barrier) to a barrier (well), an intermediate potential region exists for the electrons. The influence of finite thickness of the interfaces on the electron dispersion law is very important, since; the electron energy spectrum governs the electron transport in SLs. In addition to it, for effective mass SLs, the electronic subbands appear continually in real space [146]. In section 1.2.19 to 1.2.26 we shall study the EP in the quantum dots of the said SLs. The section 1.3 contains the results and discussion for this chapter.

1.1. Theoretical Background

1.2.1. The EP from QDs of Nonlinear Optical Materials

The form of k.p matrix for nonlinear optical compounds can be expressed extending Bodnar [66] as

$$H = \begin{bmatrix} H_1 & H_2 \\ H_2^{+} & H_1 \end{bmatrix} \tag{1.1}$$

where, and $H_2 \equiv \begin{bmatrix} 0 & -f_{,+} & 0 & f_{,-} \\ f_{,+} & 0 & 0 & 0 \\ 0 & 0 & 0 & 0 \\ f_{,+} & 0 & 0 & 0 \end{bmatrix}$

in which E_{g_0} is the band gap in the absence of any field, $P_{||}$ and are the momentum matrix elements parallel and perpendicular to the direction of crystal axis respectively, is the crystal field splitting constant, $\Delta_{||}$ and $\Delta_{\perp}$ are the spin-orbit splitting constants parallel and perpendicular to the C-axis respectively, and $i = \sqrt{-1}$. Thus, neglecting the contribution of the higher bands and the free electron term, the diagonalization of the above matrix leads to the dispersion relation of the conduction electrons in bulk specimens of nonlinear optical compounds [147] as

$$\gamma(E) = f_1(E) k_s^2 + f_2(E) k_z^2 \tag{1.2}$$

where

$$\gamma(E) \equiv E(E+E_{g_0})\left[(E+E_{g_0})(E+E_{g_0}+\Delta_{\|})+\delta\left(E+E_{g_0}+\frac{2}{3}\Delta_{\|}\right)+\frac{2}{9}\left(\Delta_{\|}^2-\Delta_{\perp}^2\right)\right], \quad k_s^2 = k_x^2 + k_y^2,$$

$$f_1(E) \equiv \frac{\hbar^2 E_{g_0}\left(E_{g_0}+\Delta_{\perp}\right)}{\left[2m_{\perp}^*\left(E_{g_0}+\frac{2}{3}\Delta_{\perp}\right)\right]}\left[\delta\left(E+E_{g_0}+\frac{1}{3}\Delta_{\|}\right)+\left(E+E_{g_0}\right)\left(E+E_{g_0}+\frac{2}{3}\Delta_{\|}\right)+\frac{1}{9}\left(\Delta_{\|}^2-\Delta_{\|}^2\right)\right],$$

$$f_2(E) \equiv \frac{\hbar^2 E_{g_0}\left(E_{g_0}+\Delta_{\|}\right)}{\left[2m_{\|}^*\left(E_{g_0}+\frac{2}{3}\Delta_{\|}\right)\right]}\left[\left(E+E_{g_0}\right)\left(E+E_{g_0}+\frac{2}{3}\Delta_{\|}\right)\right]$$

$m_{\|}^*$ and $m_{\perp}^*$ are the longitudinal and transverse effective electron masses at the edge of the conduction band respectively.

Let and be the quantized energy levels due to infinitely deep potential well along i[th]-axis with n_i $(=1,2,3,...)$ as the size quantum numbers. Therefore, from (1.2), one can write

$$\gamma(E_{n_x}) = f_1(E_{n_x})\left(\frac{\pi n_x}{d_x}\right)^2 \tag{1.3}$$

$$\gamma(E_{n_y}) = f_1(E_{n_y})\left(\frac{\pi n_y}{d_y}\right)^2 \tag{1.4}$$

$$\gamma(E_{n_z}) = f_2(E_{n_z})\left(\frac{\pi n_z}{d_z}\right)^2 \tag{1.5}$$

From (1.2), the totally quantized energy can be expressed as

$$\gamma(E_{QD1}) = f_1(E_{QD1})\left[\left(\frac{\pi n_x}{d_x}\right)^2+\left(\frac{\pi n_y}{d_y}\right)^2\right]+f_2(E_{QD1})\left[\left(\frac{\pi n_z}{d_z}\right)^2\right] \tag{1.6}$$

The total density-of-states function in this case is given by

$$N_{0DT}(E) = \frac{2g_v}{d_x d_y d_z} \sum_{n_x=1}^{n_{x_{\max}}} \sum_{n_y=1}^{n_{y_{\max}}} \sum_{n_z=1}^{n_{z_{\max}}} \delta'(E - E_{QD1}) \tag{1.7}$$

where g_v is the valley degeneracy and δ' is the Dirac's delta function.

Using (1.7) and Fermi-Dirac statistics, the total electron concentration can be written as

$$n_{0D} = \frac{2g_v}{d_x d_y d_z} \sum_{n_x=1}^{n_{x_{\max}}} \sum_{n_y=1}^{n_{y_{\max}}} \sum_{n_z=1}^{n_{z_{\max}}} F_{-1}(\eta_{31}) \tag{1.8}$$

where $\eta_{31} \equiv (k_B T)^{-1}(E_{F0D} - E_{QD1})$, E_{F0D} is the Fermi energy in the present case, and $F_j(\eta)$ is the one parameter Fermi-Dirac integral of order which can be written as [148],

$$F_j(\eta) = \left(\frac{1}{\Gamma(j+1)}\right) \int_0^\infty y^j (1 + \exp(y - \eta))^{-1} dy, \quad y > -1 \tag{1.9}$$

where is the complete Gamma function or for all j, analytically continued as a complex contour integral around the negative axis

$$F_j(\eta) = A_j \int_{-\infty}^{(0+)} y^j (1 + \exp(-y - \eta))^{-1} dy \tag{1.10}$$

in which $A_j \equiv \frac{\Gamma(-j)}{2\pi\sqrt{-1}}$.

Therefore the electron concentration per sub-band is given by

$$\Delta n_{0D} = \left(\frac{2g_v}{d_x d_y d_z}\right) F_{-1}(\eta_{31}) \tag{1.11}$$

The expression of the total photo-emitted current density in this case is

$$J_{0D} = \frac{e\alpha_0 g_v}{2} \sum_{n_x=1}^{n_{x_{\max}}} \sum_{n_y=1}^{n_{y_{\max}}} \sum_{n_{z_{\min}}}^{n_{z_{\max}}} (\Delta n_{0D}) v_z(E_{n_z}) \tag{1.12}$$

where the should be the nearest integer of the following inequality

$$n_{z_{\min}} \geq \left(d_z\sqrt{\gamma(W-h\nu)}\right)/\left(\pi\sqrt{f_2(W-h\nu)}\right) \tag{1.13}$$

and the velocity of the photo-emitted electrons in the n_z^{th} sub-band can be written as

$$v_z\left(E_{n_z}\right)=\frac{1}{\hbar}Q_1\left(E_{n_z}\right) \tag{1.14}$$

in which

$$Q_1\left(E_{n_z}\right)\equiv\left(2f_2^{3/2}\left(E_{n_z}\right)\left[\gamma\left(E_{n_z}\right)\right]^{1/2}\right)/\left(\left[f_2\left(E_{n_z}\right)\gamma'\left(E_{n_z}\right)-\gamma\left(E_{n_z}\right)f_2'\left(E_{n_z}\right)\right]\right)$$

Using the appropriate equations we get

$$J_{0D}=\left(\frac{e\alpha_0 g_v}{\hbar d_x d_y d_z}\right)\sum_{n_{x=1}}^{n_{x_{\max}}}\sum_{n_{y=1}}^{n_{y_{\max}}}\sum_{n_{z_{\min}}}^{n_{z_{\max}}}F_{-1}\left(\eta_{31}\right)Q_1\left(E_{n_z}\right) \tag{1.15}$$

1.2.2. The EP from QDs of III-V Materials

The dispersion relation of the conduction electrons of III-V semiconductors are described by the models of Kane (both three and two bands) [149], Stillman et. al. [150], Newson and Kurobe [151], Rossler [152], Palik et. al. [153], Johnson and Dickey [154] and Agafonov et. al. [155] respectively. For the purpose of complete and coherent presentation, the Einstein's photoelectric effect in QDs of III-V compounds have also been investigated in accordance with the aforementioned different dispersion relations for relative comparison as follows:

(a) The Three Band Model of KANE

Under the conditions, $\delta=0$, $\Delta_{\parallel}=\Delta_{\perp}=\Delta$ (isotropic spin orbit splitting constant) and $m_{\parallel}^{*}=m_{\perp}^{*}=m^{*}$ (isotropic effective electron mass at the edge of the conduction band), (1.2) gets simplified into the form as

$$\frac{\hbar^2k^2}{2m^*}=I(E),\ I(E)\equiv\frac{E\left(E+E_{g_0}\right)\left(E+E_{g_0}+\Delta\right)\left(E_{g_0}+\frac{1}{3}\Delta\right)}{E_{g_0}\left(E_{g_0}+\Delta\right)\left(E+E_{g_0}+\frac{2}{3}\Delta\right)} \tag{1.16}$$

which is known as the three band model of Kane [148] and is often used to study the electronic properties of III-V materials.

The quantized energy levels in QDs of III-V semiconductors in accordance with the three band model of Kane can be expressed as

$$I\left(E_{n_x}\right)=\frac{\hbar^2}{2m^*}\left(\frac{\pi n_x}{d_x}\right)^2 \tag{1.17}$$

$$I\left(E_{n_y}\right)=\frac{\hbar^2}{2m^*}\left(\frac{\pi n_y}{d_y}\right)^2 \tag{1.18}$$

and $$I\left(E_{n_z}\right)=\frac{\hbar^2}{2m^*}\left(\frac{\pi n_z}{d_z}\right)^2 \tag{1.19}$$

The totally quantized energy $\left(E_{QD2}\right)$ in this case assumes the form

$$I\left(E_{QD2}\right)=\frac{\hbar^2\pi^2}{2m^*}\left[\left(\frac{n_x}{d_x}\right)^2+\left(\frac{n_y}{d_y}\right)^2+\left(\frac{n_z}{d_z}\right)^2\right] \tag{1.20}$$

The velocity of the photoelectron in the n_z^{th} quantized level is given by

$$v_z\left(E_{n_z}\right)=\sqrt{\frac{2}{m^*}}Q_2\left(E_{n_z}\right) \tag{1.21}$$

where, $Q_2\left(E_{n_z}\right)\equiv\left[\left(I'\left(E_{n_z}\right)\right)'\right]^{-1}\left(\sqrt{I\left(E_{n_z}\right)}\right)$

Thus the photo-emitted current density can be expressed as

$$J_{0D}=\left[\frac{\alpha_0 e g_v}{\left(d_x d_y d_z\right)}\sqrt{\frac{2}{m^*}}\right]\sum_{n_{x=1}}^{n_{x_{\max}}}\sum_{n_{y=1}}^{n_{y_{\max}}}\sum_{n_{z_{\min}}}^{n_{z_{\max}}}Q_2\left(E_{n_z}\right)F_{-1}\left(\eta_{32}\right) \tag{1.22}$$

in which $n_{z_{\min}}$ is the nearest integer of the following inequality,

$$n_{z_{\min}}\geq\left(\frac{d_z}{\pi}\frac{\sqrt{2m^*}}{\hbar}\left[I\left(W-h\nu\right)\right]^{1/2}\right) \tag{1.23}$$

$\eta_{32} = (k_B T)^{-1}(E_{F0D} - E_{QD2})$. has to be determined form (1.20). The form of n_{0D} as given by (1.8) remains unchanged in which η_{31} should be replaced by η_{32}.

(b) The Two Band Model of Kane

Under the inequalities $\Delta >> E_{g_0}$ or $\Delta << E_{g_0}$, (1.16) assumes the form

$$E(1+\alpha E) = \left(\hbar^2 k^2 / 2m^*\right), \ \alpha \equiv 1/E_{g_0} \tag{1.24}$$

The (1.24) is known as the two-band model of Kane and should be as such for studying the electronic properties of the materials whose band-structures obey the above inequalities [148].

Therefore under the constraints or the photo-emitted current density in the present case assumes the form

$$J_{0D} = \frac{\alpha_0 g_v e}{d_x d_y d_z}\left(\sqrt{\frac{2}{m^*}}\right)\sum_{n_{x=1}}^{n_{x_{\max}}}\sum_{n_{y=1}}^{n_{y_{\max}}}\sum_{n_{z_{\min}}}^{n_{z_{\max}}} Q_3\left(E_{n_z}\right)F_{-1}\left(\eta_{33}\right) \tag{1.25}$$

where, $n_{z_{\min}} \geq \left(\frac{d_z}{\pi}\frac{\sqrt{2m^*}}{\hbar}\left[(W-h\nu)\left[1+\alpha(W-h\nu)\right]\right]^{1/2}\right)$,

$Q_3\left(E_{n_z}\right) \equiv \left[\sqrt{E_{n_z}\left(1+\alpha E_{n_z}\right)} / \left(1+2\alpha E_{n_z}\right)\right]$

and $\eta_{33} \equiv \frac{1}{k_B T}\left[E_{F0D} - E_{QD3}\right]$

The E_{n_z} obeys the equation

$$E_{n_z}\left(1+\alpha E_{n_z}\right) = \frac{\hbar^2 \pi^2}{2m^*}\left(\frac{n_z}{d_z}\right)^2 \tag{1.26}$$

The totally quantized energy in this case is given by

$$E_{QD3}\left(1+\alpha E_{QD3}\right) = \frac{\hbar^2 \pi^2}{2m^*}\left[\left(\frac{n_x}{d_x}\right)^2 + \left(\frac{n_y}{d_y}\right)^2 + \left(\frac{n_z}{d_z}\right)^2\right] \tag{1.27}$$

The form of n_{0D} as given by (1.8) remains unchanged in which η_{31} should be replaced by where E_{QD3} has to be determined form (1.27). For $\alpha \to 0$, the expressions for Δn_{0D}, E_{n_z}, $n_{z_{\min}}$, and total electron concentration (n_{0D}), for QDs of wide-gap materials can, respectively, be written as

$$\Delta n_{0D} = \frac{2g_v}{d_x d_y d_z} F_{-1}(\eta) \tag{1.28}$$

$$E_{n_z} = \frac{\hbar^2}{2m^*}\left(\frac{n_z \pi}{d_z}\right)^2 \tag{1.29}$$

$$n_{z_{\min}} \geq \left(\frac{d_z}{\pi}\right)\frac{\sqrt{2m^*}}{\hbar}\sqrt{(W - h\nu)} \ (1.30) \tag{1.30}$$

$$J_{0D} = \frac{\alpha_o e \hbar \pi g_v}{m^* d_x d_y d_z^2} \sum_{n_x=1}^{n_{x\max}} \sum_{n_y=1}^{n_{y\max}} \sum_{n_{z\min}}^{n_{z\max}} \left[n_z F_{-1}(\eta_6) \right] \tag{1.31}$$

$$n_{0D} = \frac{2g_v}{d_x d_y d_z} \sum_{n_x=1}^{n_{x\max}} \sum_{n_y=1}^{n_{y\max}} \sum_{n_z=1}^{n_{z\max}} F_{-1}(\eta_6) \tag{1.32}$$

where, $\eta_6 \equiv \dfrac{E_{F0D} - E_{QD}}{k_B T}$ in which, $E_{QD} = \dfrac{\hbar^2 \pi^2}{2m^*}\left[\left(\dfrac{n_x}{d_x}\right)^2 + \left(\dfrac{n_y}{d_y}\right)^2 + \left(\dfrac{n_z}{d_z}\right)^2\right]$

(c) The model of Stillman et. al.

In accordance with the model of Stillman et. al. [150], the electron dispersion law of III-V materials assumes the form

$$E = t_{11} k^2 - t_{12} k^4 \tag{1.33}$$

where, and $t_{12} \equiv \left(1 - \dfrac{m^*}{m_0}\right)^2 \left(\dfrac{\hbar^2}{2m^*}\right)^2 \left[\left(3E_{g_0} + 4\Delta + \dfrac{2\Delta^2}{E_{g_0}}\right).\left\{\left(E_{g_0} + \Delta\right)\left(2\Delta + 3E_{g_0}\right)\right\}^{-1}\right]$

The (1.33) can be expressed as

$$\frac{\hbar^2 k^2}{2m^*} = I_{11}(E) \tag{1.34}$$

where, $I_{11}(E) \equiv a_{11}\left[1-\left(1-a_{12}E\right)^{1/2}\right]$, $a_{11} \equiv \left(\frac{\hbar^2 t_{11}}{4m^* t_{12}}\right)$, $a_{12} \equiv \frac{4t_{12}}{t_{11}^{\;2}}$.

Therefore proceeding as above, the photoelectric current density in this case can be written as

$$J_{0D} = \frac{\alpha_0 g_v e}{d_x d_y d_z}\left(\frac{2}{m^*}\right)^{1/2} \sum_{n_x=1}^{n_{x\max}} \sum_{n_y=1}^{n_{y\max}} \sum_{n_{z\min}}^{n_{z\max}} Q_5\left(E_{n_z}\right) F_{-1}\left(\eta_{34}\right) \tag{1.35}$$

where $n_{z_{\min}} \geq \frac{d_z}{\pi}\frac{\sqrt{2m^*}}{\hbar}\left[I_{11}(W-h\nu)\right]^{1/2}$, $Q_5\left(E_{n_z}\right) \equiv \left[\left[I_{11}\left(E_{n_z}\right)\right]^{1/2}\left[I_{11}'\left(E_{n_z}\right)\right]^{-1}\right]$ and

$\eta_{34} \equiv \left(k_B T\right)^{-1}\left[E_{F0D} - E_{QD4}\right]$

The E_{n_z} obeys the equation

$$\left(\hbar^2/2m^*\right)\left(\pi n_z/d_z\right)^2 \equiv I_{11}\left(E_{n_z}\right) \tag{1.36}$$

The E_{QD4} in this case can be defined as

$$I_{11}\left(E_{QD4}\right) = \frac{\hbar^2 \pi^2}{2m^*}\left[\left(\frac{n_x}{d_x}\right)^2 + \left(\frac{n_y}{d_y}\right)^2 + \left(\frac{n_z}{d_z}\right)^2\right] \tag{1.37}$$

The form of n_{0D} as given by (1.8) remains unchanged in which η_{31} should be replaced by where E_{QD4} has to be determined form (1.37).

(d) The Model of Newson and Kurobe

In accordance with the model of Newson and Kurobe, the electron dispersion law in this case can be expressed as [151]

$$E = a_{13}k_z^{\;4} + \left[\frac{\hbar^2}{2m^*} + a_{14}k_s^{\;2}\right]k_z^{\;2} + \frac{\hbar^2}{2m^*}k_s^{\;2} + a_{14}k_x^{\;2}k_y^{\;2} + a_{13}\left(k_x^{\;4} + k_y^{\;4}\right) \tag{1.38}$$

where is the non-parabolicity constant, $a_{14}\left(\equiv 2a_{13} + a_{15}\right)$ and a_{15} is known as the warping parameter.

The $v_z\left(E_{n_z}\right)$ and E_{n_z} are given by

$$v_z\left(E_{n_z}\right) = 4a_{13}^{1/4}\left(E_{n_z}\right)^{3/4} / \hbar \tag{1.39}$$

$$E_{n_z} = a_{13}\left(\frac{\pi n_z}{d_z}\right)^4 \tag{1.40}$$

The totally quantized energy in this case can be written as

$$\begin{aligned} E_{QD5} &= a_{13}\left(\frac{\pi n_z}{d_z}\right)^4 + \left[\frac{\hbar^2}{2m^*} + a_{14}\left(\left(\frac{\pi n_x}{d_x}\right)^2 + \left(\frac{\pi n_y}{d_y}\right)^2\right)\right]\cdot\left(\frac{\pi n_z}{d_z}\right)^2 \\ &+ \frac{\hbar^2}{2m^*}\left[\left(\frac{\pi n_x}{d_x}\right)^2 + \left(\frac{\pi n_y}{d_y}\right)^2\right] + a_{14}\pi^4\left(\frac{n_x n_y}{d_x d_y}\right)^2 + a_{13}\pi^4\left[\left(\frac{n_x}{d_x}\right)^4 + \left(\frac{n_y}{d_y}\right)^4\right] \end{aligned} \tag{1.41}$$

The photo-emitted current density in this case is given by

$$J_{0D} = \frac{4\alpha_0 e g_v}{\hbar d_x d_y d_z}\left(a_{13}^{1/4}\right)\sum_{n_x=1}^{n_{x\max}}\sum_{n_y=1}^{n_{y\max}}\sum_{n_{z\min}}^{n_{z\max}}\left(E_{n_z}\right)^{3/4} F_{-1}\left(\eta_{35}\right) \tag{1.42}$$

where, $n_{z_{\min}} \geq \frac{d_z}{\pi}\left(a_{13}\right)^{-1/4}\left(W - h\nu\right)^{1/4}$ and $\eta_{35} \equiv \frac{E_{F0D} - E_{QD5}}{k_B T}$

The form of n_{0D} as given by (1.8) remains unchanged in which η_{31} should be replaced by where E_{QD5} has to be determined form (1.41).

(e) The Model of Rossler

The dispersion relation of the conduction electrons in this case in accordance with the model of Rossler can be written as [152]

$$\begin{aligned} E &= \frac{\hbar^2 k^2}{2m^*} + \left[\alpha_{11} + \alpha_{12}k\right]k^4 + \left(\beta_{11} + \beta_{12}k\right)\left[k_x^2 k_y^2 + k_y^2 k_z^2 + k_z^2 k_x^2\right] \\ &\pm\left[\gamma_{11} + \gamma_{12}k\right]\left[k^2\left(k_x^2 k_y^2 + k_y^2 k_z^2 + k_z^2 k_x^2\right) - 9k_x^2 k_y^2 k_z^2\right]^{1/2} \end{aligned} \tag{1.43}$$

where, α_{12}, β_{11}, β_{12}, γ_{11} and γ_{12} are energy-band constants.

The and E_{n_z} in this case are given by

$$v_z\left(E_{n_z}\right)=\frac{1}{\hbar}Q_6\left(E_{n_z}\right) \tag{1.44}$$

where and E_{n_z} obeys the equation

$$E_{n_z}=\left[\frac{\hbar^2}{2m^*}\left(\frac{\pi n_z}{d_z}\right)^2+\left[\alpha_{11}+\alpha_{12}\frac{\pi n_z}{d_z}\right]\left(\frac{\pi n_z}{d_z}\right)^4\right]. \tag{1.45}$$

The photo-emitted current density can be expressed as

$$J_{0D}=\frac{\alpha_0 e g_v}{2\hbar}\left(\frac{1}{d_x d_y d_z}\right)\sum_{n_x=1}^{n_{x\max}}\sum_{n_y=1}^{n_{y\max}}\sum_{n_{z\min}}^{n_{z\max}}Q_6\left(E_{n_z}\right)F_{-1}\left(\eta_{36,\pm}\right) \tag{1.46}$$

where, $\eta_{36_\pm}\equiv\left(E_{F0D}-E_{QD6,\pm}\right)\left(k_BT\right)^{-1}$ and in this case assumes the form

$$E_{QD6_\pm}\equiv\frac{\hbar^2\pi^2}{2m^*}\left[\left(\frac{n_x}{d_x}\right)^2+\left(\frac{n_y}{d_y}\right)^2+\left(\frac{n_z}{d_z}\right)^2\right]+\left[\alpha_{11}+\alpha_{12}\left[\left(\frac{\pi n_x}{d_x}\right)^2+\left(\frac{\pi n_y}{d_y}\right)^2+\left(\frac{\pi n_z}{d_z}\right)^2\right]^{1/2}\right].$$

$$\left[\left(\frac{\pi n_x}{d_x}\right)^2+\left(\frac{\pi n_y}{d_y}\right)^2+\left(\frac{\pi n_z}{d_z}\right)^2\right]^2+\left[\beta_{11}+\beta_{12}\left[\left(\frac{\pi n_x}{d_x}\right)^2+\left(\frac{\pi n_y}{d_y}\right)^2+\left(\frac{\pi n_z}{d_z}\right)^2\right]^{1/2}\right].$$

$$\left[\pi^4\left(\frac{n_x n_y}{d_x d_y}\right)^2+\pi^4\left(\frac{n_y n_z}{d_y d_z}\right)^2+\pi^4\left(\frac{n_z n_x}{d_z d_x}\right)^2\right]\pm\left[\gamma_{11}+\gamma_{12}\left[\left(\frac{\pi n_x}{d_x}\right)^2+\left(\frac{\pi n_y}{d_y}\right)^2+\left(\frac{\pi n_z}{d_z}\right)^2\right]^{1/2}\right].$$

$$\left[\left[\left(\frac{\pi n_x}{d_x}\right)^2+\left(\frac{\pi n_y}{d_y}\right)^2+\left(\frac{\pi n_z}{d_z}\right)^2\right]\left[\pi^4\left(\frac{n_x n_y}{d_x d_y}\right)^2+\pi^4\left(\frac{n_y n_z}{d_y d_z}\right)^2+\pi^4\left(\frac{n_z n_x}{d_z d_x}\right)^2\right]-9\pi^6\left(n_x n_y n_z/d_x d_y d_z\right)^6\right] \tag{1.47}$$

The $n_{z_{\min}}$ is the nearest integer of the equation

$$\left(W-h\nu\right)=\left(\frac{\hbar^2}{2m^*}\right)\left(\frac{\pi n_{z_{\min}}}{d_z}\right)^2+\left[\alpha_{11}+\alpha_{12}\frac{\pi n_{z_{\min}}}{d_z}\right]\times\left(\frac{\pi n_{z_{\min}}}{d_z}\right)^4 \tag{1.48}$$

The electron concentration can be written as

$$n_{0D} = \frac{g_v}{d_x d_y d_z} \sum_{n_x=1}^{n_{x_{\max}}} \sum_{n_y=1}^{n_{y_{\max}}} \sum_{n_z=1}^{n_{z_{\max}}} F_{-1}\left(\eta_{36,\pm}\right) \tag{1.49}$$

(f) The Model of Palik et. al.

In accordance with the model of Palik et. al. [153], the energy spectrum of the conduction electrons in III-V materials up to the fourth order in effective mass theory, taking into account the interactions the heavy hole, the light hole and the split-off bands can be written as [153]

$$E = \frac{\hbar^2 k^2}{2m^*} - b_{11} k^4 \tag{1.50}$$

where, $b_{11} \equiv \left[\frac{\hbar^4}{4E_{g_0}\left(m^*\right)^2}\right]\left[\frac{1+\frac{x_{11}{}^2}{2}}{1+\frac{x_{11}}{2}}\right] . (1-y_1)^2$, and $y_1 \equiv m^* / m_0$.

From (1.50) we get

$$\frac{\hbar^2 k^2}{2m^*} = I_{12}(E) \tag{1.51}$$

where, $I_{12}(E) \equiv b_{12}\left[a_{12} - \sqrt{a_{12}{}^2 - 4Eb_{11}}\right], b_{12} \equiv \frac{a_{12}}{2b_{11}}$ and $a_{12} \equiv \frac{\hbar^2}{2m^*}$.

The photoemission current density in this case is given by

$$J_{0D} = \frac{\alpha_0 e g_v}{\hbar d_x d_y d_z}\left(\frac{2}{m^*}\right)^{1/2} \sum_{n_x=1}^{n_{x\max}} \sum_{n_y=1}^{n_{y\max}} \sum_{n_{z\min}}^{n_{z\max}} Q_7\left(E_{n_z}\right) F_{-1}\left(\eta_{37}\right) \tag{1.52}$$

where, $n_{z_{\min}} \geq \left(\frac{\sqrt{2m^*}}{\hbar}\right)\left(\frac{d_z}{\pi}\right)\left[I_{12}(W - h\nu)\right]^{1/2}$,

$Q_7\left(E_{n_z}\right) \equiv \left[\left[I_{12}\left(E_{n_z}\right)\right]^{1/2} / I_{12}'\left(E_{n_z}\right)\right]$ and $\eta_{37} \equiv \frac{\left(E_{F0D} - E_{QD7}\right)}{k_B T}$

The E_{n_z} and are defined by the following equations:

$$I_{12}\left(E_{n_z}\right)=\frac{\hbar^2}{2m^*}\left(\frac{\pi n_z}{d_z}\right)^2 \tag{1.53}$$

$$I_{12}\left(E_{QD7}\right)=\frac{\hbar^2}{2m^*}\left[\left(\frac{\pi n_x}{d_x}\right)^2+\left(\frac{\pi n_y}{d_y}\right)^2+\left(\frac{\pi n_z}{d_z}\right)^2\right] \tag{1.54}$$

The form of n_{0D} as given by (1.8) remains unchanged in which η_{31} should be replaced by where E_{QD7} has to be determined form (1.54).

(g) The Model of Johnson and Dickey

In accordance with the model of Johnson and Dickey [154] the electron dispersion law in III-V materials assumes the form

$$E=-\frac{E_{g_0}}{2}+\frac{\hbar^2k^2}{2}\left[\frac{1}{m^*}-\frac{1}{m_0}\right]+\frac{E_{g_0}}{2}\left[1+a_{15}\left(E\right)k^2\right]^{1/2} \tag{1.55}$$

where, $a_{15}\left(E\right)\equiv\frac{4\hbar^2}{2m^*E_{g_0}}a_{14}\left(E\right)$, $a_{14}\left(E\right)\equiv\frac{\left(E_{g_0}+\Delta\right)\left(E+E_{g_0}+\frac{2}{3}\Delta\right)}{\left(E_{g_0}+\frac{2}{3}\Delta\right)\left(E+E_{g_0}+\Delta\right)}$, and $\left(\overline{P}\right)$ is the energy band constant.

The in this case is given by

$$E_{QD8}=-\frac{E_{g_0}}{2}+\frac{\hbar^2}{2}\left[\frac{1}{m^*}-\frac{1}{m_0}\right]\left[\left(\frac{\pi n_x}{d_x}\right)^2+\left(\frac{\pi n_y}{d_y}\right)^2+\left(\frac{\pi n_z}{d_z}\right)^2\right]+\frac{E_{g_0}}{2}\left[1+a_{15}\left(E_{QD8}\right)\left[\left(\frac{\pi n_x}{d_x}\right)^2+\left(\frac{\pi n_y}{d_y}\right)^2+\left(\frac{\pi n_z}{d_z}\right)^2\right]\right]^{1/2} \tag{1.56}$$

The can be written as, $\nu_z\left(E_{n_z}\right)=\frac{1}{\hbar}Q_8\left(E_{n_z}\right)$ (1.57)

where,

$$Q_8\left(E_{n_z}\right) \equiv \left[\left[1 - \frac{E_{g_0} a_{15}'\left(E_{n_z}\right)}{4}\left(\frac{\pi n_z}{d_z}\right)^2\left[1 + a_{15}\left(E_{n_z}\right)\left(\frac{\pi n_z}{d_z}\right)^2\right]^{-1/2}\right]^{-1}\right.$$

$$\left.\left[\hbar^2\left(\frac{1}{m^*} - \frac{1}{m_0}\right)\left(\frac{n_z\pi}{d_z}\right) + \frac{E_{g_0} a_{15}\left(E_{n_z}\right)}{4}\left(\frac{\pi n_z}{d_z}\right)\left[1 + a_{15}\left(E_{n_z}\right)\left(\frac{\pi n_z}{d_z}\right)^2\right]^{-1/2}\right]\right]$$

and $E_{n_z} = -\frac{E_{g_0}}{2} + \frac{\hbar^2}{2}\left(\frac{n_z\pi}{d_z}\right)^2\left(\frac{1}{m^*} - \frac{1}{m_0}\right) + \frac{E_{g_0}}{2}\left[1 + a_{15}\left(E_{n_z}\right)\left(\frac{\pi n_z}{d_z}\right)^2\right]^{1/2}$

The photo-emitted current density can be written as

$$J_{0D} = \frac{\alpha_0 g_v e}{\hbar d_x d_y d_z}\sum_{n_x=1}^{n_{x_{\max}}}\sum_{n_y=1}^{n_{y_{\max}}}\sum_{n_{z_{\min}}}^{n_{z_{\max}}} Q_8\left(E_{n_z}\right)F_{-1}\left(\eta_{38}\right) \tag{1.58}$$

where, $n_{z_{\min}}$ is the nearest integer of the following equation

$$(W - h\nu) = -\frac{E_{g_0}}{2} + \frac{\hbar^2}{2}\left(\frac{\pi n_{z_{\min}}}{d_z}\right)^2\left(\frac{1}{m^*} - \frac{1}{m_0}\right) + \frac{E_{g_0}}{2}\left[1 + a_{15}(W - h\nu)\left(\frac{\pi n_{z_{\min}}}{d_z}\right)^2\right]^{1/2} \tag{1.59}$$

and, $\eta_{38} \equiv \frac{E_{F0D} - E_{QD8}}{k_B T}$

The form of n_{0D} as given by (1.8) remains unchanged in which η_{31} should be replaced by where E_{QD8} has to be determined form (1.59).

(h) The Model of Agafonov et. al.

In accordance with the model of Agafonov et. al. [155], the electron dispersion law can be expressed as

$$E = \frac{\bar{y} - E_{g_0}}{2}\left[1 - T_5\left(\frac{k_x^{\,4} + k_y^{\,4} + k_z^{\,4}}{\bar{y}k^2}\right)\right] \tag{1.60}$$

where, $(\bar{y})^2 = \left[E_{g_0}^{\ 2} + \frac{8}{3}P_0^2 k^2\right]$, $T_5 = \frac{\bar{D}\sqrt{3} - 3\bar{B}}{2}$,, $\bar{D} = -40\left(\frac{\hbar^2}{2m_0}\right)$ and P_0 is the momentum matrix element.

The in this case can be written as

$$\nu_z\left(E_{n_z}\right)=\frac{1}{\hbar}Q_9\left(E_{n_z}\right) \tag{1.61}$$

where,

$$Q_9\left(E_{n_z}\right)\equiv\left[\left[\frac{1}{2}\left[\sqrt{E_{g_0}{}^2+\frac{8}{3}P^2\left(\frac{\pi n_z}{d_z}\right)^2}-E_{g_0}\right]\left[-2T_5\left(\frac{\pi n_z}{d_z}\right)\right]\cdot\left[E_{g_0}{}^2+\frac{8}{3}P_0^2\left(\frac{\pi n_z}{d_z}\right)^2\right]^{-1/2}+\left[\frac{8}{3}P^2T_5\left(\frac{\pi n_z}{d_z}\right)^3\right]\left(E_{g_0}{}^2+\frac{8}{3}P^2\left(\frac{\pi n_z}{d_z}\right)^2\right)^{-3/2}\right]\right.$$
$$\left.+\frac{8}{3}P_0^2\left(\frac{\pi n_z}{d_z}\right)\left[E_{g_0}{}^2+\frac{8}{3}P_0^2\left(\frac{\pi n_z}{d_z}\right)^2\right]^{-1/2}\left[1-\left[T_5\left(\frac{\pi n_z}{d_z}\right)^2\right]\left[\left[E_{g_0}{}^2+\frac{8}{3}P_0^2\left(\frac{\pi n_z}{d_z}\right)^2\right]^{1/2}\right]^{-1}\right]\right]$$

and E_{n_z} is defined by the following equation

$$E_{n_z}=\frac{1}{2}\left[\left(E_{g_0}{}^2+\frac{8}{3}P^2\left(\frac{\pi n_z}{d_z}\right)^2\right)^{1/2}-E_{g_0}\right]\left[1-T_5\cdot\left(E_{g_0}{}^2+\frac{8}{3}P^2\left(\frac{\pi n_z}{d_z}\right)^2\right)^{-1/2}\left(\frac{\pi n_z}{d_z}\right)^2\right] \tag{1.62}$$

The totally quantized energy can be written as

$$E_{QD9}=\left(\frac{\psi_{30}-E_{g_0}}{2}\right)\left[1-\left[T_5\left[\left(\frac{\pi n_x}{d_x}\right)^4+\left(\frac{\pi n_y}{d_y}\right)^4+\left(\frac{\pi n_z}{d_z}\right)^4\right]\right]\cdot\left[\psi_{30}\left[\left(\frac{\pi n_x}{d_x}\right)^2+\left(\frac{\pi n_y}{d_y}\right)^2+\left(\frac{\pi n_z}{d_z}\right)^2\right]\right]^{-1}\right] \tag{1.63}$$

where, $\psi_{30}=\left[E_{g_0}{}^2+\frac{8}{3}P^2\left[\left(\frac{\pi n_x}{d_x}\right)^2+\left(\frac{\pi n_y}{d_y}\right)^2+\left(\frac{\pi n_z}{d_z}\right)^2\right]\right]^{1/2}$

The photo-emitted current density is given by in which, and should be determined from the following equation

$$J_{0D}=\frac{\alpha_0 g_v e}{\hbar d_x d_y d_z}\sum_{n_x=1}^{n_{x\max}}\sum_{n_y=1}^{n_{y\max}}\sum_{n_{z\min}}^{n_{z\max}}Q_9\left(E_{n_z}\right)F_{-1}\left(\eta_{39}\right) \tag{1.64}$$

in which, and should be determined from the following equation

$$(W-h\nu)=\frac{1}{2}\left[\sqrt{E_{g_0}^{\ 2}+\frac{8}{3}P^2\left(\left(\frac{\pi n_{z_{\min}}}{d_z}\right)^2\right)}-E_{g_0}\right]\left[1-T_5\left(\frac{\pi n_{z_{\min}}}{d_z}\right)^2\left[E_{g_0}^{\ 2}+\frac{8}{3}P^2\left(\left(\frac{\pi n_{z_{\min}}}{d_z}\right)^2\right)\right]^{-1/2}\right] \tag{1.65}$$

The form of η_{0D} is given by (1.8) where η_{31} should be replaced by η_{39} .

1.2.3. The EP from QDs of II-VI Materials

The carrier energy spectra in bulk specimens of II-VI compounds in accordance with Hopfield model can be written as [77]

$$E = A_o k_s^2 + B_o k_z^2 \pm C_o k_s \tag{1.66}$$

where $A_0 \equiv \hbar^2/2m_\perp^*$, and represents the splitting of the two-spin states by the spin orbit coupling and the crystalline field.

Using (1.65), the totally quantized energy in this case can be expressed as

$$E_{QD10,\pm} = A_0\left[\left(\frac{\pi n_x}{d_x}\right)^2+\left(\frac{\pi n_y}{d_y}\right)^2\right]+B_0\left(\frac{\pi n_z}{d_z}\right)^2 \pm C_0\left[\left(\frac{\pi n_x}{d_x}\right)^2+\left(\frac{\pi n_y}{d_y}\right)^2\right]^{1/2} \tag{1.67}$$

The $\nu_z\left(E_{n_z}\right)$ and E_{n_z} are given by

$$\nu_z\left(E_{n_z}\right)=\frac{\hbar}{m_{\|}^*}\left(\frac{\pi n_z}{d_z}\right) \tag{1.68}$$

$$E_{n_z}=\frac{\hbar^2}{2m_{\|}^*}\left(\frac{\pi n_z}{d_z}\right)^2 \tag{1.69}$$

The electron concentration can be written as

$$n_{0D}=\frac{g_v}{d_x d_y d_z}\sum_{n_x=1}^{n_{x\max}}\sum_{n_y=1}^{n_{y\max}}\sum_{n_z=1}^{n_{z\max}}\left[F_{-1}\left(\eta_{40,\pm}\right)\right] \tag{1.70}$$

where, $\eta_{40,\pm} = \frac{1}{k_B T}\left[E_{F0D} - E_{QD10,\pm}\right]$

The photo-emitted current density is given by

$$J_{0D} = \frac{\alpha_0 g_v \hbar \pi e}{2 d_x d_y d_z^2 m_{\|}^*} \sum_{n_x=1}^{n_{x\max}} \sum_{n_y=1}^{n_{y\max}} \sum_{n_{z\min}}^{n_{z\max}} n_z F_{-1}\left(\eta_{40,\pm}\right) \tag{1.71}$$

where,

$$n_{z_{\min}} \geq \frac{d_z}{\pi} \frac{\sqrt{2m_{\|}^*}}{\hbar} (W - h\nu)^{1/2} \tag{1.72}$$

1.2.4. The EP from QDs of n-Gallium Phosphide

The dispersion relation of the conduction electrons in bulk specimens of n-GaP is given by [156]

$$E = \frac{\hbar^2 k_s^2}{2m_\perp^*} + \frac{\hbar^2}{2m_{\|}^*}\left[k_s^2 + k_z^2\right] - \left[\frac{\hbar^4 k_0^2}{m_{\|}^{*2}}\left(k_s^2 + k_z^2\right) + \left|V_G\right|^2\right]^{1/2} + \left|V_G\right| \tag{1.73}$$

where, and are constants of the energy spectrum. Using (1.72), the totally quantized energy in this case is given by

$$E_{QD11} = \frac{\hbar^2}{2m_\perp^*}\left[\left(\frac{\pi n_x}{d_x}\right)^2 + \left(\frac{\pi n_y}{d_y}\right)^2\right] + \frac{\hbar^2}{2m_{\|}^*}\left[\left(\frac{\pi n_x}{d_x}\right)^2 + \left(\frac{\pi n_y}{d_y}\right)^2 + \left(\frac{\pi n_z}{d_z}\right)^2\right] - \left[\frac{\hbar^4 k_0^2}{m_{\|}^{*2}}\left[\left(\frac{\pi n_x}{d_x}\right)^2 + \left(\frac{\pi n_y}{d_y}\right)^2 + \left(\frac{\pi n_z}{d_z}\right)^2\right] + \left|V_G\right|^2\right]^{1/2} + \left|V_G\right| \tag{1.74}$$

The $\nu_z\left(E_{n_z}\right)$ and E_{n_z} can be respectively written as

$$v_z\left(E_{n_z}\right) = \frac{\hbar}{m_{\|}^*}\left(\frac{\pi n_z}{d_z}\right) Q_{10}\left(E_{n_z}\right) \tag{1.75}$$

where, $Q_{10}\left(E_{n_z}\right) = \left[1 - \left(\frac{\hbar^2 k_0^2}{m_{\|}^*}\right)\left\{\left|V_G\right|^2 + \frac{\hbar^4 k_0^2}{\left(m_{\|}^*\right)^2}\left(\frac{\pi n_z}{d_z}\right)^2\right\}^{-1/2}\right]$ and

(1.76)

The electron concentration assumes the form

$$n_{0D} = \frac{2g_v}{d_x d_y d_z} \sum_{n_x=1}^{n_{x\max}} \sum_{n_y=1}^{n_{y\max}} \sum_{n_z=1}^{n_{z\max}} \left[F_{-1}(\eta_{41}) \right] \tag{1.77}$$

where, $\eta_{41} = \frac{1}{k_B T}\left[E_{F0D} - E_{QD11} \right]$ in which is determined from (1.73).

The photo-emitted current density is given by

$$J_{0D} = \frac{\alpha_0 \pi \hbar g_v e}{d_x d_y d_z^2 m_{\|}^*} \sum_{n_x=1}^{n_{x\max}} \sum_{n_y=1}^{n_{y\max}} \sum_{n_{z\min}}^{n_{z\max}} Q_{10}\left(E_{n_z}\right) F_{-1}(\eta_{41}) \tag{1.78}$$

where, $n_{z_{\min}}$ is the nearest integer of the following equation

$$(W - h\nu) = \frac{\hbar^2}{2m_{\|}^*}\left(\frac{\pi n_{z_{\min}}}{d_z}\right)^2 + |V_G| - \left[|V_G|^2 + \frac{\hbar^4 k_0^2}{\left(m_{\|}^*\right)^2}\left(\frac{\pi n_{z_{\min}}}{d_z}\right)^2 \right]^{1/2} \tag{1.79}$$

1.2.5. The EP from QDs of n-Germanium

It is well known that the conduction electrons of n-Ge obey two different types of dispersion laws since band non-parabolicity has been included in two different ways as given in the literature [157-158].

The energy spectrum of the conduction electrons in bulk specimens of n-Ge can be expressed in accordance with Cardona et. al. [157] as

$$E = -\frac{E_{g_0}}{2} + \frac{\hbar^2 k_z^2}{2m_{\|}^*} + \left[\frac{E_{g_0}^2}{4} + E_{g_0} k_s^2 \left(\frac{\hbar^2}{2m_{\perp}^*} \right) \right]^{1/2} \tag{1.80}$$

where in this case and $m_{\perp}^*$ are the longitudinal and transverse effective masses along <111> direction at the edge of the conduction band respectively. The totally quantized energy is given by

$$E_{QD12} = -\frac{E_{g_0}}{2} + \frac{\hbar^2}{2m_{\|}^*}\left(\frac{\pi n_z}{d_z}\right)^2 + \left[\frac{E_{g_0}^2}{4} + E_{g_0}\left(\frac{\hbar^2}{2m_{\perp}^*}\right)\left\{ \left(\frac{\pi n_x}{d_x}\right)^2 + \left(\frac{\pi n_y}{d_y}\right)^2 \right\} \right]^{1/2} \tag{1.81}$$

The $\nu_z\left(E_{n_z}\right)$ and E_{n_z} can be respectively written as

$$\nu_z\left(E_{n_z}\right)=\frac{\hbar}{m_{\parallel}^{*}}\left(\frac{\pi n_z}{d_z}\right) \tag{1.82}$$

$$E_{n_z}=\frac{\hbar^2}{2m_{\parallel}^{*}}\left(\frac{\pi n_z}{d_z}\right)^2 \tag{1.83}$$

The electron concentration assumes the form

$$n_{0D}=\frac{2g_v}{d_x d_y d_z}\sum_{n_x=1}^{n_{x\max}}\sum_{n_y=1}^{n_{y\max}}\sum_{n_z=1}^{n_{z\max}}\left[F_{-1}\left(\eta_{42}\right)\right] \tag{1.84}$$

where, $\eta_{42}=\frac{1}{k_B T}\left[E_{F0D}-E_{QD12}\right]$

in which is determined from (1.80).
The photo-emitted current density is given by

$$J_{0D}=\frac{\alpha_0\pi\hbar g_v e}{d_x d_y d_z^2 m_{\parallel}^{*}}\sum_{n_x=1}^{n_{x\max}}\sum_{n_y=1}^{n_{y\max}}\sum_{n_{z\min}}^{n_{z\max}} n_z F_{-1}\left(\eta_{42}\right) \tag{1.85}$$

where, $n_{z_{\min}}$ is the nearest integer of the following inequality

$$n_{z_{\min}}\geq\frac{d_z}{\pi}\frac{\sqrt{2m_{\parallel}^{*}}}{\hbar}\left(W-h\nu\right)^{1/2} \tag{1.86}$$

b) The dispersion relation of the conduction electron in bulk specimens of n-Ge can be expressed in accordance with the model of Wang and Ressler [158] can be written as

$$E=\frac{\hbar^2 k_z^2}{2m_{\parallel}^{*}}+\frac{\hbar^2 k_s^2}{2m_{\perp}^{*}}-\overline{c}_1\left(\frac{\hbar^2 k_s^2}{2m_{\perp}^{*}}\right)^2-\overline{d}_1\left(\frac{\hbar^2 k_s^2}{2m_{\perp}^{*}}\right)\left(\frac{\hbar^2 k_z^2}{2m_{\parallel}^{*}}\right)-\overline{e}_1\left(\frac{\hbar^2 k_z^2}{2m_{\parallel}^{*}}\right)^2 \tag{1.87}$$

where $\overline{c}_1\equiv\overline{C}\left(2m_{\perp}^{*}/\hbar^2\right)^2$, $\overline{C}\equiv1.4\overline{A}$, $\overline{A}\equiv\frac{1}{4}\left(\hbar^4/E_{g_0}m_{\perp}^{*2}\right)\left(1-\frac{m_{\perp}^{*}}{m_0}\right)^2$,

$$\bar{d}_1 \equiv \bar{d}\left(\frac{4m_{\perp}^{*}m_{\parallel}^{*}}{\hbar^4}\right),\ \bar{d} \equiv 0.8\bar{A}, \text{ and } \bar{e}_0 \equiv 0.005\bar{A}.$$

The totally quantized energy in this case is given by

$$E_{QD13} = \frac{\hbar^2}{2m_{\parallel}^{*}}\left(\frac{\pi n_z}{d_z}\right)^2 + \frac{\hbar^2}{2m_{\perp}^{*}}\left\{\left(\frac{\pi n_x}{d_x}\right)^2 + \left(\frac{\pi n_y}{d_y}\right)^2\right\} - \bar{c}_1\left(\frac{\hbar^2}{2m_{\perp}^{*}}\right)^2\left\{\left(\frac{\pi n_x}{d_x}\right)^2 + \left(\frac{\pi n_y}{d_y}\right)^2\right\}^2 - \bar{d}_1\left(\frac{\hbar^2}{2m_{\perp}^{*}}\right)\left\{\left(\frac{\pi n_x}{d_x}\right)^2 + \left(\frac{\pi n_y}{d_y}\right)^2\right\}\left(\frac{\hbar^2}{2m_{\parallel}^{*}}\right)\left(\frac{\pi n_z}{d_z}\right)^2 - \bar{e}_1\left(\frac{\hbar^2}{2m_{\parallel}^{*}}\right)^2\left(\frac{\pi n_z}{d_z}\right)^4 \tag{1.88}$$

The $\nu_z\left(E_{n_z}\right)$ and E_{n_z} can be respectively written as

$$\nu_z\left(E_{n_z}\right) = \frac{\hbar}{m_{\parallel}^{*}}\left(\frac{n_z\pi}{d_z}\right)Q_{11}\left(E_{n_z}\right) \tag{1.89}$$

where, $Q_{11}\left(E_{n_z}\right) = \left[1 - \bar{e}_1\frac{\hbar^2}{\left(m_{\parallel}^{*}\right)^2}\left(\frac{\pi n_z}{d_z}\right)^2\right]$

$$E_{n_z} = \frac{\hbar^2}{2m_{\parallel}^{*}}\left(\frac{\pi n_z}{d_z}\right)^2 - \bar{e}_1\left(\frac{\hbar^2}{2m_{\parallel}^{*}}\right)^2\left(\frac{\pi n_z}{d_z}\right)^4 \tag{1.90}$$

The electron concentration assumes the form

$$n_{0D} = \frac{2g_v}{d_x d_y d_z}\sum_{n_x=1}^{n_{x\max}}\sum_{n_y=1}^{n_{y\max}}\sum_{n_z=1}^{n_{z\max}}\left[F_{-1}\left(\eta_{43}\right)\right] \tag{1.91}$$

where, in which is determined from (1.87)

The photo-emitted current density is given by

$$J_{0D} = \frac{\alpha_0\pi\hbar g_v e}{d_x d_y d_z^2 m_{\parallel}^{*}}\sum_{n_x=1}^{n_{x\max}}\sum_{n_y=1}^{n_{y\max}}\sum_{n_{z\min}}^{n_{z\max}}Q_{11}\left(E_{n_z}\right)F_{-1}\left(\eta_{43}\right) \tag{1.92}$$

where, $n_{z_{\min}}$ is the nearest integer of the following inequality

$$\left(W - h\nu\right) = \frac{\hbar^2}{2m_{\parallel}^{*}}\left(\frac{\pi n_{z_{\min}}}{d_z}\right)^2 - \bar{e}_1\left(\frac{\hbar^2}{2m_{\parallel}^{*}}\right)^2\left(\frac{\pi n_{z_{\min}}}{d_z}\right)^4 \tag{1.93}$$

1.2.6. The EP from QDs of Tellurium

The carriers of Tellurium find various descriptions for the energy-wave vector dispersion relations in the literature. Among them we shall use the E-k dispersion relations as given by Bouat et. al. [159] and Ortenberg and Button [160] respectively.

The dispersion relation of the conduction electrons of Tellurium can be written in accordance with Bouat et. al. as [159]

$$E = A_6 k_z^2 + A_7 k_s^2 \pm \left[A_8 k_z^2 + A_9 k_s^2 \right]^{1/2} \tag{1.94}$$

where A_6 , A_7 , A_8 and A_9 are the energy band constants.

The $v_z\left(E_{n_{z,\pm}}\right)$ in this case is given by

$$v_z\left(E_{n_{z,\pm}}\right) = \frac{1}{\hbar} Q_{12}\left(E_{n_{z,\pm}}\right) \tag{1.95}$$

where $Q_{12}\left(E_{n_{z,\pm}}\right) = 2A_6\left(\frac{n_z \pi}{d_z}\right) \pm \sqrt{A_8}$

in which $E_{n_{z,\pm}} = A_6\left(\frac{\pi n_z}{d_z}\right)^2 \pm \left(\frac{\pi n_z}{d_z}\right)\sqrt{A_8}$

The totally quantized energy can be written as

$$E_{QD14,\pm} = A_6\left(\frac{\pi n_z}{d_z}\right)^2 + A_7\left[\left(\frac{\pi n_x}{d_x}\right)^2 + \left(\frac{\pi n_y}{d_y}\right)^2\right] \pm \left[A_8\left(\frac{\pi n_z}{d_z}\right)^2 + A_9\left[\left(\frac{\pi n_x}{d_x}\right)^2 + \left(\frac{\pi n_y}{d_y}\right)^2\right]\right]^{1/2} \tag{1.96}$$

The electron concentration is given by

$$n_{0D} = \frac{g_v}{d_x d_y d_z} \sum_{n_x=1}^{n_{x\max}} \sum_{n_y=1}^{n_{y\max}} \sum_{n_z=1}^{n_{z\max}} F_{-1}\left(\eta_{44,\pm}\right) \tag{1.97}$$

where $\eta_{44,\pm} = \left(E_{F0D} - E_{QD14,\pm}\right) / k_B T$

The photo-emitted current density is given by

$$J_{0D} = \frac{\alpha_0 e g_v}{2\hbar d_x d_y d_z} \sum_{n_x=1}^{n_{x_{\max}}} \sum_{n_y=1}^{n_{y_{\max}}} \sum_{n_z=1}^{n_{z_{\max}}} Q_{12}\left(E_{n_{z,\pm}}\right) F_{-1}\left(\eta_{44,\pm}\right) \tag{1.98}$$

where $n_{z_{\min}}$ can be determined from

$$W - h\nu = A_6\left(\frac{\pi n_{z_{\min}}}{d_z}\right)^2 \pm \sqrt{A_8}.\frac{\pi}{d_z}.n_{z_{\min}} \tag{1.99}$$

The dispersion relation of the conduction electrons of Tellurium can be written in accordance with the model of Ortenberg and Button as [160]

$$E = t_1 + t_2 k_z^2 + t_3 k_s^2 + t_4 k_s^4 + t_5 k_s^2 k_z^2 \pm \left[\left(t_1 + t_6 k_s^2\right)^2 + t_7 k_z^2\right]^{1/2} \tag{1.100}$$

where t_1, t_2, t_3, t_4, t_5, t_6 and t_7 are the energy band constants.

The $v_z\left(E_{n_{z,\pm}}\right)$ in this case is given by

$$v_z\left(E_{n_{z,\pm}}\right) = \frac{1}{\hbar} Q_{13}\left(E_{n_{z,\pm}}\right) \tag{1.101}$$

where $Q_{13}\left(E_{n_{z,\pm}}\right) = \left[2t_2\left(\frac{\pi n_z}{d_z}\right) \pm t_7\left(\frac{\pi n_z}{d_z}\right).\left[t_1^2 + t_7\left(\frac{\pi n_z}{d_z}\right)^2\right]^{-1/2}\right]$ in which $E_{n_{z,\pm}} = t_1 + t_2\left(\frac{\pi n_z}{d_z}\right)^2 \pm \left[t_1^2 + t_7\left(\frac{\pi n_z}{d_z}\right)^2\right]^{1/2}$

The totally quantized energy can be written as

$$E_{QD15,\pm} = t_1 + t_2\left(\frac{\pi n_z}{d_z}\right)^2 + t_3\left[\left(\frac{\pi n_x}{d_x}\right)^2 + \left(\frac{\pi n_y}{d_y}\right)^2\right] + t_4\left[\left(\frac{\pi n_x}{d_x}\right)^2 + \left(\frac{\pi n_y}{d_y}\right)^2\right]^2 +$$
$$t_5\left[\left(\frac{\pi n_x}{d_x}\right)^2 + \left(\frac{\pi n_y}{d_y}\right)^2\right]\left(\frac{\pi n_z}{d_z}\right)^2 \pm \left[\left[t_1 + t_6\left\{\left(\frac{\pi n_x}{d_x}\right)^2 + \left(\frac{\pi n_y}{d_y}\right)^2\right\}\right]^2 + t_7\left(\frac{\pi n_z}{d_z}\right)^2\right]^{1/2} \tag{1.102}$$

The electron concentration is given by

$$n_{0D} = \frac{g_v}{d_x d_y d_z} \sum_{n_x=1}^{n_{x_{\max}}} \sum_{n_y=1}^{n_{y_{\max}}} \sum_{n_z=1}^{n_{z_{\max}}} F_{-1}\left(\eta_{45,\pm}\right) \tag{1.103}$$

where $\eta_{45,\pm} = \left(E_{F0D} - E_{QD15,\pm}\right)/k_B T$

The photo-emitted current density can be written as

$$J_{0D} = \frac{\alpha_0 e g_v}{2\hbar d_x d_y d_z} \sum_{n_x=1}^{n_{x_{\max}}} \sum_{n_y=1}^{n_{y_{\max}}} \sum_{n_z=1}^{n_{z_{\max}}} Q_{13}\left(E_{n_{z,\pm}}\right) F_{-1}\left(\eta_{45,\pm}\right) \tag{1.104}$$

where $n_{z_{\min}}$ should be determined from the following equation

$$W - h\nu = t_1 + t_2\left(\frac{\pi n_{z_{\min}}}{d_z}\right)^2 \pm \left[t_1^2 + t_7\left(\frac{\pi n_{z_{\min}}}{d_z}\right)^2\right]^{1/2} \tag{1.105}$$

1.2.7. The EP from QDs Graphite

The carrier dispersion law in graphite can be written as [161]

$$E = \frac{1}{2}\left(E_2 + E_3\right) \pm \left[\frac{1}{4}\left(E_2 - E_3\right)^2 + \eta_{46}^2 k^2\right]^{1/2} \tag{1.106}$$

where $E_2 \equiv \left(\Delta_1 - 2\gamma_1 \cos\phi_0 + 2\gamma_5 \cos^2\phi_0\right)$, Δ_1, γ_1 and γ_5 are energy band constants.

$\phi_0 \equiv \left(\bar{c}k_z/2\right)$, $E_3 \equiv \left(2\gamma_2\cos^2\phi_0\right)$, $\eta_{46} \equiv \left(\frac{\sqrt{3}}{2}\right)\bar{a}\left(\gamma_0 + 2\gamma_4\cos\phi_0\right)$, and are constants of the spectrum.

The $v_z\left(E_{n_{z,\pm}}\right)$ in this case is given by

$$v_z\left(E_{n_{z,\pm}}\right) = \frac{1}{\hbar} Q_{14}\left(E_{n_{z,\pm}}\right) \tag{1.107}$$

where $Q_{14}\left(E_{n_{z,\pm}}\right) = \frac{1}{2}\left[\gamma_1\bar{c}\sin\left(\frac{\bar{c}(\pi n_z)}{2d_z}\right) - \gamma_5\bar{c}\sin\left(\frac{\bar{c}(\pi n_z)}{2d_z}\right) - \gamma_2\bar{c}\sin\left(\frac{\bar{c}(\pi n_z)}{2d_z}\right)\right]$

$$\pm\frac{1}{2}\left[\frac{1}{4}\left(\Delta_1 - 2\gamma_1\cos\left(\frac{\bar{c}(\pi n_z)}{2d_z}\right) + 2\gamma_5\cos^2\left(\frac{\bar{c}(\pi n_z)}{2d_z}\right) - 2\gamma_2\cos^2\left(\frac{\bar{c}(\pi n_z)}{2d_z}\right)\right)^2 + \frac{3(\bar{a})^2}{4}\left(\frac{n_z\pi}{d_z}\right)^2 \cdot \left(\gamma_0 + 2\gamma_4\cos\left(\frac{\bar{c}(\pi n_z)}{2d_z}\right)\right)^2\right]^{-1/2} \cdot$$

$$\left[\frac{1}{2}\left[\Delta_1 - 2\gamma_1\cos\left(\frac{\bar{c}(\pi n_z)}{2d_z}\right) + 2\gamma_5\cos^2\left(\frac{\bar{c}(\pi n_z)}{2d_z}\right) - 2\gamma_2\cos^2\left(\frac{\bar{c}(\pi n_z)}{2d_z}\right)\right]\right]$$

$$\left[\bar{c}\gamma_1 \sin\left(\frac{\bar{c}(\pi n_z)}{2d_z}\right) - \bar{c}\gamma_5 \sin\left(\frac{\bar{c}(\pi n_z)}{2d_z}\right) + \gamma_2 \bar{c} \sin\left(\frac{\bar{c}(\pi n_z)}{2d_z}\right)\right]$$

$$+\left(\bar{a}\right)^2 \frac{3}{4}\left(\frac{\pi n_z}{d_z}\right)\left[\gamma_0 + 2\gamma_4 \cos\left(\frac{\bar{c}(\pi n_z)}{2d_z}\right)\right]^2$$

$$-\left(\bar{a}\right)^2 \frac{3}{2}\bar{c}\gamma_4\left(\frac{\pi n_z}{d_z}\right)^2\left[\gamma_0 + 2\gamma_4 \cos\left(\frac{\bar{c}(\pi n_z)}{2d_z}\right)\right]\Bigg]$$

in which

$$E_{n_{z,\pm}} = \frac{1}{2}\left[\Delta_1 - 2\gamma_1 \cos\left(\frac{\bar{c}(\pi n_z)}{2d_z}\right) + 2\gamma_5 \cos^2\left(\frac{\bar{c}(\pi n_z)}{2d_z}\right) + 2\gamma_2 \cos^2\left(\frac{\bar{c}(\pi n_z)}{2d_z}\right)\right]$$

$$\pm\left[\frac{1}{4}\left(\Delta_1 - 2\gamma_1 \cos\left(\frac{\bar{c}(\pi n_z)}{2d_z}\right) + 2\gamma_5 \cos^2\left(\frac{\bar{c}(\pi n_z)}{2d_z}\right) - 2\gamma_2 \cos^2\left(\frac{\bar{c}(\pi n_z)}{2d_z}\right)\right)^2 + \frac{3(\bar{a})^2}{4}\left(\gamma_0 + 2\gamma_4 \cos\left(\frac{\bar{c}(\pi n_z)}{2d_z}\right)\right)^2\left(\frac{\pi n_z}{d_z}\right)^2\right]^{1/2}$$

The totally quantized energy can be written as

$$E_{QD16,\pm} = \frac{1}{2}\left[\Delta_1 - 2\gamma_1 \cos\left(\frac{\bar{c}(\pi n_z)}{2d_z}\right) + 2\gamma_5 \cos^2\left(\frac{\bar{c}(\pi n_z)}{2d_z}\right) + 2\gamma_2 \cos^2\left(\frac{\bar{c}(\pi n_z)}{2d_z}\right)\right] \pm$$

$$\left[\frac{1}{4}\left(\Delta_1 - 2\gamma_1 \cos^2\left(\frac{\bar{c}(\pi n_z)}{2d_z}\right) + 2\gamma_5 \cos^2\left(\frac{\bar{c}(\pi n_z)}{2d_z}\right) - 2\gamma_2 \cos^2\left(\frac{\bar{c}(\pi n_z)}{2d_z}\right)\right)^2 + \frac{3}{4}\left(\bar{a}\right)^2\right.$$

$$\left.\left(\gamma_0 + 2\gamma_4 \cos\left(\frac{n_z \pi \bar{c}}{2d_z}\right)\right)^2\left[\left(\frac{\pi n_x}{d_x}\right)^2 + \left(\frac{\pi n_y}{d_y}\right)^2 + \left(\frac{\pi n_z}{d_z}\right)^2\right]\right]^{1/2} \tag{1.108}$$

The electron concentration is given by

$$n_{0D} = \frac{g_v}{d_x d_y d_z} \sum_{n_x=1}^{n_{x\max}} \sum_{n_y=1}^{n_{y\max}} \sum_{n_z=1}^{n_{z\max}} F_{-1}\left(\eta_{46,\pm}\right) \tag{1.109}$$

where $\eta_{46,\pm} = \left(E_{F0D} - E_{QD16,\pm}\right)/k_B T$

The photo-emitted current density can be written as

$$J_{0D}=\frac{\alpha_0 g_v e}{2d_x d_y d_z \hbar}\sum_{n_x=1}^{n_{x_{\max}}}\sum_{n_y=1}^{n_{y_{\max}}}\sum_{n_{z_{\min}}}^{n_{z_{\max}}}F_{-1}(\eta_{46})Q_{14}\left(E_{n_{z,\pm}}\right) \tag{1.110}$$

where $n_{z_{\min}}$ should be determined from the following equation

$$(W-h\nu)=\frac{1}{2}\left[\Delta_1-2\gamma_1\cos\left(\frac{\bar{c}\left(\pi n_{z_{\min}}\right)}{2d_z}\right)+2\gamma_5\cos^2\left(\frac{\bar{c}\left(\pi n_{z_{\min}}\right)}{2d_z}\right)+2\gamma_2\cos^2\left(\frac{\bar{c}\left(\pi n_{z_{\min}}\right)}{2d_z}\right)\right]$$

$$\pm\left[\frac{1}{4}\left(\Delta_1-2\gamma_1\cos\left(\frac{\bar{c}\left(\pi n_{z_{\min}}\right)}{2d_z}\right)+2\gamma_5\cos^2\left(\frac{\bar{c}\left(\pi n_{z_{\min}}\right)}{2d_z}\right)-2\gamma_2\cos^2\left(\frac{\bar{c}\left(\pi n_{z_{\min}}\right)}{2d_z}\right)\right)^2+\frac{3(\bar{a})^2}{4}\left(\gamma_0+2\gamma_4\cos\left(\frac{\bar{c}\left(\pi n_{z_{\min}}\right)}{2d_z}\right)\right)^2\left(\frac{\pi n_{z_{\min}}}{d_z}\right)^2\right]^{1/2} \tag{1.111}$$

1.2.8. The EP from QDs of Platinum Antimonide

The dispersion relation for the n-type $PtSb_2$ can be written as [162]

$$\left(E+\lambda_0\frac{\left(\bar{\bar{a}}\right)^2}{4}k^2-lk_S^2\frac{\left(\bar{\bar{a}}\right)^2}{4}\right)\left(E+\delta_0-\upsilon\frac{\left(\bar{\bar{a}}\right)^2}{4}k^2-\bar{\bar{n}}\frac{\left(\bar{\bar{a}}\right)^2}{4}k_S^2\right)=I\frac{\left(\bar{\bar{a}}\right)^4}{16}k^4 \tag{1.112}$$

where $\omega_1\equiv\left(\lambda_0\frac{\left(\bar{\bar{a}}\right)^2}{4}-l\frac{\left(\bar{\bar{a}}\right)^2}{4}\right)$, $\omega_2\equiv\lambda_0\frac{\left(\bar{\bar{a}}\right)^2}{4}$,

$\omega_3\equiv\left(\bar{\bar{n}}\frac{\left(\bar{\bar{a}}\right)^2}{4}+\upsilon\frac{\left(\bar{\bar{a}}\right)^2}{4}\right)$, $\omega_4\equiv\upsilon\frac{\left(\bar{\bar{a}}\right)^2}{4}$, $I_1=I(\left(\bar{\bar{a}}\right)^4/16)$,

λ_0, l, δ_0, ν and $\bar{\bar{n}}$ are the band constants and $\bar{\bar{a}}$ is the lattice constant.

The $v_z(E_{n_z})$ in this case is given by

$$v_z(E_{n_z})=\frac{1}{\hbar}Q_{15}\left(E_{n_z}\right) \tag{1.113}$$

where $Q_{15}\left(E_{n_z}\right)=\left[2E_{n_z}+\delta_0-\omega_4\left(\frac{\pi n_z}{d_2}\right)^2+\omega_2\left(\frac{\pi n_z}{d_z}\right)^2\right]^{-1}.$

$$\cdot\left[4I_1\left(\frac{\pi n_z}{d_z}\right)^3+2\omega_4\left(\frac{\pi n_z}{d_z}\right)\left(E_{n_z}+\omega_2\left(\frac{\pi n_z}{d_z}\right)^2\right)-2\omega_2\left(\frac{\pi n_z}{d_z}\right)\left(E+\delta_0-\omega_4\left(\frac{\pi n_z}{d_z}\right)^2\right)\right]$$

in which E_{n_z} should be determined from the following equation

$$\left[E_{n_z}+\omega_2\left(\frac{\pi n_z}{d_z}\right)^2\right]\left[E_{n_z}+\delta_0-\omega_4\left(\frac{\pi n_z}{d_z}\right)^2\right]=I_1\left(\frac{\pi n_z}{d_z}\right)^4 \tag{1.114}$$

The totally quantized energy in this case can be written as

$$\left[E_{QD17}+\omega_1\left[\left(\frac{\pi n_x}{d_x}\right)^2+\left(\frac{\pi n_y}{d_y}\right)^2\right]+\omega_2\left(\frac{\pi n_z}{d_z}\right)^2\right].$$

$$\left[E_{QD}+\delta_0+\omega_3\left[\left(\frac{\pi n_x}{d_x}\right)^2+\left(\frac{\pi n_y}{d_y}\right)^2\right]-\omega_4\left(\frac{\pi n_z}{d_z}\right)^2\right]$$

$$=I_1\left[\left(\frac{\pi n_x}{d_x}\right)^2+\left(\frac{\pi n_y}{d_y}\right)^2+\left(\frac{\pi n_z}{d_z}\right)^2\right]^2 \tag{1.115}$$

The carrier concentration is given by

$$p_{0D}=\frac{2g_v}{d_x d_y d_z}\sum_{n_x=1}^{n_{x\max}}\sum_{n_y=1}^{n_{y\max}}\sum_{n_z=1}^{n_{z\max}}F_{-1}\left(\eta_{47}\right) \tag{1.116}$$

where $\eta_{47}=(E_{F0D}-E_{QD17})/(K_BT)$

The photo-emitted current density assumes the form

$$J_{0D}=\frac{\alpha_0 e g_v}{d_x d_y d_z\hbar}\sum_{n_x=1}^{n_{x\max}}\sum_{n_y=1}^{n_{y\max}}\sum_{n_{z\min}}^{n_{z\max}}Q_{15}\left(E_{n_z}\right)F_{-1}\left(\eta_{47}\right) \tag{1.117}$$

where $n_{z_{\min}}$ should be determined from the following equation

$$\left[(W-h\nu)+\omega_2\left(\frac{\pi n_{z_{\min}}}{d_z}\right)^2\right]\left[(W-h\nu)+\delta_0-\omega_4\left(\frac{\pi n_{z_{\min}}}{d_z}\right)^2\right]=I_1\left[\frac{\pi n_{z_{\min}}}{d_z}\right]^4 \quad (1.118)$$

1.2.9. The EP from QDs of Zero-gap Materials

The dispersion relation of the conduction electrons in gapless materials [163] is given by

$$E=\frac{\hbar^2k^2}{2m_c}+\frac{3e^2k}{128\varepsilon_{sc}}-\frac{2}{\pi}E_B\ln\left|\frac{k}{k_0}\right| \quad (1.119)$$

where ε_{sc} is the semiconductor permittivity, E_B is the Bohr electron energy and k_0 is the inverse Bohr radius.

The $v_z(E_{n_z})$ in this case is given by

$$v_z(E_{n_z})=\frac{1}{\hbar}Q_{16}\left(E_{n_z}\right) \quad (1.120)$$

where $Q_{16}\left(E_{n_z}\right)=\left[\frac{\hbar}{m_c}\left(\frac{\pi n_z}{d_z}\right)+\frac{1}{\hbar}\left[\frac{3e^2}{128\varepsilon_{sc}}-\frac{2E_Bd_z}{\pi^2n_z}\right]\right]$

in which E_{n_z} should be determined from the following equation

$$E_{n_z}=\left[\frac{\hbar^2}{2m_c}\left(\frac{\pi n_z}{d_z}\right)^2+\frac{3e^2}{128\varepsilon_{sc}}\cdot\left(\frac{\pi n_z}{d_z}\right)-\frac{2}{\pi}E_B\ln\left|\frac{\pi n_z}{d_zk_0}\right|\right] \quad (1.121)$$

The totally quantized energy is given by

$$E_{QD18}=\frac{\hbar^2\pi^2}{2m_c}\left[\left(\frac{n_x}{d_x}\right)^2+\left(\frac{n_y}{d_y}\right)^2+\left(\frac{n_z}{d_z}\right)^2\right]-\frac{3e^2}{128\varepsilon_{sc}}\left[\left(\frac{\pi n_x}{d_x}\right)^2+\left(\frac{\pi n_y}{d_y}\right)^2+\left(\frac{\pi n_z}{d_z}\right)^2\right]^{1/2}$$

$$-\frac{E_B}{\pi}\ln\left|\frac{(\pi n_x/d_x)^2+(\pi n_y/d_y)^2+(\pi n_z/d_z)^2}{k_0^2}\right| \quad (1.122)$$

The electron concentration assumes the form

$$n_{0D} = \frac{2g_v}{d_x d_y d_z} \sum_{n_x=1}^{n_{x\max}} \sum_{n_y=1}^{n_{y\max}} \sum_{n_z=1}^{n_{z\max}} F_{-1}(\eta_{48}) \tag{1.123}$$

where $\eta_{48} = \left(E_{F0D} - E_{QD18}\right) / k_B T$

The photo-emitted current density can be written as

$$J_{0D} = \frac{\alpha_0 g_v e}{\hbar d_x d_y d_z} \sum_{n_x=1}^{n_{x\max}} \sum_{n_y=1}^{n_{y\max}} \sum_{n_z=1}^{n_{z\max}} Q_{16}\left(E_{n_z}\right) F_{-1}(\eta_{48}) \tag{1.124}$$

where $n_{z_{\min}}$ should be determined from the following equation

$$\left(W - h\nu\right) = \frac{\hbar^2}{2m_c}\left[\frac{\pi n_{z_{\min}}}{d_z}\right]^2 + \frac{3e^2}{128\varepsilon_{sc}}\left(\frac{\pi n_{z_{\min}}}{d_z}\right) - \frac{2}{\pi} E_B \ln\left|\frac{\pi n_{z_{\min}}}{k_0 d_z}\right| \tag{1.125}$$

1.2.10. The EP from QDs of II-V Materials

The dispersion relation of the holes in II-V compounds in accordance with Yamada [164] can be expressed as

$$E = A_{10}k_x^2 + A_{11}k_y^2 + A_{12}k_z^2 + A_{13}k_x \pm \left[\left(A_{14}k_x^2 + A_{15}k_y^2 + A_{16}k_z^2 + A_{17}k_x\right)^2 + A_{18}k_y^2 + A_{19}^2\right]^{1/2} \tag{1.126}$$

where A_{10}, A_{11}, A_{12}, A_{13}, A_{14}, A_{15}, A_{16}, A_{17}, A_{18} and A_{19} are energy band constants.

The $v_z(n_z)$ in this case is given by

$$v_z\left(E_{n_z}\right) = \frac{1}{\hbar} Q_{24,\pm}\left(E_{n_z}\right) \tag{1.127}$$

where $Q_{24,\pm}(n_z) = \left[2A_{12}\left(\frac{\pi n_z}{d_z}\right) \pm 2A_{16}^2\left(\frac{\pi n_z}{d_z}\right)^3\left[A_{16}^2\left(\frac{\pi n_z}{d_z}\right)^4 + A_{19}^2\right]^{-1/2}\right]$

in which E_{n_z} should be determined from the following equation

$$E_{n_z} = A_{12}\left(\frac{\pi n_z}{d_z}\right)^2 \pm \left[A_{16}^2\left(\frac{\pi n_z}{d_z}\right)^4 + A_{19}^2\right]^{1/2} \tag{1.128}$$

The totally quantized energy assumes the form

$$E_{QD31,\pm} = A_{10}\left(\frac{\pi n_x}{d_x}\right)^2 + A_{11}\left(\frac{\pi n_y}{d_y}\right)^2 + A_{12}\left(\frac{\pi n_z}{d_z}\right)^2 + A_{13}\left(\frac{\pi n_x}{d_x}\right)$$

$$\pm\left[\left(A_{14}\left(\frac{\pi n_x}{d_x}\right)^2 + A_{15}\left(\frac{\pi n_y}{d_y}\right)^2 + A_{16}\left(\frac{\pi n_z}{d_z}\right)^2 + A_{17}\left(\frac{\pi n_x}{d_x}\right)\right)^2 + A_{18}\left(\frac{\pi n_y}{d_y}\right)^2 + A_{19}{}^2\right]^{1/2} \tag{1.129}$$

The hole concentration is given by

$$p_{0D} = \frac{g_v}{d_x d_y d_z}\sum_{n_x=1}^{n_{x\max}}\sum_{n_y=1}^{n_{y\max}}\sum_{n_z=1}^{n_{z\max}}\left[1+\exp\left(\eta_{60,\pm}\right)\right]^{-1}$$

where $\eta_{60,\pm} = \left(E_{F0D} - E_{QD31,\pm}\right)/k_B T$ 1.130)

The photo-emitted current density can be written as

$$J_{0D} = \frac{\alpha_0 e g_v}{2\hbar d_x d_y d_z}\sum_{n_x=1}^{n_{x\max}}\sum_{n_y=1}^{n_{y\max}}\sum_{n_{z\min}}^{n_{z\max}} Q_{24,\pm}\left(E_{n_z}\right)\left[1+\exp\left(\eta_{60,\pm}\right)\right]^{-1} \tag{1.131}$$

where $n_{z_{\min}}$ should be the nearest integer of the following equation

$$W - h\nu = A_{12}\left(\frac{\pi n_{z_{\min}}}{d_z}\right)^2 \pm\left[A_{16}^2\left(\frac{\pi n_{z_{\min}}}{d_z}\right)^4 + A_{19}^2\right]^{1/2} \tag{1.132}$$

3.2.11. The EP from QDs of Gallium Antimonide

The conduction electrons of n-GaSb obey the three dispersion relations as provided by Seiler et. al. [165], Mathur et. al. [166] and Zhang [167] respectively. The photoemission from QDs of GaSb is being presented in accordance with the aforementioned models for the joint purpose of coherent presentation and relative assessment.

a) In accordance with the model of Seiler et. al. [165] the dispersion relation of the conduction electrons in n-GaSb assume the form

$$E = \left[-\frac{E_{g_0}}{2} + \frac{E_{g_0}}{2}\left[1+\alpha_4 k^2\right]^{1/2} + \frac{\overline{\zeta_0}\hbar^2 k^2}{2m_0} + \frac{\overline{v_0}\theta_1(k)\hbar^2}{2m_0} \pm \frac{\overline{w_0}\theta_2(k)\hbar^2}{2m_0}\right] \tag{1.133}$$

where $\alpha_4 = 4P_0^2\left(E_{g_0}+\frac{2}{3}\Delta\right)\left[E_{g_0}^2\left(E_{g_0}+\Delta\right)\right]^{-1}$, represents the warping of the Fermi surface, represents the inversion asymmetry splitting of the conduction band, $\overline{\zeta_0}\left(=-2.1\right)$, $\overline{v_0}\left(=-1.49\right)$ and $\overline{w_0}\left(=0.42\right)$ represent the constants of the spectrum.

The $v_z(E_{n_z})$ in this case is given by

$$v_z\left(E_{n_{z,\pm}}\right)=\frac{1}{\hbar}Q_{18}\left(E_{n_{z,\pm}}\right) \tag{1.134}$$

where $Q_{18}\left(E_{n_{z,\pm}}\right)=\left[\alpha_4 E_{g_0}\left(\frac{\pi n_z}{d_z}\right)\left[1+\alpha_4\left(\frac{\pi n_z}{d_z}\right)^2\right]^{-1/2}+\left(\frac{\overline{\zeta_0}\hbar^2}{m_0}\right)\left(\frac{\pi n_z}{d_z}\right)\right]$

in which $E_{n_{z,\pm}}$ should be determined from the following equation

$$E_{n_{z,\pm}}=-\frac{E_{g_0}}{2}+\frac{E_{g_0}}{2}\left[1+\alpha_4\left(\frac{\pi n_z}{d_z}\right)^2\right]^{1/2}+\frac{\overline{\zeta_0}\hbar^2}{2m_0}\left(\frac{\pi n_z}{d_z}\right)^2 \tag{1.135}$$

The totally quantized energy is given by

$$E_{QD0,\pm}=\left[-\frac{E_{g_0}}{2}+\frac{E_{g_0}}{2}\left[1+\alpha_4\left[\left(\frac{\pi n_x}{d_x}\right)^2+\left(\frac{\pi n_y}{d_y}\right)^2+\left(\frac{\pi n_z}{d_z}\right)^2\right]\right]^{1/2}\right.$$
$$\left.+\left(\frac{\overline{\zeta_0}\hbar^2}{2m_0}\right)\left[\left(\frac{\pi n_x}{d_x}\right)^2+\left(\frac{\pi n_y}{d_y}\right)^2+\left(\frac{\pi n_z}{d_z}\right)^2\right]+\left(\frac{\overline{v_0}\hbar^2}{2m_0}\right)S_6\pm\left(\frac{\overline{w_0}S_7\hbar^2}{2m_0}\right)\right] \tag{1.136}$$

where

$$S_6=\left[\left(\frac{\pi n_x}{d_x}\right)^2+\left(\frac{\pi n_y}{d_y}\right)^2+\left(\frac{\pi n_z}{d_z}\right)^2\right]^{-2}\left[\left(\frac{\pi^2 n_x n_y}{d_x d_y}\right)^2+\left(\frac{\pi^2 n_y n_z}{d_y d_z}\right)^2+\left(\frac{\pi^2 n_z n_x}{d_x d_z}\right)^2\right],$$

$$S_7=\left[\left\{\left(\left(\frac{\pi n_x}{d_x}\right)^2+\left(\frac{\pi n_y}{d_y}\right)^2+\left(\frac{\pi n_z}{d_z}\right)^2\right)\left(\frac{n_x^2 n_y^2\pi^4}{d_x^2 d_y^2}+\frac{n_y^2 n_z^2\pi^4}{d_y^2 d_z^2}+\frac{n_z^2 n_x^2\pi^4}{d_z^2 d_x^2}\right)-\frac{9\pi^6 n_x^2 n_y^2 n_z^2}{d_x^2 d_y^2 d_z^2}\right\}^{1/2}\right.$$

$$\left[\left(\frac{\pi n_x}{d_x}\right)^2+\left(\frac{\pi n_y}{d_y}\right)^2+\left(\frac{\pi n_z}{d_z}\right)^2\right]^{-1/2}\Bigg]$$

The electron concentration is given by

$$n_{0D}=\frac{g_v}{d_x d_y d_z}\sum_{n_x=1}^{n_{x\max}}\sum_{n_y=1}^{n_{y\max}}\sum_{n_z=1}^{n_{z\max}}F_{-1}\left(\eta_{50,\pm}\right)$$

where $\eta_{50,\pm}=\left(E_{F0D}-E_{QD20,\pm}\right)/k_BT$ (1.137)

The photo-emitted current density can be written as

$$J_{0D}=\frac{\alpha_0 e g_v}{2\hbar d_x d_y d_z}\sum_{n_x=1}^{n_{x\max}}\sum_{n_y=1}^{n_{y\max}}\sum_{n_{z\min}}^{n_{z\max}}Q_{18}\left(E_{n_{z,\pm}}\right)F_{-1}\left(\eta_{50,\pm}\right) \tag{1.138}$$

where $n_{z_{\min}}$ should be determined from the following equation

$$(W-h\nu)=-\frac{E_{g_0}}{2}+\frac{E_{g_0}}{2}\left[1+\alpha_4\left(\frac{\pi n_{z_{\min}}}{d_z}\right)^2\right]^{1/2}+\overline{\zeta}_0\left(\frac{\hbar^2}{2m_0}\right)\left(\frac{\pi n_{z_{\min}}}{d_z}\right)^2 \tag{1.139}$$

(b) In accordance with the model of Mathur et. al [166], the electron dispersion law in n-GaSb can be written as

$$E=\alpha_9 k^2+\frac{E_{g1}}{2}\left[\sqrt{1+\alpha_{10}k^2}-1\right] \tag{1.140}$$

where $\alpha_9=\frac{\hbar^2}{2m_0}$, $E_{g1}=E_{g_0}+\left[\frac{5\times10^{-S}T^2}{2(112+T)}\right]$, $\alpha_{10}=\left(\frac{\left(2\hbar^2\right)}{\left(E_{g1}\right)}\right)\left[\frac{1}{m^*}-\frac{1}{m_0}\right]$.

From (1.139), we get

$$k^2=\frac{E}{\alpha_9}+\alpha_{11}-\left[\alpha_{12}E+\alpha_{13}\right]^{1/2} \tag{1.141}$$

where $\alpha_{11} = \frac{E_{g1}^2}{8\alpha_9^2}\left[\alpha_{10} + \frac{4\alpha_9}{E_{g1}}\right], \alpha_{12} = \left(\frac{E_{g1}^2}{\alpha_9^3}\right), \alpha_{13} = \frac{E_{g1}^4}{64\alpha_9^4}\left[\alpha_{10}^2 + \frac{10\alpha_9^2}{E_{g1}^2} - \frac{8\alpha_9\alpha_{10}}{E_{g1}}\right].$

The $v_z(E_{n_z})$ in this case is given by

$$v_z\left(E_{n_z}\right) = \frac{1}{\hbar}Q_{19}\left(E_{n_z}\right) \tag{1.142}$$

where $Q_{19}\left(E_{n_z}\right) = \left[2\alpha_9\left(\frac{\pi n_z}{d_z}\right) + \frac{E_{g1}}{2}\left[\alpha_{10}\left(\frac{\pi n_z}{d_z}\right)\left(1 + \alpha_{10}\left(\frac{\pi n_z}{d_z}\right)^2\right)^{-1/2}\right]\right]$

in which E_{n_z} should be determined from the following equation

$$E_{n_z} = \left(\frac{2n_z\pi}{d_z}\right)\left[\alpha_9 + \frac{\alpha_{10}E_{g1}}{4}\left[1 + \alpha_{10}\left(\frac{\pi n_z}{d_z}\right)^2\right]^{-1/2}\right] \tag{1.143}$$

The totally quantized energy assumes the form

$$E_{QD21} = \alpha_9\left[\left(\frac{\pi n_x}{d_x}\right)^2 + \left(\frac{\pi n_y}{d_y}\right)^2 + \left(\frac{\pi n_z}{d_z}\right)^2\right] + \frac{E_{g1}}{2}\left[\sqrt{1 + \alpha_{10}\left[\left(\frac{\pi n_x}{d_x}\right)^2 + \left(\frac{\pi n_y}{d_y}\right)^2 + \left(\frac{\pi n_z}{d_z}\right)^2\right]} - 1\right] \tag{1.144}$$

The electron concentration is given by

$$n_{0D} = \frac{2g_v}{d_x d_y d_z}\sum_{n_x=1}^{n_{x\max}}\sum_{n_y=1}^{n_{y\max}}\sum_{n_{z\min}}^{n_{z\max}} F_{-1}\left(\eta_{51}\right) \tag{1.145}$$

where $\eta_{51} = \left(E_{F0D} - E_{QD21}\right)/\left(k_B T\right)$

The photo-emitted current density can be written as

$$J_{0D} = \frac{\alpha_0 e g_v}{\hbar d_x d_y d_z}\sum_{n_x=1}^{n_{x\max}}\sum_{n_y=1}^{n_{y\max}}\sum_{n_{z\min}}^{n_{z\max}} Q_{19}\left(E_{n_z}\right)F_{-1}\left(\eta_{51}\right) \tag{1.146}$$

where $n_{z_{\min}}$ should be determined from the following equation

$$W - h\nu = \alpha_9 \left(\frac{\pi n_{z_{\min}}}{d_z}\right)^2 + \frac{E_{g1}}{2}\left[-1+\sqrt{1+\alpha_{10}\left(\frac{\pi n_{z_{\min}}}{d_z}\right)^2}\right] \tag{1.147}$$

(c) The dispersion relation of the conduction electrons in n-GaSb can be expressed in accordance with Zhang [176] as

$$E = \left[\overline{E_1} + \overline{E_2} Z_1(k)\right]k^2 + \left[\overline{E_3} + \overline{E_4} Z_1(k)\right]k^4 + k^6\left[\overline{E_5} + \overline{E_6} Z_1(k) + \overline{E_7} Z_2(k)\right] \tag{1.148}$$

where $Z_1(k) = \frac{5\sqrt{21}}{4}\left[\frac{k_x^4 + k_y^4 + k_z^4}{k^4} - \frac{3}{5}\right]$,

$$Z_2(k) = \left[\frac{639639}{32}\right]^{1/2}\left[\frac{k_x^2 k_y^2 k_z^2}{k^6} + \frac{1}{12}\left(\frac{k_x^4 + k_y^4 + k_z^4}{k^4} - \frac{3}{5}\right) - \frac{1}{105}\right],$$

the coefficients are in eV, the values of k are in $10\left(\frac{\mathrm{a}}{2\pi}\right)$ times of k in atomic units (a is lattice constant), $\overline{E_1}$, $\overline{E_2}$, $\overline{E_3}$, $\overline{E_4}$, $\overline{E_5}$, $\overline{E_6}$ and $\overline{E_7}$ are energy-band constants.

The $v_z(E_{n_z})$ in this case is given by

$$v_z\left(E_{n_z}\right) = \frac{1}{\hbar} Q_{20}\left(E_{n_z}\right) \tag{1.149}$$

where

$$Q_{20}\left(E_{n_z}\right) = 2\left(\frac{\pi n_z}{d_z}\right)\left[\overline{E_1} + \frac{\sqrt{21}}{2}.\overline{E_2}\right] + 4\left(\frac{\pi n_z}{d_z}\right)^3\left[\overline{E_3} + \frac{\sqrt{21}}{2}.\overline{E_4}\right] + 6\left(\frac{\pi n_z}{d_z}\right)^5.\left[\overline{E_5} + \frac{\sqrt{21}}{2}.\overline{E_6} + \left(\frac{\overline{E_7}}{42}\right)\left(\frac{639639}{32}\right)^{1/2}\right]$$

in which E_{n_z} should be determined from the following equation

$$E_{n_z} = \left[\overline{E_1} + \overline{E_2}\left(\frac{\sqrt{21}}{2}\right)\right]\left(\frac{\pi n_z}{d_z}\right)^2 + \left(\overline{E_3} + \overline{E_4}\left(\frac{\sqrt{21}}{2}\right)\right)\left(\frac{\pi n_z}{d_z}\right)^4 + \left(\frac{\pi n_z}{d_z}\right)^6\left[\overline{E_5} + \overline{E_6}\left(\frac{\sqrt{21}}{2}\right) + \overline{E_7}\left(\frac{639639}{32}\right)^{1/2}\left[\frac{1}{42}\right]\right] \tag{1.150}$$

The totally quantized energy is given by

$$E_{QD22} = \left[\overline{E_1} + \overline{E_2} Z_3\right]\left[\left(\frac{\pi n_x}{d_x}\right)^2 + \left(\frac{\pi n_y}{d_y}\right)^2 + \left(\frac{\pi n_z}{d_z}\right)^2\right]$$

$$+\left[\left(\frac{\pi n_x}{d_x}\right)^2 + \left(\frac{\pi n_y}{d_y}\right)^2 + \left(\frac{\pi n_z}{d_z}\right)^2\right]^2 .\left[\overline{E_3} + \overline{E_4} Z_3\right]$$

$$+\left[\left(\frac{\pi n_x}{d_x}\right)^2 + \left(\frac{\pi n_y}{d_y}\right)^2 + \left(\frac{\pi n_z}{d_z}\right)^2\right]^3 .\left[\overline{E_5} + \overline{E_6} Z_3 + \overline{E_7} Z_4\right] \tag{1.151}$$

where $Z_3 = \frac{5\sqrt{21}}{4}\left[\left[\left(\frac{n_x\pi}{d_x}\right)^4 + \left(\frac{n_y\pi}{d_y}\right)^4 + \left(\frac{n_z\pi}{d_z}\right)^4\right] . \left[\left(\frac{n_x\pi}{d_x}\right)^2 + \left(\frac{n_y\pi}{d_y}\right)^2 + \left(\frac{n_z\pi}{d_z}\right)^2\right]^{-2} - \frac{3}{5}\right]$

and $Z_4 = \left[\frac{639639}{32}\right]^{1/2}$.

$$\left[\left(\frac{n_x}{d_x}\frac{n_y}{d_y}\frac{n_z}{d_z}\right)^2\left[\left(\frac{n_x}{d_x}\right)^2 + \left(\frac{n_y}{d_y}\right)^2 + \left(\frac{n_z}{d_z}\right)^2\right]^{-3} + \frac{1}{(15)\sqrt{21}} Z_3 - \frac{1}{105}\right]$$

The electron concentration assumes the form

$$n_{0D} = \frac{g_v}{d_x d_y d_z}\sum_{n_x=1}^{n_{x_{\max}}}\sum_{n_y=1}^{n_{y_{\max}}}\sum_{n_z=1}^{n_{z_{\max}}} F_{-1}(\eta_{52}) \text{ where } \eta_{52} = \left(E_{F0D} - E_{QD22}\right)/k_B T \tag{1.152}$$

The photo-emitted current density can be written as

$$J_{0D} = \frac{\alpha_0 e g_v}{2\hbar d_x d_y d_z}\sum_{n_x=1}^{n_{x_{\max}}}\sum_{n_y=1}^{n_{y_{\max}}}\sum_{n_{z_{\min}}}^{n_{z_{\max}}} Q_{20} F_{-1}\left(E_{n_z}\right) F_{-1}(\eta_{52}) \tag{1.153}$$

where $n_{z_{\min}}$ should be determined from the following equation

$$W - h\nu = \left[\overline{E_1} + \overline{E_2}\left(\frac{\sqrt{21}}{2}\right)\right]\left(\frac{\pi n_{z_{\min}}}{d_z}\right)^2 + \left(\frac{\pi n_{z_{\min}}}{d_z}\right)^4\left[\overline{E_3} + \overline{E_4}\left(\frac{\sqrt{21}}{2}\right)\right] + \left(\frac{\pi n_{z_{\min}}}{d_z}\right)^6\left[\overline{E_5} + \overline{E_6}\left[\frac{\sqrt{21}}{2}\right] + \overline{E_7} . \left(\frac{1}{42}\right)\left(\frac{639639}{32}\right)^{1/2}\right] \tag{1.154}$$

3.2.12. The EP from QDs of Stressed Materials

The electron dispersion law in bulk specimens of stressed materials can be written as [168]

$$\frac{k_x^2}{[a^*(E)]^2}+\frac{k_y^2}{[b^*(E)]^2}+\frac{k_z^2}{[c^*(E)]^2}=1 \tag{1.155}$$

where, $\left[a^*(E)\right]^2=\dfrac{\bar{K}_0(E)}{\bar{A}_0(E)+\frac{1}{2}\bar{D}_0(E)}$, $\bar{K}_0(E)=\left[E-C_1\varepsilon-\dfrac{2C_2^2\varepsilon_{xy}^2}{3E'_{g_0}}\right]\left(\dfrac{3E'_{g_0}}{2B_2^2}\right)$, is the conduction band deformation potential, is the trace of the strain tensor which can be written as, C_2 is a constant which describes the strain interaction between the conduction and valance bands, $E'_{g_0}=E_{g_0}+E-C_1\varepsilon$, E_{g_0} is the band gap in the absence of stress, B_2 is the momentum matrix element, $\bar{A}_0(E)=\left[1-\dfrac{(\bar{a}_0+C_1)}{E'_{g_0}}+\dfrac{3\bar{b}_0\varepsilon_{xx}}{2E'_{g_0}}-\dfrac{\bar{b}_0\varepsilon}{2E'_{g_0}}\right]$, $\bar{a}_0=-\frac{1}{3}\left(\bar{b}_0+2\bar{m}\right)$, $\bar{b}_0=\frac{1}{3}\left(\bar{l}-\bar{m}\right)$, $\bar{d}_0=\dfrac{2\bar{n}}{\sqrt{3}}$, $\bar{l},\bar{m},\bar{n}$ are the matrix elements of the strain perturbation operator, $\bar{D}_0(E)=\left(\bar{d}_0\sqrt{3}\right)\dfrac{\varepsilon_{xy}}{E'_{g_0}}$, $\left[b^*(E)\right]^2=\dfrac{\bar{K}_0(E)}{\bar{A}_0(E)-\frac{1}{2}\bar{D}_0(E)}$,

and $\bar{L}_0(E)=\left[1-\dfrac{(\bar{a}_0+C_1)}{E'_{g_0}}+\dfrac{3\bar{b}_0\varepsilon_{zz}}{E'_{g_0}}-\dfrac{\bar{b}_0\varepsilon}{2E'_{g_0}}\right]$

The $v_z\left(E_{n_z}\right)$ for QDs of stressed materials in this case is given by

$$v_z\left(E_{n_z}\right)=[Q_{21}\left(E_{n_z}\right)/\hbar] \tag{1.156}$$

where $Q_{21}\left(E_{n_z}\right)=[\left[c^*\left(E_{n_z}\right)\right]']^{-1}$

The E_{n_z} can be expressed through the equation

$$c^*\left(E_{n_z}\right)=(n_z\pi/d_z) \tag{1.157}$$

The totally quantized energy in this case assumes the form

$$\left(\frac{\pi n_x}{d_x}\right)^2 \left[a^*\left(E_{QD23}\right)\right]^{-2} + \left(\frac{\pi n_y}{d_y}\right)^2 \left[b^*\left(E_{QD23}\right)\right]^{-2} + \left(\frac{\pi n_z}{d_z}\right)^2 \left[c^*\left(E_{QD23}\right)\right]^{-2} = 1$$

(1.158)

The electron concentration is given by

$$n_{0D} = \frac{2g_v}{d_x d_y d_z} \sum_{n_x=1}^{n_{x_{\max}}} \sum_{n_y=1}^{n_{y_{\max}}} \sum_{n_z=1}^{n_{z_{\max}}} F_{-1}\left(\bar{\eta}_{53}\right) \tag{1.159}$$

where $\bar{\eta}_{53} = \left(k_B T\right)^{-1}\left(E_{F0D} - E_{QD23}\right)$

The photo-emitted current density can be written as

$$J_{0D} = \frac{\alpha_0 e g_v}{d_x d_y d_z \hbar} \sum_{n_x=1}^{n_{x_{\max}}} \sum_{n_y=1}^{n_{y_{\max}}} \sum_{n_z=1}^{n_{z_{\max}}} Q_{21}\left(E_{n_z}\right) F_{-1}\left(\bar{\eta}_{53}\right) \tag{1.160}$$

where $n_{z_{\min}}$ satisfies the following inequality

$$n_{z_{\min}} \geq \left(\frac{d_z}{\pi}\right)\left[c^*\left(W - h\nu\right)\right] \tag{1.161}$$

1.2.13. The EP from Qds of Bismuth

(a) The McClure and Choi Model

The dispersion relation of the carriers in Bi can be written, following the McClure and Choi [105], as

$$E(1+\alpha E) = \frac{p_x^2}{2m_1} + \frac{p_y^2}{2m_2} + \frac{p_z^2}{2m_3} + \frac{p_y^2}{2m_2}\alpha E\left\{1 - \left(\frac{m_2}{m_2'}\right)\right\} + \frac{p_y^4 \alpha}{4m_2 m_2'} - \frac{\alpha p_x^2 p_y^2}{4m_1 m_2} - \frac{\alpha p_y^2 p_z^2}{4m_2 m_3}$$

(1.162)

where $p_i \equiv \hbar k_i$, $i = x, y, z$, are the effective carrier masses at the band-edge along x, y and z directions respectively and m_2' is the effective-mass tensor component at the top of the valence band (for electrons) or at the bottom of the conduction band (for holes).

The conduction electrons in QDs of Bi in accordance with McClure and Choi (1.162) can be expressed as

$$v_z\left(E_{n_z}\right)=\frac{\pi\hbar n_z}{d_z m_3\left(1+2\alpha E_{n_z}\right)} \tag{1.163}$$

E_{n_z} should be determined from the following equation

$$E_{n_z}\left(1+\alpha E_{n_z}\right)=\frac{\hbar^2}{2m_3}\left(\frac{\pi n_z}{d_z}\right)^2 \tag{1.164}$$

The totally quantized energy is given by

$$E_{QD24}\left(1+\alpha E_{QD24}\right)=\frac{\hbar^2}{2m_1}\left(\frac{\pi n_x}{d_x}\right)^2+\frac{\hbar^2}{2m_2}\left(\frac{\pi n_y}{d_y}\right)^2+\frac{\hbar^2}{2m_3}\left(\frac{\pi n_z}{d_z}\right)^2$$
$$+\frac{\hbar^2\alpha E_{QD24}}{2m_2}\left(\frac{\pi n_y}{d_y}\right)^2\left(1-\frac{m_2}{m_2'}\right)+\frac{\alpha\hbar^4}{4m_2m_2'}\left(\frac{\pi n_y}{d_y}\right)^4-\frac{\alpha\hbar^4}{4m_1m_2}\left(\frac{\pi^2 n_x^2}{d_x^2}\cdot\frac{\pi^2 n_y^2}{d_y^2}\right)-\frac{\alpha\hbar^4\pi^4}{4m_2m_3}\left(\frac{n_y n_z}{d_y d_z}\right)^2 \tag{1.165}$$

The electron concentration is given by

$$n_{0D}=\frac{2g_v}{d_x d_y d_z}\sum_{n_x=1}^{n_{x_{\max}}}\sum_{n_y=1}^{n_{y_{\max}}}\sum_{n_z=1}^{n_{z_{\max}}}F_{-1}\left(\eta_{54}\right) \tag{1.166}$$

where $\eta_{54}=\left(k_BT\right)^{-1}\left(E_{F0D}-E_{QD24}\right)$

The photo-emitted current density can be written as

$$J_{0D}=\frac{aeg_v\hbar\pi}{m_3.d_x d_y d_z^2}\sum_{n_x=1}^{n_{x_{\max}}}\sum_{n_y=1}^{n_{y_{\max}}}\sum_{n_{z_{\min}}}^{n_{z_{\max}}}\frac{n_z F_{-1}\left(\eta_{54}\right)}{\left(1+2\alpha E_{n_z}\right)} \tag{1.167}$$

where $n_{z_{\min}}$ should be determined from the following equation

$$\left(W-hv\right)\left[1+\alpha\left(W-hv\right)\right]=\frac{\hbar^2\pi^2}{2m_3}\left(\frac{n_{z_{\min}}}{d_z}\right)^2 \tag{1.168}$$

(b) The Hybrid Model

The dispersion relation of the carriers in bulk specimens of Bi in accordance with the Hybrid model can be represented as [104]

$$E(1+\alpha E)=\frac{\theta_0(E)(\hbar k_y)^2}{2M_2}+\frac{\alpha\gamma_0\hbar^4k_y^4}{4M_2^2}+\frac{\hbar^2k_x^2}{2m_1}+\frac{\hbar^2k_z^2}{2m_3}, \quad (1.169)$$

$\theta_0(E)\equiv\left[1+\alpha E(1-\gamma_0)+\bar{\delta}_0\right]$, $\gamma_0\equiv\frac{M_2}{m_2}$, and the other notations are defined in [104].

The $v_z(E_{n_z})$ for QDs of Bi in accordance with Hybrid model (1.167) can be written as

$$v_z\left(E_{n_z}\right)=(\pi\hbar n_z/(d_z m_3(1+2\alpha E_{n_z}))) \quad (1.170)$$

in which E_{n_z} should be determined from the following equation

$$E_{n_z}\left(1+\alpha E_{n_z}\right)=\left(\hbar^2/2m_3\right)\left(\pi n_z/d_z\right)^2 \quad (1.171)$$

The totally quantized energy is given by

$$E_{QD25}\left(1+\alpha E_{QD25}\right)=\frac{\hbar^2}{2m_1}\left(\frac{\pi n_z}{d_z}\right)^2+\frac{\hbar^2}{2m_3}\left(\frac{\pi n_z}{d_z}\right)^2$$

$$+\frac{\hbar^2}{2M_2}\left(\frac{\pi n_y}{d_y}\right)^2\left[1+\bar{\delta}_0+\alpha E_{QD25}(1-\bar{\gamma}_0)\right]+\frac{\bar{\gamma}_0\hbar^4}{4M_2^2E_{g_0}}\left(\frac{\pi n_y}{d_y}\right)^4 \quad (1.172)$$

The electron concentration is given by

$$n_{0D}=\frac{2g_v}{d_xd_yd_z}\sum_{n_x=1}^{n_{x\max}}\sum_{n_y=1}^{n_{y\max}}\sum_{n_z=1}^{n_{z\max}}F_{-1}(\eta_{55}) \quad (1.173)$$

where $\eta_{55}=(k_BT)^{-1}\left(E_{F0D}-E_{QD25}\right)$

The photo-emitted current density can be written as

$$J_{0D}=\frac{aeg_v\hbar\pi}{m_3.d_xd_yd_z^2}\sum_{n_x=1}^{n_{x\max}}\sum_{n_y=1}^{n_{y\max}}\sum_{n_{z\min}}^{n_{z\max}}\frac{n_zF_{-1}(\eta_{55})}{\left(1+2\alpha E_{n_z}\right)} \quad (1.174)$$

where $n_{z_{\min}}$ should be determined from the following equation

$$(W-hv)\left[1+\alpha(W-hv)\right]=[\hbar^2/2m_3][\pi n_{z_{\min}}/d_z]^2 \quad (1.175)$$

(c) The Cohen Model

In accordance with the Cohen model [103], the dispersion law of the carriers in Bi is given by

$$E(1+\alpha E)=\frac{p_x^{\,2}}{2m_1}+\frac{p_z^{\,2}}{2m_3}-\frac{\alpha E p_y^{\,2}}{2m_2'}+\frac{p_y^{\,2}(1+\alpha E)}{2m_2}+\frac{\alpha p_y^{\,4}}{4m_2m_2'} \tag{1.176}$$

The $v_z(E_{n_z})$ for QDs of Bi in accordance with the model of Cohen (1.174) can be written as

$$v_z\left(E_{n_z}\right)=\frac{\pi\hbar n_z}{\left(1+2\alpha E_{n_z}\right)d_z m_3} \tag{1.177}$$

in which E_{n_z} should be determined from the following equation

$$E_{n_z}\left(1+\alpha E_{n_z}\right)=\left(\hbar^2/2m_3\right)\left(\pi n_z/d_z\right)^2 \tag{1.178}$$

The totally quantized energy E_{QD26} is given by

$$E_{QD26}\left(1+\alpha E_{QD26}\right)=\frac{\hbar^2}{2m_1}\left(\frac{\pi n_x}{d_x}\right)^2+\frac{\hbar^2}{2m_3}\left(\frac{\pi n_z}{d_z}\right)^2+\frac{\alpha\hbar^4}{4m_2m_2'}\left(\frac{\pi n_y}{d_y}\right)^4+\frac{\hbar^2}{2m_2}\left(\frac{\pi n_y}{d_y}\right)^2\left[1+\alpha E_{QD26}\left(1-\frac{m_2}{m_2'}\right)\right] \tag{1.179}$$

The electron concentration is given by

$$n_{0D}=\frac{2g_v}{d_x d_y d_z}\sum_{n_x=1}^{n_{x_{\max}}}\sum_{n_y=1}^{n_{y_{\max}}}\sum_{n_z=1}^{n_{z_{\max}}}F_{-1}\left(\eta_{56}\right) \tag{1.180}$$

where $\eta_{56}=\left(E_{F0D}-E_{QD26}\right)/k_BT$

The photo-emitted current density can be written as

$$J_{0D}=\frac{\hbar\pi a e g_v}{m_3.d_x d_y d_z^{\,2}}\sum_{n_x=1}^{n_{x_{\max}}}\sum_{n_y=1}^{n_{y_{\max}}}\sum_{n_{z_{\min}}}^{n_{z_{\max}}}\frac{n_z F_{-1}\left(\eta_{56}\right)}{\left(1+2\alpha E_{n_z}\right)} \tag{1.181}$$

where $n_{z_{\min}}$ should be determined from the following equation

$$(W-h\nu)\left[1+\alpha(W-h\nu)\right]=\left(\hbar^2/2m_3\right)\left(\pi n_{z_{\min}}/d_z\right)^2 \quad (1.182)$$

(d) The Lax Model

The electron energy spectra in bulk specimens of Bi in accordance with the Lax model is given by [97]

$$E(1+\alpha E)=\left(p_x^{\,2}/2m_1\right)+\left(p_y^{\,2}/2m_2\right)+\left(p_z^{\,2}/2m_3\right) \quad (1.183)$$

The $v_z(E_{n_z})$ for QDs of Bi in accordance with the model of Lax (1.181) can be written as

$$v_z\left(E_{n_z}\right)=\left(\pi\hbar n_z/(d_z.m_3)\left(1+2\alpha E_{n_z}\right)\right) \quad (1.184)$$

in which E_{n_z} should be determined from the following equation

$$E_{n_z}\left(1+\alpha E_{n_z}\right)=\left(\hbar^2/2m_3\right)\left(\pi n_z/d_z\right)^2 \quad (1.185)$$

The totally quantized energy E_{QD27}

$$E_{QD27}\left(1+\alpha E_{QD27}\right)=\frac{\hbar^2}{2m_1}\left(\frac{\pi n_x}{d_x}\right)^2+\frac{\hbar^2}{2m_2}\left(\frac{\pi n_y}{d_y}\right)^2+\frac{\hbar^2}{2m_3}\left(\frac{\pi n_z}{d_z}\right)^2 \quad (1.186)$$

The electron concentration is given by

$$n_{0D}=\frac{2g_v}{d_x d_y d_z}\sum_{n_x=1}^{n_{x\max}}\sum_{n_y=1}^{n_{y\max}}\sum_{n_z=1}^{n_{z\max}}F_{-1}(\eta_{57}) \quad (1.187)$$

where $\eta_{57}=\left(E_{F0D}-E_{QD27}\right)/k_BT$

The photo-emitted current density can be written as

$$J_{0D}=\frac{\hbar\pi aeg_v}{m_3.d_x d_y d_z^{\,2}}\sum_{n_x=1}^{n_{x\max}}\sum_{n_y=1}^{n_{y\max}}\sum_{n_{z\min}}^{n_{z\max}}\frac{n_z F_{-1}(\eta_{57})}{\left(1+2\alpha E_{n_z}\right)} \quad (1.188)$$

where $n_{z_{\min}}$ should be determined from the (a) following equation

$$(W-hv)\left[1+\alpha(W-hv)\right]=\left(\hbar^2/2m_3\right)\left[\pi n_{z_{\min}}/d_z\right]^2 \quad (1.189)$$

1.2.14. The EP from QDs of IV-VI Materials

In this section, the photoemission from QDs of IV-VI materials has been presented in accordance with the models of Dimmock [169], Bangert and Kastner [170] and Foley et. al. [171] respectively, although the carrier dispersion laws in this compound also follows Cohen and Lax models.

a) The dispersion relation of the conduction electrons in IV-VI semiconductors can be expressed in accordance with Dimmock [169] as

$$\left[\varepsilon-\frac{E_{g_0}}{2}-\frac{\hbar^2k_s{}^2}{2m_t^-}-\frac{\hbar^2k_z{}^2}{2m_l^-}\right]\left[\varepsilon+\frac{E_{g_0}}{2}+\frac{\hbar^2k_s{}^2}{2m_t^-}+\frac{\hbar^2k_z{}^2}{2m_l^-}\right]=P_\perp{}^2k_s{}^2+P_\parallel^2k_z{}^2 \quad (1.190)$$

where ε is the energy as measured from the center of the band gap E_{g_0} and $m_t^{\pm}$ and $m_l^{\pm}$ represent the contributions to the transverse and longitudinal effective masses of the external L_6^+ and bands arising from the $\vec{k}.\vec{p}$ perturbations with the other bands taken to the second order.

Substituting $\varepsilon=E+\left(\frac{E_{g_0}}{2}\right)$, $P_\perp{}^2=\hbar^2E_{g_0}/2m_t^*$ and $P_\parallel^2=\hbar^2E_{g_0}/2m_l^*$ (m_t^* and m_l^* are the transverse and longitudinal effective masses at k = 0), in (1.188), the following equation is obtained

$$\left[E-\frac{\hbar^2k_s^2}{2m_t^-}-\frac{\hbar^2k_z^2}{2m_l^-}\right]\left[1+\alpha E+\alpha\left(\frac{\hbar^2k_s^2}{2m_t^+}\right)+\alpha\left(\frac{\hbar^2k_z^2}{2m_l^+}\right)\right]=\frac{\hbar^2k_s^2}{2m_t^*}+\frac{\hbar^2k_z^2}{2m_l^*} \quad (1.191)$$

The $v_z\left(E_{n_z}\right)$ in given by

$$v_z\left(E_{n_z}\right)=\frac{\pi\hbar n_z}{d_z m_l^*}Q_{21}\left(E_{n_z}\right) \quad (1.192)$$

where

$$Q_{21}\left(E_{n_z}\right)=\left[1+\frac{\alpha\hbar^2m_l^*}{m_l^+m_l^-}\left(\frac{\pi n_z}{d_z}\right)^2+\frac{m_l^*}{m_l^-}\left(1+\alpha E_{n_z}\right)\right]\left[1+2\alpha E_{n_z}+\frac{\alpha\hbar^2}{2m_l^+}\left(\frac{\pi n_z}{d_z}\right)^2\left(1-\frac{m_l^+}{m_l^-}\right)\right]^{-1}$$

in which E_{n_z} should be determined from the following equation

$$E_{n_z}(1+\alpha E_{n_z})+\alpha E_{n_z}\left(\frac{\hbar^2}{2m_l^+}\left(\frac{\pi n_z}{d_z}\right)^2\right)-\frac{\hbar^2}{2m_l^-}\left(\frac{\pi n_z}{d_z}\right)^2.(1+\alpha E_{n_z})-\frac{\alpha\hbar^4}{4m_l^+m_l^-}\left(\frac{\pi n_z}{d_z}\right)^4=\frac{\hbar^2}{2m_l^*}\left(\frac{\pi n_z}{d_z}\right)^2 \quad (1.193)$$

The totally quantized energy assumes the form

$$\left[E_{QD28}-\frac{\hbar^2}{2m_t^-}\left\{\left(\frac{\pi n_x}{d_x}\right)^2+\left(\frac{\pi n_y}{d_y}\right)^2\right\}-\frac{\hbar^2}{2m_l^-}\left(\frac{\pi n_z}{d_z}\right)^2\right]\left[1+\alpha E_{QD28}+\frac{\alpha\hbar^2}{2m_t^+}\left\{\left(\frac{\pi n_x}{d_x}\right)^2+\left(\frac{\pi n_y}{d_y}\right)^2\right\}\right.$$
$$\left.+\alpha\frac{\hbar^2}{2m_l^+}\left(\frac{\pi n_z}{d_z}\right)^2\right]=\frac{\hbar^2}{2m_t^*}\left[\left(\frac{\pi n_x}{d_x}\right)^2+\left(\frac{\pi n_y}{d_y}\right)^2\right]+\frac{\hbar^2}{2m_l^*}\left(\frac{\pi n_z}{d_z}\right)^2 \quad (1.194)$$

The electron concentration is given by

$$n_{0D}=\frac{2g_v}{d_xd_yd_z}\sum_{n_x=1}^{n_{x\max}}\sum_{n_y=1}^{n_{y\max}}\sum_{n_z=1}^{n_{z\max}}F_{-1}(\overline{\eta_{57}}) \quad (1.195)$$

where $\overline{\eta_{57}}=(E_{F0D}-E_{QD28})/k_BT$

The photo-emitted current density can be written as

$$J_{0D}=\frac{\pi\hbar\alpha_0eg_v}{d_xd_yd_z^2m_l^*}\sum_{n_x=1}^{n_{x\max}}\sum_{n_y=1}^{n_{y\max}}\sum_{n_{z\min}}^{n_{z\max}}Q_{21}(E_{n_z})F_{-1}(\overline{\eta_{57}}) \quad (1.196)$$

where $n_{z_{\min}}$ should be the nearest integer of the (b) following equation

$$(W-h\nu)[1+\alpha(W-h\nu)]+\alpha(W-h\nu)\frac{\hbar^2}{2m_l^+}\left(\frac{\pi n_{z_{\min}}}{d_z}\right)^2-$$
$$\frac{\hbar^2}{2m_l^-}\left(\frac{\pi n_{z_{\min}}}{d_z}\right)^2(1+\alpha(W-h\nu))$$
$$-\frac{\alpha\hbar^4}{4m_l^+m_l^-}\left(\frac{\pi n_{z_{\min}}}{d_z}\right)^4=\left(\frac{\hbar^2}{2m_l^*}\right)\left[\frac{\pi n_{z_{\min}}}{d_z}\right]^2 \quad (1.197)$$

b) The electron dispersion law in IV-VI compounds can be written in accordance with Bangert and Kastner [170] as

$$\Gamma(E)=\overline{F}_1(E)k_s^2+\overline{F}_2(E)k_z^2 \quad (1.198)$$

where $\Gamma(E) \equiv 2E$, $\overline{F}_1(E) \equiv \left[\frac{(\overline{R})^2}{E+E_{g_0}} + \frac{(\overline{s})^2}{E+\Delta_c'} + \frac{(\overline{Q})^2}{E+\Delta_c''}\right]$

$\overline{F}_2(E) \equiv \left[\frac{2(\overline{A})^2}{E+E_{g_0}} + \frac{(\overline{s}+\overline{Q})^2}{E+\Delta_c''}\right]$, $\overline{R}$, $\bar{s}$, Δ_c', $\overline{Q}$, and $\overline{A}$ are the spectrum constants.

The $v_z(E_{n_z})$ in given by

$$v_z(E_{n_z}) = \left(Q_{22}(E_{n_z})/\hbar\right) \tag{1.199}$$

where $Q_{22}(E_{n_z}) = 2\left(\frac{\pi n_z}{d_z}\right)\overline{F}_2(E_{n_z})\left[\Gamma'(E_{n_z}) - \left(\frac{\pi n_z}{d_z}\right)^2 \left(\overline{F}_2(E_{n_z})\right)'\right]^{-1}$

in which E_{n_z} should be determined from the following equation

$$\Gamma(E_{n_z}) = \overline{F}_2(E_{n_z})\left(\frac{\pi n_z}{d_z}\right)^2 \tag{1.200}$$

The totally quantized energy can be expressed as

$$\Gamma(E_{QD29}) = F_1(E_{QD29})\left[\left(\frac{\pi n_x}{d_x}\right)^2 + \left(\frac{\pi n_y}{d_y}\right)^2\right] + F_2(E_{QD29})\left(\frac{\pi n_z}{d_z}\right)^2 \tag{1.201}$$

The electron concentration is given by

$$n_{0D} = \frac{2g_v}{d_x d_y d_z}\sum_{n_x=1}^{n_{x\max}}\sum_{n_y=1}^{n_{y\max}}\sum_{n_z=1}^{n_{z\max}} F_{-1}(\eta_{58}) \tag{1.202}$$

where $\eta_{58} = (k_B T)^{-1}(E_{F0D} - E_{QD29})$

The photo-emitted current density can be written as

$$J_{0D} = \frac{\alpha_0 e g_v}{\hbar d_x d_y d_z}\sum_{n_x=1}^{n_{x\max}}\sum_{n_y=1}^{n_{y\max}}\sum_{n_{z\min}}^{n_{z\max}} Q_{22}(E_{n_z}) F_{-1}(\eta_{58}) \tag{1.203}$$

where $n_{z_{\min}}$ should be the nearest integer of the following equation

$$\Gamma(W - h\nu) = \overline{F}_2(W - h\nu)\left(\pi n_{z_{\min}} / d_z\right)^2 \tag{1.204}$$

It may be noted that under the condition $\overline{S} = \overline{Q} = 0, \left(\overline{R}\right)^2 = \left(\overline{A}\right)^2 = \left(\hbar^2 E_{g_0} / m^*\right)$, the (1.196) reduces to well-known two band model of Kane.

(c) In accordance with Foley et. al [171], the electron dispersion relation assumes the form

$$E + \frac{E_{g_0}}{2} = \frac{\hbar^2 k_s^{\,2}}{2m_\perp^-} + \frac{\hbar^2 k_z^{\,2}}{2m_\parallel^-} + \left[\left[\frac{\hbar^2 k_s^{\,2}}{2m_\perp^+} + \frac{\hbar^2 k_z^{\,2}}{2m_\parallel^+} + \frac{E_{g_0}}{2}\right]^2 + P_\parallel^2 k_z^{\,2} + P_\perp^{\,2} k_s^{\,2}\right]^{1/2} \tag{1.205}$$

where $\frac{1}{m_\perp^\pm} = \frac{1}{2}\left[\frac{1}{m_{tc}} \pm \frac{1}{m_{tv}}\right], \frac{1}{m_\parallel^\pm} = \frac{1}{2}\left[\frac{1}{m_{lc}} \pm \frac{1}{m_{lv}}\right]$, m_{tc} and m_{lc} are the transverse and longitudinal effective electron masses of the conduction electrons at the edge of the conduction band, m_{tv} and m_{lv} are the transverse and longitudinal effective hole masses of the holes at the edge of the valence band.

The $v_z\left(E_{n_z}\right)$ in given by

$$v_z\left(E_{n_z}\right) = \left(Q_{23}\left(E_{n_z}\right) / \hbar\right) \tag{1.206}$$

where

$$Q_{23}\left(E_{n_z}\right) = \frac{\hbar}{m_\parallel^-}\left(\frac{\pi n_z}{d_z}\right) + \frac{1}{\hbar}\left[\frac{E_{g_0}}{2} + \frac{\hbar}{2m_\parallel^+}\left(\frac{\pi n_z}{d_z}\right)^2 + P_\parallel^2\left(\frac{\pi n_z}{d_z}\right)^2\right]^{-1/2}\left[\left(\frac{E_{g_0}}{2} + \frac{\hbar^2}{2m_\parallel^+}\left(\frac{\pi n_z}{d_z}\right)\right)\left(\frac{\hbar^2 \pi n_z}{m_\parallel^+ d_z}\right) + P_\parallel^2\left(\frac{\pi n_z}{d_z}\right)\right]$$

in which E_{n_z} should be determined from the following equation

$$E_{n_z} = -\frac{E_{g_0}}{2} + \frac{\hbar^2}{2m_\parallel^-}\left(\frac{\pi n_z}{d_z}\right)^2 + \left[\left(\frac{E_{g_0}}{2} + \frac{\hbar^2}{2m_\parallel^+}\left(\frac{\pi n_z}{d_z}\right)^2\right)^2 + P_\parallel^2\left(\frac{\pi n_z}{d_z}\right)^2\right]^{1/2} \tag{1.207}$$

The totally quantized energy for this model is given by

$$E_{QD30} = -\frac{E_{g_0}}{2} + \frac{\hbar^2}{2m_\perp^-}\left[\left(\frac{n_x\pi}{d_x}\right)^2 + \left(\frac{n_y\pi}{d_y}\right)^2\right] + \frac{\hbar^2}{2m_\parallel^-}\left(\frac{n_z\pi}{d_z}\right)^2 + \left[\left[\frac{E_{g_0}}{2} + \frac{\hbar}{2m_\parallel^+}\left(\frac{n_z\pi}{d_z}\right)^2 + \frac{\hbar^2}{2m_\perp^+}\left\{\left(\frac{n_x\pi}{d_x}\right)^2 + \left(\frac{n_y\pi}{d_y}\right)^2\right\}\right]^2\right.$$

$$\left. + P_\parallel^2\left(\frac{n_z\pi}{d_z}\right)^2 + P_\perp^2\left\{\left(\frac{n_x\pi}{d_x}\right)^2 + \left(\frac{n_y\pi}{d_y}\right)^2\right\}\right]^{1/2} \quad (1.208)$$

The electron concentration assumes the form

$$n_{0D} = \frac{2g_v}{d_x d_y d_z}\sum_{n_x=1}^{n_{x\max}}\sum_{n_y=1}^{n_{y\max}}\sum_{n_z=1}^{n_{z\max}} F_{-1}(\eta_{59}) \quad (1.209)$$

where $\eta_{59} = \left(E_{F0D} - E_{QD30}\right)/\left(k_B T\right)$

The photo-emitted current density can be written as

$$J_{0D} = \frac{\alpha_0 e g_v}{d_x d_y d_z \hbar}\sum_{n_x=1}^{n_{x\max}}\sum_{n_y=1}^{n_{y\max}}\sum_{n_{z\min}}^{n_{z\max}} Q_{23}\left(E_{n_z}\right)F_{-1}(\eta_{59}) \quad (1.210)$$

where $n_{z_{\min}}$ should be the nearest integer of the following equation

$$(W - hv) = -\frac{E_{g_0}}{2} + \frac{\hbar^2}{2m_\parallel^-}\left(\frac{\pi n_{z_{\min}}}{d_z}\right)^2 + \left[\left(\frac{E_{g_0}}{2} + \frac{\hbar^2}{2m_\parallel^+}\left(\frac{\pi n_{z_{\min}}}{d_z}\right)^2\right)^2 + P_\parallel^2\left(\frac{\pi n_{z_{\min}}}{d_z}\right)^2\right]^{1/2} \quad (1.211)$$

1.2.15. The EP from QDs of Lead Germanium Telluride

The dispersion law of n-type $Pb_{1-x}Ge_xTe$ with x = 0.01 can be expressed following Vassilev [172] as

$$\left[E - 0.606k_s^2 - 0.722k_z^2\right]\left[E + \bar{E}_{g_0} + 0.411k_s^2 + 0.377k_z^2\right] =$$

$$= 0.23k_s^2 + 0.02k_z^2 \pm \left[0.06\bar{E}_{g_0} + 0.061k_s^2 + 0.0066k_z^2\right]k_s \quad (1.212)$$

where $\bar{E}_{g_0} = 0.21$ eV, k_x, k_y and k_z are in the units of m^{-1}

The $v_z(E_{n_z})$ in this case is given by

$$v_z\left(E_{n_z}\right)=(1/\hbar)Q_{17}\left(E_{n_z}\right) \tag{1.213}$$

where

$$Q_{17}\left(E_{n_z}\right)=\left[2E_{n_z}-0.0345\left(\frac{\pi n_z}{d_z}\right)^2\right]^{-1}\cdot\left[0.04\left(\frac{\pi n_z}{d_z}\right)+0.1444\left(\frac{\pi n_z}{d_z}\right)\left[E_{n_z}+0.0377\left(\frac{\pi n_z}{d_z}\right)^2\right]-0.0754\left(\frac{\pi n_z}{d_z}\right)\left[E_{n_z}-0.0722\left(\frac{\pi n_z}{d_z}\right)^2\right]\right]$$

in which E_{n_z} should be determined from the following equation

$$\left[0.0722\left(\frac{\pi n_z}{d_z}\right)^2-E_{n_z}\right]\left[-E_{n_z}-\overline{E_{g_0}}-0.0377\left(\frac{\pi n_z}{d_z}\right)^2\right]=0.02\left(\frac{\pi n_z}{d_z}\right)^2\pm\left[\left(\frac{\pi n_x}{d_x}\right)^2+\left(\frac{\pi n_y}{d_y}\right)^2\right]^{1/2}\left[0.06\overline{E_{g_0}}+0.0066\left(\frac{\pi n_z}{d_z}\right)^2\right] \tag{1.214}$$

The totally quantized energy $\left(E_{QD19,\pm}\right)$ is given by

$$\begin{aligned}
&\left[E_{QD19,\pm}-0.606\left(\left(\frac{\pi n_x}{d_x}\right)^2+\left(\frac{\pi n_y}{d_y}\right)^2\right)-0.0722\left(\frac{\pi n_z}{d_z}\right)^2\right]\\
&\left[E_{QD19,\pm}+0.471\left(\left(\frac{\pi n_x}{d_x}\right)^2+\left(\frac{\pi n_y}{d_y}\right)^2\right)+\overline{E_{g_0}}+0.0377\left(\frac{\pi n_z}{d_z}\right)^2\right]\\
&=0.23\left(\left(\frac{\pi n_x}{d_x}\right)^2+\left(\frac{\pi n_y}{d_y}\right)^2\right)+0.02\left(\frac{\pi n_z}{d_z}\right)^2\pm\left[\left(\frac{\pi n_x}{d_x}\right)^2+\left(\frac{\pi n_y}{d_y}\right)^2\right]^{1/2}\\
&\left[0.06\overline{E_{g_0}}+0.061\left(\left(\frac{\pi n_x}{d_x}\right)^2+\left(\frac{\pi n_y}{d_y}\right)^2\right)+0.0066\left(\frac{\pi n_z}{d_z}\right)^2\right]
\end{aligned} \tag{1.215}$$

The electron concentration is given by

$$n_{0D}=\frac{g_v}{d_x d_y d_z}\sum_{n_x=1}^{n_{x_{max}}}\sum_{n_y=1}^{n_{y_{max}}}\sum_{n_z=1}^{n_{z_{max}}}F_{-1}\left(\eta_{49}\right) \tag{1.216}$$

where $\eta_{49}=[(E_{F0D}-E_{QD19,\pm})/k_B T]$

The photo-emitted current density can be written as

$$J_{0D}=\frac{\alpha_0 g_v e}{2\hbar d_x d_y d_z}\sum_{n_x=1}^{n_{x_{max}}}\sum_{n_y=1}^{n_{y_{max}}}\sum_{n_z=1}^{n_{z_{max}}} Q_{17}\left(E_{n_z}\right)F_{-1}\left(\eta_{49}\right) \tag{1.217}$$

where $n_{z_{min}}$ should be determined from the following equation

$$\left[\left(W-h\nu\right)-0.0722\left(\frac{\pi n_{z_{min}}}{d_z}\right)^2\right]\left[\left(W-h\nu\right)+\overline{E_{g_0}}+0.0377\left(\frac{\pi n_{z_{min}}}{d_z}\right)^2\right]=0.02\left(\frac{\pi n_{z_{min}}}{d_z}\right)^2 \tag{1.218}$$

1.2.16. The EP from QDs of Zinc and Cadmium diphosphides

The dispersion relation of the holes of Cadmium and Zinc diphosphides can approximately be written following Chuiko [173] as

$$E=\left[\beta_1+\frac{\beta_2\beta_3(k)}{8\beta_4}\right]k^2\pm\left\{\left[\beta_4\beta_3(k)\left(\beta_5-\frac{\beta_2\beta_3(k)}{8\beta_4}\right)k^2\right]+8\beta_4^2\left(1-\frac{\beta_3^2(k)}{4}\right)-\beta_2\left(1-\frac{\beta_3^2(k)}{4}\right)k^2\right\}^{1/2} \tag{1.219}$$

where and are system constants and $\beta_3\left(k\right)=[(k_x^{\,2}+k_y^{\,2}-2k_x^{\,2})/k^2]$

The $v_z\left(n_z\right)$ in this case is given by

$$v_z\left(E_{n_z}\right)=\left(1/\hbar\right)Q_{25}\left(E_{n_z}\right) \tag{1.220}$$

where

$$Q_{25}(n_z)=\left[2\left(\frac{\pi n_z}{d_z}\right)\left(\beta_1-\frac{\beta_2}{4\beta_4}\right)\mp\left(\frac{\pi n_z}{d_z}\right)\left[\left(-2\beta_4\left(\beta_5+\frac{\beta_2}{4\beta_4}\right)-\frac{3}{2}\beta_2\right)\right]\left[12\beta_4^2+\left[\left(+2\beta_4\left(\beta_5+\frac{\beta_2}{4\beta_4}\right)+\frac{3}{2}\beta_2\right)\right]\left(\frac{\pi n_z}{d_z}\right)^2\right]^{-1/2}\right]$$

in which E_{n_z} should be determined from the following equation

$$E_{n_z}=\left(\beta_1-\frac{\beta_2}{4\beta_4}\right)\left(\frac{\pi n_z}{d_z}\right)^2\pm\left\{\left(-2\beta_4\left(\beta_5+\frac{\beta_2}{4\beta_4}\right)-\frac{3}{2}\beta_2\right)\left(\frac{\pi n_z}{d_z}\right)^2+12\beta_4^2\right\}^{1/2} \tag{1.221}$$

The totally quantized energy is given by

$$E_{QD32,\pm} = \left[\beta_1 + \frac{\beta_2\beta_6}{8\beta_4}\right]\beta_7^2 \pm \left\{\left[\beta_4\beta_6\left(\beta_5 - \frac{\beta_2\beta_6}{8\beta_4}\right)\beta_7^2\right] + 8\beta_4^2\left(1 - \frac{\beta_6^2}{4}\right) - \beta_2\left(1 - \frac{\beta_6^2}{4}\right)\beta_7^2\right\}^{1/2} \quad (1.222)$$

where $\beta_6 = \frac{1}{\beta_7^2}\left[\left(\frac{\pi n_x}{d_x}\right)^2 + \left(\frac{\pi n_y}{d_y}\right)^2 - 2\left(\frac{\pi n_z}{d_z}\right)^2\right]$ and

$$\beta_7^2 = \left[\left(\frac{\pi n_x}{d_x}\right)^2 + \left(\frac{\pi n_y}{d_y}\right)^2 + \left(\frac{\pi n_z}{d_z}\right)^2\right]$$

The hole concentration assumes the form

$$p_{0D} = \frac{g_v}{d_x d_y d_z} \sum_{n_x=1}^{n_{x\max}} \sum_{n_y=1}^{n_{y\max}} \sum_{n_z=1}^{n_{z\max}} \left[1 + \exp\left(\eta_{61,\pm}\right)\right]^{-1} \quad (1.223)$$

where $\eta_{61,\pm} = \left(E_{F0D} - E_{QD32,\pm}\right)/\left(k_B T\right)$

The photo-emitted current density can be written as

$$J_{0D} = \frac{\alpha_0 e g_v}{2\hbar d_x d_y d_z} \sum_{n_x=1}^{n_{x\max}} \sum_{n_y=1}^{n_{y\max}} \sum_{n_{z\min}}^{n_{z\max}} Q_{25}\left(E_{n_z}\right)\left[1 + \exp\left(\eta_{61,\pm}\right)\right]^{-1} \quad (1.224)$$

where $n_{z_{\min}}$ should be the nearest integer of the following equation

$$W - h\nu = \left(\beta_1 - \frac{\beta_2}{4\beta_4}\right)\left(\frac{\pi n_{z_{\min}}}{d_z}\right)^2 \pm \left\{\left(-2\beta_4\left(\beta_5 + \frac{\beta_2}{4\beta_4}\right) - \frac{3}{2}\beta_2\right)\left(\frac{\pi n_{z_{\min}}}{d_z}\right)^2 + 12\beta_4^2\right\}^{1/2} \quad (1.225)$$

1.2.17. The EP from QDs of Bismuth Telluride

The dispersion relation of the holes in Bi_2Te_3 can be expressed as [174]

$$E\left(1 + \alpha E\right) = \frac{\hbar^2}{2m_0}\left(\alpha_{11}k_x^2 + \alpha_{22}k_y^2 + \alpha_{33}k_z^2 + 2\alpha_{23}k_y k_z\right) \quad (1.226)$$

where are spectrum constants.

The is given by

$$v_z\left(E_{n_z}\right)=\left(\pi\hbar\alpha_{33}n_z/\left(d_z m_0\left(1+2\alpha E_{n_z}\right)\right)\right) \tag{1.227}$$

in which E_{n_z} should be determined from the following equation

$$E_{n_z}\left(1+\alpha E_{n_z}\right)=\frac{\hbar^2}{2m_0}\left(\alpha_{33}\left(\frac{\pi n_z}{d_z}\right)^2\right) \tag{1.228}$$

The totally quantized energy E_{QD33} can be written as

$$E_{QD33}\left(1+\alpha E_{QD33}\right)=\frac{\hbar^2}{2m_0}\left[\alpha_{11}\left(\frac{n_x\pi}{d_x}\right)^2+\alpha_{22}\left(\frac{n_y\pi}{d_y}\right)^2+\alpha_{33}\left(\frac{n_z\pi}{d_z}\right)^2+2\alpha_{23}\left(\frac{\pi^2 n_y n_z}{d_y d_z}\right)\right] \tag{1.229}$$

The hole concentration is given by

$$p_{0D}=\frac{2g_v}{d_x d_y d_z}\sum_{n_x=1}^{n_{x\max}}\sum_{n_y=1}^{n_{y\max}}\sum_{n_z=1}^{n_{z\max}}\left[1+\exp\left(\eta_{62}\right)\right]^{-1} \tag{1.230}$$

where $\eta_{62}=\left[E_{F0D}-E_{QD33}\right]\left(k_B T\right)^{-1}$

The photo-emitted current density assumes the form

$$J_{0D}=\frac{\pi\alpha_0 e g_v\hbar\alpha_{33}}{m_0 d_x d_y d_z^2}\sum_{n_x=1}^{n_{x\max}}\sum_{n_y=1}^{n_{y\max}}\sum_{n_{z\min}}^{n_{z\max}}\frac{n_z\left[1+\exp\left(\eta_{62}\right)\right]^{-1}}{\left(1+2\alpha E_{n_z}\right)} \tag{1.231}$$

where $n_{z_{\min}}$ should be the nearest integer of the following equation

$$\left(W-hv\right)\left[1+\alpha\left(W-hv\right)\right]=\frac{\hbar^2\alpha_{33}}{2m_0}\left[\pi n_{z_{\min}}/d_z\right]^2 \tag{1.232}$$

1.2.18. The EP from QDs Antimony

The dispersion relation of the conduction electrons in Antimony (Sb) can be written following Ketterson [175] as

$$\frac{2m_0E}{\hbar^2} = \overline{\alpha_{11}}k_x^2 + \overline{\alpha_{22}}k_y^2 + \overline{\alpha_{33}}k_z^2 + 2\overline{\alpha_{23}}k_yk_z \tag{1.233}$$

$$\frac{2m_0E}{\hbar^2} = a_1k_x^2 + a_2k_y^2 + a_3k_z^2 + a_4k_yk_z + a_5k_zk_x + a_6k_xk_y \tag{1.234}$$

$$\frac{2m_0E}{\hbar^2} = a_1k_x^2 + a_2k_y^2 + a_3k_z^2 + a_4k_yk_z - a_5k_zk_x - a_6k_xk_y \tag{1.235}$$

where

$a_1 = \frac{1}{4}\left(\overline{\alpha_{11}} + 3\overline{\alpha_{22}}\right), a_2 = \frac{1}{4}\left(\overline{\alpha_{22}} + 3\overline{\alpha_{11}}\right), a_3 = \overline{\alpha_{33}}, a_4 = \overline{\alpha_{33}}, a_5 = \sqrt{3}$

$, a_6 = \sqrt{3}\left(\overline{\alpha_{22}} - \overline{\alpha_{11}}\right), \overline{\alpha_{11}}, \overline{\alpha_{22}}, \overline{\alpha_{33}}$ and $\overline{\alpha_{23}}$ are the system constants.

For the QDs, whose bulk carrier dispersion law is described by (1.231), the is given by

$$v_{z1}\left(E_{n_{z1}}\right) = \overline{\alpha_{33}}\frac{\hbar}{m_0}.\frac{\pi n_{z1}}{d_z} \tag{1.236}$$

where $E_{n_{z1}} = \bar{\alpha}_{33}\frac{\hbar^2}{2m_0}.\left(\frac{\pi n_{z1}}{d_z}\right)^2$ (1.237)

The totally quantized energy can be written as

$$\frac{2m_0\overline{E_{QD1}}}{\hbar^2} = \bar{\alpha}_{11}\left(\frac{\pi n_{x1}}{d_x}\right)^2 + \bar{\alpha}_{22}\left(\frac{\pi n_{y1}}{d_y}\right)^2 + \bar{\alpha}_{33}\left(\frac{\pi n_{z1}}{d_y}\right)^2 + 2\bar{\alpha}_{23}\left(\frac{\pi^2 n_{y1}.n_{z1}}{d_yd_z}\right) \tag{1.238}$$

The electron concentration for the QDs, with bulk carrier dispersion law is described by (1.231) is given by

$$n_{0D1} = \frac{2}{d_xd_yd_z}\sum_{n_{x1}=1}^{n_{x1\max}}\sum_{n_{y1}=1}^{n_{y1\max}}\sum_{n_{z1}=1}^{n_{z1\max}} F_{-1}(\eta_{63}) \tag{1.239}$$

where $\eta_{63} = \frac{1}{k_BT}\left[E_{F0D} - \overline{E_{QD1}}\right]$

For the QDs, whose bulk carrier dispersion law is described by (1.232), the photo-emitted current density can be expressed as

$$J_{0D1} = \frac{\alpha_0 e\hbar\pi\bar{\alpha}_{33}}{m_0 . d_x d_y d_z^{\,2}} \sum_{n_{x1}=1}^{n_{x1_{\max}}} \sum_{n_{y1}=1}^{n_{y1_{\max}}} \sum_{n_{z1_{\min}}}^{n_{z1_{\max}}} n_{z1} F_{-1}(\eta_{63}) \tag{1.240}$$

where $n_{z1_{\min}}$ should be the nearest integer of the following equation

$$W - hv = \bar{\alpha}_{33} \frac{\hbar^2}{2m_0} . \left(\frac{\pi n_{z1_{\min}}}{d_z} \right)^2 \tag{1.241}$$

Similarly for QDs, whose bulk carrier dispersion law is given by (1.232), the corresponding photo-current density and the electron concentration can, respectively, be expressed as

$$J_{0D2} = \frac{\pi\hbar\alpha_0 e a_3}{m_0 . d_x d_y d_z^{\,2}} \sum_{n_{x2}=1}^{n_{x2_{\max}}} \sum_{n_{y2}=1}^{n_{y2_{\max}}} \sum_{n_{z2_{\min}}}^{n_{z2_{\max}}} n_{z2} F_{-1}(\eta_{64}) \tag{1.242}$$

and $$n_{0D2} = \frac{2}{d_x d_y d_z} \sum_{n_{x2}=1}^{n_{x2_{\max}}} \sum_{n_{y2}=1}^{n_{y2_{\max}}} \sum_{n_{z2}=1}^{n_{z2_{\max}}} F_{-1}(\eta_{64}) \tag{1.243}$$

where $\eta_{64} = \frac{1}{k_B T}\left[E_{F0D} - \overline{E_{QD2}} \right]$, is given by

$$\overline{E_{QD2}} = \frac{\hbar^2}{2m_0} \left[a_1 \left(\frac{n_{x2}\pi}{d_x} \right)^2 + a_2 \left(\frac{n_{y2}\pi}{d_y} \right)^2 + a_3 \left(\frac{n_{z2}\pi}{d_z} \right)^2 + a_4 \frac{n_{y2} n_{z2} \pi^2}{d_y d_z} + a_5 \frac{n_{x2} n_{z2} \pi^2}{d_x d_z} + a_6 \frac{\pi^2 n_{x2} n_{y2}}{d_x d_y} \right] \tag{1.244}$$

The $n_{z2_{\min}}$ should be the nearest integer of the following equation

$$W - hv = a_3 \left(\frac{\hbar^2}{2m_0} \right) \left(\frac{\pi n_{z2_{\min}}}{d_z} \right)^2 \tag{1.245}$$

Similarly for QDs, whose bulk carrier dispersion law is given by (1.233), the corresponding photocurrent density and the electron concentration can, respectively, be written as

$$J_{0D3} = \frac{\pi\hbar\alpha_0 e a_3}{m_0 . d_x d_y d_z^{\,2}} \sum_{n_{x3}=1}^{n_{x3_{\max}}} \sum_{n_{y3}=1}^{n_{y3_{\max}}} \sum_{n_{z3_{\min}}}^{n_{z3_{\max}}} n_{z3} F_{-1}(\eta_{65}) \tag{1.246}$$

and $n_{0D3} = \frac{2}{d_x d_y d_z} \sum_{n_{x3}=1}^{n_{x3_{\max}}} \sum_{n_{y3}=1}^{n_{y3_{\max}}} \sum_{n_{z3}=1}^{n_{z3_{\max}}} F_{-1}(\eta_{65})$ (1.247)

where $\eta_{65} = \frac{1}{k_B T}\left[E_{F0D} - \overline{E_{QD3}}\right]$, is given by

$$\overline{E_{QD3}} = \frac{\hbar^2}{2m_0}\left[a_1\left(\frac{n_{x3}\pi}{d_x}\right)^2 + a_2\left(\frac{n_{y3}\pi}{d_y}\right)^2 + a_3\left(\frac{n_{z3}\pi}{d_z}\right)^2 + a_4\frac{n_{y3}n_{z3}\pi^2}{d_y d_z} + a_5\frac{n_{x3}n_{z3}\pi^2}{d_x d_z} + a_6\frac{\pi^2 n_{x3}n_{y3}}{d_x d_y}\right]$$ (1.248)

The should be the nearest integer of the following equation

$$W - h\upsilon = a_3\left(\hbar^2/2m_0\right)\left(\pi n_{z3_{\min}}/d_z\right)^2$$ (1.249)

Therefore, the respective simplified expressions of the total electron concentration and the current density from QDs of antimony can be written as

$$J_{0D} = J_{0D1} + J_{0D2} + J_{0D3}$$ (1.250)

and $n_{0D} = n_{0D1} + n_{0D2} + n_{0D3}$ (1.251)

1.2.19. The EP from III-V Quantum Dot Superlattices with Graded Interfaces

The electron dispersion law in bulk specimens of the constituent materials of III-V SLs whose energy band structures are defined by three band model of Kane can be expressed following (1.16) as

$$(\hbar^2 k^2)/(2m_i^*) = EG(E, E_{g_{0i}}, \Delta_i)$$ (1.252)

$i = 1, 2, \ldots$ and $G(E, E_{g_{0i}}, \Delta_i) \equiv \frac{\left(E_{g_{0i}} + \frac{2}{3}\Delta_i\right)\left(E + E_{g_{0i}} + \Delta_i\right)\left(E + E_{g_{0i}}\right)}{\left[E_{g_{0i}}\left(E_{g_{0i}} + \Delta_i\right)\left(E + E_{g_{0i}} + \frac{2}{3}\Delta_i\right)\right]}$.

Therefore, the dispersion law of the electrons of III-V SLs with graded interfaces can be expressed, following Jiang and Lin [176], as

$$\cos(L_o k) = \frac{1}{2}\Phi(E,k_s) \tag{1.253}$$

where is the period length, a_0 and are the widths of the barrier and the well respectively,

$$\Phi(E,k_s) \equiv \Big[2\cosh\{\beta(E,k_s)\}\cos\{\gamma(E,k_s)\} + \varepsilon(E,k_s)\sinh\{\beta(E,k_s)\}\sin\{\gamma(E,k_s)\}$$

$$+\Delta_0\Bigg[\left(\frac{K_1^2(E,k_s)}{K_2(E,k_s)} - 3K_2(E,k_s)\right)\cosh\{\beta(E,k_s)\}\sin\{\gamma(E,k_s)\}$$

$$+\left(3K_1(E,k_s) - \frac{\{K_2(E,k_s)\}^2}{K_1(E,k_s)}\right)\sinh\{\beta(E,k_s)\}\cos\{\gamma(E,k_s)\}\Bigg]$$

$$+\frac{1}{12}\left[\frac{5\{K_2(E,k_s)\}^3}{K_1(E,k_s)} + \frac{5\{K_1(E,k_s)\}^3}{K_2(E,k_s)} - 34K_2(E,k_s)K_1(E,k_s)\right]\sinh\{\beta(E,k_s)\}\sin\{\gamma(E,k_s)\}\Bigg]\Bigg],$$

$$\beta(E,k_s) \equiv K_1(E,k_s)[a_0 - \Delta_0],\ K_1(E,k_s) \equiv \left[\frac{2m_2^* E'}{\hbar^2}G(E - V_o,\alpha_2,\Delta_2) + k_s^2\right]^{1/2},$$

$E' \equiv V_0 - E$, is the potential barrier encountered by the electron $(V_0 \equiv |E_{g_2} - E_{g_1}|)$, $\alpha_i \equiv 1/E_{g_{0i}}$, Δ_0 is the interface width, $k_s^2 = k_x^2 + k_y^2$, $\gamma(E,k_s) = K_2(E,k_s)[b_0 - \Delta_0]$ and

$$\varepsilon(E,k_s) \equiv \left[\frac{K_1(E,k_s)}{K_2(E,k_s)} - \frac{K_2(E,k_s)}{K_1(E,k_s)}\right].$$

Therefore the total electron concentration per unit volume in III-V quantum dot SL is given by

$$n_0 = \frac{2g_v}{d_x d_y d_z}\sum_{n_x=1}^{n_{x_{max}}}\sum_{n_y=1}^{n_{y_{max}}}\sum_{n_z=1}^{n_{z_{max}}} F_{-1}(\overline{\eta_9}) \tag{1.254}$$

where $\overline{\eta_9} \equiv \dfrac{E_{FQDSLGI} - \overline{E_{QD10}}}{k_B T}$ in which $E_{FQDSLGI}$ is the Fermi energy in this case, is the root

$$\left(\frac{\pi n_z}{d_z}\right)^2 = \left[\frac{1}{L_0^2}\left[\cos^{-1}\left[\frac{1}{2}f_{10}(E,n_x,n_y)\right]\right|_{E=\overline{E_{QD10}}}\right]^2 - \left(\frac{\pi n_x}{d_x}\right)^2 - \left(\frac{\pi n_y}{d_y}\right)^2\Bigg] \tag{1.255}$$

and

$$f_{10}(E,n_x,n_y) \equiv \Big[2\cosh\{\bar{\beta}(E,n_x,n_y)\}\cos\{\bar{\gamma}(E,n_x,n_y)\} + \bar{\varepsilon}(E,n_x,n_y)\sinh\{\bar{\beta}(E,n_x,n_y)\}$$
$$\sin\{\bar{\gamma}(E,n_x,n_y)\}$$
$$+\Delta_0\Bigg[\Bigg(\frac{\{\bar{K}_1(E,n_x,n_y)\}^2}{\bar{K}_2(E,n_x,n_y)} - 3\bar{K}_2(E,n_x,n_y)\Bigg)\cosh\{\bar{\beta}(E,n_x,n_y)\}\sin\{\bar{\gamma}(E,n_x,n_y)\}$$
$$+\Bigg(3\bar{K}_1(E,n_x,n_y) - \frac{\{\bar{K}_2(E,n_x,n_y)\}^2}{\bar{K}_1(E,n_x,n_y)}\Bigg)\sinh\{\bar{\beta}(E,n_x,n_y)\}\cos\{\bar{\gamma}(E,n_x,n_y)\}\Bigg]$$
$$+\Delta_0\Big[2\big(\{\bar{K}_1(E,n_x,n_y)\}^2 - \{\bar{K}_2(E,n_x,n_y)\}^2\big)\cosh\{\bar{\beta}(E,n_x,n_y)\}\cos\{\bar{\gamma}(E,n_x,n_y)\}$$
$$+\frac{1}{12}\Bigg[\frac{5\{\bar{K}_2(E,n_x,n_y)\}^3}{\bar{K}_1(E,n_x,n_y)} + \frac{5\{\bar{K}_1(E,n_x,n_y)\}^3}{\bar{K}_2(E,n_x,n_y)} - 34\bar{K}_2(E,n_x,n_y)\bar{K}_1(E,n_x,n_y)\Bigg]\sinh\{\bar{\beta}(E,n_x,n_y)\}\sin\{\bar{\gamma}(E,n_x,n_y)\}\Big]\Big],$$

$$\bar{\beta}(E,n_x,n_y) \equiv \bar{K}_1(E,n_x,n_y)[a_0 - \Delta_0]$$

$$,\ \bar{K}_1(E,n_x,n_y) \equiv \left[\frac{2m_2^* E'}{\hbar^2}G(E-V_o,\alpha_2,\Delta_2) + \left\{\left(\frac{\pi n_x}{d_x}\right)^2 + \left(\frac{\pi n_y}{d_y}\right)^2\right\}\right]^{1/2}$$

$$,\ \bar{\gamma}(E,n_x,n_y) = \bar{K}_2(E,n_x,n_y)[b_0 - \Delta_0],$$

$$\bar{K}_2(E,n_x,n_y) \equiv \left[\frac{2m_1^* E}{\hbar^2}G(E,\alpha_1,\Delta_1) - \left\{\left(\frac{\pi n_x}{d_x}\right)^2 + \left(\frac{\pi n_y}{d_y}\right)^2\right\}\right]^{1/2}$$

and $\bar{\varepsilon}(E,n_x,n_y) \equiv \left[\frac{\bar{K}_1(E,n_x,n_y)}{\bar{K}_2(E,n_x,n_y)} - \frac{\bar{K}_2(E,n_x,n_y)}{\bar{K}_1(E,n_x,n_y)}\right].$

The photo-emitted current density is given by

$$J_1 = \frac{2\alpha_0 e L_0 g_v}{d_x d_y d_z \hbar}\sum_{n_x=1}^{n_{x_{\max}}}\sum_{n_y=1}^{n_{y_{\max}}}\sum_{n_{z_{\min}}}^{n_{z_{\max}}} F_{-1}(\eta_9)\left|\frac{\sin\left(\frac{L_0 n_z \pi}{d_z}\right)}{\phi'(E_{n_z},0)}\right| \qquad (1.256)$$

where E_{n_z} is determined from

$$\phi\left(E_{n_z},0\right)=2\cos\left[\frac{L_0\pi n_z}{d_z}\right] \tag{1.257}$$

$$\phi(E_{n_z},0)=\phi(E,0)|_{E=E_{n_z}} \tag{1.258}$$

in which, $\varepsilon(E,0)\equiv\left[\frac{K_1(E,0)}{K_2(E,0)}-\frac{K_2(E,0)}{K_1(E,0)}\right]$,

the n_z should be the nearest integer of the following inequality

$$n_{z_{\min}}\geq\left(\frac{d_z}{\pi L_0}\right)\cos^{-1}\left[\frac{1}{2}\phi(W-h\upsilon,0)\right] \tag{1.259}$$

$$\phi(E,0)=\Big[\big(2\beta(E,0)\big)\sinh\big(\beta(E,0)\big)\cos\big(\gamma(E,0)\big)-(2\gamma'(E,0))\cosh\big(\beta(E,0)\big)\sin\big(\gamma(E,0)\big)+\big[\zeta_1(E,0)\big]+$$
$$\Delta_0\Bigg[\Bigg(\frac{2K_1(E,0)K_1'(E,0)}{K_2(E,0)}-\frac{K_1^2(E,0)K_2'(E,0)}{K_2^2(E,0)}-3K_2'(E,0)\Bigg)$$
$$.\cosh\big(\beta(E,0)\big)\sin\big(\gamma(E,0)\big)+\left(\frac{K_1^2(E,0)}{K_2(E,0)}-3K_2(E,0)\right)$$
$$\Big\{\big(\beta'(E,0)\big)\sinh\big(\beta(E,0)\big)\sin\big(\gamma(E,0)\big)+\gamma'(E,0)\cos\big(\gamma(E,0)\big)\cosh\big(\beta(E,0)\big)\Big\}$$
$$+\sinh\big(\beta(E,0)\big)\cos\big\{\gamma(E,0)\big\}$$
$$\left\{3K_1'(E,0)-\frac{2K_2(E,0)K_2'(E,0)}{K_1(E,0)}+\frac{K_2^2(E,0)K_1'(E,0)}{K_1^2(E,0)}\right\}$$
$$+\left(3K_1(E,0)-\frac{K_2^2(E,0)}{K_1(E,0)}\right)\Big\{\beta'(E,0)\cosh\{\beta(E,0)\}\cos\{\gamma(E,0)\}-\gamma'(E,0)\sin\{\gamma(E,0)\}\sinh\{\beta(E,0)\}\Big\}\Bigg]$$
$$+\Delta_0\Big[4\big\{K_1(E,0)K_1'(E,0)-K_2(E,0)K_2'(E,0)\big\}\cosh\{\beta(E,0)\}\cos\{\gamma(E,0)\}+2\big(K_1^2(E,0)-K_2^2(E,0)\big)$$
$$\big\{\beta'(E,0)\sinh(E,0)\cos\{\gamma(E,0)\}-\gamma'(E,0)\sin\{\gamma(E,0)\}\cosh\{\beta(E,0)\}\big\}$$
$$+\frac{1}{12}\Bigg[\frac{15K_2^2(E,0)K_2'(E,0)}{K_1(E,0)}-\frac{5K_2^3(E,0)K_1'(E,0)}{K_1^2(E,0)}+\frac{15K_1^2(E,0)K_1'(E,0)}{K_2(E,0)}$$

$$-\frac{5K_1^3(E,0)K_2'(E,0)}{K_2^2(E,0)}-34K_2'(E,0)K_1(E,0)-34K_2(E,0)K_1'(E,0)\Big]$$

$$\left[\sinh(\beta(E,0))\sin(\gamma(E,0))\right]+\frac{1}{12}\left[\frac{5K_2^3(E,0)}{K_1(E,0)}+\frac{5K_1^3(E,0)}{K_2(E,0)}\right.$$

$$-34K_1(E,0)K_2(E,0)\Big]\Big[\beta'(E,0)\cosh\{\beta(E,0)\}\sin\{\gamma(E,0)\}$$

$$+\gamma'(E,0)\cos\{\gamma(E,0)\}\sinh\{\beta(E,0)\}\Big]\Big],$$

$$\beta'(E,0)=K_1'(E,0)(a_0-\Delta_0),$$

$$K_1'(E,0)=\left(m_2^*\left(K_1(E,0)\hbar^2\right)^{-1}\right)\left[-G(E-V_0,\alpha_2,\Delta_2)\right]$$

$$[1+(E-V_0)[[1/E+E_{g_{02}}]+[1/E+E_{g_{02}}+\Delta_2]-[1/E+E_{g_{02}}+\frac{2}{3}\Delta_2]]],\ \gamma'(E,0)=K_2'(E,0)(b_0-\Delta_0),$$

$$K_2'(E,0)=\left[\frac{m_1^*}{K_2(E,0)\hbar^2}\right]\left[G(E,\alpha_1,\Delta_1)\right]\left[1+E\left(\frac{1}{E+E_{g_{01}}}+\frac{1}{E+E_{g_{01}}+\Delta_1}-\frac{1}{E+E_{g_{01}}+\frac{2}{3}\Delta_1}\right)\right]$$

$$,\zeta_1(E,0)=$$

$$\left[\varepsilon'(E,0)\sinh\{\beta(E,0)\}\sin\{\gamma(E,0)\}+\varepsilon(E,0)\beta'(E,0)\cosh\{\beta(E,0)\}\sin\{\gamma(E,0)\}\right.$$

$$\left.+\gamma'(E,0)\varepsilon(E,0)\sinh(\beta(E,0))\cos\{\gamma(E,0)\}\right]$$

and $\varepsilon'(E,0)=\left[\frac{K_1'(E,0)}{K_2(E,0)}-\frac{K_1(E,0)K_2'(E,0)}{K_2^2(E,0)}-\frac{K_2'(E,0)}{K_1(E,0)}+\frac{K_2(E,0)K_1'(E,0)}{K_1^2(E,0)}\right].$

1.2.20. The EP from II-VI Quantum dot Superlattices with Graded Interfaces

The electron dispersion laws of the constituent materials of II-VI SLs are given by

$$E=\frac{\hbar^2k_s^2}{2m_{\perp,1}^*}+\frac{\hbar^2k_z^2}{2m_{\parallel,1}^*}\pm C_0k_s \tag{1.260}$$

and

$$\frac{\hbar^2k^2}{2m_2^*}=EG(E,E_{g_{02}},\Delta_2) \tag{1.261}$$

where and are the transverse and longitudinal effective electron masses respectively at the edge of the conduction band for the first material. The energy-wave vector dispersion relation of the conduction electrons in II-VI SLs with graded interfaces can be expressed as

$$\cos(L_o k) = \frac{1}{2}\Phi_1(E,k_s) \tag{1.262}$$

where

$$\Phi_1(E,k_s) \equiv \left[2\cosh\{\beta_1(E,k_s)\}\cos\{\gamma_1(E,k_s)\} + \varepsilon_1(E,k_s)\sinh\{\beta_1(E,k_s)\}\sin\{\gamma_1(E,k_s)\}\right.$$

The total electron concentration per unit volume is given by

$$n_0 = \frac{g_v}{d_x d_y d_z}\sum_{n_x=1}^{n_{x_{\max}}}\sum_{n_y=1}^{n_{y_{\max}}}\sum_{n_z=1}^{n_{z_{\max}}} F_{-1}(\overline{\eta_{10}}) \tag{1.263}$$

where $\overline{\eta_{10}} \equiv \frac{E_{FQDSLGI} - \overline{E_{QD11}}}{k_B T}$ and is the root of the equation

$$\left(\frac{\pi n_z}{d_z}\right)^2 = \left[\frac{1}{L_0^2}\left[\cos^{-1}\left[\frac{1}{2}f_{11}(E,n_x,n_y)\right]\right|_{E=\overline{E_{QD11}}}\right]^2 - \left(\frac{\pi n_x}{d_x}\right)^2 - \left(\frac{\pi n_y}{d_y}\right)^2\Bigg] \tag{1.264}$$

where,

$$+\Delta_0\left[\left[\left(\frac{\{\bar{K}_3(E,n_x,n_y)\}^2}{\bar{K}_4(E,n_x,n_y)} - 3\bar{K}_4(E,n_x,n_y)\right)\cosh\{\bar{\beta}_1(E,n_x,n_y)\}\sin\{\bar{\gamma}_1(E,n_x,n_y)\}\right.\right.$$

$$\left.+\left(3\bar{K}_3(E,n_x,n_y) - \frac{\{\bar{K}_4(E,n_x,n_y)\}^2}{\bar{K}_3(E,n_x,n_y)}\right)\sinh\{\bar{\beta}_1(E,n_x,n_y)\}\cos\{\bar{\gamma}_1(E,n_x,n_y)\}\right]$$

$$+\Delta_0\left[2\left(\{\bar{K}_3(E,n_x,n_y)\}^2 - \{\bar{K}_4(E,n_x,n_y)\}^2\right)\cosh\{\bar{\beta}_1(E,n_x,n_y)\}\cos\{\bar{\gamma}_1(E,n_x,n_y)\}\right.$$

The photo-emitted current density is given by

$$J_1 = \frac{2\alpha_0 e L_0 g_v}{d_x d_y d_z \hbar}\sum_{n_x=1}^{n_{x_{\max}}}\sum_{n_y=1}^{n_{y_{\max}}}\sum_{n_{z_{\min}}}^{n_{z_{\max}}} F_{-1}(\overline{\eta_{10}})\left|\frac{\sin\left(\frac{L_0 n_z \pi}{d_z}\right)}{\phi_1'(E_{n_z},0)}\right| \tag{1.265}$$

where E_{n_z} is determined from

$$\phi_1\left(E_{n_z},0\right)=2\cos\left[\frac{L_0\pi n_z}{d_z}\right] \tag{1.266}$$

$$\phi_1(\,E_{n_z},0)=\phi_1(E,0)\,|_{E=E_{n_z}} \tag{1.267}$$

in which, $\varepsilon_1(E,0)\equiv\left[\frac{K_3(E,0)}{K_4(E,0)}-\frac{K_4(E,0)}{K_3(E,0)}\right]$,

, $K_3(E,0)=\left[\frac{2m_2^*}{\hbar^2}(V_0-E)G(E-V_0,\alpha_2,\Delta_2)\right]^{1/2}$,

$\beta_1(E,0)=K_3(E,0)(a_0-\Delta_0)$, $\gamma_1(E,0)=K_4(E,0)(b_0-\Delta_0)$,

$K_4(E,0)=\left[\frac{2m_{\parallel,1}^*E}{\hbar^2}\right]^{1/2}$,

the n_z should be the nearest integer of the following inequality

$$n_{z_{\min}}\geq\left(\frac{d_z}{\pi L_0}\right)\cos^{-1}\left[\frac{1}{2}\phi_1\left(W-h\upsilon,0\right)\right] \tag{1.268}$$

in which, $\varepsilon_1(E,0)\equiv\left[\frac{K_3(E,0)}{K_4(E,0)}-\frac{K_4(E,0)}{K_3(E,0)}\right]$,

$$\phi_1'(E,0)=\left[(2\beta_1'(E,0))\sinh\left(\beta_1(E,0)\right)\cos\left(\gamma_1(E,0)\right)-(2\gamma_1'(E,0))\cosh\left(\beta_1(E,0)\right)\sin\left(\gamma_1(E,0)\right)+\left[\zeta_2(E,0)\right]+\right.$$

$$\Delta_0\left[\left(\frac{2K_3(E,0)K_3'(E,0)}{K_4(E,0)}-\frac{K_3^2(E,0)K_4'(E,0)}{K_4^2(E,0)}-3K_4'(E,0)\right)\cosh\left[\beta_1(E,0)\right]\sinh\left[\gamma_1(E,0)\right]+\right.$$

$$\left(\frac{K_3^2(E,0)}{K_4(E,0)}-3K_4(E,0)\right)$$

$$\left\{(\beta_1'(E,0))\sinh\left(\beta_1(E,0)\right)\sin\left(\gamma_1(E,0)\right)+\gamma_1'(E,0)\cos\left(\gamma_1(E,0)\right)\cosh\left(\beta_1(E,0)\right)\right\}$$

$$+\sinh\left(\beta_1(E,0)\right)\cos\left\{\gamma_1(E,0)\right\}$$

$$\left\{3K_3'(E,0)-\frac{2K_4(E,0)K_4'(E,0)}{K_3(E,0)}+\frac{K_4^2(E,0)K_3'(E,0)}{K_3^2(E,0)}\right\}$$

$$+\left(3K_3(E,0)-\frac{K_4^2(E,0)}{K_3(E,0)}\right)\left\{\beta_1'(E,0)\cosh\{\beta_1(E,0)\}\cos\{\gamma_1(E,0)\}-\gamma_1'(E,0)\sin\{\gamma_1(E,0)\}\sinh\{\beta_1(E,0)\}\}\right]$$

$$+\Delta_0\left[4\{K_3(E,0)K_3'(E,0)-K_4(E,0)K_4'(E,0)\}\cosh\{\beta_1(E,0)\}\cos\{\gamma_1(E,0)\}+2\left(K_3^2(E,0)-K_4^2(E,0)\right)\right.$$

$$\{\beta_1'(E,0)\sinh(E,0)\cos\{\gamma_1(E,0)\}-\gamma_1'(E,0)\sin\{\gamma_1(E,0)\}\cosh\{\beta_1(E,0)\}\}$$

$$+\frac{1}{12}\left[\frac{15K_4^2(E,0)K_4'(E,0)}{K_3(E,0)}-\frac{5K_4^3(E,0)K_3'(E,0)}{K_3^2(E,0)}+\frac{15K_3^2(E,0)K_3'(E,0)}{K_4(E,0)}\right.$$

$$\left.-\frac{5K_3^3(E,0)K_4'(E,0)}{K_4^2(E,0)}-34K_4'(E,0)K_3(E,0)-34K_4(E,0)K_3'(E,0)\right]$$

$$\left[\sinh(\beta_1(E,0))\sin(\gamma_1(E,0))\right]$$

$$+\frac{1}{12}\left[\frac{5K_4^3(E,0)}{K_3(E,0)}+\frac{5K_3^3(E,0)}{K_4(E,0)}-34K_3(E,0)K_4(E,0)\right]$$

$$\left[\beta_1'(E,0)\cosh\{\beta_1(E,0)\}\sin\{\gamma_1(E,0)\}+\right.$$

$$\left.\gamma_1'(E,0)\cos\{\gamma_1(E,0)\}\sinh\{\beta_1(E,0)\}\right]\Big],$$

$$\beta_1'(E,0)=K_3'(E,0)(a_0-\Delta_0),$$

$$K_3'(E,0)=\left(m_2^*\left(K_3(E,0)\hbar^2\right)^{-1}\right)\left[-G(E-V_0,\alpha_2,\Delta_2)\right]$$

$$\left[1+(E-V_0)\left[\frac{1}{E+E_{g02}}+\frac{1}{E+E_{g02}+\Delta_2}-\frac{1}{E+E_{g02}+\frac{2}{3}\Delta_2}\right]\right],$$

$$,\ \gamma_1'(E,0)=K_4'(E,0)(b_0-\Delta_0),\ K_4'(E,0)=\left[\frac{m_{1,1}^*}{K_4(E,0)\hbar^2}\right],$$

$$\zeta_2(E,0)=\left[\varepsilon_1'(E,0)\sinh\{\beta_1(E,0)\}\sin\{\gamma_1(E,0)\}+\varepsilon_1(E,0)\beta_1'(E,0)\cosh\{\beta_1(E,0)\}\sin\{\gamma_1(E,0)\}\right.$$

$$\left.\gamma_1'(E,0)\varepsilon_1(E,0)\sinh[\beta_1(E,0)]\cos[\gamma_1(E,0)]\right]$$

and $\varepsilon_1'(E,0)=\left[\frac{K_3'(E,0)}{K_4(E,0)}-\frac{K_3(E,0)K_4'(E,0)}{K_4^2(E,0)}-\frac{K_4'(E,0)}{K_3(E,0)}+\frac{K_4(E,0)K_3'(E,0)}{K_3^2(E,0)}\right].$

1.2.21. The EP from IV-VI Quantum Dot Superlattices with Graded Interfaces

The E-k dispersion relation of the conduction electrons of the constituent materials of the IV-VI SLs can be expressed as

$$E = a_i k_s^2 + b_i k_z^2 + \left[\left[c_i k_s^2 + d_i k_z^2 \right] + \left(e_i k_s^2 + f_i k_y^2 + \frac{E_{g0i}}{2} \right)^2 \right]^{1/2} - \frac{E_{g0i}}{2} \quad (1.269)$$

where, $a_i \equiv \left[\frac{\hbar^2}{2m_{\perp,i}^-} \right]$, $b_i \equiv \left(\frac{\hbar^2}{2m_{\parallel,i}^-} \right)^2$, $C_i \equiv P_{\perp,i}^2$, $d_i \equiv P_{\perp,i}^2$, $e_i \equiv \left[\frac{\hbar^2}{2m_{\perp,i}^+} \right]$ and $f_i \equiv \left(\frac{\hbar^2}{2m_{\parallel,i}^+} \right)^2$.

The electron dispersion law in IV-VI SLs with graded interfaces can be expressed as

$$\cos(L_o k) = \frac{1}{2} \Phi_2(E, k_s) \quad (1.270)$$

where

$$\Phi_2(E,k_s) \equiv \Big[2\cosh\{\beta_2(E,k_s)\}\cos\{\gamma_2(E,k_s)\} + \varepsilon_2(E,k_s)\sinh\{\beta_2(E,k_s)\}\sin\{\gamma_2(E,k_s)\}$$
$$+ \frac{1}{12}\left[\frac{5\{K_5(E,k_s)\}^3}{K_6(E,k_s)} + \frac{5\{K_6(E,k_s)\}^3}{K_5(E,k_s)} - 34 K_6(E,k_s) K_5(E,k_s) \right] \sinh\{\beta_2(E,k_s)\}\sin\{\gamma_2(E,k_s)\} \Big]\Big]$$

$$\beta_2(E,k_s) \equiv K_5(E,k_s)[a_0 - \Delta_0],$$

$$K_5(E,k_x,k_y) \equiv \left[\left[(E-V_0)^2 H_{32} + (E-V_0) H_{42}(k_x,k_y) + H_{52}(k_x,k_y) \right]^{1/2} - \left[(E-V_0) H_{12} + H_{22}(k_x,k_y) \right] \right]^{1/2}$$

$$, \gamma_2(E,k_s) = K_6(E,k_s)[b_0 - \Delta_0],$$

$$K_6(E,k_x,k_y) \equiv \left[\left[EH_{11} + H_{21}(k_x,k_y) \right] - \left[H_{31}E^2 + EH_{41}(k_x,k_y) + H_{51}(k_x,k_y) \right]^{1/2} \right]^{1/2}$$

$$, H_{1i} = b_i (b_i^2 - f_i^2)^{-1}, i = 1 \text{ and } 2,$$

$$, H_{2i}(k_x,k_y) = \left[2(b_i^2 - f_i^2) \right]^{-1} \left[E_{g_i} b_i + d_i + f_i E_{g_i} + 2(e_i f_i - a_i b_i)(k_x^2 + k_y^2) \right]$$

$$, H_{3i} = \frac{f_i^2}{(b_i^2 - f_i^2)^2},$$

$$H_{4i}(k_x,k_y) = \left[4(b_i^2 - f_i^2)^2 \right]^{-1} \left[4b_i^2 E_{gi} + 4b_i d_i + 4b_i f_i E_{g_i} + 4f_i^2 E_{g_i} + 8(k_x^2 + k_y^2)\left[b_i^2 a_i + C_i f_i b_i - a_i^2 b_i \right] \right]$$

$$H_{5i}(k_x,k_y) \equiv \left[4(b_i^2 - f_i^2)^2 \right]^{-1} \Big[(k_x^2 + k_y^2)^2 \left[-8a_i b_i C_i f_i + 4b_i^2 C_i^2 + 4f_i^2 a_i^2 - 4f_i^2 C_i^2 \right]$$
$$+ (k_x^2 + k_y^2)\left[8d_i C_i f_i - 4a_i b_i d_i - 4a_i b_i f_i E_{gi} + 4b_i^2 C_i + 4b_i^2 e_i E_{gi} - 4a_i f_i^2 E_{gi} - 4f_i^2 e_i E_{gi} \right]$$
$$+ \left[E_{g_i}^2 b_i^2 + d_i^2 + f_i^2 E_{gi}^2 + 2d_i f_i E_{g_i} \right] \Big]$$

and $\varepsilon_2(E,k_s) \equiv \left[\frac{K_5(E,k_s)}{K_6(E,k_s)} - \frac{K_6(E,k_s)}{K_5(E,k_s)}\right]$.

The total electron concentration per unit volume is given by

$$n_0 = \frac{2g_v}{d_x d_y d_z} \sum_{n_x=1}^{n_{x_{\max}}} \sum_{n_y=1}^{n_{y_{\max}}} \sum_{n_{z_{\min}}}^{n_{z_{\max}}} F_{-1}\left(\overline{\eta_{11}}\right) \tag{1.271}$$

where $\overline{\eta_{11}} \equiv \frac{E_{FQDSLGI} - \overline{E_{QD12}}}{k_B T}$ and is the root of the equation

$$\left(\frac{\pi n_z}{d_z}\right)^2 = \left[\left[\frac{1}{L_0^2}\left[\cos^{-1}\left[\frac{1}{2} f_{12}(E,n_x,n_y)\Big|_{E=\overline{E_{QD12}}}\right]\right]^2 - \left(\frac{\pi n_x}{d_x}\right)^2 - \left(\frac{\pi n_y}{d_y}\right)^2\right] \tag{1.272}$$

$$f_{12}(E,n_x,n_y) \equiv \left[2\cosh\{\bar{\beta}_2(E,n_x,n_y)\}\cos\{\bar{\gamma}_2(E,n_x,n_y)\} + \bar{\varepsilon}_2(E,n_x,n_y)\sinh\{\bar{\beta}_2(E,n_x,n_y)\}\sin\{\bar{\gamma}_2(E,n_x,n_y)\}\right.$$
$$\left. + \frac{1}{12}\left[\frac{5\{\bar{K}_5(E,n_x,n_y)\}^3}{\bar{K}_6(E,n_x,n_y)} + \frac{5\{\bar{K}_6(E,n_x,n_y)\}^3}{\bar{K}_5(E,n_x,n_y)} - 34\bar{K}_6(E,n_x,n_y)\bar{K}_5(E,n_x,n_y)\right]\sinh\{\bar{\beta}_2(E,n_x,n_y)\}\sin\{\bar{\gamma}_2(E,n_x,n_y)\}\right]\right],$$

$$\bar{\beta}_2(E,n_x,n_y) \equiv \bar{K}_5(E,n_x,n_y)[a_0 - \Delta_0],$$

$$\bar{K}_5(E,n_x,n_y) \equiv \left[\left[(E-V_0)^2 H_{32} + (E-V_0)\bar{H}_{42}(n_x,n_y) + \bar{H}_{52}(n_x,n_y)\right]^{1/2} - \left[(E-V_0)H_{12} + \bar{H}_{22}(n_x,n_y)\right]\right]^{1/2}$$

$$\bar{H}_{2i}(n_x,n_y) = \left[2\left(b_i^2 - f_i^2\right)\right]^{-1}\left[E_{g_i} b_i + d_i + f_i E_{g_i} + 2(C_i f_i - a_i b_i)\Psi_1(n_x,n_y)\right],$$

$$\Psi_1(n_x,n_y) = \left\{\left(\frac{\pi n_x}{d_x}\right)^2 + \left(\frac{\pi n_y}{d_y}\right)^2\right\}$$

$$\bar{H}_{4i}(n_x,n_y) = \left[4\left(b_i^2 - f_i^2\right)^2\right]^{-1}\left[4b_i d_i + 4b_i f_i E_{g_i} + 4f_i^2 E_{g_i} + 4b_i^2 E_{g_i} + 8\Psi_1(n_x,n_y)\left[a_i b_i^2 + C_i f_i b_i - b_i a_i^2\right]\right]$$

$$\bar{H}_{5i}(n_x,n_y) \equiv \left[4\left(b_i^2 - f_i^2\right)^2\right]^{-1}\left[\{\Psi_1(n_x,n_y)\}^2\left[-8a_i b_i C_i f_i + 4b_i^2 C_i^2 + 4f_i^2 d_i^2 - 4f_i^2 C_i^2\right] + \Psi_1(n_x,n_y)[8d_i C_i f_i\right.$$
$$-4a_i b_i d_i - 4a_i f_i E_{g_i} b_i + 4b_i^2 C_i + 4b_i^2 e_i E_{g_i} - 4a_i f_i^2 E_{g_i} - 4f_i^2 e_i E_{g_i}]$$
$$\left. + \left[E_{g_i}^2 b_i^2 + d_i^2 + f_i^2 E_{gi}^2 + 2E_{g_i} f_i d_i\right]\right]$$

$$\bar{\gamma}_2(E,n_x,n_y) = \bar{K}_6(E,n_x,n_y)[b_0 - \Delta_0],$$

$$\bar{\varepsilon}_2\left(E,n_x,n_y\right)\equiv\left[\frac{\bar{K}_5\left(E,n_x,n_y\right)}{\bar{K}_6\left(E,n_x,n_y\right)}-\frac{\bar{K}_6\left(E,n_x,n_y\right)}{\bar{K}_5\left(E,n_x,n_y\right)}\right] \text{ and}$$

$$\bar{K}_6\left(E,n_x,n_y\right)\equiv\left[\left[EH_{11}+\bar{H}_{21}\left(n_x,n_y\right)\right]-\left[H_{31}E^2+E\bar{H}_{41}\left(n_x,n_y\right)+\bar{H}_{51}\left(n_x,n_y\right)\right]^{1/2}\right]^{1/2}$$

The photo-emitted current density is given by

$$J_1=\frac{2\alpha_0 eL_0 g_v}{d_x d_y d_z \hbar}\sum_{n_x=1}^{n_{x_{\max}}}\sum_{n_y=1}^{n_{y_{\max}}}\sum_{n_{z_{\min}}}^{n_{z_{\max}}}F_{-1}\left(\overline{\eta_{11}}\right)\left|\frac{\sin\left(\frac{L_0 n_z\pi}{d_z}\right)}{\phi_2'\left(E_{n_z},0\right)}\right| \tag{1.273}$$

where E_{n_z} is determined from

$$\phi_2\left(E_{n_z},0\right)=2\cos\left[\frac{L_0\pi n_z}{d_z}\right] \tag{1.274}$$

$$\phi_2(E_{n_z},0)=\phi_2(E,0)\,|_{E=E_{n_z}} \tag{1.275}$$

in which, $\varepsilon_2(E,0)\equiv\left[\frac{K_5(E,0)}{K_6(E,0)}-\frac{K_6(E,0)}{K_5(E,0)}\right]$,

$\beta_2(E,0)=K_5(E,0)(a_0-\Delta_0)$,

$$k_5(E,0)=\left[\left[(E-V_0)^2 H_{32}+(E-V_0)H_{42}(0,0)+H_{52}(0,0)\right]^{1/2}-\left[(E-V_0)H_{12}+H_{22}(0,0)\right]\right]^{1/2}$$

$$H_{4i}(0,0)=\left[\left(b_i^2-f_i^2\right)^2\right]^{-1}\left[b_i^2E_{gi}+b_id_i+b_if_iE_{g_i}+f_i^2E_{g_i}\right],$$

$$H_{5i}(0,0)\equiv\left[4\left(b_i^2-f_i^2\right)^2\right]^{-1}\left[E_{g_i}^2b_i^2+d_i^2+f_i^2E_{gi}^2+2d_if_iE_{g_i}\right],$$

$$H_{2i}(0,0)=\left[2\left(b_i^2-f_i^2\right)\right]^{-1}\left[E_{g_i}b_i+d_i+f_iE_{g_i}\right],\quad \gamma_2(E,0)=K_6(E,0)(b_0-\Delta_0),$$

$$k_6(E,0)=\left[\left[EH_{11}+H_{21}(0,0)\right]-\left[H_{31}E^2+EH_{41}(0,0)+H_{51}(0,0)\right]^{1/2}\right]^{1/2}$$

the n_z should be the nearest integer of the following inequality

$$n_{z_{\min}} \geq \left(\frac{d_z}{\pi L_0}\right)\cos^{-1}\left[\frac{1}{2}\phi_2\left(W-h\upsilon,0\right)\right] \tag{1.276}$$

in which, $\varepsilon_2(E,0) \equiv \left[\frac{K_5(E,0)}{K_6(E,0)} - \frac{K_6(E,0)}{K_5(E,0)}\right]$,

$$\phi_2'(E,0)=\Big[\left(2\beta_2'(E,0)\right)\sinh\left(\beta_2(E,0)\right)\cos\left(\gamma_2(E,0)\right)-\left(2\gamma_2'(E,0)\right)\cosh\left(\beta_2(E,0)\right)\sin\left(\gamma_2(E,0)\right)+\left[\zeta_3(E,0)\right]+$$

$$\Delta_0\Bigg[\left(\frac{2K_5(E,0)K_5'(E,0)}{K_6(E,0)}-\frac{K_5^2(E,0)K_6'(E,0)}{K_6^2(E,0)}-3K_6'(E,0)\right)\cosh\left[\beta_2(E,0)\right]\sinh\left[\gamma_2(E,0)\right]+$$

$$\left(\frac{K_5^2(E,0)}{K_6(E,0)}-3K_6(E,0)\right)$$

$$\left\{\left(\beta_2'(E,0)\right)\sinh\left(\beta_2(E,0)\right)\sin\left(\gamma_2(E,0)\right)+\gamma_2'(E,0)\cos\left(\gamma_2(E,0)\right)\cosh\left(\beta_2(E,0)\right)\right\}$$

$$+\sinh\left(\beta_2(E,0)\right)\cos\left\{\gamma_2(E,0)\right\}$$

$$\left\{3K_5'(E,0)-\frac{2K_6(E,0)K_6'(E,0)}{K_5(E,0)}+\frac{K_6^2(E,0)K_5'(E,0)}{K_5^2(E,0)}\right\}$$

$$+\left(3K_5(E,0)-\frac{K_6^2(E,0)}{K_5(E,0)}\right)\left\{\beta_2'(E,0)\cosh\left\{\beta_2(E,0)\right\}\cos\left\{\gamma_2(E,0)\right\}-\gamma_2'(E,0)\sin\left\{\gamma_2(E,0)\right\}\sinh\left\{\beta_2(E,0)\right\}\right\}\Bigg]$$

$$+\Delta_0\Big[4\left\{K_5(E,0)K_5'(E,0)-K_6(E,0)K_6'(E,0)\right\}\cosh\left\{\beta_2(E,0)\right\}\cos\left\{\gamma_2(E,0)\right\}+2\left(K_5^2(E,0)-K_6^2(E,0)\right)$$

$$\left\{\beta_2'(E,0)\sinh(E,0)\cos\left\{\gamma_2(E,0)\right\}-\gamma_2'(E,0)\sin\left\{\gamma_2(E,0)\right\}\cosh\left\{\beta_2(E,0)\right\}\right\}$$

$$+\frac{1}{12}\Bigg[\frac{15K_6^2(E,0)K_6'(E,0)}{K_5(E,0)}-\frac{5K_6^3(E,0)K_5'(E,0)}{K_5^2(E,0)}$$

$$+\frac{15K_5^2(E,0)K_5'(E,0)}{K_6(E,0)}-\frac{5K_5^3(E,0)K_6'(E,0)}{K_6^2(E,0)}-34K_6'(E,0)K_5(E,0)$$

$$-34K_6(E,0)K_5'(E,0)\Bigg]\left[\sinh\left(\beta_2(E,0)\right)\sin\left(\gamma_2(E,0)\right)\right]+$$

$$\frac{1}{12}\left[\frac{5K_6^3(E,0)}{K_5(E,0)}+\frac{5K_5^3(E,0)}{K_6(E,0)}-34K_5(E,0)K_6(E,0)\right]$$

$$\left[\beta_2'(E,0)\cosh\left\{\beta_2(E,0)\right\}\sin\left\{\gamma_2(E,0)\right\}\right.$$

$$\left.+\gamma_2'(E,0)\cos\left\{\gamma_2(E,0)\right\}\sinh\left\{\beta_2(E,0)\right\}\right]\Big]\text{,}\quad \beta_2'(E,0)=K_5'(E,0)\left(a_0-\Delta_0\right),$$

$$K_5'(E,0)=\left[2K_5(E,0)\right]^{-1}\left[-H_{12}+\frac{\left[2(E-V_0)H_{32}+H_{42}(0,0)\right]}{2\left[(E-V_0)^2H_{32}+(E-V_0)H_{42}(0,0)+H_{52}(0,0)\right]^{1/2}}\right],$$

$$\gamma_2'(E,0)=K_6'(E,0)(b_0-\Delta_0),$$

$$K_6'(E,0)=\left[2K_6(E,0)\right]^{-1}\left[H_{11}-\frac{\left[2EH_{31}+H_{41}(0,0)\right]}{2\left[E^2H_{31}+EH_{41}(0,0)+H_{51}(0,0)\right]^{1/2}}\right]$$

$$\zeta_3(E,0)=\left[\varepsilon_2'(E,0)\sinh\{\beta_2(E,0)\}\sin\{\gamma_2(E,0)\}+\varepsilon_2(E,0)\beta_2'(E,0)\cosh\{\beta_2(E,0)\}\sin\{\gamma_2(E,0)\}\right.$$
$$\gamma_2'(E,0)\varepsilon_2(E,0)\sinh\left[\beta_2(E,0)\right]\cos\left[\gamma_2(E,0)\right]$$

and

$$\varepsilon_2'(E,0)=\left[\frac{K_5'(E,0)}{K_6(E,0)}-\frac{K_5(E,0)K_6'(E,0)}{K_6^2(E,0)}-\frac{K_6'(E,0)}{K_5(E,0)}+\frac{K_6(E,0)K_5'(E,0)}{K_5^2(E,0)}\right].$$

1.2.22. The EP from HgTe/CdTe Quantum dot Superlattices with Graded Interfaces

The dispersion relation of the conduction electrons of the constituent materials of HgTe/CdTe SLs can be expressed as

$$E=\frac{\hbar^2k^2}{2m_1^*}+\frac{3|e|^2k}{128\varepsilon_{sc}} \tag{1.277}$$

$$\frac{\hbar^2k^2}{2m_2^*}=EG\left(E,E_{g_{02}},\Delta_2\right) \tag{1.278}$$

The electron energy dispersion law in HgTe/CdTe SL is given by

$$\cos(L_0k)=\frac{1}{2}\Phi_3(E,k_s) \tag{1.279}$$

where,

$$\Phi_3(E,k_s)\equiv\left[2\cosh\{\beta_3(E,k_s)\}\cos\{\gamma_3(E,k_s)\}+\varepsilon_3(E,k_s)\sinh\{\beta_3(E,k_s)\}\sin\{\gamma_3(E,k_s)\}\right.$$

$$+\Delta_0\left[\left(\frac{\{K_7(E,k_s)\}^7}{K_8(E,k_s)}-3K_8(E,k_s)\right)\cosh\{\beta_3(E,k_s)\}\sin\{\gamma_3(E,k_s)\}+\left(3K_7(E,k_s)-\frac{\{K_8(E,k_s)\}^2}{K_7(E,k_s)}\right)\sinh\{\beta_3(E,k_s)\}\cos\{\gamma_3(E,k_s)\}\right]$$

$$+\Delta_0\left[2\left(\{K_7(E,k_s)\}^2-\{K_8(E,k_s)\}^2\right)\cosh\{\beta_3(E,k_s)\}\cos\{\gamma_3(E,k_s)\}\right.,$$

$$\beta_3(E,k_s)\equiv K_7(E,k_s)[a_0-\Delta_0],$$

$$K_7(E,k_s)\equiv\left[\frac{2m_2^*E'}{\hbar^2}G(E-V_0,E_{g_{02}},\Delta_2)+k_s^2\right]^{1/2},$$

$$\gamma_3(E,k_s)=K_8(E,k_s)[b_0-\Delta_0],$$

$$K_8(E,k_x,k_y)\equiv\left[\frac{B_0^2+2AE-B_0\sqrt{B_0^2+4AE}}{2A^2}-k_s^2\right]^{1/2},\quad B_0=\frac{3|e|^2}{128\varepsilon_{sc}},$$

and $\varepsilon_3(E,k_s)\equiv\left[\frac{K_7(E,k_s)}{K_8(E,k_s)}-\frac{K_8(E,k_s)}{K_7(E,k_s)}\right].$

The total electron concentration per unit volume is given by

$$n_0=\frac{2g_v}{d_xd_yd_z}\sum_{n_x=1}^{n_{x_{max}}}\sum_{n_y=1}^{n_{y_{max}}}\sum_{n_z=1}^{n_{z_{max}}}F_{-1}\left(\overline{\eta_{12}}\right) \tag{1.280}$$

where, $\overline{\eta_{12}}\equiv\frac{E_{FQDSLGI}-\overline{E_{QD13}}}{k_BT}$ and $\overline{E_{QD13}}$ is the root of the equation

$$\left(\frac{\pi n_z}{d_z}\right)^2=\left[\left[\frac{1}{L_0^2}\left[\cos^{-1}\left[\frac{1}{2}f_{13}(E,n_x,n_y)\Big|_{E=\overline{E_{QD13}}}\right]\right]^2-\left(\frac{\pi n_x}{d_x}\right)^2-\left(\frac{\pi n_y}{d_y}\right)^2\right] \tag{1.281}$$

where

$$f_{13}(E,n_x,n_y)\equiv\left[2\cosh\{\bar\beta_3(E,n_x,n_y)\}\cos\{\bar\gamma_3(E,n_x,n_y)\}+\bar\varepsilon_3(E,n_x,n_y)\sinh\{\bar\beta_3(E,n_x,n_y)\}\sin\{\bar\gamma_3(E,n_x,n_y)\}\right.$$

$$+\Delta_0\left[\left(\frac{\{\bar K_7(E,n_x,n_y)\}^7}{\bar K_8(E,n_x,n_y)}-3\bar K_8(E,n_x,n_y)\right)\cosh\{\bar\beta_3(E,n_x,n_y)\}\sin\{\bar\gamma_3(E,n_x,n_y)\}\right.$$

$$+\left(3\bar{K}_7(E,n_x,n_y)-\frac{\{\bar{K}_8(E,n_x,n_y)\}^2}{\bar{K}_7(E,n_x,n_y)}\right)\sinh\{\bar{\beta}_3(E,n_x,n_y)\}\cos\{\bar{\gamma}_3(E,n_x,n_y)\}\Bigg]$$

$$+\Delta_0\Bigg[2\left(\{\bar{K}_7(E,n_x,n_y)\}^2-\{\bar{K}_8(E,n_x,n_y)\}^2\right)\cosh\{\bar{\beta}_3(E,n_x,n_y)\}\cos\{\bar{\gamma}_3(E,n_x,n_y)\}$$

$$+\frac{1}{12}\left[\frac{5\{\bar{K}_8(E,n_x,n_y)\}^3}{\bar{K}_7(E,n_x,n_y)}+\frac{5\{\bar{K}_7(E,n_x,n_y)\}^3}{\bar{K}_8(E,n_x,n_y)}-34\bar{K}_7(E,n_x,n_y)\bar{K}_8(E,n_x,n_y)\right]\sinh\{\bar{\beta}_3(E,n_x,n_y)\}\sin\{\bar{\gamma}_3(E,n_x,n_y)\}\Bigg]\Bigg],$$

$$\bar{\beta}_3(E,n_x,n_y)\equiv\bar{K}_7(E,n_x,n_y)[a_0-\Delta_0],$$

$$\bar{K}_7(E,n_x,n_y)\equiv\left[\frac{2m_2^*E'}{\hbar^2}G(E-V_0,E_{g_{02}},\Delta_2)+\left\{\left(\frac{\pi n_x}{d_x}\right)^2+\left(\frac{\pi n_y}{d_y}\right)^2\right\}\right]^{1/2}$$

$$,\bar{\gamma}_3(E,n_x,n_y)=\bar{K}_8(E,n_x,n_y)[b_0-\Delta_0]$$

$$,\bar{K}_8(E,n_x,n_y)\equiv\left[\frac{B_0^2+2AE-B_0\sqrt{B_0^2+4AE}}{2A^2}-\left\{\left(\frac{\pi n_x}{d_x}\right)^2+\left(\frac{\pi n_y}{d_y}\right)^2\right\}\right]^{1/2}$$

and $\bar{\varepsilon}_3(E,n_x,n_y)\equiv\left[\dfrac{\bar{K}_7(E,n_x,n_y)}{\bar{K}_8(E,n_x,n_y)}-\dfrac{\bar{K}_8(E,n_x,n_y)}{\bar{K}_7(E,n_x,n_y)}\right]$.

The photo-emitted current density is given by

$$J_1=\frac{2\alpha_0 eL_0 g_v}{d_x d_y d_z\hbar}\sum_{n_x=1}^{n_{x_{\max}}}\sum_{n_y=1}^{n_{y_{\max}}}\sum_{n_{z_{\min}}}^{n_{z_{\max}}}F_{-1}(\eta_{12})\left|\frac{\sin\left(\dfrac{L_0 n_z\pi}{d_z}\right)}{\phi_3'(E_{n_z},0)}\right| \tag{1.282}$$

where E_{n_z} is determined from

$$\phi_3(E_{n_z},0)=2\cos\left[\frac{L_0\pi n_z}{d_z}\right] \tag{1.283}$$

$$\phi_3(E_{n_z},0)=\phi_3(E,0)|_{E=E_{n_z}} \tag{1.284}$$

in which, $\varepsilon_3(E,0) \equiv \left[\frac{K_7(E,0)}{K_8(E,0)} - \frac{K_8(E,0)}{K_7(E,0)}\right]$,

$$\beta_3(E,0) = K_7(E,0)(a_0 - \Delta_0), \quad K_7(E,0) \equiv \left[\frac{2m_2^* E'}{\hbar^2} G(E - V_0, E_{g_{02}}, \Delta_2)\right]^{1/2}$$

$$\gamma_3(E,0) = K_8(E,0)(b_0 - \Delta_0), \quad K_8(E,0) \equiv \left[\frac{B_0^2 + 2AE - B_0\sqrt{B_0^2 + 4AE}}{2A^2}\right]^{1/2}$$

the n_z should be the nearest integer of the following inequality

$$n_{z_{\min}} \geq \left(\frac{d_z}{\pi L_0}\right)\cos^{-1}\left[\frac{1}{2}\phi_3(W - h\upsilon, 0)\right] \tag{1.285}$$

$$\phi_3'(E,0) = \left[(2\beta_3'(E,0))\sinh(\beta_3(E,0))\cos(\gamma_3(E,0)) - (2\gamma_3'(E,0))\cosh(\beta_3(E,0))\sin(\gamma_3(E,0)) + [\zeta_4(E,0)] + \right.$$

$$\Delta_0\left[\left(\frac{2K_7(E,0)K_7'(E,0)}{K_8(E,0)} - \frac{K_7^2(E,0)K_8'(E,0)}{K_8^2(E,0)} - 3K_8'(E,0)\right)\cosh[\beta_3(E,0)]\sinh[\gamma_3(E,0)] + \right.$$

$$\left(\frac{K_7^2(E,0)}{K_8(E,0)} - 3K_8(E,0)\right)$$

$$\{(\beta_3'(E,0))\sinh(\beta_3(E,0))\sin(\gamma_3(E,0)) + \gamma_3'(E,0)\cos(\gamma_3(E,0))\cosh(\beta_3(E,0))\}$$

$$+ \sinh(\beta_3(E,0))\cos\{\gamma_3(E,0)\}$$

$$\left\{3K_7'(E,0) - \frac{2K_8(E,0)K_8'(E,0)}{K_7(E,0)} + \frac{K_8^2(E,0)K_7'(E,0)}{K_7^2(E,0)}\right\}$$

$$+\left(3K_7(E,0) - \frac{K_8^2(E,0)}{K_7(E,0)}\right)\left\{\beta_3'(E,0)\cosh\{\beta_3(E,0)\}\cos\{\gamma_3(E,0)\} - \gamma_3'(E,0)\sin\{\gamma_3(E,0)\}\sinh\{\beta_3(E,0)\}\right\}\Big]$$

$$+\Delta_0\left[4\{K_7(E,0)K_7'(E,0) - K_8(E,0)K_8'(E,0)\}\cosh\{\beta_3(E,0)\}\cos\{\gamma_3(E,0)\} + 2\left(K_7^2(E,0) - K_8^2(E,0)\right)\right.$$

$$\{\beta_3'(E,0)\sinh(E,0)\cos\{\gamma_3(E,0)\} - \gamma_3'(E,0)\sin\{\gamma_3(E,0)\}\cosh\{\beta_3(E,0)\}\}$$

$$+\frac{1}{12}\left[\frac{15K_8^2(E,0)K_8'(E,0)}{K_7(E,0)} - \frac{5K_8^3(E,0)K_7'(E,0)}{K_7^2(E,0)} + \frac{15K_7^2(E,0)K_7'(E,0)}{K_8(E,0)}\right.$$

$$\left. - \frac{5K_7^3(E,0)K_8'(E,0)}{K_8^2(E,0)} - 34K_8'(E,0)K_7(E,0) - 34K_8(E,0)K_7'(E,0)\right]$$

$$\left[\sinh\left(\beta_3(E,0)\right)\sin\left(\gamma_3(E,0)\right)\right]+\frac{1}{12}\left[\frac{5K_8^3(E,0)}{K_7(E,0)}+\frac{5K_7^3(E,0)}{K_8(E,0)}-34K_7(E,0)K_8(E,0)\right]$$

$$\left[\beta_3{}'(E,0)\cosh\left\{\beta_3(E,0)\right\}\sin\left\{\gamma_3(E,0)\right\}+\right.$$

$$\left.\gamma_3{}'(E,0)\cos\left\{\gamma_3(E,0)\right\}\sinh\left\{\beta_3(E,0)\right\}\right]\Big]_{,}\ \beta_3{}'(E,0)=K_7'(E,0)(a_0-\Delta_0),$$

$$K_7'(E,0)=\left(m_2^*\left(K_7(E,0)\hbar^2\right)^{-1}\right)\left[-G(E-V_0,\alpha_2,\Delta_2)\right]$$

$$\left[1+(E-V_0)\left[\frac{1}{E+E_{g02}}+\frac{1}{E+E_{g02}+\Delta_2}-\frac{1}{E+E_{g02}+\frac{2}{3}\Delta_2}\right]\right]\Big]_{,}$$

$$\gamma_2{}'(E,0)=K_6'(E,0)(b_0-\Delta_0),$$

$$K_8{}'(E,0)=\left[2AK_8(E,0)\right]^{-1}\left[1-\frac{B_0}{\sqrt{B_0^2+4AE}}\right]$$

,

$$\zeta_4(E,0)=\left[\varepsilon_3{}'(E,0)\sinh\left\{\beta_3(E,0)\right\}\sin\left\{\gamma_3(E,0)\right\}+\varepsilon_3(E,0)\beta_3{}'(E,0)\cosh\left\{\beta_3(E,0)\right\}\sin\left\{\gamma_3(E,0)\right\}\right.$$

$$\left.\gamma_3{}'(E,0)\varepsilon_3(E,0)\sinh\left[\beta_3(E,0)\right]\cos\left[\gamma_3(E,0)\right]\right]$$

and $\varepsilon_3{}'(E,0)=\left[\frac{K_7'(E,0)}{K_8(E,0)}-\frac{K_7(E,0)K_8'(E,0)}{K_8^2(E,0)}-\frac{K_8'(E,0)}{K_7(E,0)}+\frac{K_8(E,0)K_7'(E,0)}{K_7^2(E,0)}\right].$

1.2.23. The EP from III-V Quantum Dot Effective Mass Superlattices

Following Sasaki the electron dispersion law in III-V effective mass superlattices (EMSLs) can be written as

$$k_x^2=\left[\frac{1}{L_0^2}\left\{\cos^{-1}\left(f\left(E,k_y,k_z\right)\right)\right\}^2-k_\perp^2\right] \tag{1.286}$$

in which,

$$f\left(E,k_y,k_z\right)\equiv a_1\cos\left[a_0C_1\left(E,k_\perp\right)+b_0D_1\left(E,k_\perp\right)\right]-a_2\cos\left[a_0C_1\left(E,k_\perp\right)-b_0D_1\left(E,k_\perp\right)\right]$$

, $k_\perp^2\equiv k_y^2+k_z^2$,

$$a_1 \equiv \left[\sqrt{\frac{m_2^*}{m_1^*}}+1\right]^2 \left[4\left(\frac{m_2^*}{m_1^*}\right)^{1/2}\right]^{-1}$$

$$, C_1(E,k_\perp) \equiv \left[\left(\frac{2m_1^* E}{\hbar^2}\right) G(E, E_{g_{01}}, \Delta_1) - k_\perp^2\right]^{1/2}$$

$$, \text{and } a_2 \equiv \left[-1+\sqrt{\frac{m_2^*}{m_1^*}}\right]^2 \left[4\left(\frac{m_2^*}{m_1^*}\right)^{1/2}\right]^{-1}$$

The electron concentration is given by

$$N_0 = \frac{2g_v}{d_x d_y d_z} \sum_{n_x=1}^{n_{x_{\max}}} \sum_{n_y=1}^{n_{y_{\max}}} \sum_{n_z=1}^{n_{z_{\max}}} F_{-1}(\eta_{13}) \tag{1.287}$$

where, $\eta_{13} = \frac{E_{FQDSLEM} - E_{QD14}}{k_B T}$, $E_{FQDSLEM}$ is the Fermi energy in this case and E_{QD14} is the root of the equation

$$\left(\frac{\pi n_x}{d_x}\right)^2 = \left[\frac{1}{L_0^2}\left[\cos^{-1}\left[f_{14}(E, n_y, n_z)\Big|_{E=E_{QD14}}\right]\right]^2 - \left(\frac{\pi n_y}{d_y}\right)^2 - \left(\frac{\pi n_z}{d_z}\right)^2\right] \tag{1.288}$$

in which

$$f_{14}(E,n_y,n_z)\Big|_{E=E_{QD14}} = \left[a_1 \cos\left[a_0 g_5(E,n_y,n_z) + b_0 h_5(E,n_y,n_z)\right] - a_2 \cos\left[a_0 g_5(E,n_y,n_z) - b_0 h_5(E,n_y,n_z)\right]\right]\Big|_{E=E_{QD14}} \tag{1.289}$$

$$g_5(E,n_y,n_z) \equiv \left[\frac{2m_1^* E G(E, E_{g_{01}}, \Delta_1)}{\hbar^2} - \left(\frac{\pi n_y}{d_y}\right)^2 - \left(\frac{\pi n_z}{d_z}\right)^2\right]^{1/2}$$

$$, h_5(E,n_y,n_z) \equiv \left[\frac{2m_2^* E G(E, E_{g_{02}}, \Delta_2)}{\hbar^2} - \left(\frac{\pi n_y}{d_y}\right)^2 - \left(\frac{\pi n_z}{d_z}\right)^2\right]^{1/2},$$

The photo-emitted current density is given by

$$J_1 = \frac{\alpha_0 e g_v}{d_x d_y d_z \hbar} \sum_{n_y=1}^{n_{y_{\max}}} \sum_{n_z=1}^{n_{z_{\max}}} \sum_{n_{x_{\min}}}^{n_{x_{\max}}} L_5\left(E_{n_x},0\right) F_{-1}\left(\eta_{13}\right) \tag{1.290}$$

where $L_5\left(E_{n_x},0\right) \equiv \left[I_5\left(E_{n_x},0\right)\right]\left[1-\left\{\bar{f}_5\left(E_{n_x},0\right)\right\}^2\right]^{-1/2}$,

$$I_5\left(E_{n_x},0\right) \equiv \left[a_2\left[a_0 g_1'\left(E_{n_x},0\right)-b_0 h_1'\left(E_{n_x},0\right)\right]\right.$$
$$\left[\sin\left[a_0 g_1\left(E_{n_x},0\right)-b_0 h_1\left(E_{n_x},0\right)\right]\right]$$
$$\left.-a_1\left[a_0 g_1'\left(E_{n_x},0\right)+b_0 h_1'\left(E_{n_x},0\right)\right]\left[\sin\left[a_0 g_1\left(E_{n_x},0\right)+b_0 h_1\left(E_{n_x},0\right)\right]\right]\right],$$

$$g_1'\left(E_{n_x},0\right) \equiv \left[\frac{m_1^*}{\hbar^2 g_1\left(E_{n_x},0\right)}\right]\left[G\left(E_{n_x},E_{g_{01}},\Delta_1\right)+E_{n_x} G'\left(E_{n_x},E_{g_{01}},\Delta_1\right)\right],$$

$$G'\left(E_{n_x},E_{g_{0i}},\Delta_i\right) \equiv \left[G\left(E_{n_x},E_{g_{0i}},\Delta_i\right)\right]\left[\left(E_{n_x}+E_{g_{0i}}+\Delta_i\right)^{-1}+\left(E_{n_x}+E_{g_{0i}}\right)^{-1}-\left(E_{n_x}+E_{g_{0i}}+\frac{2}{3}\Delta_i\right)^{-1}\right],$$

$$G\left(E_{n_x},E_{g_{0i}},\Delta_i\right) \equiv \left[\left(E_{n_x}+E_{g_{0i}}\right)\left(E_{n_x}+E_{g_{0i}}+\Delta_i\right)\left(E_{g_{0i}}+\frac{2}{3}\Delta_i\right)\right]\left[E_{g_{0i}}\left(E_{g_{0i}}+\Delta_i\right)\left(E_{n_x}+E_{g_{0i}}+\frac{2}{3}\Delta_i\right)\right]^{-1},$$

$$h_1'\left(E_{n_x},0\right) \equiv \left[\frac{m_2^*}{\hbar^2 h_1\left(E_{n_x},0\right)}\right]\left[G\left(E_{n_x},E_{g_{02}},\Delta_2\right)+E_{n_x} G'\left(E_{n_x},E_{g_{02}},\Delta_2\right)\right],$$

$$g_1\left(E_{n_x},0\right) \equiv \left[\frac{2m_1^* E_{n_x}}{\hbar^2} G\left(E_{n_x},E_{g_{01}},\Delta_1\right)\right]^{1/2},$$

$$h_1\left(E_{n_x},0\right) \equiv \left[\frac{2m_2^* E_{n_x}}{\hbar^2} G\left(E_{n_x},E_{g_{02}},\Delta_2\right)\right]^{1/2}$$

and

$$\bar{f}_5\left(E_{n_x},0\right) \equiv \left[a_1 \cos\left[a_0 g_1\left(E_{n_x},0\right)+b_0 h_1\left(E_{n_x},0\right)\right]-a_2 \cos\left[a_0 g_1\left(E_{n_x},0\right)-b_0 h_1\left(E_{n_x},0\right)\right]\right] \tag{1.291}$$

E_{n_x} is the root of the equation

$$\bar{f}_5\left(E_{n_x}\right) = \cos\left[\frac{L_0 \pi n_x}{d_x}\right] \tag{1.292}$$

The should be the nearest integer of the following inequality

$$n_{x_{\min}} \geq \left(\frac{d_x}{\pi L_0}\right)\cos^{-1}\left[\overline{f}_5\left(W - h\upsilon, 0\right)\right] \tag{1.293}$$

1.2.24. The EP from II-VI Quantum Dot Effective Mass Superlattices

The electron concentration is given by

$$N_0 = \frac{g_v}{d_x d_y d_z} \sum_{n_x=1}^{n_{x_{\max}}} \sum_{n_y=1}^{n_{y_{\max}}} \sum_{n_z=1}^{n_{z_{\max}}} F_{-1}\left(\eta_{14}\right) \tag{1.294}$$

where $\eta_{14} = \dfrac{E_{FQDSLEM} - E_{QD15}}{k_B T}$ and

E_{QD15} is the root of the equation

$$\left(\frac{\pi n_z}{d_z}\right)^2 = \left[\frac{1}{L_0^2}\left[\cos^{-1}\left[f_{15}\left(E, n_x, n_y\right)\Big|_{E=E_{QD15}}\right]\right]^2 - \left(\frac{\pi n_x}{d_x}\right)^2 - \left(\frac{\pi n_y}{d_y}\right)^2\right] \tag{1.295}$$

in which

$$f_{15}\left(E, n_x, n_y\right)\Big|_{E=E_{QD15}} \equiv \left[a_3 \cos\left[a_0 g_6\left(E, n_x, n_y\right) + b_0 h_6\left(E, n_x, n_y\right)\right] - a_4 \cos\left[a_0 g_6\left(E, n_x, n_y\right) - b_0 h_6\left(E, n_x, n_y\right)\right]\right]\Big|_{E=E_{QD15}} \tag{1.296}$$

$$g_6\left(E, n_x, n_y\right) \equiv \left(\frac{2m_{\parallel,1}^*}{\hbar^2}\right)^{1/2}\left[E - \frac{\hbar^2}{2m_{\perp,1}^*}\left\{\left(\frac{\pi n_x}{d_x}\right)^2 + \left(\frac{\pi n_y}{d_y}\right)^2\right\} \mp C_0\left\{\left(\frac{\pi n_x}{d_x}\right)^2 + \left(\frac{\pi n_y}{d_y}\right)^2\right\}^{1/2}\right]^{1/2}$$

and

$$h_6\left(E, n_x, n_y\right) \equiv \left[\frac{2m_2^* E}{\hbar^2} G\left(E, E_{g_{02}}, \Delta_2\right) - \left\{\left(\frac{\pi n_x}{d_x}\right)^2 + \left(\frac{\pi n_y}{d_y}\right)^2\right\}\right]^{1/2}$$

The photo-emitted current density is given by

$$J = \frac{\alpha_0 e g_v}{2 d_x d_y d_z \hbar} \sum_{n_x=1}^{n_{x_{\max}}} \sum_{n_y=1}^{n_{y_{\max}}} \sum_{n_{z_{\min}}}^{n_{z_{\max}}} L_6\left(E_{n_z},0\right) F_{-1}\left(\eta_{14}\right) \tag{1.297}$$

$$L_6\left(E_{n_z},0\right) \equiv \left[I_6\left(E_{n_z},0\right)\right]\left[1-\left\{f_7\left(E_{n_z},0\right)\right\}^2\right]^{-1/2},$$

$$I_6\left(E_{n_z},0\right) \equiv \left[a_4\left[a_0 g_2'\left(E_{n_z},0\right)-b_0 h_2'\left(E_{n_z},0\right)\right]\right.$$

$$\left[\sin\left[a_0 g_2\left(E_{n_z},0\right)-b_0 h_2\left(E_{n_z},0\right)\right]\right]-a_3\left[a_0 g_2'\left(E_{n_z},0\right)+b_0 h_2'\left(E_{n_z},0\right)\right]$$

$$\left.\left[\sin\left[a_0 g_2\left(E_{n_z},0\right)+b_0 h_2\left(E_{n_z},0\right)\right]\right]\right], g_2'\left(E_{n_z},0\right) \equiv \left[\frac{2m_{\|,1}^*}{\hbar^2}\right]^{1/2}\left(\sqrt{4E_{n_z}}\right)^{-1},$$

$$h_2'\left(E_{n_z},0\right) \equiv \left[\frac{m_2^*}{\hbar^2 h_2\left(E_{n_z},0\right)}\right]\left[G\left(E_{n_z},E_{g_{02}},\Delta_2\right)+E_{n_z} G'\left(E_{n_z},E_{g_{02}},\Delta_2\right)\right],$$

$$G\left(E_{n_z},E_{g_{0i}},\Delta_i\right) \equiv \left[\left(E_{n_z}+E_{g_{0i}}\right)\left(E_{n_z}+E_{g_{0i}}+\Delta_i\right)\left(E_{g_{0i}}+\frac{2}{3}\Delta_i\right)\right]\left[E_{g_{0i}}\left(E_{g_{0i}}+\Delta_i\right)\left(E_{n_z}+E_{g_{0i}}+\frac{2}{3}\Delta_i\right)\right]^{-1},$$

$$G'\left(E_{n_z},E_{g_{02}},\Delta_2\right) \equiv \left[G\left(E_{n_z},E_{g_{02}},\Delta_2\right)\right]\left[\left(E_{n_z}+E_{g_{02}}+\Delta_2\right)^{-1}+\left(E_{n_z}+E_{g_{02}}\right)^{-1}-\left(E_{n_z}+E_{g_{02}}+\frac{2}{3}\Delta_2\right)^{-1}\right],$$

$i = 2$ in this case.

$$g_2\left(E_{n_z},0\right) \equiv \left[\frac{2m_{\|,1}^*}{\hbar^2}\right]^{1/2}\left(E_{n_z}\right)^{1/2},$$

$$h_2\left(E_{n_z},0\right) \equiv \left[\frac{2m_2^*}{\hbar^2} E_{n_z} G\left(E_{n_z},E_{g_{02}},\Delta_2\right)\right]^{1/2},$$

$$f_7\left(E_{n_z},0\right) \equiv \left[a_3\cos\left[a_0 g_2\left(E_{n_z},0\right)+b_0 h_2\left(E_{n_z},0\right)\right]-a_4\cos\left[a_0 g_2\left(E_{n_z},0\right)-b_0 h_2\left(E_{n_z},0\right)\right]\right] \tag{1.298}$$

The is the root of the equation

$$\cos\left[L_0\left(\frac{\pi n_z}{d_z}\right)\right] = f_7\left(E_{n_z},0\right) \tag{1.299}$$

The should be the nearest integer of the following inequality

$$n_{z_{\min}} \geq \left(\frac{d_z}{\pi L_0}\right)\left[\cos^{-1}\left[f_7\left(W - h\upsilon, 0\right)\right]\right] \tag{1.300}$$

1.2.25. The EP from IV-VI Quantum Dot Effective Mass Superlattices

The electron concentration is given by

$$N_0 = \frac{2g_v}{d_x d_y d_z} \sum_{n_x=1}^{n_{x_{\max}}} \sum_{n_y=1}^{n_{y_{\max}}} \sum_{n_{z_{\min}}}^{n_{z_{\max}}} F_{-1}\left(\eta_{15}\right) \tag{1.301}$$

where $\eta_{15} = \dfrac{E_{FQDSLEM} - E_{QD16}}{k_B T}$ and

E_{QD16} is the root of the equation

$$\left(\frac{\pi n_z}{d_z}\right)^2 = \left[\frac{1}{L_0^2}\left[\cos^{-1}\left[f_{16}\left(E, n_x, n_y\right)\Big|_{E=E_{QD16}}\right]\right]^2 - \left(\frac{\pi n_x}{d_x}\right)^2 - \left(\frac{\pi n_y}{d_y}\right)^2\right] \tag{1.302}$$

where

$$f_{16}\left(E, n_x, n_y\right)\Big|_{E=E_{QD16}} \equiv \left[a_5 \cos\left[a_0 g_7\left(E, n_x, n_y\right) + b_0 h_7\left(E, n_x, n_y\right)\right] - a_6 \cos\left[a_0 g_7\left(E, n_x, n_y\right) - b_0 h_7\left(E, n_x, n_y\right)\right]\right]_{E=E_{QD16}} \tag{1.303}$$

$$g_7\left(E, n_x, n_y\right) \equiv \left[\left[H_{11}E + H_{21}\left(n_x, n_y\right)\right] - \left(H_{31}E^2 + H_{41}\left(n_x, n_y\right)E + H_{51}\left(n_x, n_y\right)\right)^{1/2}\right]^{1/2},$$

$$h_7\left(E, n_x, n_y\right) \equiv \left[\left[H_{12}E + H_{22}\left(n_x, n_y\right)\right] - \left[H_{32}E^2 + H_{42}\left(n_x, n_y\right)E + H_{52}\left(n_x, n_y\right)\right]^{1/2}\right]^{1/2}$$

The photo-emitted current density is given by

$$J_1 = \frac{\alpha_0 e g_v}{d_x d_y d_z \hbar} \sum_{n_x=1}^{n_{x_{\max}}} \sum_{n_y=1}^{n_{y_{\max}}} \sum_{n_{z_{\min}}}^{n_{z_{\max}}} L_7\left(E_{n_z}, 0\right) F_{-1}\left(\eta_{15}\right) \tag{1.304}$$

where $L_7\left(E_{n_z}, 0\right) \equiv \left[I_7\left(E_{n_z}, 0\right)\right]\left[1 - \left\{f_8\left(E_{n_z}, 0\right)\right\}^2\right]^{-1/2}$,

$$I_7\left(E_{n_z},0\right) \equiv \left[a_6\left[a_0 g_3'\left(E_{n_z},0\right)-b_0 h_3'\left(E_{n_z},0\right)\right]\left[\sin\left[a_0 g_3\left(E_{n_z},0\right)-b_0 h_3\left(E_{n_z},0\right)\right]\right]\right.$$

$$\left.-a_5\left[a_0 g_3'\left(E_{n_z},0\right)+b_0 h_3'\left(E_{n_z},0\right)\right]\left[\sin\left[a_0 g_3\left(E_{n_z},0\right)+b_0 h_3\left(E_{n_z},0\right)\right]\right]\right],$$

$$g_3'\left(E_{n_z},0\right) \equiv \left[2g_3\left(E_{n_z},0\right)\right]^{-1}\left[H_{11}-\frac{1}{2}\left[2E_{n_z}H_{31}+H_{41}(0,0)\right]\left[H_{31}E_{n_z}{}^2+H_{41}(0,0)E_{n_z}+H_{51}(0,0)\right]^{-1/2}\right],$$

$$g_3\left(E_{n_z},0\right) \equiv \left[\left[H_{11}E_{n_z}+H_{21}(0,0)\right]-\left[H_{31}E_{n_z}{}^2+H_{41}(0,0)E_{n_z}+H_{51}(0,0)\right]^{1/2}\right]^{1/2},$$

$$h_3'\left(E_{n_z},0\right) \equiv \left[2h_3\left(E_{n_z},0\right)\right]^{-1}\left[H_{12}-\frac{1}{2}\left[2E_{n_z}H_{32}+H_{42}(0,0)\right]\right.$$

$$\left.\left[H_{32}E_{n_z}{}^2+H_{42}(0,0)E_{n_z}+H_{52}(0,0)\right]^{-1/2}\right]$$

$$, h_3\left(E_{n_z},0\right) \equiv \left[\left[H_{12}E_{n_z}+H_{22}(0,0)\right]-\right.$$

$$\left.\left[H_{32}E_{n_z}{}^2+H_{42}(0,0)E_{n_z}+H_{52}(0,0)\right]^{1/2}\right]^{1/2} \text{ and}$$

$$f_8\left(E_{n_z},0\right) \equiv \left[a_5\cos\left[a_0 g_3\left(E_{n_z},0\right)+b_0 h_3\left(E_{n_z},0\right)\right]-a_6\cos\left[a_0 g_3\left(E_{n_z},0\right)-b_0 h_3\left(E_{n_z},0\right)\right]\right].$$

The E_{n_z} can be determined from the equation

$$\cos\left[L_0\left(\frac{\pi n_z}{d_z}\right)\right] = f_8\left(E_{n_z},0\right) \tag{1.305}$$

The should be the nearest integer of the following inequality

$$n_{z_{\min}} \geq \left(\frac{d_z}{\pi L_0}\right)\left[\cos^{-1}\left[f_8\left(W-h\upsilon,0\right)\right]\right] \tag{1.306}$$

1.2.26. The EP from HgTe/CdTe Quantum Dot Effective Mass Superlattices

The electron concentration is given by

$$N_0 = \frac{2g_v}{d_x d_y d_z}\sum_{n_x=1}^{n_{x_{\max}}}\sum_{n_y=1}^{n_{y_{\max}}}\sum_{n_z=1}^{n_{z_{\max}}} F_{-1}\left(\eta_{16}\right) \tag{1.307}$$

where $\eta_{16} = \dfrac{E_{FQDSLEM} - E_{QD17}}{k_B T}$ and

E_{QD17} is the root of

$$\left(\frac{\pi n_z}{d_z}\right)^2 = \left[\frac{1}{L_0^2}\left[\cos^{-1}\left[f_{17}\left(E, n_x, n_y\right)\Big|_{E=E_{QD17}}\right]\right]^2 - \left(\frac{\pi n_y}{d_y}\right)^2 - \left(\frac{\pi n_x}{d_x}\right)^2\right] \quad (1.308)$$

where

$$f_{17}(E,n_x,n_y)\Big|_{E=E_{QD17}} \equiv \Big[a_3 \cos\left[a_0 g_8(E,n_x,n_y) + b_0 h_8(E,n_x,n_y)\right] - a_4 \cos\left[a_0 g_8(E,n_x,n_y) - b_0 h_8(E,n_x,n_y)\right]\Big]\Big|_{E=E_{QD17}} \quad (1.309)$$

in which

$$g_8\left(E, n_x, n_y\right) \equiv \left[\frac{2m_1^* \omega_1(E)}{\hbar^2} - \left[\left(\frac{\pi n_x}{d_x}\right)^2 + \left(\frac{\pi n_y}{d_y}\right)^2\right]\right]^{1/2},$$

$$\omega_1(E) \equiv \left[\frac{9m_1^* e^4}{\left(128\varepsilon_{sc}\hbar\right)^2} + E - \left(\frac{3e^2 m_1^*}{\left(128\varepsilon_{sc}\hbar^2\right)}\right)\left[\left(\frac{3e^2}{128\varepsilon_{sc}}\right)^2 + \frac{2\hbar^2 E}{m_1^*}\right]^{1/2}\right] \text{ and}$$

$$h_8\left(E, n_x, n_y\right) \equiv \left[\frac{2m_2^*}{\hbar^2} G\left(E, E_{g_{02}}, \Delta_2\right) - \left[\left(\frac{n_x \pi}{d_x}\right)^2 + \left(\frac{n_y \pi}{d_y}\right)^2\right]\right]^{1/2}$$

The photo-emitted current density is given by

$$J_1 \equiv \frac{\alpha_0 e g_v}{d_x d_y d_z \hbar} \sum_{n_x=1}^{n_{x_{\max}}} \sum_{n_y=1}^{n_{y_{\max}}} \sum_{n_{z_{\min}}}^{n_{z_{\max}}} L_8\left(E_{n_z}, 0\right) F_{-1}\left(\eta_{16}\right) \quad (1.310)$$

in which

$$L_8(E_{n_z},0) \equiv \left[1 - \left\{f_9(E_{n_z},0)\right\}^2\right]^{-1/2}\Big[a_4\left[a_0 g_4'(E_{n_z},0) - b_0 h_4'(E_{n_z},0)\right]\left[\sin\left[a_0 g_4(E_{n_z},0) - b_0 h_4(E_{n_z},0)\right]\right]$$
$$- a_3\left[a_0 g_4'(E_{n_z},0) + b_0 h_4'(E_{n_z},0)\right]\left[\sin\left[a_0 g_4'(E_{n_z},0) + b_0 h_4'(E_{n_z},0)\right]\right]\Big],$$

$$f_9(E_{n_z},0)\equiv\left[a_3\cos\left[a_0g_4(E_{n_z},0)+b_0h_4(E_{n_z},0)\right]-a_4\cos\left[a_0g_4(E_{n_z},0)-b_0h_4(E_{n_z},0)\right]\right],$$

$$g_4(E_{n_z},0)\equiv\left[2m_1^*\omega_1(E_{n_z})/\hbar^2\right]^{1/2}$$

$$,\ h_4(E_{n_z},0)\equiv\left[\frac{2m_2^*}{\hbar^2}G(E_{n_z},E_{g_{02}},\Delta_2)\right]^{1/2},$$

$$g_4'(E_{n_z},0)\equiv\left[g_4(E_{n_z},0)\right]^{-1}\left(\frac{m_1^*}{\hbar^2}\right)\left[1-\frac{3e^2}{128\varepsilon_{sc}}\left[\left(\frac{3e^2}{128\varepsilon_{sc}}\right)^2+\frac{2\hbar^2E_{n_z}}{m_1^*}\right]^{-1/2}\right]$$

and $h_4'(E_{n_z},0)\equiv\left[h_4(E_{n_z},0)\right]^{-1}\left[\frac{m_2^*}{\hbar^2}G'(E_{n_z},E_{g_{02}},\Delta_2)\right]$.

The E_{n_z} should be determined from the equation

$$\cos\left[L_0\left(\frac{\pi n_z}{d_z}\right)\right]=f_9(E_{n_z},0) \tag{1.311}$$

The should be the nearest integer of the following inequality

$$n_{z_{\min}}\geq\left(\frac{d_z}{\pi L_0}\right)\left[\cos^{-1}\left[f_9(W-h\upsilon,0)\right]\right] \tag{1.312}$$

The process of finding $f_9(W-h\upsilon,0)$ is same as $f_7(W-h\upsilon,0)$.

The content of this chapter finds two applications in the field of quantized structures

1. Carrier Contribution to the Elastic Constants

The knowledge of the carrier contribution to the elastic constants is useful in studying the mechanical properties of the materials and has been investigated in the literature [177-180]. The electronic contribution to the second- and third- order elastic constants can be written as [177-180]

$$\Delta C_{44}=-\frac{(\overline{G_0})^2}{9d_z}\frac{\partial n_0}{\partial E_{F2D}}, \tag{1.313}$$

and

$$\Delta C_{456} = \frac{\left(\overline{G_0}\right)^3}{27 d_z} \frac{\partial^2 n_0}{\partial E_{F2D}^2}, \tag{1.314}$$

where is the deformation potential constant. Thus, using appropriate equations for the surface electron concentration we can study and ΔC_{456} for quantum wells, inversion layers and nipi structures of Kane type semiconductors in the presence of strong light waves.

2. Third Order Nonlinear Optical Susceptibility

This particular susceptibility can be written as [181]

$$\chi_{NP}\left(\omega_1, \omega_2, \omega_3\right) = \frac{n_0 e^4 \left\langle \varepsilon^4 \right\rangle}{24 \omega_1 \omega_2 \omega_3 \left(\omega_1 + \omega_2 + \omega_3\right) \hbar^4} \tag{1.315}$$

where and the other notations are defined in [181]. The term $\left(\frac{\partial^4 E}{\partial k_z^4}\right)$ can be formulated by using the dispersion relations of different materials as given in appropriate sections of this chapter. Thus one can investigate the $\chi_{NP}\left(\omega_1, \omega_2, \omega_3\right)$ for all materials as considered in this chapter.

1.3. Result and Discussions

Using (1.8) and (1.15) and the band constants from the table 1, we have plotted the normalized photo-emitted current density from QDs of $CdGeAs_2$ as a function of film thickness as shown in plot (a) of Figure 1.1, where the plot (b) indicates the case for $\delta = 0$ .The plots (c) and (d) represent the photoemission in this case in accordance with three (using (1.22)) and two (using (1.32)) band models of Kane. In Figure 1.2 and 1.3, the afore-mentioned variable has been plotted as functions of normalized incident photon energy and normalized electron degeneracy respectively for all cases of Figure 1.1. In Figure 1.4, the normalized photo-emitted current density from QDs of n-InSb has been plotted as a function of in accordance with the (a) three band model of Kane, (b) two band model of Kane, the models of (c) Stillman et. al. (using (1.35), (d) Palik et. al. (using (1.52)), (e) Johnson et. al. (using (1.58)) and (f) Agafonov et. al (using (1.63)) respectively. The plots (g) and (h) refer to QDs of GaAs in accordance with the models of Newson et.al. (using (1.42)) and Rossler. (using (1.46)) respectively in the same figure. In Figure 1.5 and .6, the normalized photoemission for all cases of Figure 1.4 has been plotted as functions of normalized incident photon energy and the normalized electron degeneracy respectively. The Figure 1.7 exhibits the normalized photo-emitted current density from QDs of CdS (using (1.70)) as a function of d_z as shown in plot (a), in which the plot (b) is valid for $C_0 = 0$. The plot (c) in the same figure is valid for QDs of GaP and has been drawn by using (1.77) and (1.76) respectively. In Figure 1.8 and 1.9, the aforementioned variable has been plotted as functions of normalized incident photon energy and normalized electron degeneracy respectively for all cases of Figure 1.7.

Table 1. The numerical values of the constants of the energy-wave vector dispersion relations of few materials

	Materials	Numerical Values of the Energy Band Constants
1	The conduction electrons of n-Cadmium Germanium Arsenide can be described by three types of band models	1. The values of the energy band constants in accordance with the generalized electron dispersion relation of nonlinear optical materials are as follows $E_{g_0} = 0.57eV$, $\Delta_{\parallel} = 0.30eV$, $\Delta_{\perp} = 0.36eV$, $m_{\parallel}^* = 0.034m_0$, $m_{\perp}^* = 0.039m_0$, $T = 4K$, $\delta = -0.21eV$ and . 2. In accordance with the three band model of Kane, the spectrum constants are given by $\Delta = (\Delta_{\parallel} + \Delta_{\perp})/2 = 0.33eV$, $E_{g_0} = 0.57eV$, $m^* = (m_{\parallel}^* + m_{\perp}^*)/2 = 0.0365m_0$ and $\delta = 0eV$. 3. In accordance with two band model of Kane, $E_{g_0} = 0.57eV$ and $m^* = 0.0365m_0$.
2	n-Indium Arsenide	The values $E_{g_0} = 0.36eV$, $\Delta = 0.43eV$, $m^* = 0.026m_0$ and are valid for three band model of Kane.
3	n-Gallium Arsenide	The values $E_{g_0} = 1.55eV$, $\Delta = 0.35eV$, $m^* = 0.07m_0$ and $g_v = 1$ are valid for three band model of Kane [125]. The values $\alpha_{13} = -1.97 \times 10^{-37} eVm^4$ and are valid for the model of Newson and Kurobe. The values $\alpha_{11} = -2132 \times 10^{-40} eVm^4$, $\alpha_{12} = 9030 \times 10^{-50} eVm^5$, $\beta_{11} = -2493 \times 10^{-40} eVm^4$, $\beta_{12} = 12594 \times 10^{-50} eVm^5$, $\gamma_{11} = 30 \times 10^{-30} eVm^3$, are valid for the model of Rossler .
4	n-Gallium Alumi-nium Arsenide	$E_{g_0} = (1.424 + 1.266x + 0.26x^2)eV$, $\Delta = (0.34 - 0.5x)eV$, $m^* = [0.066 + 0.088x]m_0$ and .
5	n-Mercury Cadmium Telluride	$E_{g_0} = (-0.302 + 1.93x + 5.35 \times 10^{-4}(1-2x)T - 0.810x^2 + 0.832x^3)eV$, $\Delta = (0.63 + 0.24x - 0.27x^2)eV$, $m^* = 0.1m_0 E_{g_0}(eV)^{-1}$ and .
6	n-Indium Gal-lium Arsenide Phosphide lattice matched to Indium Phosphide	$E_{g_0} = (1.337 - 0.73y + 0.13y^2)eV$, $\Delta = (0.114 + 0.26y - 0.22y^2)eV$, $m^* = (0.08 - 0.039y)m_0$, $y = (0.1896 - 0.4052x)/(0.1896 - 0.0123x)$ and $g_v = 1$.
7	n-Indium Antimo-nide	$E_{g_0} = 0.2352eV$, $\Delta = 0.81eV$, $m^* = 0.01359m_0$ and .

Table 1. (Continued).

	Materials	Numerical Values of the Energy Band Constants
8	n-Gallium Antimo-nide	The values of $E_{g_0} = 0.81eV$, $\Delta = 0.80eV$, $P = 9.48\times10^{-10} eVm$, $\overline{\varsigma_0} = -2.1$, $\overline{v_0} = -1.49$, $\overline{\omega_0} = 0.42$ and $g_v = 1$ are valid for the model of Seiler et.al.The values $\overline{E_1} = 1.024eV$, $\overline{E_2} = 0eV$, $\overline{E_3} = -1.132eV$, $\overline{E_4} = 0.05eV$, $\overline{E_5} = 1.107eV$, and are valid for the model of Zhang .
9	p-Cadmium Sulphide	$m_{\|}^* = 0.7m_0$, $m_{\perp}^* = 1.5m_0$, $\overline{\lambda_0} = 1.4\times10^{-10} eVm$ and $g_v = 1$.
10	n-Lead Telluride	The values $m_t^- = 0.070m_0$, $m_l^- = 0.54m_0$, $m_t^+ = 0.010m_0$, $m_l^+ = 1.4m_0$, $P_{\|} = 141meVnm$, $P_{\perp} = 486meVnm$, $E_{g_0} = 190meV$ and are valid for the Dimmock model . The values $\left(\overline{R}\right)^2 = 2.3\times10^{-19}\left(eVm\right)^2$, $E_{g_0} = 0.16eV$, $\left(\overline{s}\right)^2 = 4.6\left(\overline{R}\right)^2$, $\Delta_c' = 3.07eV$, $\left(\overline{Q}\right)^2 = 1.3\left(\overline{R}\right)^2$, $\Delta_c'' = 3.28eV$, are valid for the model of Bangert and Kastner . The values $m_{tv} = 0.0965m_0$, $m_{lv} = 1.33m_0$, $m_{tc} = 0.088m_0$, are valid for the model of Foley et. al. The values $m_1 = 0.0239m_0$, $m_2 = 0.024m_0$, $m_2' = 0.31m_0$, $m_3 = 0.24m_0$ [184] are valid for the Cohen model.
11	n-Lead Tin Telluride	$m_t^- = 0.070m_0$, $m_l^- = 0.54m_0$, $m_t^+ = 0.010m_0$, $m_l^+ = 1.4m_0$, $P_{\|} = 141meV\ nm$, $P_{\perp} = 486meV\ nm$, $E_g = 190meV$, and $\varepsilon_{sc} = 33\varepsilon_0$.
12	n-Lead Tin Selenide	$m_t^- = 0.063m_0$, $m_l^- = 0.41m_0$, $m_t^+ = 0.089m_0$, $m_l^+ = 1.6m_0$, $P_{\|} = 137meV\ nm$, $P_{\perp} = 464meV\ nm$, $E_g = 90meV$, and $\varepsilon_{sc} = 60\varepsilon_0$.
13	Stressed n-Indium Antimo-nide	The values $m^* = 0.048m_o$, $E_{g_0} = 0.081eV$, $B_2 = 9\times10^{-10} eVm$, $C_1^c = 3eV$, $C_2^c = 2eV$, $a_0 = -10eV$, $b_0 = -1.7eV$, $d = -4.4eV$, $S_{xx} = 0.6\times10^{-3}(kbar)^{-1}$, $S_{yy} = 0.42\times10^{-3}(kbar)^{-1}$, $S_{zz} = 0.39\times10^{-3}(kbar)^{-1}$, $S_{xy} = 0.5\times10^{-3}(kbar)^{-1}$, $\varepsilon_{xx} = \sigma S_{xx}$, $\varepsilon_{yy} = \sigma S_{yy}$, $\varepsilon_{zz} = \sigma S_{zz}$, $\varepsilon_{xy} = \sigma S_{xy}$, is the stress in kilobar, and $g_v = 1$ are valid for the model of Seiler et. al.

14	Bismuth	$E_{g_0} = 0.0153eV$, $m_1 = 0.00194m_0$, $m_2 = 0.313m_0$, $m_3 = 0.00246m_0$, $m_2' = 0.36m_0$, $g_v = 3$, [157], $M_2 = 0.128m_0$ and $M_2' = 0.80m_0$.
15	Mercury Telluride	$m_v^* = 0.028m_0$, $g_v = 1$ and $\varepsilon_\infty = 15.2\varepsilon_0$.
16	Platinum Antimonide	For valence bands, along <111> direction, $\lambda_1 = 0.33eV$, $l_1 = 1.09eV$, $\nu_1 = 0.17eV$, $\overline{\overline{n}} = 0.22eV$, $\overline{\overline{a}} = 0.643nm$, $I_0 = 0.30(eV)^2$, $\delta_0' = 0.33eV$ and $g_v = 8$.
17	n-Gallium Phosphide	$m_\parallel^* = 0.92m_0$, $m_\perp^* = 0.25m_0$, $k_0 = 1.7\times10^{19}m^{-1}$, $\lvert V_G\rvert = 0.21eV$, $g_v = 6$ and $g_s = 2$.
18	Germanium	$E_{g_0} = 0.785eV$, $m_\parallel^* = 1.57m_0$ and $m_\perp^* = 0.0807m_0$.
19	Tellurium	The values $A_6 = 6.7\times10^{-16} meVm^2$, $A_7 = 4.2\times10^{-16} meVm^2$, and are valid for the model of Bouat et. al. The values $t_1 = 0.06315eV$, $t_2 = -10.0\hbar^2/2m_0$, $t_3 = -5.55\hbar^2/2m_0$, $t_4 = 0.3\times10^{-36}eVm^4$, $t_5 = 0.3\times10^{-36}eVm^4$, $t_6 = -5.55\hbar^2/2m_0$ and are valid for the model of Ortenberg and Button .
20	Graphite	The values $\Delta_1 = -0.0002eV$, $\gamma_1 = 0.392eV$, $\gamma_5 = 0.194eV$, $\overline{c} = 0.674nm$, $\gamma_2 = -0.019eV$, $\overline{a} = 0.246nm$, $\gamma_0 = 3eV$, are valid for the model of Brandt et. al.
21	Lead German-ium Telluride	The values $\overline{E_{g_0}} = 0.21eV$ and $g_v = 4$ are valid for the model of Vassilev.
22	Cadmium Antimo-nide	The values $A_{10} = -4.65\times10^{-19}eVm^2$, $A_{11} = -2.035\times10^{-19}eVm^2$, $A_{12} = -5.12\times10^{-19}eVm^2$, $A_{13} = -0.25\times10^{-10}eVm$, $A_{14} = 1.42\times10^{-19}eVm^2$, $A_{15} = 0.405\times10^{-19}eVm^2$, $A_{16} = -4.07\times10^{-19}eVm^2$, $A_{17} = 3.22\times10^{-10}eVm$, and $A_{19} = 0.070eV$ are valid for the model of Yamada.
23	Cadmium Diphosphide	The values $\beta_1 = 8.6\times10^{-21}eVm^2$, $\beta_2 = 1.8\times10^{-21}(eVm)^2$, $\beta_4 = 0.0825eV$ and are valid for the model of Chuiko .
24	Zinc Dipho-sphide	The values $\beta_1 = 8.7\times10^{-21}eVm^2$, $\beta_2 = 1.9\times10^{-21}(eVm)^2$, $\beta_4 = 0.0875eV$ and are valid for the model of Chuiko .
25	Bismuth Telluride	The values $\overline{\overline{E_{g_0}}} = 0.145eV$, $\alpha_{11} = 3.25$, $\alpha_{22} = 4.81$, $\alpha_{33} = 9.02$, $\alpha_{23} = 4.15$, $g_s = 2$ and $g_v = 6$ are valid for the model of Stordeur et. al.

Table 1. (Continued).

	Materials	Numerical Values of the Energy Band Constants
26	Antimony	The values $\bar{\alpha}_{11} = 16.7$, $\bar{\alpha}_{22} = 5.98$, $\bar{\alpha}_{33} = 11.61$ and $\bar{\alpha}_{23} = 7.54$ are valid for the model of Ketterson .
27	Zinc Selenide	$m_2^* = 0.16m_0$, $\Delta_2 = 0.42eV$ and $E_{g_{02}} = 2.82eV$.
28	Lead Selenide	$m_t^- = 0.23m_0$, $m_l^- = 0.32m_0$, $m_t^+ = 0.115m_0$, $m_l^+ = 0.303m_0$, $P_{\parallel} \approx 138meVnm$, $P_{\perp} = 471meVnm$ and $E_{g_0} = 0.28eV$.

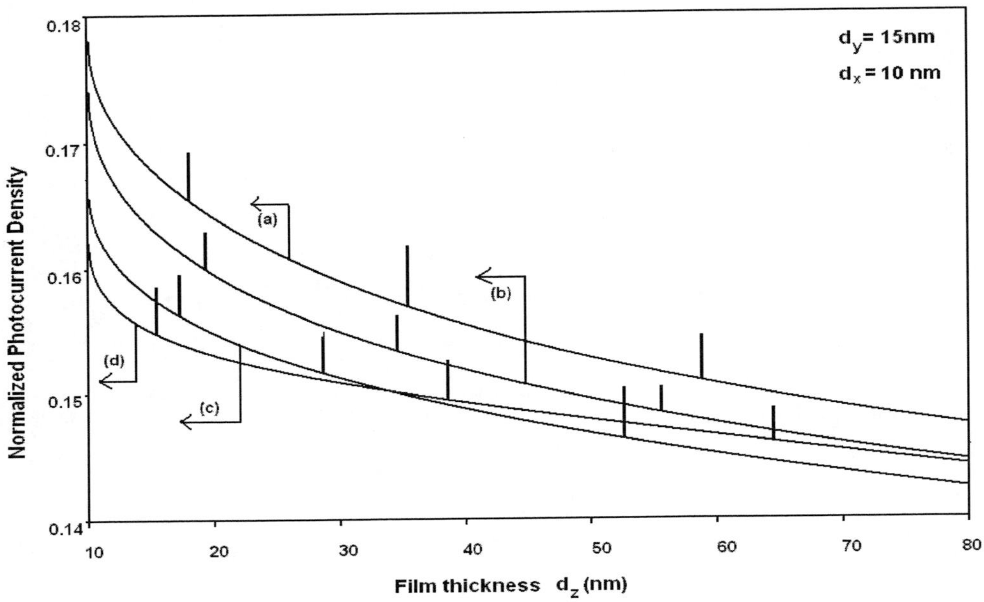

Figure 1.1. Plot of the normalized photocurrent density from QDs of $CdGeAs_2$ as a function of in accordance with (a) generalized band model (b) $\delta = 0$, (c) the three-band model of Kane and (d) the two band model of Kane.

The Figure 1.10 exhibits the plot of the normalized photo-emitted current density as a function of film thickness from QDs of n-Ge in accordance with the models of (a) Cardona et. al. (using (1.84) and (1.83) and (b) Wang et. al. (using (1.91) and (1.90)) respectively. The curves (c) and (d) in the same figure refer to the photoemission from QDs of Tellurium in accordance with the models of Bouat et. al. (using (1.97) and (1.96)) and Ortenberg et. al. (using (1.103) and (1.102)) respectively. Besides, the curve (e) of the same figure has been plotted for QDs of Graphite (using (1.109) and (1.108)). In Figure 1.11 and 1.12, the normalized photo-emitted current density has been plotted as functions of normalized incident photon energy and normalized electron degeneracy respectively for all cases of Figure 1.10. In Figure 1.13, we have drawn the plots of the normalized photoemission as a function of film thickness from QDs of (a) n-$PtSb_2$ (using (1.115) and (1.116)) (b) zerogap materials, taking HgTe as an example (using (1.123) and (1.122)) and (c) $Pb_{1-x}Ge_xTe$ (using (1.214) and (1.215)). In Figures 1.14 and 1.15, the normalized photoemission in this case has been plotted as functions of normalized incident photon energy and normalized carrier degeneracy respectively for all cases of Figure 1.13. In Figure 1.16, we have drawn the plots of the normalized photoemission as a function of film thickness from QDs of GaSb in accordance with the models of (a) Seiler et. al. (using (1.137) and (1.136)), (b) Mathur et. al. (using (1.145) and (1.144)) and (c) Zhang et. al. (using (1.152) and (1.151)) respectively. The curve (d) in the same figure refers to the QDs of stressed materials, where stressed n-InSb has been considered as an example (using (1.158) and (1.157)). In Figure 1.17 and 1.18, the normalized photo-current density in this case has been plotted as functions of normalized incident photon energy and the normalized electron degeneracy respectively for all cases of Figure 1.16.

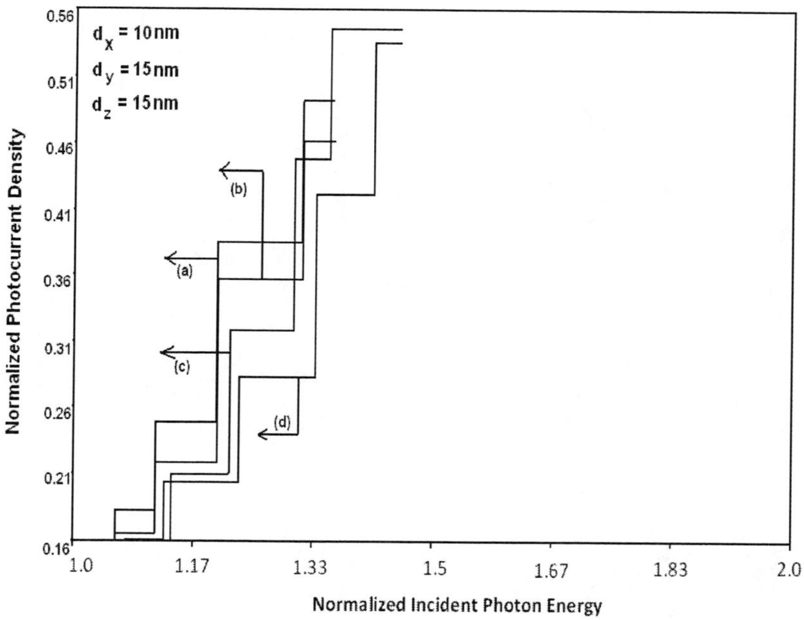

Figure 1.2. Plot of the normalized photocurrent density from QDs of $CdGeAs_2$ as a function of normalized incident photon energy for all the cases of Figure 3.1.

The Figure 1.19 exhibits the plot of the normalized photo-current density as a function of film thickness from QDs of Bismuth in accordance with the models of (a) McClure and Choi (using (1.164) and (1.165)), (b) the Cohen (using (1.179) and (1.178)), (c) the Hybrid (using (1.173) and (1.171)) and (d) the Lax (using (1.185) and (1.186)) respectively.

In Figure 1.20 and 1.21, the same variable has been plotted as functions of normalized incident photon energy and normalized electron degeneracy respectively for all cases of Figure 1.19. The Figure 1.22 depicts the plot of the normalized photoemission current density as a function of film thickness from QDs of IV-VI materials taking n-PbTe as an example in accordance with the models of (a) Dimmock (using (1.194) and (1.193)) (b) Cohen (using (1.179) and (1.178)), (c) Bangert et. al. (using (1.201) and (1.200), (d) Foley et. al. (using (1.208) and (1.207)) respectively. Besides, the plot (e) of Figure 1.22 exhibits the photoemission current density from QDs of II-V materials taking CdSb as an example (using (1.128) and (1.129)). In Figure 1.23 and 1.24, the normalized photoemission in this case has been plotted as functions of normalized incident photon energy and normalized carrier degeneracy respectively for all cases of Figure 1.22.

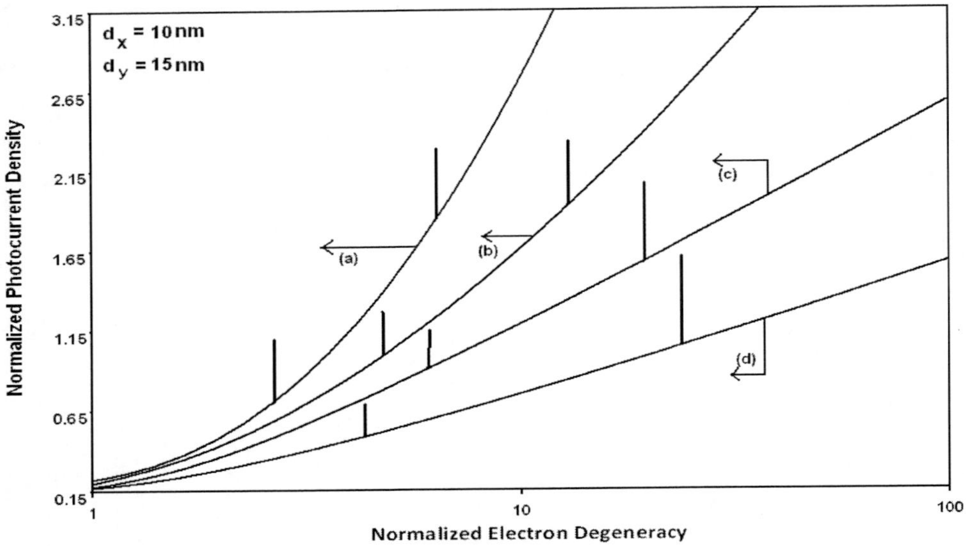

Figure 1.3. Plot of the normalized photocurrent density from QDs of $CdGeAs_2$ as a function of normalized electron degeneracy for all the cases of Figure 3.1.

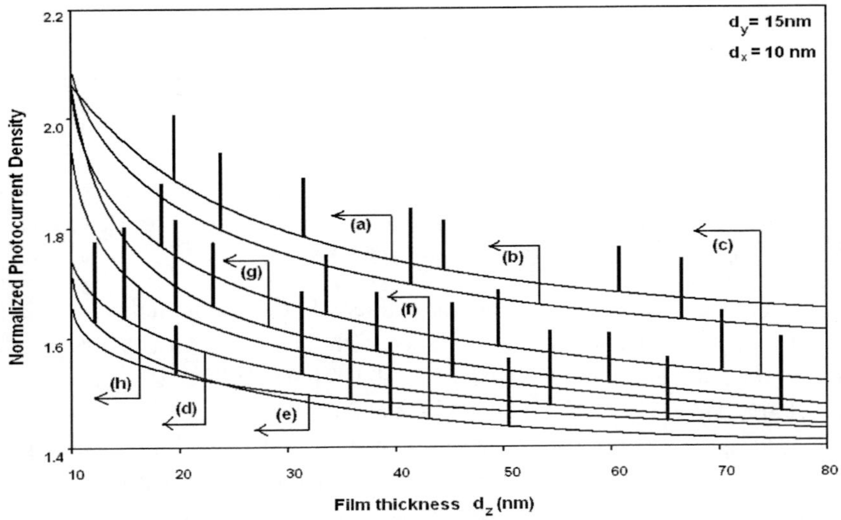

Figure 1.4. Plot of the normalized photocurrent density as a function of d_z from QDs of n-InSb in accordance with the (a) three band model of Kane, (b) two band model of Kane, (c) model of Stillman et. al. (d) model of Palik et. al (e) model of Johnson et. al. (f) model of Agafonov et. al. The plots (g) and (h) refer to QDs of GaAs in accordance with the models of Newson et.al. and Rossler respectively.

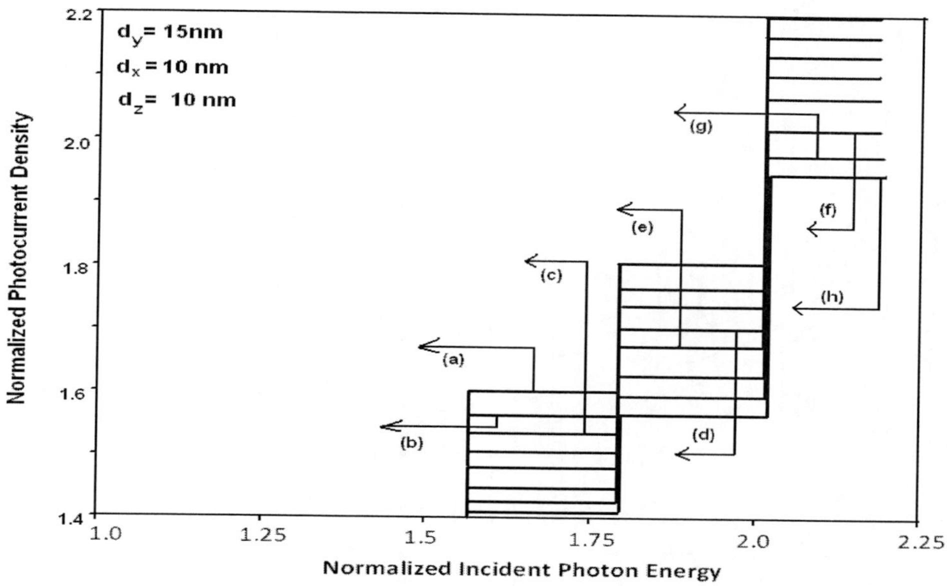

Figure 1.5. Plot of the normalized photocurrent density from QDs of n-InSb and n-GaAs as a function of normalized incident photon energy for all the cases of Figure 3.4.

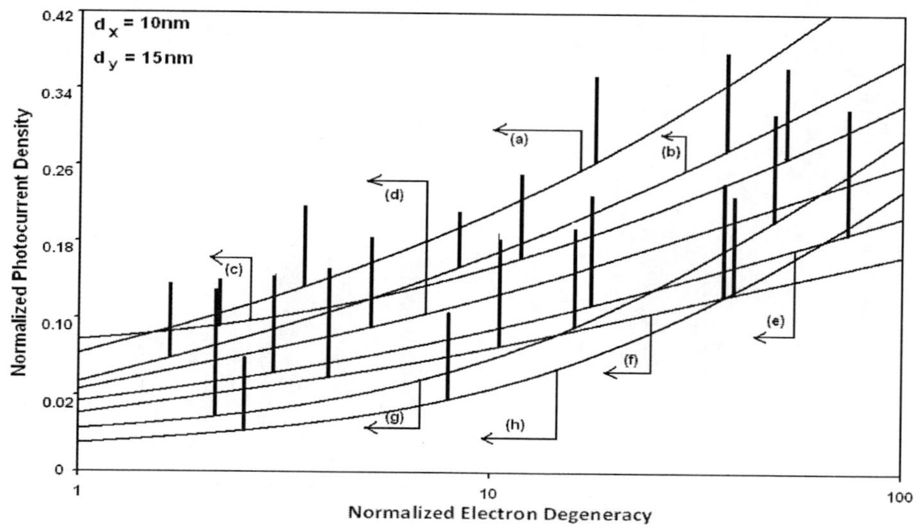

Figure 1.6. Plot of the normalized photocurrent density from QDs of n-InSb and n-GaAs as a function of normalized electron degeneracy for all the cases of Figure 3.4.

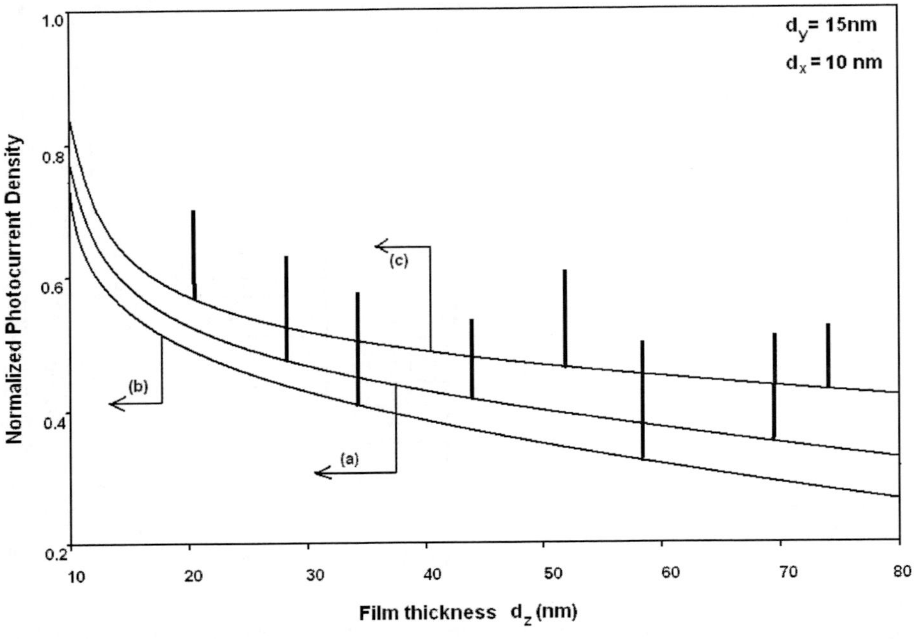

Figure 1.7. Plot of the normalized photocurrent density from QDs of (a) CdS with $C_0 \neq 0$, (b) $C_0 = 0$ respectively and (c) GaP as a function of d_z.

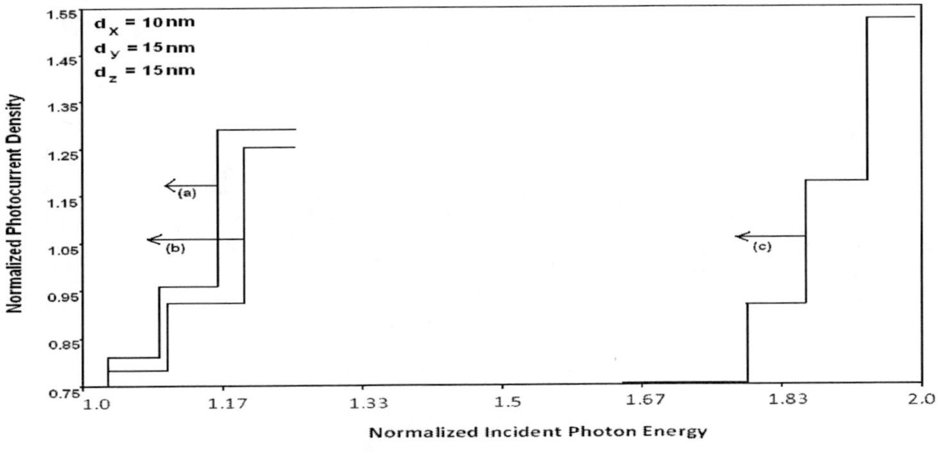

Figure 1.8. Plot of the normalized photocurrent density from QDs of (a) CdS with $C_0 \neq 0$, (b) $C_0 = 0$ respectively and (c) GaP as a function of normalized incident photon energy.

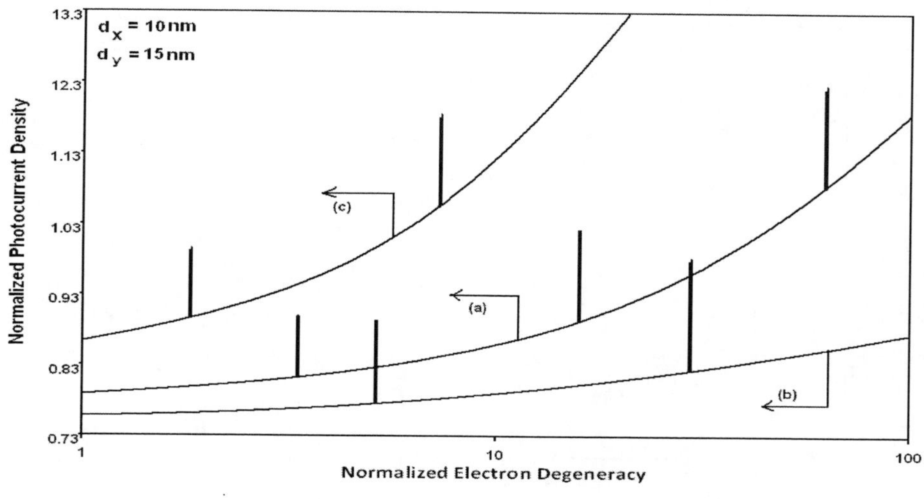

Figure 1.9. Plot of the normalized photocurrent density from QDs of (a) CdS with $C_0 \neq 0$, (b) $C_0 = 0$ respectively and (c) GaP as a function of normalized electron degeneracy.

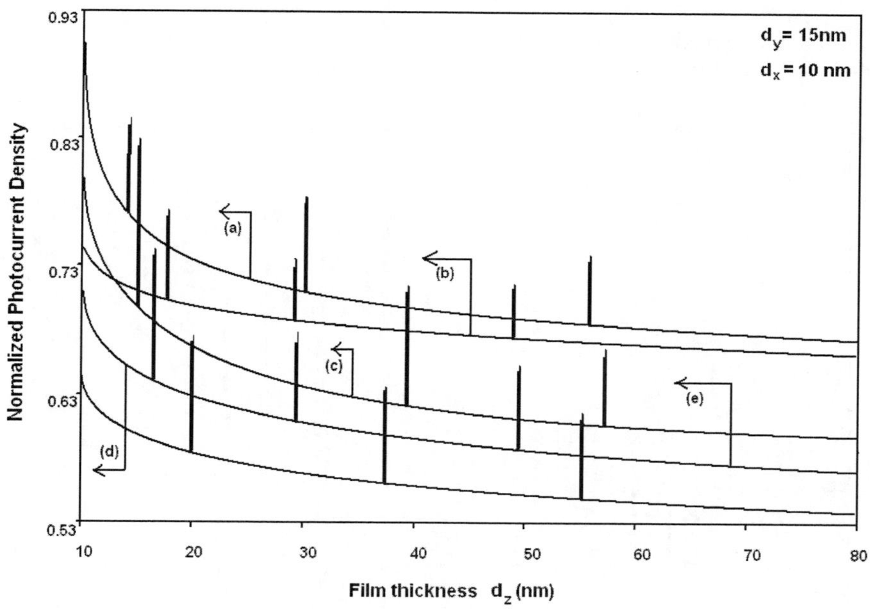

Figure 1.10. Plot of the normalized photocurrent density from QDs of (a) n-Ge in accordance with the model of Cardona et. al., (b) n-Ge in accordance with the model of Wang et. al. as a function of film thickness The curves (c) and (d) refer to the same plot for QDs of Tellurium in accordance with the model of Bouat et. al. and Ortenberg et. al. respectively. The curve (e) stands for the QDs of Graphite.

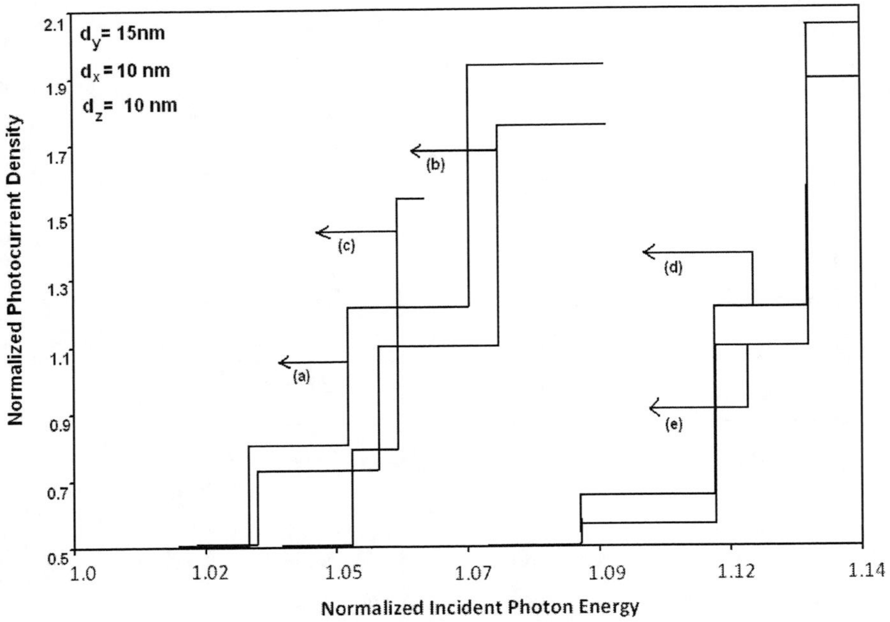

Figure 1.11. Plot of the normalized photocurrent density as a function of normalized incident photon energy for all the cases of Figure 1.10.

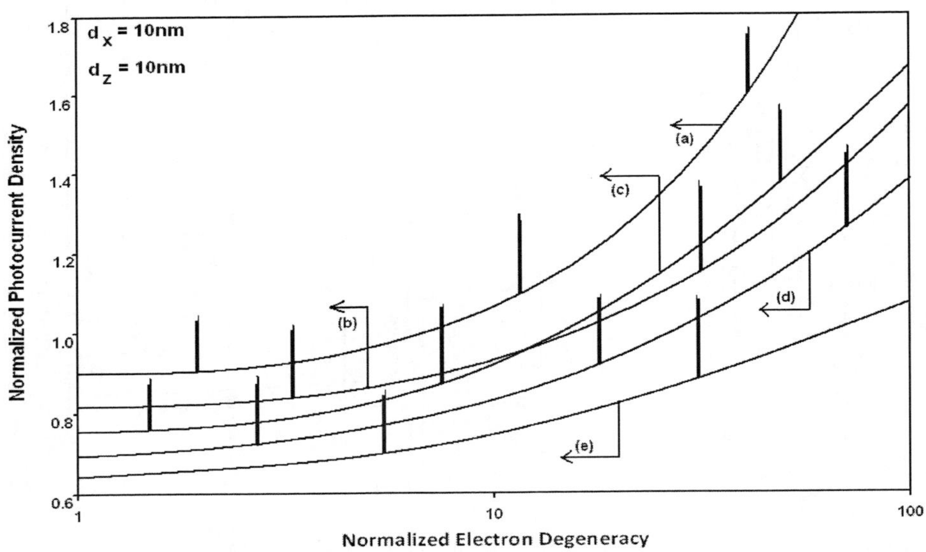

Figure 1.12. Plot of the normalized photocurrent density as a function of normalized electron degeneracy for all the cases of Figure 1.10.

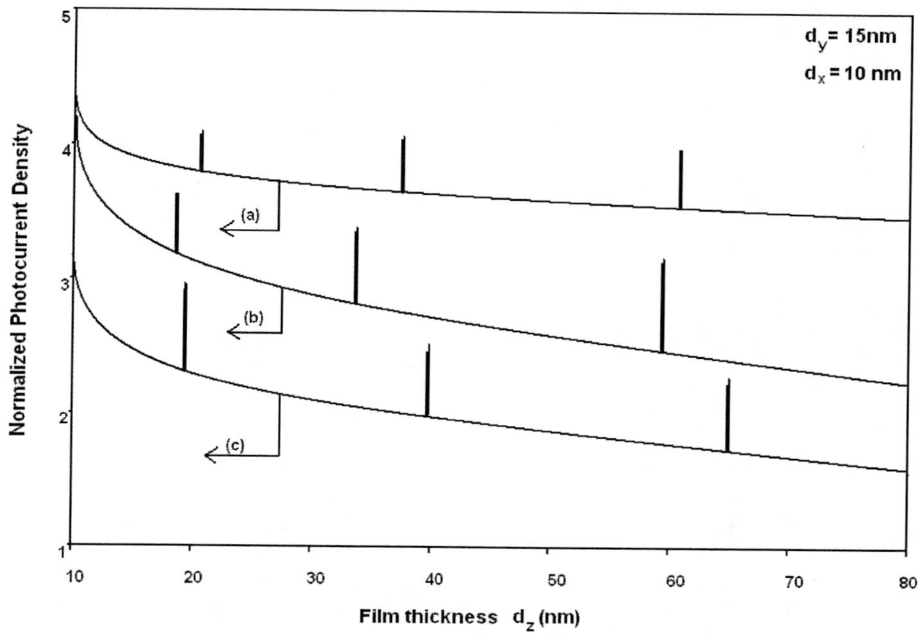

Figure 1.13. Plot of the normalized photocurrent density from QDs of (a) $PtSb_2$, (b) HgTe and (c) $Pb_{1-x}Ge_xTe$ as a function of film thickness.

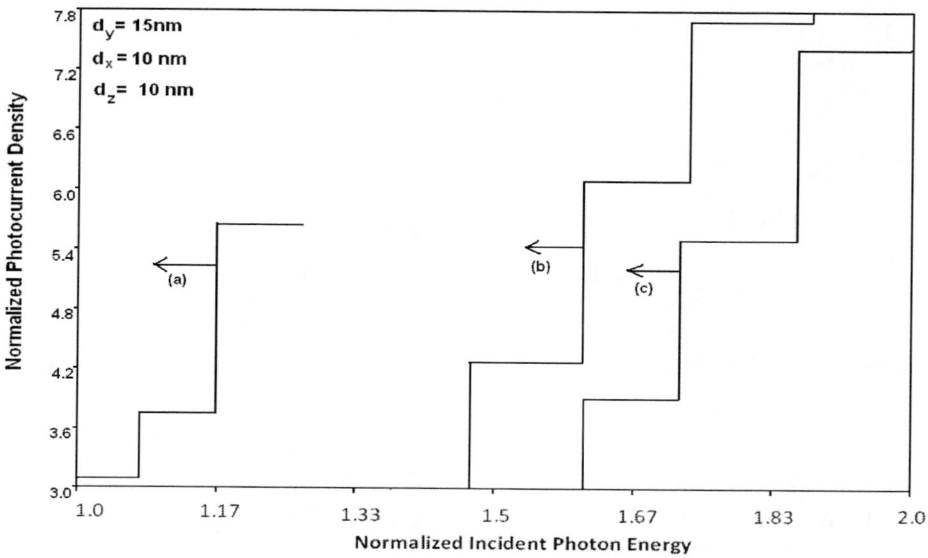

Figure 1.14. Plot of the normalized photocurrent density from QDs of (a) $PtSb_2$, (b) HgTe and (c) $Pb_{1-x}Ge_xTe$ as a function of normalized incident photon energy.

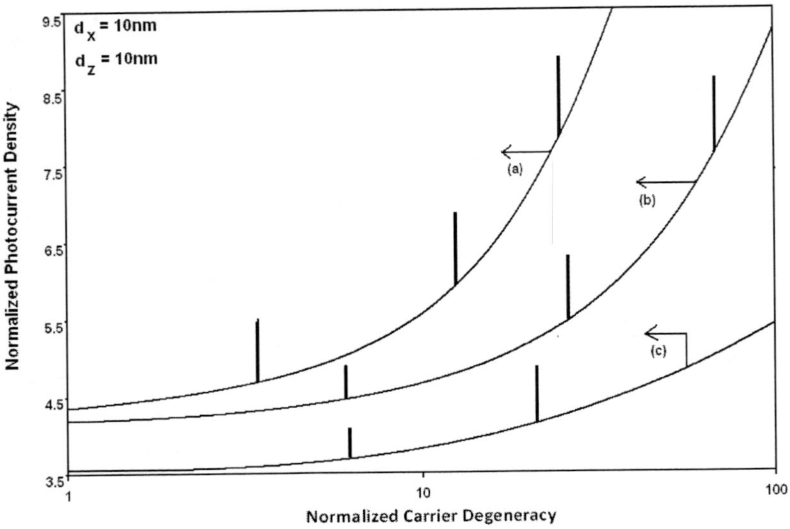

Figure 1.15. Plot of the normalized photocurrent density from QDs of (a) $PtSb_2$, (b) HgTe and (c) $Pb_{1-x}Ge_xTe$ as a function of normalized carrier degeneracy.

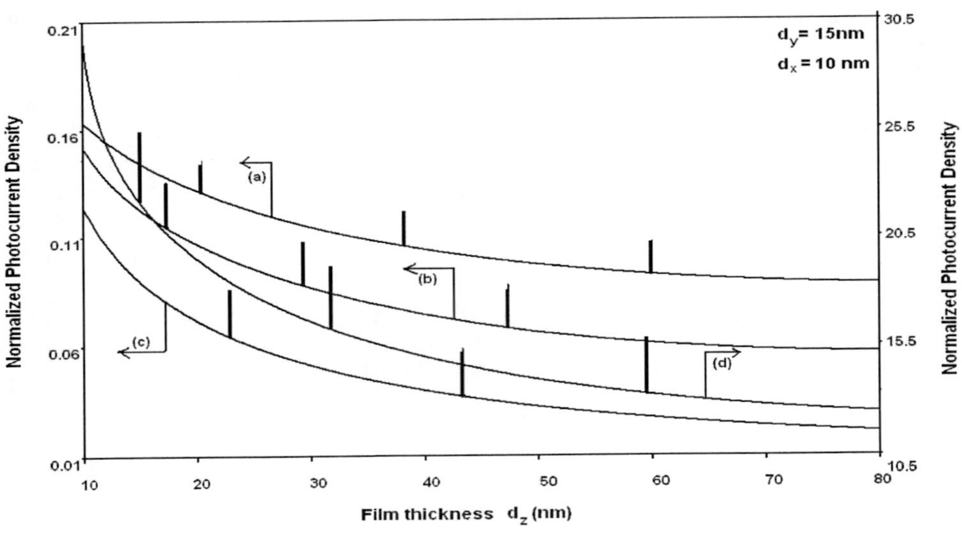

Figure 1.16. Plot of the normalized photocurrent density from QDs of GaSb in accordance with the models of (a) Seiler et. al., (b) Mathur et. al. and (c) Zhang et. al. respectively as a function of film thickness. The plot (d) refers to QDs of stressed InSb.

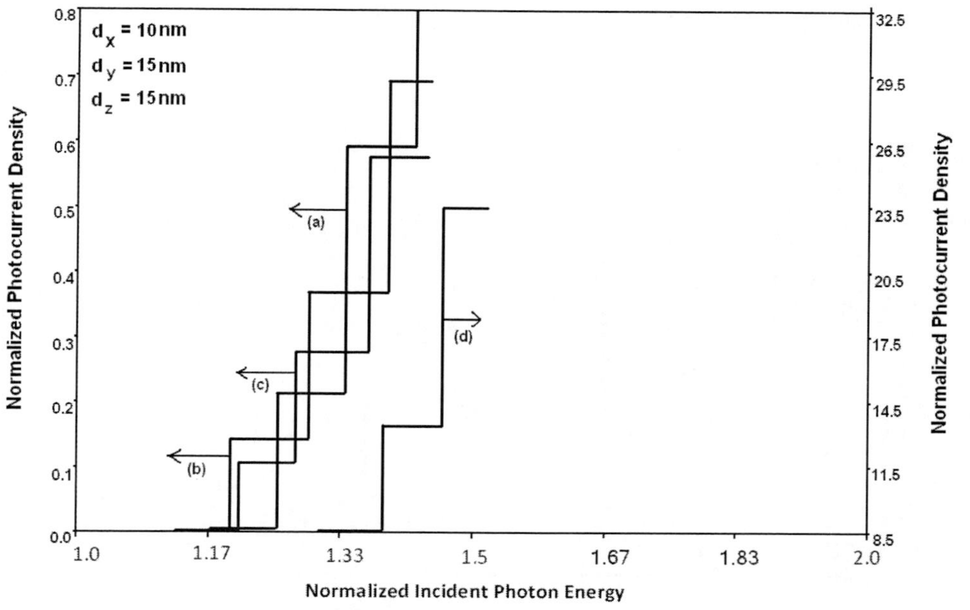

Figure 1.17. Plot of the normalized photocurrent density as a function of normalized incident photon energy for all the cases of Figure 1.16.

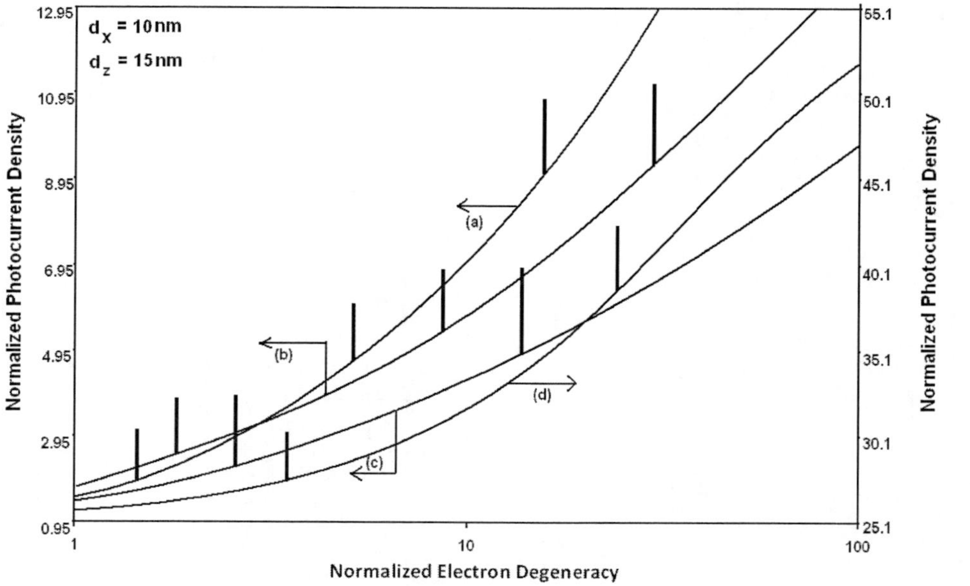

Figure 1.18. Plot of the normalized photocurrent density as a function of normalized electron degeneracy for all the cases of Figure 1.16.

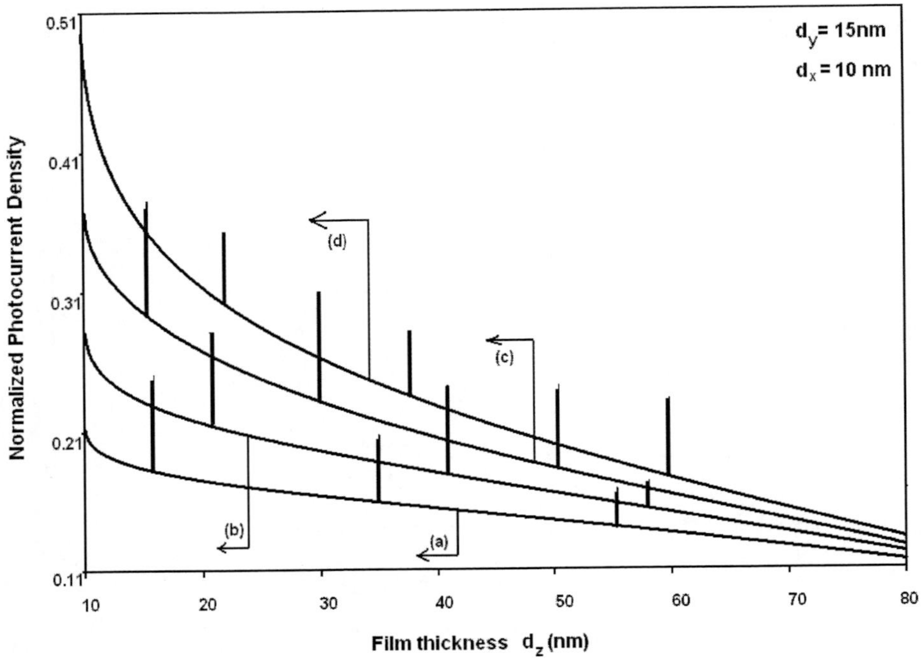

Figure 1.19. Plot of the normalized photocurrent density as a function of film thickness from QDs of Bismuth in accordance with the models of (a) McClure and Choi, (b) the Cohen, (c) the Hybrid and (d) the Lax respectively.

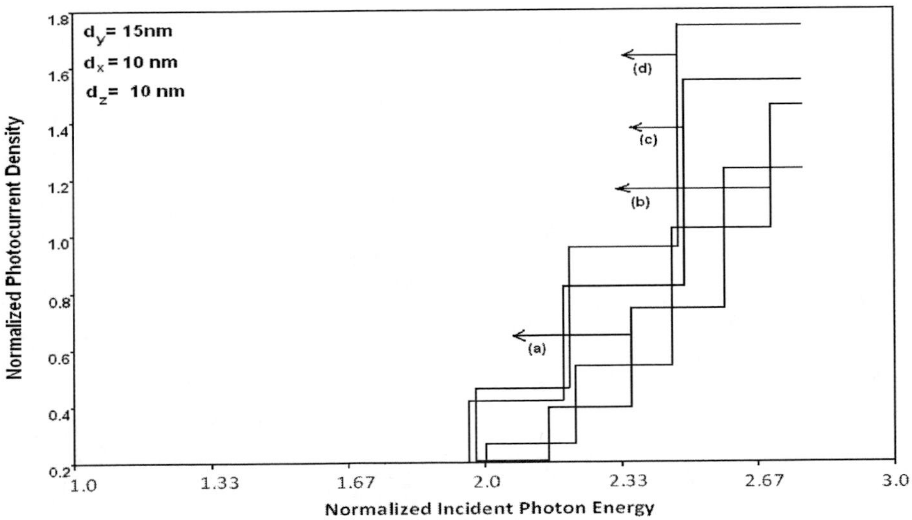

Figure 1.20. Plot of the normalized photocurrent density as a function of normalized incident photon energy for all the cases of Figure 1.19.

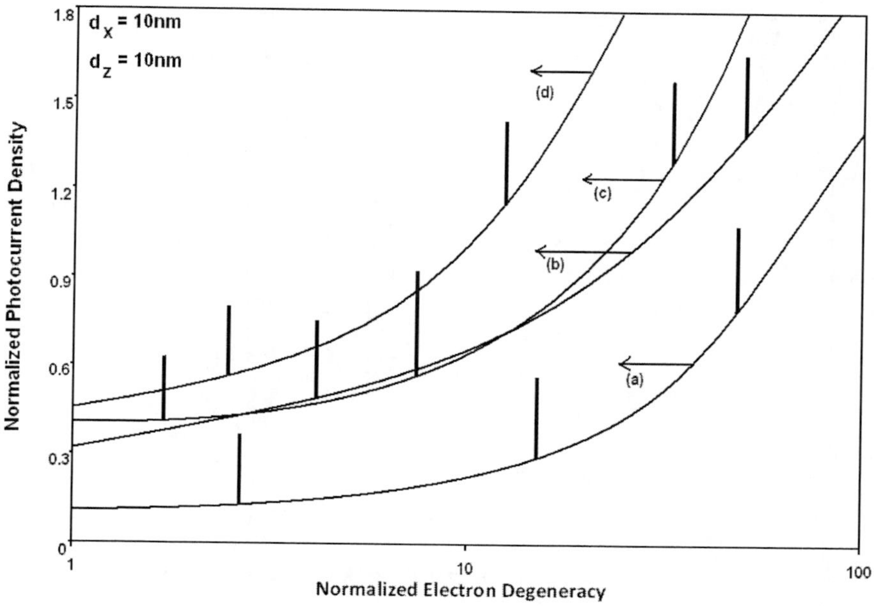

Figure 1.21. Plot of the normalized photocurrent density as a function of normalized electron degeneracy for all the cases of Figure 1.19.

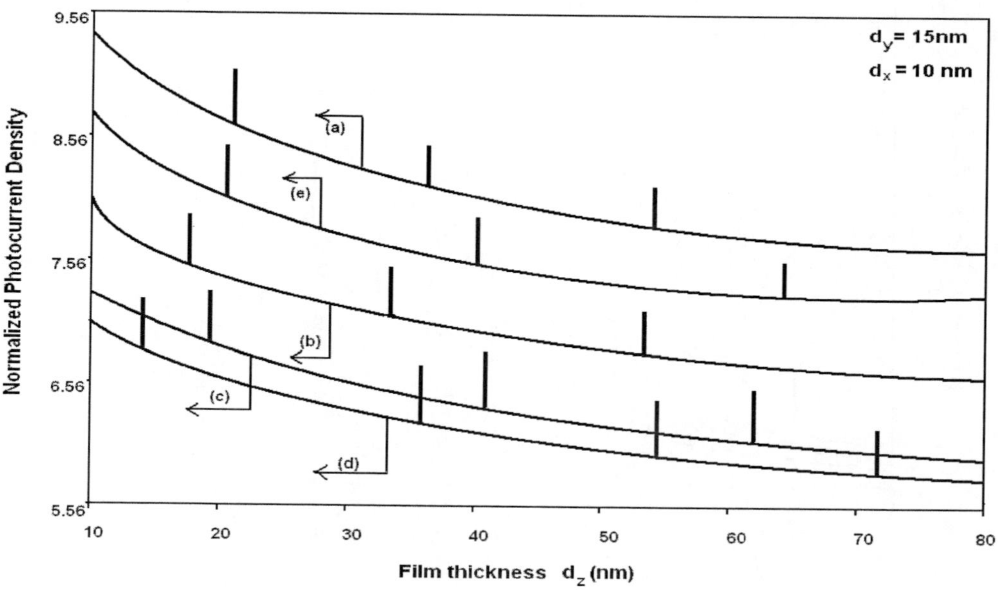

Figure 1.22. Plot of the normalized photocurrent density from QDs of PbTe in accordance with the models of (a) Dimmock (b) Cohen, (c) Bangert et. al. and (d) Foley et. al. respectively as a function of film thickness. The plot (e) refers to QDs of CdSb.

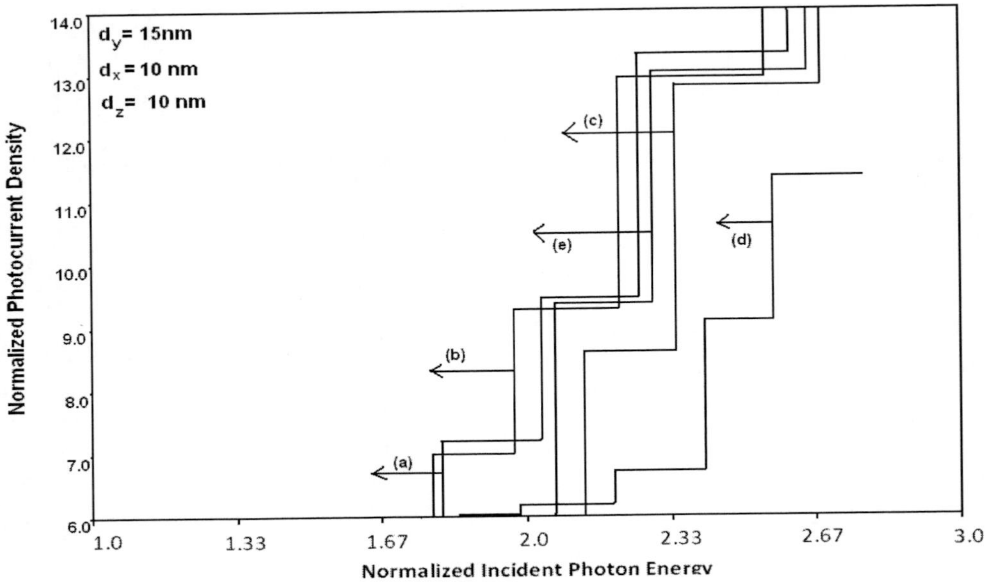

Figure 1.23. Plot of the normalized photocurrent density as a function of normalized incident photon energy for all the cases of Figure 1.22.

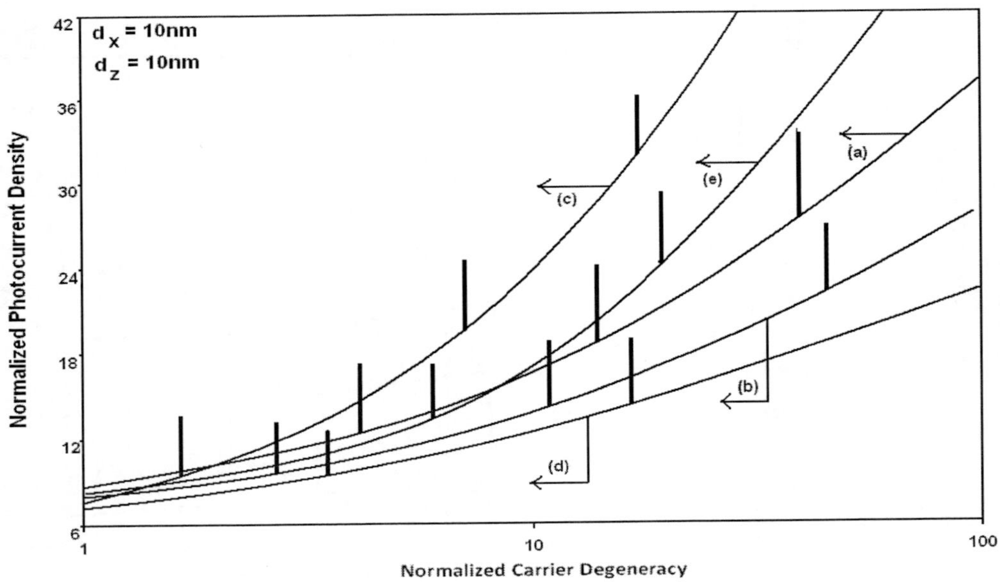

Figure 1.24. Plot of the normalized photocurrent density as a function of normalized carrier degeneracy for all the cases Figure 1.22.

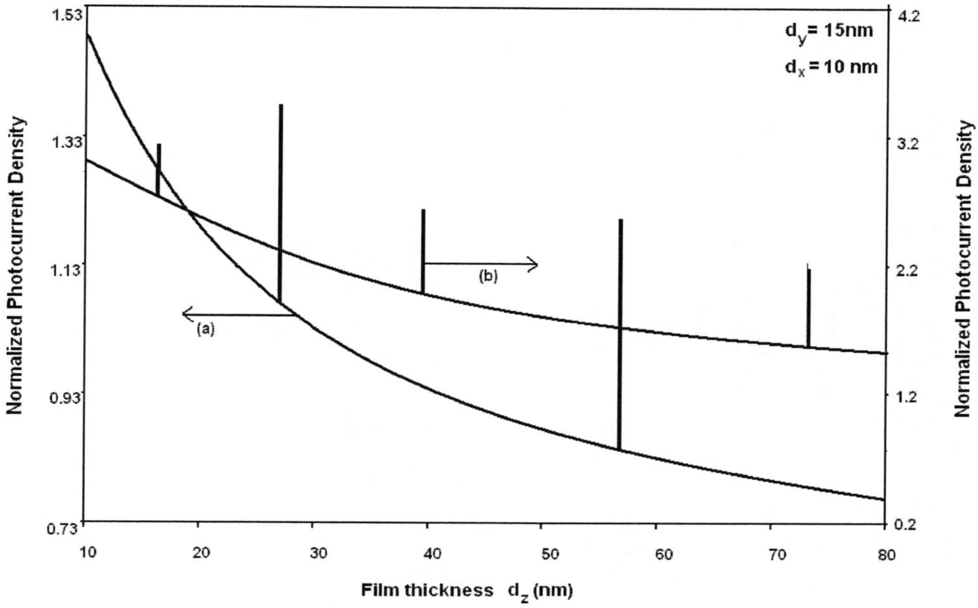

Figure 1.25. Plot of the normalized photocurrent density from QDs of (a) cadmium and (b) zinc diphosphides as a function of film thickness.

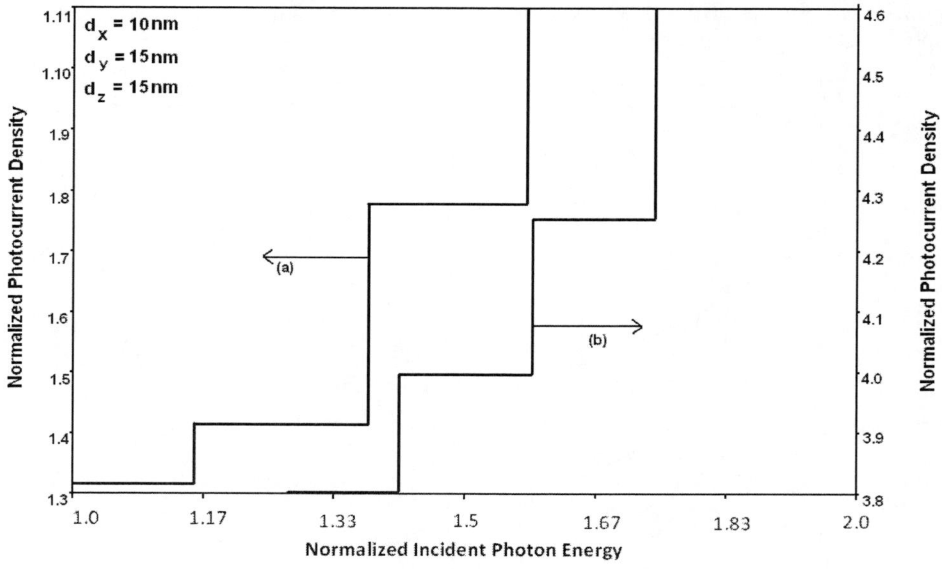

Figure 1.26. Plot of the normalized photocurrent density from QDs of (a) cadmium and (b) zinc diphosphides as a function of normalized incident photon energy.

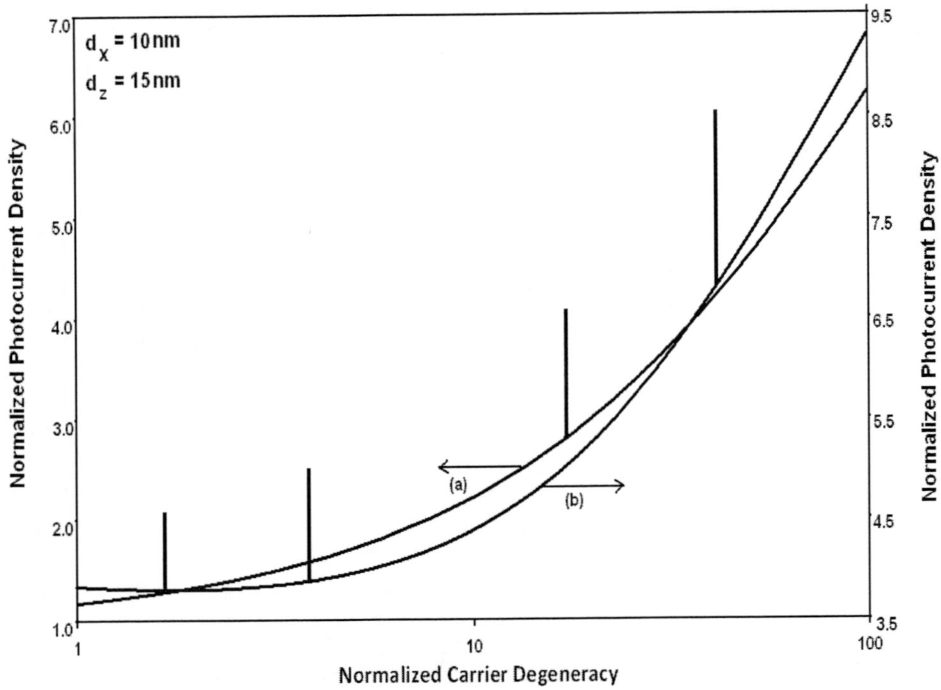

Figure 1.27. Plot of the normalized photocurrent density from QDs of (a) Cadmium and (b) zinc diphosphides as a function of normalized carrier degeneracy.

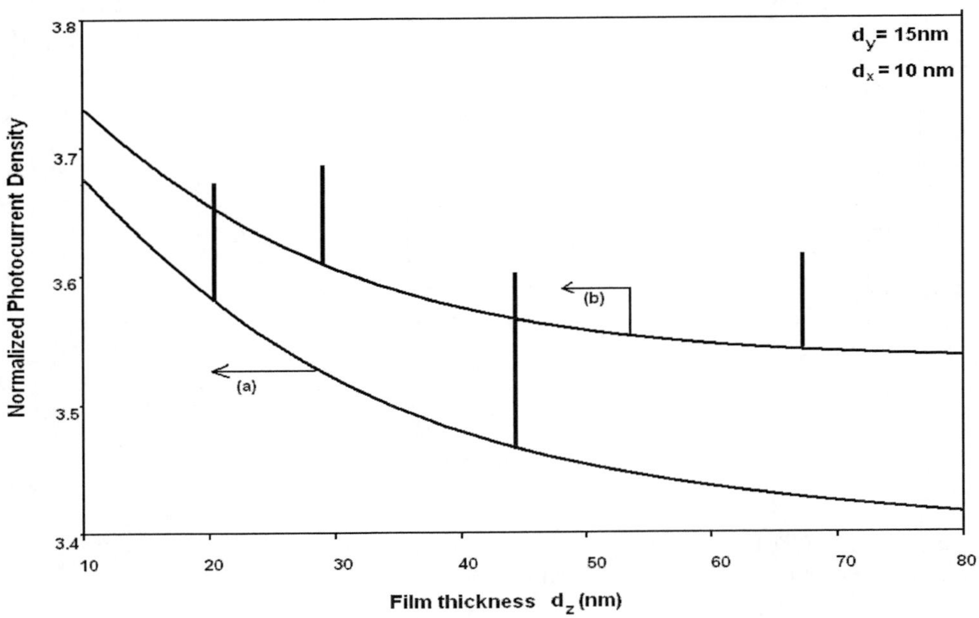

Figure 1.28. Plot of the normalized photocurrent density from QDs of (a) Bi_2Te_3 and (b) Sb as a function of film thickness.

The Figure 1.25 exhibits the plot of the normalized photoemission current density as a function of film thickness from QDs of (a) cadmium and (b) zinc diphosphides respectively which has been obtained by using (1.222) and (1.221). In Figure 1.26 and 1.27, the normalized photoemission in this case has been plotted as functions of normalized photon energy and normalized carrier degeneracy respectively for all cases of Figure 1.25. The Figure 1.28 shows the plot of the normalized photoemission as a function of film thickness from QDs of (a) Bi_2Te_3 (using (1.229) and (1.228)) and (b) Sb (using (1.248) and (1.249)) respectively. In Figure 1.29 and 1.30, the normalized photoemission in this case has been plotted as functions of normalized incident photon energy and normalized carrier degeneracy respectively for all cases of Figure 1.28.

It appears from the Figures (1.1), (1.4), (1.7), (1.10), (1.13), (116), (1.19), (1.22), (1.25) and (1.28) that the normalized photo-emitted current density increases with decreasing film thickness and exhibits spikes for various values of d_z which are totally band structure dependent. The Figures (1.2), (1.5), (1.8), (1.11), (1.14), (1.17), (1.20), (1.23), (1.26) and (1.29) exhibit the step-functional dependence of the normalized photo-emitted current density from the QDs of different materials with the incident photon energy. It is apparent from the Figures (1.3), (1.6), (1.9), (1.12), (1.15), (1.18), (1.21), (1.24), (1.27) and (1.30) that the normalized photo-emitted current density from the QDs of various materials increases with increasing normalized carrier degeneracy and exhibits spikes for different values of carrier concentration which are again band structure dependent. It may be noted that the QDs lead to the discrete energy levels, somewhat like atomic energy levels, which produce very large changes. This follows from the inherent nature of the quantum confinement of the carrier gas dealt with here. In QDs, there remain no free carrier states in between any two allowed sets of size-quantized levels unlike that for quantum wells in ultrathin films (QWs in UFs) and quantum well wires (QWWs) where the quantum confinements are 1D and 2D, respectively. Consequently, the crossing of the Fermi level by the size-quantized levels in QDs would have much greater impact on the redistribution of the carriers among the allowed levels, as compared to that for QWs in UFs and QWWs respectively. Although the photoemission varies in various manners with all the variables in all the limiting cases as evident from all the figures, the rates of variations are totally band-structure dependent. The quantum signature of QDs for Einstein's photo-electric effect is rather prominent as compared to the same from QWs in UFs and QWWs.

The photoemission from QWs in UFs, QWWs and QDs will further be investigated in details in Chapter 7 in the presence of external photo-excitation with the realization that it is the band structure which changes in a fundamental way and consequently alters the photoemission together with the fact that in general all the physical properties of all the electronic materials changes radically leading to new physical concepts.

The plot (a) of Figure 1.31 exhibits the variation of the normalized photoemission as a function of film thickness for QD SLs with graded interfaces of (a) HgTe/CdTe (using (1.278) and (1.280)), (b) $HgTe/Hg_{1-x}Cd_xTe$ (an example of III-V QD SLs and using (1.252) and (1.254)), (c) CdS/ZnSe (an example of II-VI QD SLs and using (1.261) and (1.263) with $C_0 \neq 0$) and (d) PbSe/PbTe (an example of IV-VI QD SLs and using (1.269) and (1.271) respectively. The above plots of Fig.1.31 have been obtained by using the energy band constants of the said super lattices from table 1.1. The Figures 1.32 and 1.33 demonstrate the plots for normalized photoemission as functions of normalized electron degeneracy and the normalized incident photon energy respectively for all the cases of Figure 1.31. The plot (a) of Figure 1.34 exhibits the variation of the normalized photoemission as a function of film thickness for QD effective SLs of (a) HgTe/CdTe (using (1.305) and (1.308)), (b) HgTe/$Hg_{1-x}Cd_xTe$ (an example of III-V SLs and using (1.285) and (1.288)), (c) CdS/ZnSe (an example of II-VI QD SLs and using (1.292) and (1.295) with $C_0 \neq 0$) and (d) PbSe/PbTe (an example of IV-VI QD SLs and using (1.299) and (1.302) respectively. The Figures 1.35 and 1.36 exhibit the plots for photoemission as functions of normalized electron degeneracy and the normalized incident photon energy respectively for all the cases of Figure 1.34. Finally, it is logical to conclude that the numerical values of the photo-emitted current density are totally different in all cases which exhibit the signature of the respective band structure under different physical conditions and the rates of variation are again totally energy spectrum dependent.

The theoretical results as presented here will be useful in determining the mobility even for relatively wide gap compounds whose energy band structures can be approximated by the parabolic energy bands both in the presence and absence of light waves. It is worth remarking that our basic Equation (1) covers various materials having different energy band structures. Under certain limiting conditions, all the results of the EEM for different materials having various band structures lead to the well-known expression of the EEM for degenerate compounds following the two band model of Kane. This indirect test not only exhibits the mathematical compatibility of our formulation but also shows the fact that our simple analysis is a more generalized one, since one can obtain the corresponding results for the relatively wide gap materials having parabolic energy bands under certain limiting conditions from our present derivation.

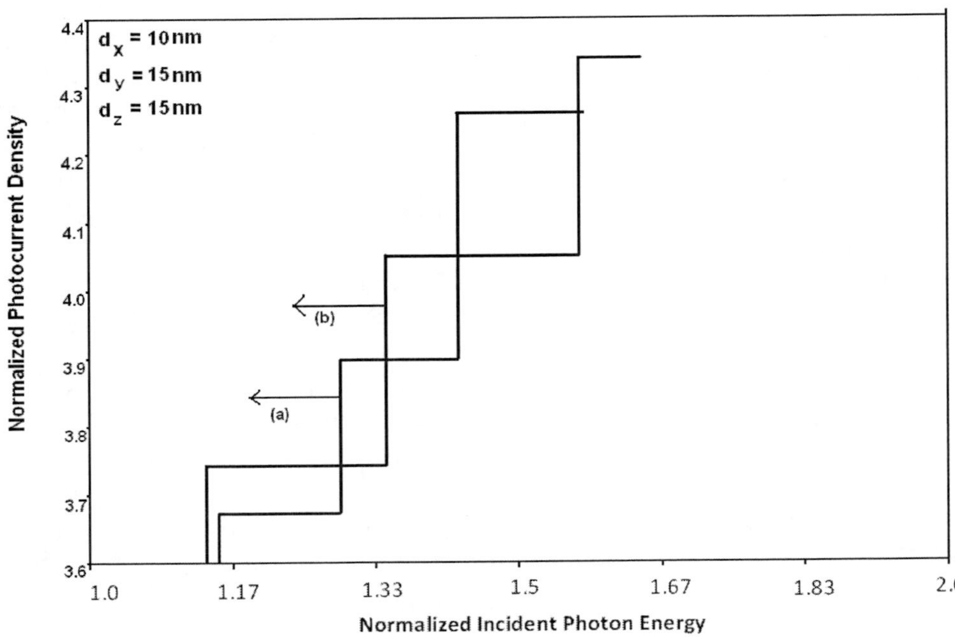

Figure 1.29. Plot of the normalized photocurrent density from QDs of (a) Bi_2Te_3 and (b) Sb as a function of normalized incident photon energy.

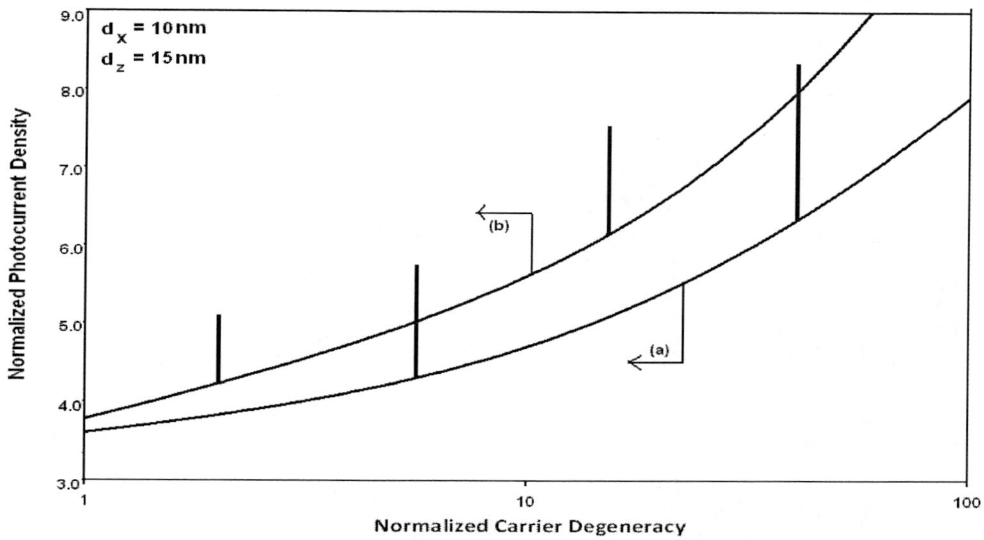

Figure 1.30. Plot of the normalized photocurrent density from QDs of (a) Bi_2Te_3 and (b) Sb as a function of normalized carrier degeneracy.

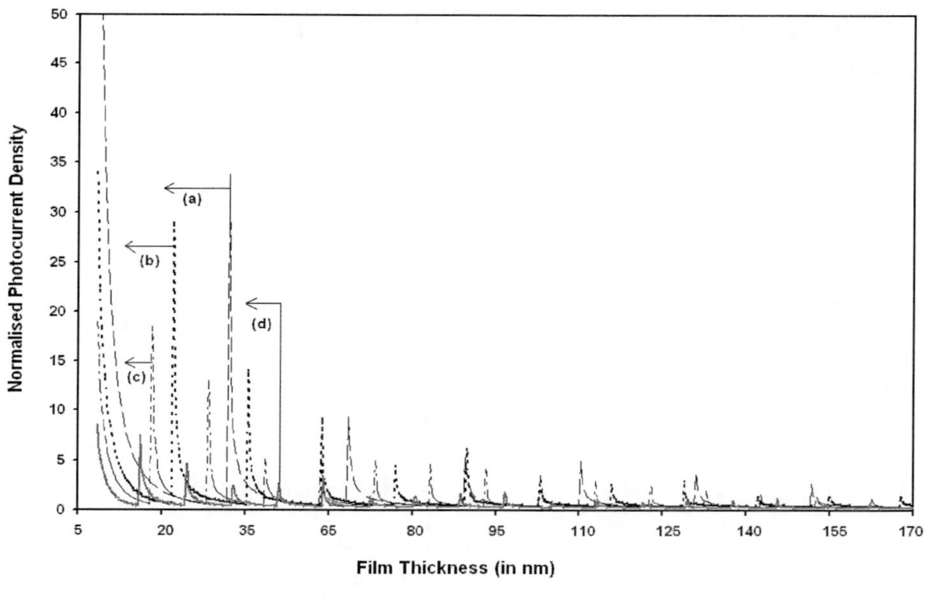

Figure 1.31. Plot of the normalized photocurrent density from (a) HgTe/CdTe (b) $HgTe/Hg_{1-x}Cd_xTe$, (c) CdS/ZnSe with $C_o \neq 0$ and (d) PbSe/PbTe quantum dots of superlattices with graded interfaces as a function of film thickness.

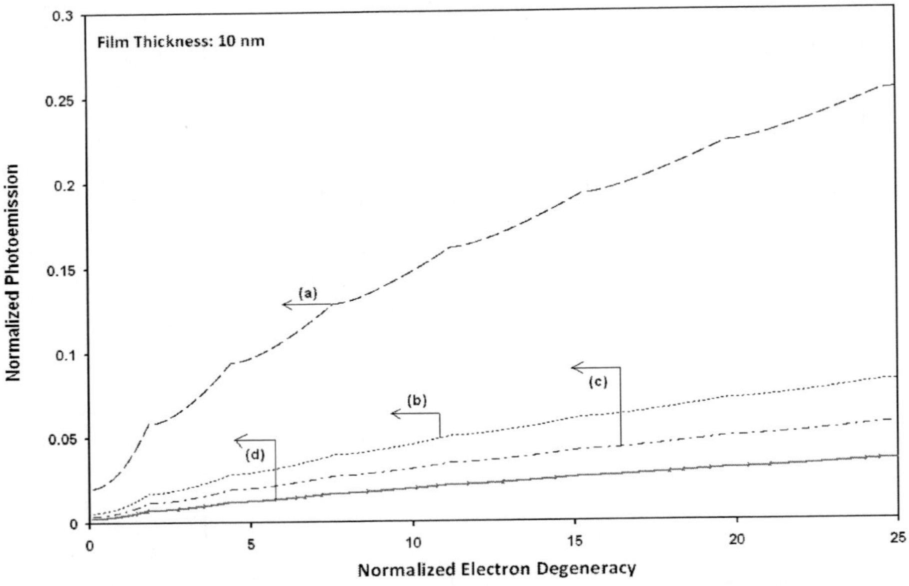

Figure 1.32. Plot of the normalized photo-emitted current density from (a) HgTe/CdTe (b) HgTe/$Hg_{1-x}Cd_xTe$, (c) CdS/ZnSe with $C_o \neq 0$ and (d) PbSe/PbTe quantum dots of superlattices with graded interfaces as a function of normalized electron degeneracy.

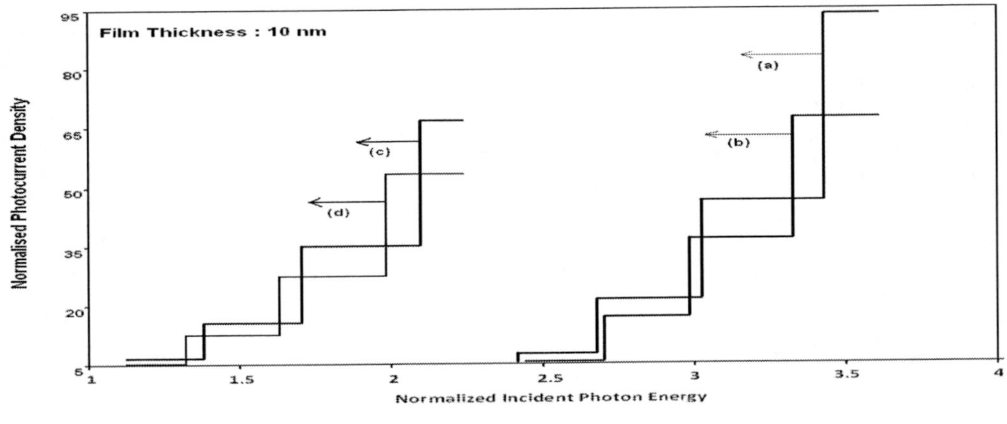

Figure 1.33. Plot of the normalized photocurrent density from (a) HgTe/CdTe (b) HgTe/$Hg_{1-x}Cd_xTe$, (c) CdS/ZnSe with $C_o \neq 0$ and (d) PbSe/PbTe quantum dots of superlattices with graded interfaces as a function of normalized incident photon energy.

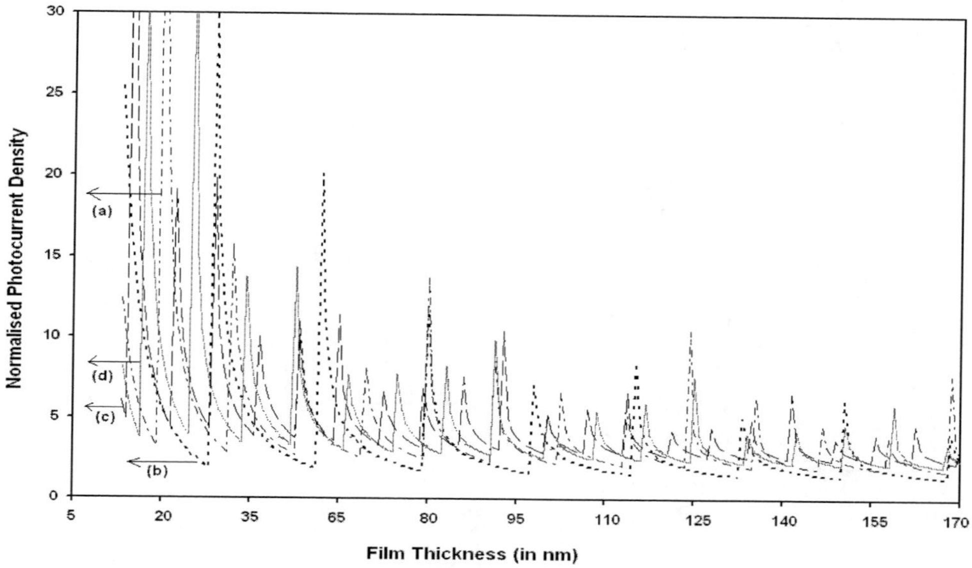

Figure 1.34. Plot of the normalized photocurrent density from (a) HgTe/CdTe (b) $HgTe/Hg_{1-x}Cd_xTe$, (c) CdS/ZnSe with $C_o \neq 0$ and (d) PbSe/PbTe quantum dots effective mass superlattices as a function of film thickness.

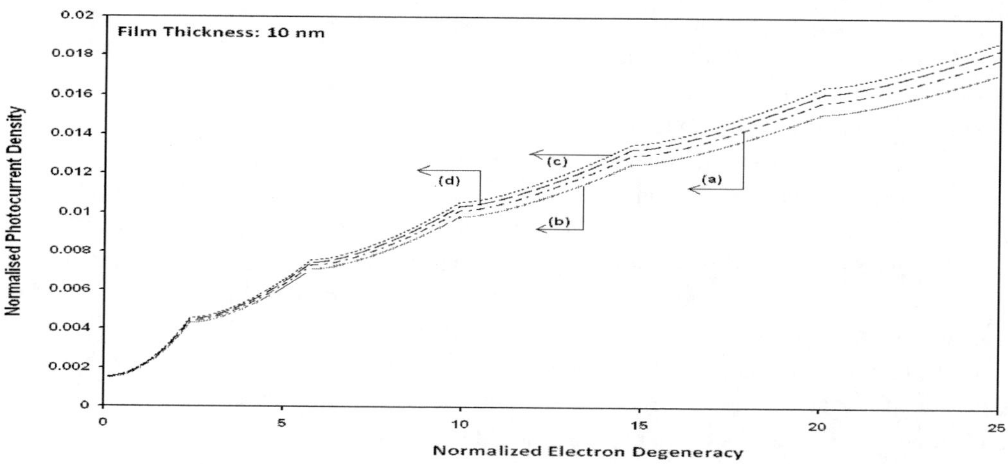

Figure 1.35. Plot of the normalized photocurrent density from (a) HgTe/CdTe (b) $HgTe/Hg_{1-x}Cd_xTe$, (c) CdS/ZnSe with $C_o \neq 0$ and (d) PbSe/PbTe quantum dots effective mass superlattices as a function of normalized electron degeneracy.

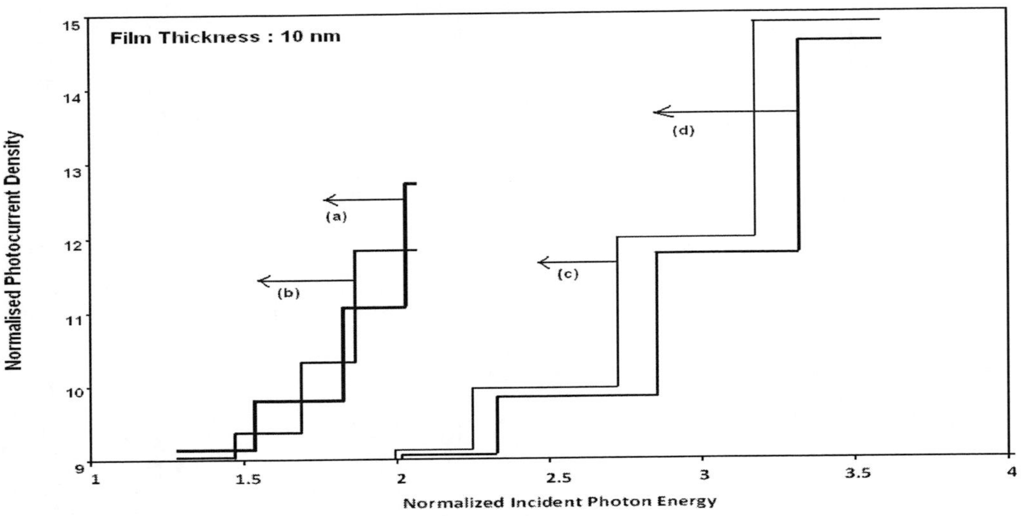

Figure 1.36. Plot of the normalized photocurrent density from (a) HgTe/CdTe (b) $HgTe/Hg_{1-x}Cd_xTe$, (c) CdS/ZnSe with $C_o \neq 0$ and (d) PbSe/PbTe quantum dots effective mass superlattices as a function of normalized incident photon energy.

It is worth remarking that the influence of an external photo-excitation is to change radically the original band structure of the material. Our method is not at all related to the DOS technique as used in the literature. From the **E-k** dispersion relation, we can obtain the DOS, but the DOS technique as used in the literature cannot provide the **E-k** dispersion relation. Therefore, our study is more fundamental than those of the existing literature because the Boltzmann transport equation, which controls the study of the charge transport properties of semiconductor devices, can be solved if and only if the **E-k** dispersion relation is known. Since the experimental results in the present case are not available in the literature to the best of our knowledge, we cannot compare our generalized theoretical analysis with the corresponding experimental data. Our formalism will be useful in probing the band structure when the experimental results for our generalized systems would appear. We have not considered other types of optoelectronic materials and other external variables in order to keep the presentation brief.

The numerical results presented in this paper would be different for other materials but the nature of variation would be unaltered. The theoretical results as given here would be useful in analyzing various other experimental data related to this phenomenon. Finally, we can write that this theory can be used to investigate the gate capacitance of nano-scale transistors, the magnetic susceptibilities, the Burstien Moss shift, plasma frequency, the Hall coefficient, the specific heat and other different transport coefficients of modern semiconductor evices operated in the presence of light waves.

ACKNOWLEDGMENTS

This chapter is dedicated to Professor Dilip Kumar Basu, Ex- Vice-Chancellor of Burdwan and Tripura Universities, India on the occasion of his 70th birth day. S. Bhattacharya and D. De together with K.P Ghatak are grateful to DST, India for sanctioning the project and the fellowship under ''SERC Fast Track Proposal of Young Scientist'' Scheme SR/FTP/ETA-37/08 and SERC/ET-0213/2011 respectively. The Senior and Corresponding author K.P. Ghatak and D.De are further grateful to UGC for sanctioning the research project having the No-F 40-469/2011(SR) under which this paper has been completed.

REFERENCES

[1] J. Cibert, P. M. Petroff, G. J. Dolan, S. J. Pearton, A. C. Gossard, J. H. English, *Appl. Phys. Letts.* 49, 1275 (1988).

[2] P. M. Petroff, A. C. Gossard, W. Wiegmann, *Appl. Phys. Letts.* 45, 620 (1984); J. M. Gaines, P. M. Petroff, H. Kroemer, R. J. Simes, R. S. Geels, J. H. English, *J. Vac. Sci. Tech. B* 6, 1378 (1988).

[3] T. Fukui, H. Saito, *Appl. Phys. Letts.* 50, 824 (1987).

[4] H. Sakaki, *Jpn. J. Appl. Phys.* 19, L735 (1980); P. M. Petroff, A. C. Gossard, R. A. Logan, W. Wiegmann, *Appl. Phys. Letts.* 41, 635 (1982).

[5] S. Bhattacharya, R. Sarkar, D. De, S. Mukherjee, S. Pahari, A. Saha, S. Roy, N. C. Paul, S. Ghosh, K. P. Ghatak, *J. Comput. and Theor. Nanosci.* 6, 112 (2009); K. P. Ghatak, S. Bhattacharya, D. De, R. Sarkar, S. Pahari, A. Dey, A. K. Dasgupta, S. N. Biswas, *J. Comput. Theor. Nanosci.* 5, 1345 (2008).

[6] K. P. Ghatak, S. Dutta, D. K. Basu, B. Nag, *Nuovo Cimento della Societa Italiana di Fisica D - Condensed Matter, Atomic, Molecular and Chemical Physics, Biophysics* 20, 227 (1998); B. Nag, K. P. Ghatak, *J. Phys. Chem. Solids* 59, 713 (1998); K. P. Ghatak, B. Nag, D. Bhattacharyya, *J. Low Temp. Phys.* 101, 983 (1995); K. P. Ghatak, D. Bhattacharyya, *Phys. Letts. A* 184, 366 (1994); K. P. Ghatak, S. N. Biswas, *Solid-State Electronics* 37, 1437 (1994); K. P. Ghatak, S. N. Biswas, *Molecular Crystals and Liquid Crystals Science and Technology Section B: Nonlinear Optics* 4, 39 (1993); K. P. Ghatak, B. Mitra, *Int. Jour. of Electronics* 72, 541 (1992); K. P. Ghatak, B. De, M. Mondal, *Phys. Stat. Sol. (b)* 105, K53 (1991); K. P. Ghatak, B. Mitra, M. Mondal, *Annalen der Physik* 48, 283 (1993).

[7] D. Bimberg, M. Grundmann, N. N. Ledentsov, *Quantum Dot Heterostructures*, John Wiley and Sons, (1999); G. Konstantatos, I. Howard, A. Fischer, S. Howland, J. Clifford, E. Klem, L. Levina, E. H. Sargent, *Nature* 442, 180 (2006).

[8] J. K. Jaiswal, H. Mattoussi, J. M. Mauro, S. M. Simon, *Nat Biotechnol* 21, 47 (2003).

[9] A. Watson, X. Wu, M. Bruchez, *Biotechniques* 34, 296 (2003); J. Nakanishi, Y. Kikuchi, T. Takarada, H. Nakayama, K. Yamaguchi, M. Maeda, *J. Am. Chem. Soc.* 126, 16314 (2004).

[10] X. Michalet, F. F. Pinaud, L. A. Bentolila, J. M. Tsay, S. Doose, J. J. Li, G. Sundaresan, A. M. Wu, S. S. Gambhir, S. Weiss, *Science* 307, 538 (2005).

[11] W. G. J. H. M. van Sark, P. L. T. M. Frederix, D. J. Van den Heuvel, H. C. G. A. Bol, J. N. J. van Lingen, C. de Mello Donegá, A. Meijerink, *J. Phys. Chem. B* 105, 8281 (2001).

[12] E. J. Sánchez, L. Novotny, X. S. Xie, *Phys Rev Lett* 82, 4014 (1999); B. Bailey, D. L. Farkas, D. L. Taylor, F. Lanni, *Nature* 366, 44 (1993).

[13] L.V. Asryan, R.A. Suris in *Selected Topics in Electronics and Systems*, 25, Ed. by E. Borovitskaya, M.S. Shur, (World Scientific, Singapore, 2002); L.V. Asryan, R.A. Suris, *Int. J. High Speed Electron. Syst.*, Special Issue on "Quantum Dot Heterostructures - Fabrication, Application, Theory," 12 [1], 111 (2002); L.V. Asryan, S. Luryi, *Future Trends in Microelectronics: The Nano Millennium*, Ed. by S. Luryi, J.M. Xu, A. Zaslavsky, Wiley Interscience, New York, 219 (2002).

[14] R. A. Freitas, Jr., *J. Comput. Theor. Nanosci.* 2, 1 (2005); A. Ferreira, C. Mavroidis, *IEEE Robotics and Automation Magazine* 13, 78 (2006); A. Dubey, G. Sharma, C. Mavroidis, S. M. Tomassone, K. Nikitczuk, M. L. Yarmush, *J. Comput. Theor. Nanosci.* 1, 18 (2004); C. Mavroidis, A. Dubey, M. L. Yarmush, *Annual Reviews of Biomedical Engineering* 6, 363 (2004).

[15] Y. Liu, J. A. Starzyk, Z. Zhu, *IEEE Trans. on Neural Networks* (2008) [In the press]; J. A. Starzyk, H. He, *IEEE Trans. on Neural Networks*, 18(2), 344 (2007); J. A. Starzyk, H. He, *IEEE Trans. Circuits Syst. II* 54(2), 176 (2007).

[16] E. –S. Hasaneen, E. Heller, R. Bansal, W. Huang, F. Jain, *Solid State Electronics* 48, 2055 (2004).

[17] T. Kawazoe, S. Tanaka, M. Ohtsu, *J. Nanophoton.* 2, 029502 (2008).

[18] H. J. Krenner, S. Stufler, M. Sabathil, E. C Clark, P. Ester, M. Bichler, G. Abstreiter, J. J Finley, A. Zrenner, *New J. Phys.* 7, 184 (2005).

[19] A. E. Zhukov, A. R. Kovsh, *Quantum Electron.* 38, 409 (2008).

[20] M Sugawara, T Akiyama, N Hatori, Y Nakata, H Ebe, H Ishikawa, *Meas. Sci. Technol.* 13, 1683 (2002); M. van der Poel, D. Birkedal, J. Hvam, M. Laemmlin, D. Bimberg, (2004) *Conference on Lasers and Electro-Optics (CLEO)* 1, 16.

[21] J. M. Costa-Fernandez, (2006) *Anal. Bioanal. Chem.* vol.384, 37; H. S. Djie, C. E. Dimas, D. –N. Wang, B. –S. Ooi, J. C. M. Hwang, G. T. Dang, W. H. Chang, (2007) *IEEE Sensors Jour.* vol.7, 251.

[22] X. –X. Zhu, Y. –C. Cao, X. Jin, J. Yang, X. –F. Hua, H. –Q. Wang, B. Liu, Z. Wang, J. –H. Wang, L. Yang, Y. –D. Zhao, (2008) *Nanotechnology* vol.19, 025708; X. Gao, W. C. W. Chan, S. Nie, (2002) *J. Biomed. Opt.* 7, 532.

[23] X. Michalet, F. F. Pinaud, L. A. Bentolila, J. M. Tsay, S. Doose, J. J. Li, G. Sundaresan, A. M. Wu, S. S. Gambhir, S. Weiss, (2005) *Science* 307, 538 ;J. K. Jaiswal, E. R. Goldman, H. Mattoussi, S. M. Simon, (2004). *Nature Methods* vol.1, 73.

[24] H. Matsueda, *Internat. Jour. Circuit Theo. and Appl.* 31, 23 (2003); X. Hu, S. Das Sarma, (2003) *Phys. Stat. Sol. (b) vol.*238, 360.

[25] G. –L. Chen, D. M. T. Kuo, W. –T. Lai, P. –W. Li, (2007) *Nanotechnology* vol.18, 475402; A. G. Pogosov, M. V. Budantsev, A. A. Shevyrin, A. E. Plotnikov, A. K. Bakarov, A. I. Toropov, (2008) *JETP Letts.* vol.87, 150.

[26] K. W. Johnston, A. G. Pattantyus-Abraham, J. P. Clifford, S. H. Myrskog, D. D. MacNeil, L. Levina, E. H. Sargent, *Appl. Phys. Letts.vol.* 92, 151115 (2008); K. S. Leschkies, R. Divakar, J. Basu, E. E. –Pommer, J. E. Boercker, C. B. Carter, U. R. Kortshagen, D. J. Norris, E. S. Aydil, *Nano Lett.* vol.7, 1793 (2007).

[27] I. –S. Liu, H. –H. Lo, C. –T. Chien, Y. –Y. Lin, C. –W. Chen, Y. –F. Chen, W. –F. Su, S. –C. Liou(2008), *J. Mater. Chem.* vol.18, 675.

[28] N. Hitoshi, Y. Sugimoto, K. Nanamoto, N. Ikeda, Y. Tanaka, Y. Nakamura, S. Ohkouchi, Y. Watanabe, K. Inoue, H. Ishikawa, K. Asakawa, (2004) *Optics Express* 12, 6606; N. Yamamoto, T. Matsuno, H. Takai, N. Ohtani, (2005) *Jpn. J. Appl. Phys.* 44, 4749; T. Yamada, Y. Kinoshita, S. Kasai, H. Hasegawa, Y. Amemiya,(2001)*Jpn. J. Appl. Phys.* 40, 4485; K. Asakawa, Y. Sugimoto, Y. Watanabe, N. Ozaki, A. Mizutani, Y. Takata, Y. Kitagawa, H. Ishikawa, N. Ikeda, K. Awazu, X. Wang, A. Watanabe, S. Nakamura, S. Ohkouchi, K. Inoue, M. Kristensen, O. Sigmund, P. I Borel, R. Baets, (2006) *New J. Phys.* vol.8, 208.

[29] A. R. Clapp, I. L. Medintz, B. R. Fisher, G. P. Anderson, H. Mattoussi, (2005) *J. Am. Chem. Soc.* 127, 1242; L. Shi, B. Hernandez, M. Selke, (2006) *J. Am. Chem. Soc.* 128, 6278; C. Wu, J. Zheng, C. Huang, J. Lai, S. Li, C. Chen, Y. Zhao, (2007) *Angewandte Chemie International Edition* vol.46, 5393.

[30] M. Cardona, L. Ley, (1978) *Photoemission in Solids 1 and 2, Topics in Applied Physics*, vols. 26, 27, (Springer-Verlag, Germany,; D. J. Lockwood, *Light Emission in Silicon in silicon based materials and devices*, vol 2, ed H. S. Nalwa (2001) (Academic Press, San Diego, USA).

[31] R. K. Pathria, (1996) *Statistical Mechanics, 2nd ed.* (Butterworth-Heinemann, Oxford.

[32] K. P. Ghatak, S. Bhattacharya, K. M. Singh, S. Choudhury, S. Pahari, (2008) Physica B vol.403, 2116.

[33] K. P. Ghatak, S. N. Biswas, (1991) in *Proceedings of the Society of Photo-optical and Instrumentation Engineers (SPIE),* Nonlinear Optics II, USA, vol. 1409, p. 28.

[34] K. P. Ghatak, (1992) SPIE, Process Module Metrology, USA, vol. 1594, p. 110; K. P. Ghatak, (1991) SPIE, International Conference on the Application and Theory of Periodic Structures, vol. 1545, p. 282.

[35] K. P. Ghatak, M. Mondal, (1988) Solid State Electron. 31, 1561.

[36] K. P. Ghatak, M. Mondal, (1991) J. Appl. Phys. 69, 1666.

[37] K. P. Ghatak, D. Bhattacharyya, B. Nag, S. N. Biswas, (1995) J. Nonlin. Optics and Quant. Optics 13, 267.

[38] M. Mondal, S. Banik, K. P. Ghatak, (1989) *J. Low Temp. Phys*. 74, 423.

[39] B. Mitra, A. Ghoshal, K. P. Ghatak, (1988) *Phys. Stat. Sol. (b)* 150, K67.

[40] B. Mitra, K. P. Ghatak, (1989) *Physica Scripta* 40, 776.

[41] K. P. Ghatak, M. Mondal, S. N. Biswas, (1990). *J. Appl. Phys*. 68, 3032 .

[42] K. P. Ghatak, S. N. Biswas, (1993). *Nonlinear Optics 4*, 39 .

[43] K. P. Ghatak, S. N. Biswas, (1991) SPIE, *Growth and Characterization of Materials for Infrared Detectors and Nonlinear Optical Switches*, USA, vol. 1484, p. 136.

[44] K. P. Ghatak, B. De,(1991) *Polymeric materials for Integrated Optics and Information Storage, Materials Research Society (MRS) Symposium Proceedings, MRS Spring Meeting,* vol. 228, p. 237.

[45] K. P. Ghatak, B. Nag, G. Majumdar, (1995) *Strained Layer Expitaxy – Materials, Processing, and Device Applications, MRS Symposium Proceedings, MRS Spring Meeting,* vol. 379, p.85.

[46] K. P. Ghatak,(1990) SPIE, *High Speed Phenomena in Photonic Materials and Optical Bistability*, USA, vol. 1280, p. 53.

[47] K. P. Ghatak, (1990) *Long Wave Length Semiconductor devices, Materials and Processes Symposium Proceedings, MRS Symposium Proceedings, MRS Spring Meeting*, vol. 216, p. 469.
[48] K. P. Ghatak, A. Ghoshal, S. Bhattacharyya, (1990) SPIE, *Nonlinear Optical Materials and Devices for Photonic Switching,* USA, vol. 1216, p. 282.
[49] K. P. Ghatak, (1992)SPIE, *Nonlinear Optics III*, USA, vol. 1626, p. 115.
[50] K. P. Ghatak, A. Ghoshal, B. De, (1990) SPIE, *Optoelectronic Devices and Applications*, USA,, vol. 1338, p. 111.
[51] R. Houdré, C. Hermann, G. Lampel, P. M. Frijlink, (1986) *Surface Sci.* vol.168, 538.
[52] T. C. Chiang, R. Ludeke, D. E. Eastman, (1982) *Phys. Rev*. B. vol.25, 6518 .
[53] S. P. Svensson, J. Kanski, T. G. Andersson, (1984) P. O. Nilsson, *J. Vacuum Sci. Technol*. B vol. 2, 235; S. F. Alvarado, F. Ciccacci, M. Campagna, (1981) *Appl. Phys. Letts*. vol39, 615.
[54] C. Majumdar, A. B. Maity, A. N. Chakravarti, (1987) *Phys. Stat. Sol. (b)* vol.140, K7.
[55] C. Majumdar, A. B. Maity, A. N. Chakravarti, (1987) *Phys. Stat. Sol. (b)* vol.141, K35.
[56] N. R. Das, K. K. Ghosh, D. Ghoshal, *Phys. Stat. Sol. (b)* vol.197, 97 (1996).
[57] C. Majumdar, A. B. Maity, A. N. Chakravarti, *Phys. Stat. Sol. (b),* vol.144, K13, (1987).
[58] N. R. Das, A. N. Chakravarti, *Phys. Stat. Sol. (b)* vol.176, 335 (1993).
[59] S. Sen, N. R. Das and A. N. Chakravarti, *Jour. of Phys: Conden. Mat.* vol.19, 186205 (2007); N. R. Das, S. Ghosh, A. N. Chakravarti, *Phys. Stat. Sol. (b)* vol.174, 45 (1992).
[60] A. B. Maity, C. Majumdar, A. N. Chakravarti, *Phys. Stat. Sol. (b)* vol.144, K93, (1987).
[61] A. B. Maity, C. Majumdar, A. N. Chakravarti, *Phys. Stat. Sol. (b)* vol.149, 565 (1988).
[62] N. R. Das, A. N. Chakravarti, *Phys. Stat. Sol. (b)* vol.169, 97 (1992).
[63] A. Modinos, *Field, Thermionic and Secondary Electron Emission Spectroscopy*, (Plenum Press, New York, 1984); A. V. D. Ziel, *Solid State Physical Electronics*, (Prentice Hall, Inc. Eaglewood Cliffs, 1957).
[64] J. E. Rowe, J. L. Shay, *Phys. Rev. B* vol.3, 451 (1973).
[65] H. Kildal, *Phys. Rev. B* vol.10, 5082 (1974).
[66] J. Bodnar, in *Proc. Int. Conf. of the Physics of Narrow-gap Semiconductors (Polish Science Publishers, Warsaw, 1978)*; G. P. Chuiko, N. N. Chuiko, *Sov. Phys. Semicond.* vol.15, 739 (1981); K. P. Ghatak and S. N. Biswas, *Proc. of SPIE* 1484, 149 (1991).
[67] A. Rogalski, *Jour. of Alloys and Comp.* vol.371, 53 (2004).
[68] A. Baumgartner, A. Chaggar, A. Patanè, L. Eaves, M. Henini, *Appl. Phys. Lett.vol.* 92, 091121 (2008).
[69] J. Devenson, R. Teissier, O. Cathabard, A. N. Baranov, *Proc. SPIE* 6909, 69090U (2008).
[70] B. S. Passmore, J. Wu, M. O. Manasreh, G. J. Salamo, *Appl. Phys. Lett.* vol.91, 233508 (2007).
[71] M. Mikhailova, N. Stoyanov, I. Andreev, B. Zhurtanov, S. Kizhaev, E. Kunitsyna, K. Salikhov, Y. Yakovlev, *Proc. SPIE* 6585, 658526 (2007).
[72] W. Kruppa, J.B. Boos, B.R. Bennett, N.A. Papanicolaou, D. Park, R. Bass, *Electron. Lett.vol.* 42, 688 (2006).
[73] E. O. Kane, *J. Phys. Chem. Sol.* 1, 249 (1957).
[74] J. A. Zapien, Y. K. Liu, Y. Y. Shan, H. Tang, C. S. Lee, S. T. Lee, *Appl. Phys. Lett.* vol.90, 213114 (2007).

[75] R. M. Park, *Proc. SPIE* 2524, 142 (1995).

[76] S. -G. Hur, E. -T. Kim, J. -H. Lee, G. –H. Kim, S. -G. Yoon, *Electrochem. Solid-State Lett.* 11, H176 (2008); H. Kroemer, *Rev. of Mod. Phys.* 73, 783 (2001); T. Nguyen Duy, J. Meslage, G. Pichard, *Jour. of Crys. Growth* 72, 490 (1985); T. Aramoto, F. Adurodija, Y. Nishiyama, T. Arita, A. Hanafusa, K. Omura, A. Morita, *Solar Energy Mat. and Solar Cells* 75, 211 (2003); H. B. Barber, *Jour. of Elect. Mat.* 25, 1232 (1996); S. Taniguchi, T.Hino, S. Itoh, K. Nakano, N. Nakayama, A. Ishibashi, M.Ikeda, *Elect. Lett.* 32, 552 (1996).

[77] J. J. Hopfield, *J. Appl. Phys.* 32, 2277 (1961).

[78] F. Hatami, V. Lordi, J. S. Harris, H. Kostial, W. T. Masselink, *J. Appl. Phys.* 97, 096106 (2005).

[79] B. W. Wessels, *J. Electrochem. Soc.* 122, 402 (1975); D. W. L. Tolfree, *J. Sci. Instrum.* **41,** 788 (1964); P.B. Hart, ***Proc. the IEEE* 61, 880 (1973).**

[80] H. Choi, M. Chang, M. Jo, S.J. Jung, H. Hwang, *Electrochem. Solid-State Lett.* 11, H154 (2008).

[81] S. Cova, M. Ghioni, A. Lacaita, C. Samori, F. Zappa, *Appl. Opt.* 35, 1956 (1996); H.W.H. Lee, B.R. Taylor, S.M. Kauzlarich, *Nonlinear Optics: Materials, Fundamentals, and Applications*, (Technical Digest, 12, 2000); E. Brundermann, U. Heugen, A. Bergner, R. Schiwon, G.W. Schwaab, S. Ebbinghaus, D.R. Chamberlin, E.E. Haller, M. Havenith, *29th International Conference on Infrared and Millimeter Waves and 12th International Conference on Terahertz Electronics* 283 (2004).

[82] P. K. Weimer, *Proc. IEEE* 52, 608 (1964).

[83] G. Ribakovs, A. A. Gundjian, *IEEE J. Quant. Electron.* QE-14, 42 (1978).

[84] S. K. Dey, (1973) *J. Vac. Sc. Tech.* 10, 227; S. J. Lynch, (1983) *Thin Solid Films* 102, 47.

[85] V. V. Kudzin, V. S. Kulakov, D. R. Pape', S. V. Kulakov, V. V. Molotok, (1997) *IEEE Ultrasonics Symposium*, 1, 749.

[86] M.S. Dresselhaus, J.G. Mavroides, (1964) *IBM Jour. of Red. and Develop.* 8, 262; P.R. Schroeder, M.S. Dresselhaus, A. Javan, (1968) *Phys. Rev. Lett.* 20, 1292; M.S. Dresselhaus(1970) in *Proceedings of the Conference on the Physic of Semimetals and Narrow Gap Semiconductors*, Ed. by D.L. Carter, R.T. Bate (Pergamon Press, New York); M. S. Dresselhaus, G. Dresselhaus(2002) *Adv. in Phys.* 51, 1.

[87] M.A. Hines, G.D. Scholes, (2003) *Adv. Mater.* **15**, 1844; C. A. Wang, R. K. Huang, D. A. Shiau, M. K. Connors, P. G. Murphy, P. W. O'Brien, A. C. Anderson, D. M. DePoy, G. Nichols, M. N. Palmisiano, (2003) *Appl. Phys. Lett* 83, 1286; C. W. Hitchcock, R. J. Gutmann, J. M. Borrego, I. B. Bhat, G. W. Charache, (1999) *IEEE Trans. Electron. Devices* 46, 2154.

[88] S. Matsushita, K. Kawai, H. Uchida, (1985) *IEEE J. Lightwave Technol.* LT-3, 533; N. K. Ailawadi, R. C. Alferness, G. D. Bergland, R. A. Thompson, (1991) *IEEE Mag. Lightwave Telecommun. Sys.* 2, 38; K. Kubota, J. Noda, O. Mikami, (1980) *IEEE J. Quantum Electron.* QE-16, 754; R. C. Alferness, C. H. Joyner, L. L. Buhl, S. K. Korotky, (1983) *IEEE J. Quantum Electron.* QE-19, 1339; R. G. Walker, (1991) *IEEE J. Quantum Electron.* 27, 654.

[89] G. C. Young, W. W. Anderson, L.B. Anderson, (1977) *IEEE Trans. Electron Dev.* 24, 492.

[90] R. L. Gordon, V. I. Neeley, (1966) H. R. Curtin, *Proc. IEEE* 54, 2014.

[91] P. K. Weimer, (1962) *Proc. IRE* 50, 1462.
[92] M.J. Lee, S.W. Wright, C.P. Judge, P.Y. Cheung, (1991) *Display Research Conference, International Conference Record*, 211.
[93] A. N. Baranov, T. I. Voronina, N. S. Zimogorova, L. M. Kauskaya, Y. P. Yakoviev, (1985) *Sov. Phys. Semicond.* 19, 1676; M. Yano, Y. Suzuki, T. Ishii, Y. Matsushima, M. Kimata, (1978) *Jpn. J. Appl. Phys.* 17, 2091; F. S. Yuang, Y.K. Su, N.Y. Li, (1991) *Jpn. J. Appl. Phys.* 30, 207; F. S. Yuang, Y. K. Su, N.Y. Li, K. J. Gan, (1990) *J. Appl. Phys.* 68, 6383; Y. K. Su, S. M. Chen, (1993) *J. Appl. Phys.* 73, 8349; S. K. Haywood, A. B. Henriques, N. J. Mason, R. J. Nicholas, P. J. Walker, (1988) *Semicond. Sci. Technol.* 3, 315.
[94] F. Hüe, M. Hÿtch, H. Bender, F. Houdellier, A. Claverie, (2008) *Phys. Rev. Lett.* 100, 156602; S. Banerjee, K.A. Shore, C.J. Mitchell, J.L. Sly, M. Missous, (2005) *IEE Proc. Circuits Devices Syst.* 152, 497; M. Razeghi, A. Evans, S. Slivken, J. S. Yu, J. G. Zheng, V. P. Dravid, (2005) *Proc. SPIE* 5840, 54; R. A. Stradling, (1991) *Semicond. Sci. Technol.* 6, C52.
[95] R. V. Belosludov, A. A. Farajian, H. Mizuseki, K. Miki, Y. Kawazoe, (2007) *Phys. Rev. B* 75, 113411; J. Heremans, C. M. Thrush, Y. -M Lin, S. Cronin, Z. Zhang, M. S. Dresselhaus, J. F. Mansfield, (2000) *Phys. Rev. B.* 61, 2921.
[96] D. Shoenberg, (1939) *Proc. Roy. Soc.* (London) 170, 341; B. Abeles, S. Meiboom, (1956) *Phys. Rev.* 101, 544.
[97] B. Lax, J. G. Mavroides, H. J. Zieger, R. J. Keyes, (1960) *Phys. Rev. Letts.* 5, 241.
[98] M. Maltz, M.S. Dresselhaus, (1970) *Phys. Rev. B* 2, 2877.
[99] M. Cankurtaran, H. Celik, T. Alper, (1986) *J. Phys. F: Metal Phys.* 16, 853.
[100] Y. -H. Kao, (1963) *Phys. Rev.* 129, 1122.
[101] R. J. Dinger, A.W. Lawson, (1971) *Phys. Rev. B* 3, 253.
[102] J. F. Koch, J. D. Jensen, (1969) *Phys. Rev.* 184, 643.
[103] M. H. Cohen, (1961) *Phys. Rev.* 121, 387.
[104] S. Takaoka, H. Kawamura, K. Murase, S. Takano, (1976) *Phys. Rev. B* 13, 1428.
[105] J. W. McClure, K.H. Choi, (1977) *Solid State Comm.* 21, 1015.
[106] G. P. Agrawal, N. K. Dutta, (1993)Semiconductor Lasers, (Van Nostrand Reinhold, New York; S. Chatterjee, (1993) U. Pal, *Opt. Eng.* 32, 2923; T. K. Chaudhuri, *Int. J. Energy Res.* 16, 481 (1992); J. H. Dughaish, (2002) *Physica B* 322, 205; C. Wood, (1988) *Rep. Prog. Phys.* 51, 459.
[107] K.-F. Hsu, S. Loo, F. Guo, W. Chen, J. S. Dyck, C. Uher, T. Hogan, E. K. Polychroniadis, M. G. Kanatzidis, (2004) *Science* 303, 818; J. Androulakis, K. F. Hsu, R. Pcionek, H. Kong, C. Uher, J. J. D'Angelo, A. Downey, T. Hogan, M. G. Kanatzidis, (2006) *Adv. Mater.* 18, 1170; P. F. P. Poudeu, J. D'Angelo, A. D. Downey, J. L. Short, T. P. Hogan, M. G. Kanatzidis, (2006) *Angew. Chem., Int. Ed.* 45, 3835; P. F. Poudeu, J. D'Angelo, H. Kong, A. Downey, J. L. Short, R. Pcionek, T. P. Hogan, C. Uher, M. G. Kanatzidis, (2006)*J. Am. Chem. Soc.* 128, 14347; J. R. Sootsman, R. J. Pcionek, H. Kong, C. Uher, M. G. Kanatzidis, (2006)*Chem. Mater.* 18, 4993.
[108] A. J. Mountvala, G. Abowitz, (1965) *J. Am. Ceram. Soc.* 48, 651; E. I. Rogacheva, I. M. Krivulkin, O. N. Nashchekina, A. Yu. Sipatov, V. A. Volobuev, M. S. Dresselhaus, (2001) *Appl. Phys. Lett.* 78, 3238; H. S. Lee, B. Cheong, T. S. Lee, K. S. Lee, W. M. Kim, J. W. Lee, S. H. Cho, J. Y. Huh, (2004) *Appl. Phys. Lett.* 85, 2782; K. Kishimoto, M. Tsukamoto, T. Koyanagi, (2002) *J. Appl. Phys.* 92, 5331.

[109] E. I. Rogacheva, O. N. Nashchekina, S. N. Grigorov, M. A. Us, M. S. Dresselhaus, S. B. Cronin, (2003)*Nanotechnology* 14, 53.

[110] E. I. Rogacheva, O. N. Nashchekina, A. V. Meriuts, S. G. Lyubchenko, M. S. Dresselhaus, G. Dresselhaus, (2005) *Appl. Phys. Lett.* 86, 063103; E. I. Rogacheva, S. N. Grigorov, O. N. Nashchekina, T. V. Tavrina, S. G. Lyubchenko, A. Yu. Sipatov, V. V. Volobuev, A. G. Fedorov, M. S. Dresselhaus, (2005)*Thin Solid Films* 493, 41.

[111] X. Qiu, Y. Lou, A. C. S. Samia, A. Devadoss, J. D. Burgess, S. Dayal, C. Burda, (2005) *Angew. Chem. Int. Ed.* 44, 5855.

[112] C. Wang, G. Zhang, S. Fan, Y. Li, (2001)*J. Phys. Chem. Solids* 62, 1957; B. Poudel, W. Z. Wang, D. Z. Wang, J. Y. Huang, Z. F. Ren, (2006) *J. Nanosci. Nanotechnol.* 6, 1050; B. Zhang, J. He, T. M. Tritt, (2006) *Appl. Phys. Lett.* 88, 043119; W. Heiss, H. Groiss, E. Kaufmann, G. Hesser, M. Böberl, G. Springholz, F. Schäffler, K. Koike, H. Harada, (2006) M. Yano, *Appl. Phys. Lett.* 88, 192109.

[113] C. Wang, G. Zhang, S. Fan, Y. Li, (2001) *J. Phys. Chem. Solids* vol.62, 1957; B. Poudel, W. Z. Wang, D. Z. Wang, J. Y. Huang, Z. F. Ren, (2006) *J. Nanosci. Nanotechnol.vol.* 6, 1050; B. Zhang, J. He, T. M. Tritt, (2006) *Appl. Phys. Lett.* vol.88, 043119; W. Heiss, H. Groiss, E. Kaufmann, G. Hesser, M. Böberl, G. Springholz, F. Schäffler, K. Koike, H. Harada, M. Yano, (2006) *Appl. Phys. Lett.* vol.88, 192109.

[114] B. A. Volkov, L. I. Ryabova, D. R. Khokhlov, (2002) *Phys. Usp.* vol.45, 819, and references therein.

[115] J. M. Thornton, (1981) *J Mol. Biol.* 151, 261; T. E. Creighton, (1986) *Methods Enzymol.* vol.131, 83; T. E. Creighton, (1988) *Bio Essaysvol.* 8, 57.

[116] V. J. Hruby, (1982) *Life Sciences* vol.31,189; V. J. Hruby, F. Al-Obeidi, W. Kazmierski, (1990) *Biochem J.* vol.268, 249.

[117] R. Wetzel, (1987) *Trends Biochem. Sci.* vol.12, 478.

[118] T. Kimura, R. Matsueda, Y. Nakagawa, E. T. Kaiser, (1982) *Anal Biochem.* vol.122, 274.

[119] D. Andreu, F. Albericio, Nuria A. Sole, M. C. Munson, M. Ferrer, G. Barany, (1994) *Methods in Molecular Biology, 35 Peptide Synthesis Protocols*, Ed. M. W. Pennington, B. M. Dunn (Humana Press Inc., New Jersey).

[120] H. J. Goldsmid, R. W. Douglas, (1954)*Brit. J. Appl. Phys.* vol.5, 386 .

[121] F. D. Rosi, B. Abeles, R. V. Jensen, (1959) *J. Phys. Chem. Sol.* vol.10, 191; T. M. Tritt (ed.), (2000) *Semiconductors and Semimetals, Vols. 69, 70 and 71: Recent Trends in Thermoelectric Materials Research I, II and III* (Academic Press, USA); D. M. Rowe (Ed.), (1995) *CRC Handchapter of Thermoelectrics* (CRC Press, USA); D. M. Rowe, C. M. Bhandari, (1983) *Modern Thermoelectrics* (Reston Publishing Company, Virginia); D. M. Rowe (ed.), (2006) *Thermoelectrics Handchapter: Macro to Nano* (CRC Press, USA).

[122] F. Smektala, C. Quemard, V. Couderc, A. Barthélémy, (2000) *J. Non-Cryst. Sol.* vol.274, 232; G. Boudebs, F. Sanchez, J. Troles, F. Smektala, (2001) *Opt. Comm.* vol.199, 425; J.M. Harbold, F.O. Ilday, F.W. Wise, J.S. Sanghera, V.Q. Nguyen, L.B. Shaw, I.D. Aggarwal, (2002) *Opt. Lett.* vol.27, 119; J.M. Harbold, F.O. Ilday, F.W. Wise, B. Aitken, (2002) *IEEE Photon. Technol. Lett.* vol.14, 822; V.T. Plaksiy, V.M. Svetlichniy, (2001) *Symposium on Physics and Engineering of Millimeter and Sub-Millimeter Waves, 2001. The Fourth International Kharkov* vol.1, 331; Z. Helin, D.M.

Rowe, S. G. K. Williams, (2001) Proc. XX International Conference on Thermoelectrics, 314.

[123] L.V. Keldysh(1962), *Sov. Phys. Solid State* vol.4, 1658.

[124] L. Esaki, R. Tsu, (1970) *IBM J. Research and Develop.* vol.14, 61; G. Bastard, (1990) *Wave mechanics applied to heterostructures*, (Editions de Physique, Les Ulis, France; E.L. Ivchenko, G. Pikus, (1995) *Superlattices and other heterostructures*, (Springer-Berlin); R. Tsu, (2005) *Superlattices to nanoelectronics*, (Elsevier, The Netherlands).

[125] P. Fürjes, Cs. Dücs, M. Ádám, J. Zettner, I. Bársony, (2004) *Superlattices and Microstructures* vol.35, 455; T. Borca-Tasciuc, D. Achimov, W. L. Liu, G. Chen, H. –W. Ren, C. –H. Lin, S. S. Pei, (2001) *Microscale Thermophysical Engg.* vol.5, 225.

[126] B. S. Williams, (2007) *Nat. Photonics* vol.1, 517; A. Kosterev, G. Wysocki, Y. Bakhirkin, S. So, R. Lewicki, F. Tittel, R. F. Curl, (2008) *Appl. Phys. B* 90, 165; M. A. Belkin, F. Capasso, F. Xie, A. Belyanin, M. Fischer, A. Wittmann, J. Faist, (2008) *Appl. Phys. Lett.* vol.92, 201101.

[127] G. J. Brown, F. Szmulowicz, R. Linville, A. Saxler, K. Mahalingam, C. –H. Lin, C. H. Kuo, W. Y. Hwang, (2000) *IEEE Photonics Technology Letts.* vol.12, 684; H. J. Haugan, G. J. Brown, L. Grazulis, K. Mahalingam, D. H. Tomich, (2004) *Physics E: Low-dimensional Systems and Nanostructures* vol.20, 527.

[128] S. A. Nikishin, V. V. Kuryatkov, A. Chandolu, B. A. Borisov, G. D. Kipshidze, I. Ahmad, M. Holtz, H. Temkin, (2003) *Jpn. J. Appl. Phys.* vol.42, L1362; Y. –K. Su, H. –C. Wang, C. –L. Lin, W. –B. Chen, S. –M. Chen, (2003) *Jpn. J. Appl. Phys.* vol.42, L751; C. H. Liu, Y. K. Su, L. W. Wu, S. J. Chang, R. W. Chuang, (2003) *Semicond. Sci. Technol. vol.*18, 545; S. –B. Che, I. Nomura, A. Kikuchi, K. Shimomura, K. Kishino(2002), *Phys. Stat. Sol. (b)* vol.229, 1001.

[129] C. P. Endres, F. Lewen, T. F. Giesen, S. Schlemmer, D. G. Paveliev, Y. I. Koschurinov, V. M. Ustinov, A. E. Zhucov, (2007) *Rev. Sci. Instrum.* vol.78, 043106.

[130] F. Klappenberger, K. F. Renk, P. Renk, B. Rieder, Y. I. Koshurinov, D. G. Pavelev, V. Ustinov, A. Zhukov, N. Maleev, A. Vasilyev, *Appl. Phys. Letts.* vol.84, 3924 (2004).

[131] X. Jin, Y. Maeda, T. Saka, M. Tanioku, S. Fuchi, T. Ujihara, Y. Takeda, N. Yamamoto, Y. Nakagawa, A. Mano, S. Okumi, M. Yamamoto, T. Nakanishi, H. Horinaka, T. Kato, T. Yasue, T. Koshikawa, (2008) *J. of Crystal Growth* vol.310, 5039; X. Jin, N. Yamamoto, Y. Nakagawa, A. Mano, T. Kato, M. Tanioku, T. Ujihara, Y. Takeda, S. Okumi, M. Yamamoto, T. Nakanishi, T. Saka, H. Horinaka, T. Kato, T. Yasue, T. Koshikawa, (2008) *Appl. Phys. Express* vol.1, 045002.

[132] B. H. Lee, K. H. Lee, S. Im, M. M. Sung, (2008) *Organic Electronics* vol.9, 1146.

[133] P. –H. Wu, Y. –K. Su, I. –L. Chen, C. –H. Chiou, J. –T. Hsu, W. –R. Chen, (2006) *Jpn. J. Appl. Phys.* vol.45, L647; A. C. Varonides, (2008) *Renewable Energy* vol.33, 273.

[134] M. Walther, G. Weimann, (2006) *Phys. Stat. Sol. (b)* vol.203, 3545.

[135] R. Rehm, M. Walther, J. Schmitz, J. Fleiβner, F. Fuchs, J. Ziegler, W. Cabanski, (2006) *Opto-Electronics Rev.* vol.14, 19; R. Rehm, M. Walther, J. Scmitz, J. Fleissner, J. Ziegler, W. Cabanski, R. Breiter, (2006) *Electronics Letts.* vol.42, 577.

[136] G. J. Brown, F. Szmulowicz, H. Haugan, K. Mahalingam, S. Houston, (2005) *Microelectronics Jour.* vol.36, 256.

[137] K. V. Vaidyanathan, R. A. Jullens, C. L. Anderson, H. L. Dunlap, (1983) *Solid State Electron.* vol.26, 717.

[138] B. A. Wilson, (1988) IEEE, *J. Quantum Electron.* vol.24, 1763.

[139] M. Krichbaum, P. Kocevar, H. Pascher, G. Bauer, (1988) *IEEE, J. Quantum Electron.* vol.24, 717.
[140] J. N. Schulman, T. C. McGill, (1979)*Appl. Phys. Letts.* vol.34, 663.
[141] H. Kinoshita, T. Sakashita, H. Fajiyasu, (1981) *J. Appl. Phys.* vol.52, 2869.
[142] L. Ghenin, R. G. Mani, J. R. Anderson, J. T. Cheung, (1989) *Phys. Rev. B* vol.39, 1419.
[143] C. A. Hoffman, J. R. Mayer, F. J. Bartoli, J. W. Han, J. W. Cook, J. F. Schetzina, J. M. Schubman, (1989) *Phys. Rev. B.* vol.39, 5208.
[144] V. A. Yakovlev, (1979) *Sov. Phys. Semicon.* vol.13, 692.
[145] E. O. Kane. (1957), *J. Phys. Chem. Solids* vol.1, 249.
[146] H. Sasaki, (1984) *Phys. Rev. B* vol.30, 7016.
[147] K. P. Ghatak, S. Bhattacharya, D. De, (2008) *Einstein Relation in Compound Semiconductors and Their Nanostructures*, Springer Series in Materials Science, Vol. 116 (Springer-Verlag, Germany).
[148] J.S. Blakemore, (1987) *Semiconductor Statistics*, Dover, New York; K P Ghatak, S Bhattacharya, S K Biswas, A Dey and A K Dasgupta, (2007) *Phys. Scr.*, vol.75, 820.
[149] B.R Nag, (1980) *Electron Transport in Compound Semiconductors*, Springer-Verlag (Springer series in Solid State Sciences, Vol-11.
[150] G. E. Stillman, C. M. Wolfe and J. O. Dimmock(1977) in *semiconductors and semimetals*, Ed. R. K. Willardon and A. C. Beer12, p 169 Academic Press, IV,V, USA.
[151] D. J. Newson and A. Karobe, (1988) *Semiconductor. Sci. and Tech.* vol.3, 786.
[152] U. Rossler, (1984) *solid state communn.*vol. 49, 943.
[153] E. D. Palik, G. S. Picus, S. Teither and R. E. Wallis, (1961) *Phys. Rev.* vol.122, 475.
[154] E. J. Johnson and D. H. Dickey, (1970) *Phys. Rev.* vol.1, 2676.
[155] V. G. Agafonov, P. M. Valov, B. S. Ryvkin and I. D. Yarashetskin, (1978) *Sov. Phys. Semiconductors*. vol.12, 1182.
[156] G. J. Rees, (1976) *Phys. of Compounds*, Proc. of the 13th Inter. Nat. Conf. Ed. by F.G. Fumi, 1166 (North Holland Company, The Netherlands,.
[157] M. Cardona, W. Paul, H. Brooks Helv, (1960) *Acta Phys.* vol.33, 329; A.F.Gibson in *"Proceeding of International school of physics "ENRICO FERMI"* course XIII, Ed. R.A Smith, (1963) Academic Press, 171.
[158] C. C. Wang, N.W. Ressler, (1970) *Phys. Rev.vol.* 2, 1827.
[159] J. Bouat and J. C. Thuillier, (1978) *Surface Sci.* vol.73, 528.
[160] M. V. Ortenberg and K. J. Button, (1977) *Phys. Rev.* B, vol.16, 2618.
[161] H. Ushio, T.Dau and Y. Uemura, (1972) *J. Phys. Soc.* Jap. vol.33, 1551.
[162] P. R. Emtage, (1965) *Phys. Rev.* vol.138, A246.
[163] V. I. Ivanov-Omskii, A. Sh. Mekhtisev, S. A. Rustambekova, E. N. Ukraintsev, (1983) *Phys. Stat. Sol. (b)* vol.119, 159.
[164] Y. Yamada, (1973) *J. Phys.* Japan, vol.35, 1600; M. Singh, P. R. Wallace, S. D. Jog and E. Arushanov, (1984) *J. Phys. Chem. Solids,* vol.45, 409.
[165] D. G. Seiler, W. M. Beeker and K. M. Roth, (1970) *Phys. Rev.* vol.1, 764.
[166] P. C. Mathur and S. Jain, (1979) *Phys. Rev.* vol.19, 1359.
[167] H. I. Zhang, (1970) *Phys. Rev. B*, vol.1, 3450.
[168] D. G. Seiler, B. D. Bajaj and A. E. Stephens, (1977) *Phys. Rev. B* vol.16, 2822; A.V. Germaneko and G.M. Minkov, (1994) *Phys. Stat. Sol. (b)* 184, 9; G.L. Bir and G.E. Pikus, (1972) *Symmetry and Strain –Induced effects in Semiconductors* Nauka, Russia. (in Russian); M. Mondal and K. P. Ghatak, (1986) *Phys. Stat. Sol. (b)* vol.135, K21.

[169] J. O. Dimmock in "*The Physics of semimetals and narrowgap semiconductors*" edited by D. L. Carter and R. T. Bates, (1971) Pergamon Press, Oxford, 319.

[170] E. Bangert and P. Kastner, (1974) *Phys. Stat. Sol. (b)*, vol.61, 503.

[171] G.M.T. Foley and P.N. Langenberg, (1977) *Phys. Rev. B,* vol.15, 4850.

[172] L. A. Vassilev, (1984) *Phys. Stat. Sol. (b)*, vol.121, 203.

[173] G. P. Chuiko, (1985) Sov. *Phys. Semi.*, vol.19 (12), 1381.

[174] M. Stordeur and W. Kuhnberger, (1975) *Phys. Stat. Sol. (b),* vol.69, 377; D. R. Lovett, (1977) *Semimetals and Narrow-Bandgap Semiconductor*, Pion Limited, 185; H. Köhler, (1976) *Phys. Stat. Sol. (b*), vol.74, 591.

[175] J. B. Ketterson, (1963) *Phys. Rev. vol.*129, 18.

[176] H. X. Jiang, J. Y. Lin, (1987) *J. Appl. Phys.* vol.61, 624.

[177] A. K. Sreedhar, S.C. Gupta, (1972) *Phys. Rev. B* vol.5, 3160; R. W. Keyes, (1961) IBM. J. Res. Devlopm. vol.5, 266; *Solid State Phys*. vol.20, 37, (1967).

[178] S. Bhattacharya, S. Chowdhury and K. P. Ghatak, (2006) *J. Compu. Theor. Nanosc.*, vol.3, 423; S. Choudhury, L.J. Singh and K. P. Ghatak, (2005) *Physica B*, vol.5,365; L. J. Singh, S. Choudhary, A. Mallik and K. P. Ghatak, (2005) *J. Compu. Theor. Nanosc.*, vol.2, 287; K. P. Ghatak, J. Y. Siddiqui and B. Nag, (2001) *Phys. Letts. A*, vol.282, 428; K. P. Ghatak, J.P. Banerjee and B. Nag, (1998) *J. Appl. Phys.,* vol.83, 1420; B. Nag, and K. P. Ghatak, (1998) *Nonlin. Opt.,* vol.19, 1; K. P. Ghatak and B. Nag, (1998) *Nanostruc. Material*., vol.10, 923; B. Nag and K. P. Ghatak, (1997) *J. Phys. Chem. Sol*., vol.58, 427;

[179] K. P. Ghatak, D. K. Basu and B. Nag, (1997) *J. Phys. Chem. Sol*., vol.58, 133; K. P. Ghatak, J.P. Banerjee, B. Goswami and B. Nag, (1996) in *Nonlin. Opt. Quant. Opt*., vol.16, 241; K. P. Ghatak, J.P. Banerjee, D. Bhattacharyya and B. Nag, (1996) *Nanotechnology, vol.***7,** 110 ; K. P. Ghatak, J.P. Banerjee, M. Mitra and B. Nag, (1996) *Nonlin. Opt*. vol.17, 193; B. Nag and K. P. Ghatak, (1996) *Phys. Scr., vol.*54, 657; K. P. Ghatak and B. Mitra, (1992) *Phys. Scr*., vol.46, 182; K. P. Ghatak,(1991) *Inter. J. Electron*., vol.71, 239.

[180] K. P. Ghatak, B. De, S. N. Biswas and M. Mondal,(1991) *Mechanical behavior of materials and structures in microelectronics*, MRS Symposium Proceedings, Spring Meeting, 2216,191; K. P. Ghatak and B. De, (1991) MRS Symposium Proceedings, 226, 191; K. P. Ghatak, B. Nag and G. Majumdar, MRS, 379, 109, (1995); **D**. Baruah, S. Choudhury, K.M. Singh and K.P. Ghatak, (2007) *J. Phys., Conf. Series*, vol.61, 80.

[181] M. Wegener, (2005) *Extreme Nonlinear Optics* (Springer-Verlag, Germany).

In: Quantum Dots and Quantum Cellular Automata
Editors: D. De, S. Bhattacharaya and K.P. Ghatak
ISBN: 978-1-62257-920-4

Chapter 2

INFLUENCE OF MICROWAVES ON THE PHOTOEMISSION FROM QUANTUM DOTS AND QUANTUM DOT SUPER LATTICES OF OPTO-ELECTRONIC MATERIALS

***P. K. Bose*[1], *S. Ghosh*[2], *M. Mitra*[2], *S. Ghatak*[3], *S. Bhattacharya*[3], *N. Paitya*[4,5], *P. K. Sarkar*[4], *S. Choudhury*[6], *S. Bhattacharya*[7], *D. De*[8,9] *and K. P. Ghatak*[4]**

[1] National Institute of Technology,Agartala,Tripura, India
[2]Department of Electronics and Telecommunication Engineering, Bengal Engineering and Science University, Howrah, India
[3]Department of Computer Science and Engineering, Sikkim Manipal Institute of Technology, Majitar, Rangpo, East Sikkim, India
[4] Department of Electronic Science, University of Calcutta, Kolkata, India
[5] Department of Applied Electronics and Instrumentation Engineering, Sikkim Manipal Institute of Technology, Majitar, Rangpo, East Sikkim, India
[6] School of Technology, North Eastern Hill University, Shillong, India
[7] Nanoscale Device Research Laboratory, Center for Electronics Design and Technology, Indian Institute Science, Bangalore, India
[8] Department of Computer Science and Engineering, West Bengal University of Technology, Kolkata, India
[9] School of Physics, University of Western Australia, The University of Western Australia, Perth, Western Australia

ABSTRACT

In this chapter, we study the photoemission from Quantum Dots (QDs) and Quantum Dot Super Lattices (QDSLs) of Opto-Electronic Materials on the basis of newly formulated electron dispersion laws in the presence of microwaves within the framework of *k.p* formalism. It appears taking QDs of $Hg_{1-x}Cd_xTe$ and $In_{1-x}Ga_xAs_yP_{1-y}$ lattice matched to InP as examples that the photoemission from QDs of optoelectronic materials decreases with increasing film thickness exhibiting prominent trapezoidal variation for relatively large thickness bandwidth. These types of variations are the special signatures of 3D confinement in QDs of optoelectronic materials respectively in the presence of microwaves. The normalized photo-emitted current density from the QD structures increases with increasing carrier degeneracy and for relatively large values of the same variable; it exhibits quantum jumps for all types of band models when the size quantum number changes from one fixed value to another. It appears that the photoemission from $HgTe/Hg_{1-x}Cd_xTe$ and $In_xGa_{1-x}As/InP$ QDSLs with graded interfaces increases with decreasing film thickness with larger oscillations again exhibiting the quantization effect. The photoemission from the said systems increases with decreasing intensity and decreasing wavelength respectively in various manners. The content of this chapter finds two applications in the field of semiconductor science.

2.1. INTRODUCTION

With the advent of semiconductor optoelectronics, there has been a considerable interest in studying the optical processes in semiconductors and their nanostructures [1]. It appears from the literature that the investigations have been carried out on the assumption that the carrier energy spectra are invariant quantities in the presence of intense microwaves, which is not fundamentally true. The physical properties of semiconductors in the presence of microwaves which change the basic dispersion relation have relatively less investigated in the literature [2-4]. In this chapter, we shall study the photo emission from quantum dots (QD) and quantum dot super lattices (QDSL) of ternary and quaternary materials in the presence of strong external photo-excitation on the basis of newly formulated electron dispersion laws.

Ternary and quaternary compounds enjoy the singular position in the entire spectrum of optoelectronic materials. It is well-known that the ternary alloy $Hg_{1-x}Cd_xTe$ is a classic narrow gap compound. By adjusting the alloy composition, the band gap of this ternary alloy can be varied to cover the spectral range from 0.8 to over 30μm [5]. $Hg_{1-x}Cd_xTe$ finds extensive applications in infrared detector materials and photovoltaic detector arrays in the 8-12μm wave bands [6]. The above uses have generated the $Hg_{1-x}Cd_xTe$ technology for the experimental realization of high mobility single crystal with specially prepared surfaces. The same compound is the optimum choice for illuminating the narrow sub-band physics because the relevant material constants are within easy experimental reach [7]. It may be mentioned that the quaternary alloy $In_{1-x}Ga_xAs_yP_{1-y}$ lattice matched to InP, also finds wide use in the fabrication of avalanche photo-detectors [8], hetero-junction lasers [9], light emitting diodes [10] and avalanche photodiodes [11]. Besides, the field effect transistors, detectors, switches, modulators, solar cells, filters and new types of integrated optical devices such as are made from the quaternary systems [12].

In Section 2.2.1, the photo emission from the QDs of the said materials has been investigated in the presence of microwaves whose unperturbed electron energy spectra are defined by the three and two band models of Kane together with parabolic energy bands by formulating respective dispersion relation in the presence of microwaves. In Section 2.2.2 and 2.2.3, the same has been investigated for quantum dot effective mass super lattices and quantum dot super lattices with graded interfaces respectively. For the purpose of relative assessment we also formulate the expression of photoemission from bulk specimens of optoelectronic materials under microwaves. The two different applications of the results of this chapter in the field of materials science in general have also been discussed in this context. The Sections 2.3 contains result and discussions pertinent to this chapter.

2.2. Theoretical Background

2.2.1. The Photo Emission from QDs of Optoelectronic Materials

The doubly degenerate wave functions $u_1(\vec{k},\vec{r})$ and $u_2(\vec{k},\vec{r})$ can be expressed as [13, 14]

$$u_1(\vec{k},\vec{r}) = a_{k+}\left[(is)\downarrow'\right] + b_{k+}\left[\frac{X'-iY'}{\sqrt{2}}\uparrow'\right] + c_{k+}\left[Z'\downarrow'\right] \tag{2.1a}$$

and

$$u_2(\vec{k},\vec{r}) = a_{k-}\left[(is)\uparrow'\right] - b_{k-}\left[\frac{X'+iY'}{\sqrt{2}}\downarrow'\right] + c_{k-}\left[Z'\uparrow'\right] \tag{2.1b}$$

s is the s-type atomic orbital in both unprimed and primed coordinates, $\downarrow'$ indicates the spin down function in the primed coordinates,

$$a_{k\pm} \equiv \beta\left[E_{g_0} - (\gamma_{0k\pm})^2(E_{g_0} - \delta')\right]^{1/2}(E_{g_0} + \delta')^{-1/2}\Big],\ \beta \equiv \left[\left(6(E_{g_0} + 2\Delta/3)(E_{g_0} + \Delta)\right)/\chi\right]^{1/2},$$

$$\chi \equiv \left(6E_{g_0}^2 + 9E_{g_0}\Delta + 4\Delta^2\right),\ \gamma_{0k\pm} \equiv \left[\frac{\left(\xi_{1k} \mp E_{g_0}\right)}{2(\xi_{1k} + \delta')}\right]^{1/2},$$

$$\xi_{1k} \equiv E_c(\vec{k}) - E_v(\vec{k}) = E_{g_0}\left[1 + 2\left(1 + \frac{m_c}{m_v}\right)\frac{\gamma(E)}{E_{g_0}}\right]^{1/2},\ \delta' \equiv \left(E_{g_0}^2\Delta\right)(\chi)^{-1},\ X',\ Y',$$

and Z' are the p-type atomic orbitals in the primed coordinates, $\uparrow'$ indicates the spin-up function in the primed coordinates, $b_{k\pm} \equiv \rho\gamma_{0k\pm}$, $\rho \equiv \left(4\Delta^2/3\chi\right)^{1/2}$, $c_{k\pm} \equiv t\gamma_{0k\pm}$ and $t \equiv \left[6(E_{g_0} + 2\Delta/3)^2/\chi\right]^{1/2}$.

We can, therefore, write the expression for the optical matrix element (OME) as

$$\text{OME} = \hat{p}_{cv}\left(\vec{k}\right) = <u_1(\vec{k},\vec{r})\left|\hat{p}\right|u_2(\vec{k},\vec{r})> \tag{2.1c}$$

Since the photon vector has no interaction in the same band for the study of inter-band optical transition, we can therefore write

$$\langle S|\hat{p}|S\rangle = \langle X|\hat{p}|X\rangle = \langle Y|\hat{p}|Y\rangle = \langle Z|\hat{p}|Z\rangle = 0$$

and $\langle X|\hat{p}|Y\rangle = \langle Y|\hat{p}|Z\rangle = \langle Z|\hat{p}|X\rangle = 0$

There are finite interactions between the conduction band (CB) and the valance band (VB) and we can obtain

$$\langle S\,|\,\hat{P}\,|\,X\rangle = \hat{i}\bullet\hat{P} = \hat{i}\bullet\hat{P}_x$$
$$\langle S\,|\,\hat{P}\,|\,Y\rangle = \hat{j}\bullet\hat{P} = \hat{j}\bullet\hat{P}_y$$
$$\langle S\,|\,\hat{P}\,|\,Z\rangle = \hat{k}\bullet\hat{P} = \hat{k}\bullet\hat{P}_z$$

where, $\hat{i}$, $\hat{j}$ and $\hat{k}$ are the unit vectors along x, y and z axes respectively.

It is well known that

$$\begin{bmatrix}\uparrow' \\ \downarrow'\end{bmatrix} = \begin{bmatrix} e^{-i\phi/2}\cos(\theta/2) & e^{i\phi/2}\sin(\theta/2) \\ -e^{-i\phi/2}\sin(\theta/2) & e^{i\phi/2}\cos(\theta/2)\end{bmatrix}\begin{bmatrix}\uparrow \\ \downarrow\end{bmatrix}$$

$$and \begin{bmatrix} X' \\ Y' \\ Z'\end{bmatrix} = \begin{bmatrix}\cos\theta\cos\phi & \cos\theta\sin\phi & -\sin\theta \\ -\sin\phi & \cos\phi & 0 \\ \sin\theta\cos\phi & \sin\theta\sin\phi & \cos\theta\end{bmatrix}\begin{bmatrix} X \\ Y \\ Z\end{bmatrix}$$

Besides, the spin vector can be written as

$$\vec{S} = \frac{\hbar}{2}\vec{\sigma} \text{ where } \sigma_x = \begin{bmatrix} 0 & 1 \\ 1 & 0\end{bmatrix},\ \sigma_y = \begin{bmatrix} 0 & -i \\ i & 0\end{bmatrix} \text{ and } \sigma_z = \begin{bmatrix} 1 & 0 \\ 0 & -1\end{bmatrix},$$

From above, we can write

$$\hat{p}_{CV}(\vec{k}) = \langle u_1(\vec{k},\vec{r})\,|\,\hat{P}\,|\,u_2(\vec{k},\vec{r})\rangle$$
$$= \left\langle \left\{ a_{k_+}\left[(iS)\downarrow'\right] + b_{k_+}\left[\left(\frac{X'-iY'}{\sqrt{2}}\right)\uparrow'\right] + c_{k_+}\left[Z'\downarrow'\right]\right\} |\,\hat{P}\,| \left\{ a_{k_-}\left[(iS)\uparrow'\right]\right.\right.$$
$$\left.\left. -b_{k_-}\left[\left(\frac{X'+iY'}{\sqrt{2}}\right)\downarrow' + c_{k_-}\left[Z'\uparrow'\right]\right]\right\}\right\rangle .$$

Using above relations, we get

$$\hat{p}_{CV}\left(\vec{k}\right)=\left\langle u_1\left(\vec{k},\vec{r}\right)|\hat{P}|u_2\left(\vec{k},\vec{r}\right)\right\rangle$$

$$=\frac{b_{k_+}a_{k_-}}{\sqrt{2}}\left\{\left\langle (X'-iY')|\hat{P}|iS\right\rangle\left\langle \uparrow'|\uparrow'\right\rangle\right\}+c_{k_+}a_{k_-}\left\{\left\langle Z'|\hat{P}|iS\right\rangle\left\langle \downarrow'|\uparrow'\right\rangle\right\}-\frac{a_{k_+}b_{k_-}}{\sqrt{2}}\left\{\left\langle iS|\hat{P}|(X'+iY')\right\rangle\left\langle \downarrow'|\downarrow'\right\rangle\right\}$$

$$+a_{k_+}c_{k_-}\left\{\left\langle iS|\hat{P}|Z'\right\rangle\left\langle \downarrow'|\uparrow'\right\rangle\right\} \tag{2.1d}$$

From (2.1d), one can write

$$\left\langle (X'-iY')|\hat{P}|iS\right\rangle=\left\langle (X')|\hat{P}|iS\right\rangle-\left\langle (iY')|\hat{P}|iS\right\rangle$$

$$=i\int u_{X'}^{*}\hat{P}S-\int -iu_{Y'}^{*}\hat{P}iu_X=i\left\langle X'|\hat{P}|S\right\rangle-\left\langle Y'|\hat{P}|S\right\rangle$$

From the above relations, for X', Y' and Z', we get

$$\left|X'\right\rangle=\cos\theta\cos\phi\left|X\right\rangle+\cos\theta\sin\phi\left|Y\right\rangle-\sin\theta\left|Z\right\rangle$$

Thus,

$$\left\langle X'|\hat{P}|S\right\rangle=\cos\theta\cos\phi\left\langle X|\hat{P}|S\right\rangle+\cos\theta\sin\phi\left\langle Y|\hat{P}|S\right\rangle-\sin\theta\left\langle Z|\hat{P}|S\right\rangle=\hat{P}\hat{r}_1$$

where $\hat{r}_1=\hat{i}\cos\theta\cos\phi+\hat{j}\cos\theta\sin\phi-\hat{k}\sin\theta$

$$\left|Y'\right\rangle=-\sin\phi\left|X\right\rangle+\cos\phi\left|Y\right\rangle+0\left|Z\right\rangle$$

Thus $\left\langle Y'|\hat{P}|S\right\rangle=-\sin\phi\left\langle X|\hat{P}|S\right\rangle+\cos\phi\left\langle Y|\hat{P}|S\right\rangle+0\left\langle Z|\hat{P}|S\right\rangle=\hat{P}\hat{r}_2$

where $\hat{r}_2=-\hat{i}\sin\phi+\hat{j}\cos\phi$

so that $\left\langle (X'-iY')|\hat{P}|S\right\rangle=\hat{P}\left(i\hat{r}_1-\hat{r}_2\right)$

Thus $$\frac{a_{k_-}b_{k_+}}{\sqrt{2}}\left\langle (X'-iY')|\hat{P}|S\right\rangle\left\langle \uparrow'|\uparrow'\right\rangle=\frac{a_{k_-}b_{k_+}}{\sqrt{2}}\hat{P}\left(i\hat{r}_1-\hat{r}_2\right)\left\langle \uparrow'|\uparrow'\right\rangle \tag{2.1e}$$

Now since,

$$\left\langle iS|\hat{P}|(X'+iY')\right\rangle=i\left\langle S|\hat{P}|X'\right\rangle-\left\langle S|\hat{P}|Y'\right\rangle=\hat{P}\left(i\hat{r}_1-\hat{r}_2\right)$$

We can write $$-\left[\frac{a_{k_+}b_{k_-}}{\sqrt{2}}\left\{\left\langle iS|\hat{P}|(X'+iY')\right\rangle\left\langle \downarrow'|\downarrow'\right\rangle\right\}\right]=-\left[\frac{a_{k_+}b_{k_-}}{\sqrt{2}}\hat{P}\left(i\hat{r}_1-\hat{r}_2\right)\left\langle \downarrow'|\downarrow'\right\rangle\right] \tag{2.1f}$$

Similarly, we get

$$|Z'\rangle = \sin\theta\cos\phi|X\rangle + \sin\theta\sin\phi|Y\rangle + \cos\theta|Z\rangle$$

So that, $\langle Z'|\hat{P}|iS\rangle = i\langle Z'|\hat{P}|S\rangle = i\hat{P}\{\sin\theta\cos\phi\hat{i} + \sin\theta\sin\phi\hat{j} + \cos\theta\hat{k}\} = i\hat{P}\hat{r}_3$

where $\hat{r}_3 = \hat{i}\sin\theta\cos\phi + \hat{j}\sin\theta\sin\phi + \hat{k}\cos\theta$

Thus, $c_{k_+}a_{k_-}\langle Z'|\hat{P}|iS\rangle\langle\downarrow'|\uparrow'\rangle = c_{k_+}a_{k_-}i\hat{P}\hat{r}_3\langle\downarrow'|\uparrow'\rangle$ (2.1g)

Similarly, we can write,

$$c_{k_-}a_{k_+}\langle iS|\hat{P}|Z'\rangle\langle\downarrow'|\uparrow'\rangle = c_{k_-}a_{k_+}i\hat{P}\hat{r}_3\langle\downarrow'|\uparrow'\rangle \quad (2.1h)$$

Therefore, we obtain

$$\frac{a_{k_-}b_{k_+}}{\sqrt{2}}\{\langle(X'-iY')|\hat{P}|S\rangle\langle\uparrow'|\uparrow'\rangle\}$$
$$-\frac{a_{k_+}b_{k_-}}{\sqrt{2}}\{\langle iS|\hat{P}|(X'+iY')\rangle\langle\downarrow'|\downarrow'\rangle\} = \frac{\hat{P}}{\sqrt{2}}\left(-a_{k_+}b_{k_-}\langle\downarrow'|\downarrow'\rangle + a_{k_-}b_{k_+}\langle\uparrow|\uparrow'\rangle\right)(i\hat{r}_1 - \hat{r}_2) \quad (2.1i)$$

Also, we can write,

$$c_{k_+}a_{k_-}\langle Z'|\hat{P}|iS\rangle\langle\downarrow'|\uparrow'\rangle + c_{k_-}a_{k_+}\langle iS|\hat{P}|Z'\rangle\langle\downarrow'|\uparrow'\rangle$$
$$= i\hat{P}\left(c_{k_+}a_{k_-} + c_{k_-}a_{k_+}\right)\hat{r}_3\left[\langle\downarrow'|\downarrow'\rangle\right] \quad (2.1j)$$

Combining the appropriate equations, we can further write

$$\hat{p}_{CV}(\vec{k}) = \frac{\hat{P}}{\sqrt{2}}(i\hat{r}_1 - \hat{r}_2)\{(b_{k_+}a_{k_-})\langle\uparrow'|\uparrow'\rangle - (b_{k_-}a_{k_+})\langle\downarrow'|\downarrow'\rangle\} + i\hat{P}\hat{r}_3\left(c_{k_+}a_{k_-} - c_{k_-}a_{k_+}\right)\langle\downarrow'|\uparrow'\rangle \quad (2.1k)$$

From the above relations, we obtain,

$$\left.\begin{array}{l}\uparrow' = e^{-i\phi/2}\cos(\theta/2)\uparrow + e^{i\phi/2}\sin(\theta/2)\downarrow \\ \downarrow' = -e^{-i\phi/2}\sin(\theta/2)\uparrow + e^{i\phi/2}\cos(\theta/2)\downarrow\end{array}\right\} \quad (2.1l)$$

Therefore,

$$\langle\downarrow'|\uparrow'\rangle_x = -\sin(\theta/2)\cos(\theta/2)\langle\uparrow|\uparrow\rangle_x + e^{-i\phi}\cos^2(\theta/2)\langle\downarrow|\uparrow\rangle_x$$
$$-e^{i\phi}\sin^2(\theta/2)\langle\uparrow|\downarrow\rangle_x + \sin(\theta/2)\cos(\theta/2)\langle\downarrow|\downarrow\rangle_x \tag{2.1m}$$

But we know from above that

$$\langle\uparrow|\uparrow\rangle_x = 0, \langle\uparrow|\downarrow\rangle = \frac{1}{2}, \langle\downarrow|\uparrow\rangle_x = \frac{1}{2} \text{ and } \langle\downarrow|\downarrow\rangle_x = 0$$

Thus, from (2.1m), we get

$$\langle\downarrow'|\uparrow'\rangle_x = \frac{1}{2}\left[e^{-i\phi}\cos^2(\theta/2) - e^{i\phi}\sin^2(\theta/2)\right]$$
$$= \frac{1}{2}\left[(\cos\phi - i\sin\phi)\cos^2(\theta/2) - (\cos\phi + i\sin\phi)\sin^2(\theta/2)\right] = \frac{1}{2}[\cos\phi\cos\theta - i\sin\phi] \tag{2.1n}$$

Similarly, we obtain

$$\langle\downarrow'|\uparrow'\rangle_y = \frac{1}{2}[i\cos\phi + \sin\phi\cos\theta] \text{ and } \langle\downarrow'|\uparrow'\rangle_z = \frac{1}{2}[-\sin\theta]$$

Therefore $\langle\downarrow'|\uparrow'\rangle = \hat{i}\langle\downarrow'|\uparrow'\rangle_x + \hat{j}\langle\downarrow'|\uparrow'\rangle_y + \hat{k}\langle\downarrow'|\uparrow'\rangle_z = \frac{1}{2}\{(\cos\theta\cos\phi - i\sin\phi)\hat{i} + (i\cos\phi + \sin\phi\cos\theta)\hat{j} - \sin\theta\hat{k}\}$

$$= \frac{1}{2}\left[\{(\cos\theta\cos\phi)\hat{i} + (\sin\phi\cos\theta)\hat{j} - \sin\theta\hat{k}\} + i\{-\hat{i}\sin\phi + \hat{j}\cos\phi\}\right] = \frac{1}{2}[\hat{r}_1 + i\hat{r}_2] = -\frac{1}{2}i[i\hat{r}_1 - \hat{r}_2]$$

Similarly, we can write

$$\langle\uparrow'|\uparrow'\rangle = \frac{1}{2}\left[\hat{i}\sin\theta\cos\phi + \hat{j}\sin\theta\sin\phi + \hat{k}\cos\theta\right] = \frac{1}{2}\hat{r}_3 \text{ and } \langle\downarrow'|\downarrow'\rangle = -\frac{1}{2}\hat{r}_3$$

Thus combining the above results we can write

$$\hat{p}_{CV}(\vec{k}) = \frac{\hat{P}}{\sqrt{2}}(i\hat{r}_1 - \hat{r}_2)\left\{(a_{k_-}b_{k_+})\langle\uparrow'|\uparrow'\rangle - (b_{k_-}a_{k_+})\langle\downarrow'|\downarrow'\rangle\right\} + i\hat{P}\hat{r}_3\left\{(c_{k_+}a_{k_-} - c_{k_-}a_{k_+})\langle\downarrow'|\uparrow'\rangle\right\}$$
$$= \frac{\hat{P}}{2}\hat{r}_3(i\hat{r}_1 - \hat{r}_2)\left\{\left(\frac{a_{k_-}b_{k_+}}{\sqrt{2}} + \frac{b_{k_-}a_{k_+}}{\sqrt{2}}\right)\right\} + \frac{\hat{P}}{2}\hat{r}_3(i\hat{r}_1 - \hat{r}_2)\left\{(c_{k_+}a_{k_-} + c_{k_-}a_{k_+})\right\}$$

Thus, $\hat{p}_{CV}\left(\vec{k}\right)=\frac{\hat{P}}{2}\hat{r}_3\left(i\hat{r}_1-\hat{r}_2\right)\left\{a_{k_+}\left(\frac{b_{k_-}}{\sqrt{2}}+c_{k_-}\right)+a_{k_-}\left(\frac{b_{k_+}}{\sqrt{2}}+c_{k_+}\right)\right\}$ (2.1o)

We can write that,

$|\hat{r}_1|=|\hat{r}_2|=|\hat{r}_3|=1$, also $\hat{P}\hat{r}_3=\hat{P}_x\sin\theta\cos\phi\hat{i}+\hat{P}_y\sin\theta\sin\phi\hat{j}+\hat{P}_z\cos\theta\hat{k}$,

$$\text{where } \hat{P}=\left\langle S|\hat{P}|X\right\rangle=\left\langle S|\hat{P}|Y\right\rangle=\left\langle S|\hat{P}|Z\right\rangle,$$
$$\left\langle S|\hat{P}|X\right\rangle=\int u_C^*(0,\vec{r})\hat{P}u_{VX}(0,\vec{r})d^3r=\hat{P}_{CVX}(0),\left\langle S|\hat{P}|Y\right\rangle=\hat{P}_{CVY}(0)\text{ and }\left\langle S|\hat{P}|Z\right\rangle=\hat{P}_{CVZ}(0),$$
$$\text{Thus } \hat{P}=\hat{P}_{CVX}(0)=\hat{P}_{CVY}(0)=\hat{P}_{CVZ}(0)=\hat{P}_{CV}(0)$$
$$\text{where } \hat{P}_{CV}(0)\equiv\int u_c^*(0,\vec{r})\hat{P}u_V(0,\vec{r})d^3r\equiv\hat{P}$$

For a plane polarized microwaves, we have the polarization vector $\vec{\varepsilon}_s=\hat{k}$, when the microwaves vector is traveling along the z-axis. Therefore, for a plane polarized light-wave, we have considered $\vec{\varepsilon}_s=\hat{k}$.

Then, from (2.1o),we get

$$\left(\vec{\varepsilon}\cdot\hat{p}_{CV}\left(\vec{k}\right)\right)=\vec{k}\cdot\frac{\hat{P}}{2}\hat{r}_3\left(i\hat{r}_1-\hat{r}_2\right)\left[A\left(\vec{k}\right)+B\left(\vec{k}\right)\right]\cos\omega t \tag{2.1p}$$

and

$$\left.\begin{aligned}A(\vec{k})&=a_{k_-}\left(\frac{b_{k_+}}{\sqrt{2}}+c_{k_+}\right)\\B(\vec{k})&=a_{k+}\left(\frac{b_{k_-}}{\sqrt{2}}+c_{k-}\right)\end{aligned}\right\} \tag{2.1q}$$

Thus,

$$\left|\vec{\varepsilon}\cdot\hat{p}_{cv}\left(\vec{k}\right)\right|^2=\left|\hat{k}\cdot\frac{\hat{P}}{2}\hat{r}_3\right|^2\left|i\hat{r}_1-\hat{r}_2\right|^2\left[A(\vec{k})+B(\vec{k})\right]^2\cos^2\omega t$$
$$=\frac{1}{4}\left|\hat{P}_z\cos\theta\right|^2\left[A\left(\vec{k}\right)+B\left(\vec{k}\right)\right]^2\cos^2\omega t \tag{2.1r}$$

So, the average value of $\left|\vec{\varepsilon}\cdot\hat{p}_{cv}\left(\vec{k}\right)\right|^2$ for a plane polarized light-wave is given by

$$\left\langle \left|\vec{\varepsilon}\cdot\hat{p}_{cv}\left(\vec{k}\right)\right|^2\right\rangle_{av} = \frac{2}{4}\left|\hat{P}_z\right|^2\left[A(\vec{k})+B(\vec{k})\right]^2\left(\int_0^{2\pi}d\phi\int_0^{\pi}\cos^2\theta\sin\theta d\theta\right)\left(\frac{1}{2}\right) = \frac{2\pi}{3}\left|\hat{P}_z\right|^2\left[A\left(\vec{k}\right)+B\left(\vec{k}\right)\right]^2$$

(2.1s)

where $\left|\hat{P}_z\right|^2 = \left(\frac{1}{2}\right)\left|\vec{k}\bullet\hat{p}_{cv}\left(0\right)\right|^2$ and

$$\left|\vec{k}\bullet\hat{p}_{cv}\left(0\right)\right|^2 = \frac{m^2}{4m_r}\frac{E_{g_0}\left(E_{g_0}+\Delta\right)}{\left(E_{g_0}+\frac{2}{3}\Delta\right)} \quad (2.1t)$$

We can express $A(\vec{k})$ and $B\left(\vec{k}\right)$ in terms of constants of the energy spectra in the following way:

Substituting $a_{k_\pm}$, $b_{k_\pm}$, $c_{k_\pm}$ and $\gamma_{0k_\pm}$ in $A(\vec{k})$ and $B\left(\vec{k}\right)$ in (8.1q) we get

$$A\left(\vec{k}\right) = \beta\left(t+\frac{\rho}{\sqrt{2}}\right)\left\{\left(\frac{E_{g_0}}{E_{g_0}+\delta'}\right)\gamma_{0k_+}^2 - \gamma_{0k_+}^2\gamma_{0k_-}^2\left(\frac{E_{g_0}-\delta'}{E_{g_0}+\delta'}\right)\right\}^{1/2} \quad (2.1u)$$

$$B\left(\vec{k}\right) = \beta\left(t+\frac{\rho}{\sqrt{2}}\right)\left\{\left(\frac{E_{g_0}}{E_{g_0}+\delta'}\right)\gamma_{0k_-}^2 - \gamma_{0k_+}^2\gamma_{0k_-}^2\left(\frac{E_{g_0}-\delta'}{E_{g_0}+\delta'}\right)\right\}^{1/2}$$

(2.1v)

in which,

$$\gamma_{0k_+}^2 \equiv \frac{\xi_{1k}-E_{g_0}}{2(\xi_{1k}+\delta')} \equiv \frac{1}{2}\left[1-\left(\frac{E_{g_0}+\delta'}{\xi_{1k}+\delta'}\right)\right] \; and \; \gamma_{0k_-}^2 \equiv \frac{\xi_{1k}+E_{g_0}}{2(\xi_{1k}+\delta')} \equiv \frac{1}{2}\left[1+\left(\frac{E_{g_0}-\delta'}{\xi_{1k}+\delta'}\right)\right]$$

Substituting $x \equiv \xi_{1k}+\delta'$ in $\gamma_{0k_\pm}^2$, we can write,

$$A\left(\vec{k}\right) = \beta\left(t+\frac{\rho}{\sqrt{2}}\right)\left\{\left(\frac{E_{g_0}}{E_{g_0}+\delta'}\right)\frac{1}{2}\left(1-\frac{E_{g_0}+\delta'}{x}\right) - \frac{1}{4}\left(\frac{E_{g_0}-\delta'}{E_{g_0}+\delta'}\right)\left(1-\frac{E_{g_0}+\delta'}{x}\right)\left(1+\frac{E_{g_0}-\delta'}{x}\right)\right\}^{1/2}$$

$$Thus\; A\left(\vec{k}\right) = \frac{\beta}{2}\left(t+\frac{\rho}{\sqrt{2}}\right)\left\{1-\frac{2a_0}{x}+\frac{a_1}{x^2}\right\}^{1/2}$$

where, $a_0 \equiv \left(E_{g_0}^2+\delta'^2\right)\left(E_{g_0}+\delta'\right)^{-1}$ and $a_1 \equiv \left(E_{g_0}-\delta'\right)^2$ and.

After tedious algebra, one can show that

$$A(\vec{k})=\frac{\beta}{2}\left(t+\frac{\rho}{\sqrt{2}}\right)(E_{g_0}-\delta')\left[\frac{1}{\xi_{1k}+\delta'}-\frac{1}{E_{g_0}+\delta'}\right]^{1/2}\left[\frac{1}{\xi_{1k}+\delta'}-\frac{(E_{g_0}+\delta')}{(E_{g_0}-\delta')^2}\right]^{1/2} \tag{2.1w}$$

Similarly, from (2.1v) can write,

$$B(\vec{k})=\beta\left(t+\frac{\rho}{\sqrt{2}}\right)\left\{\left(\frac{E_{g_0}}{E_{g_0}+\delta'}\right)\frac{1}{2}\left(1+\frac{E_{g_0}-\delta'}{x}\right)-\frac{1}{4}\left(\frac{E_{g_0}-\delta'}{E_{g_0}+\delta'}\right)\left(1-\frac{E_{g_0}+\delta'}{x}\right)\left(1+\frac{E_{g_0}-\delta'}{x}\right)\right\}^{1/2}$$

So that, finally we get,

$$B(\vec{k})=\frac{\beta}{2}\left(t+\frac{\rho}{\sqrt{2}}\right)\left(1+\frac{E_{g_0}-\delta'}{\xi_{1k}+\delta'}\right) \tag{2.1x}$$

In the presence of microwaves, the Hamiltonian $(\hat{H})$ of an electron characterized by the vector potential $\vec{A}$ can be written following [14] as

$$\hat{H}=\left[\left|\left(\hat{p}+|e|\vec{A}\right)\right|^2/2m\right]+V(\vec{r}) \tag{2.1y}$$

in which, $\hat{p}$ is the momentum operator, $V(\vec{r})$ is the crystal potential and m is the free electron mass.

Equation (2.1y) can be expressed as

$$\hat{H}=\hat{H}_0+\hat{H}' \tag{2.1A}$$

where, $\hat{H}_0=\frac{\hat{p}^2}{2m}+V(\vec{r})$

and $\hat{H}'=\frac{|e|}{2m}\vec{A}\cdot\hat{p}$ (2.1B)

The perturbed Hamiltonian $\hat{H}'$ can be written as

$$\hat{H}' = \left(\frac{-i\hbar|e|}{2m}\right)\left(\vec{A}\cdot\nabla\right) \tag{2.1C}$$

where, $i = \sqrt{-1}$ and $\hat{p} = -i\hbar\nabla$

The vector potential ($\vec{A}$) of the monochromatic light of plane wave can be expressed as

$$\vec{A} = A_0\vec{\varepsilon}_s \cos(\vec{s}_0 \cdot \vec{r} - \omega t) \tag{2.1D}$$

where A_0 is the amplitude of the microwaves, $\vec{\varepsilon}_s$ is the polarization vector, $\vec{s}_0$ is the momentum vector of the incident photon, $\vec{r}$ is the position vector, ω is the angular frequency of microwaves and t is the time scale. The matrix element of $\hat{H}'_{nl}$ between initial state, $\psi_l(\vec{q},\vec{r})$ and final state $\psi_n(\vec{k},\vec{r})$ in different bands can be written as

$$\hat{H}'_{nl} = \frac{|e|}{2m}\left\langle n\vec{k}\left|\vec{A}\bullet\hat{p}\right|l\vec{q}\right\rangle \tag{2.1E}$$

Using (2.1C) and (2.1D), we can re-write (2.1E) as

$$\hat{H}'_{nl} = \left(\frac{-i\hbar|e|A_0}{4m}\right)\vec{\varepsilon}_s\bullet\left[\left\{\left\langle n\vec{k}\left|e^{(i\vec{s}_0\cdot\vec{r})}\nabla\right|l\vec{q}\right\rangle e^{-i\omega t}\right\} + \left\{\left\langle n\vec{k}\left|e^{(-i\vec{s}_0\cdot\vec{r})}\nabla\right|l\vec{q}\right\rangle e^{i\omega t}\right\}\right] \tag{2.1F}$$

The first matrix element of (2.1F) can be written as

$$\left\langle n\vec{k}\left|e^{(i\vec{s}_0\cdot\vec{r})}\nabla\right|l\vec{q}\right\rangle$$

$$= \int e^{\left(i\left[\vec{q}+\vec{s}_0-\vec{k}\right]\cdot\vec{r}\right)} i\vec{q}u_n^*\left(\vec{k},\vec{r}\right)u_l\left(\vec{q},\vec{r}\right)d^3r + \int e^{\left(i\left[\vec{q}+\vec{s}_0-\vec{k}\right]\cdot\vec{r}\right)} u_n^*\left(\vec{k},\vec{r}\right)\nabla u_l\left(\vec{q},\vec{r}\right)d^3r \tag{2.1G}$$

The functions $u_n^* u_l$ and $u_n^*\nabla u_l$ are periodic. The integral over all space can be separated into a sum over unit cells times an integral over a single unit cell. It is assumed that the wave length of the electromagnetic wave is sufficiently large so that if $\vec{k}$ and $\vec{q}$ are within the Brillouin zone, $\left(\vec{q}+\vec{s}_0-\vec{k}\right)$ is not a reciprocal lattice vector.

Therefore, we can write equation (2.1G) as

$$<n\vec{k}\left|e^{(i\vec{s}_0\cdot\vec{r})}\nabla\right|l\vec{q}>=\left[\frac{(2\pi)^3}{\Omega}\right]\left\{i\vec{q}\delta\left(\vec{q}+\vec{s}_0-\vec{k}\right)\delta_{nl}+\delta\left(\vec{q}+\vec{s}_0-\vec{k}\right)\int_{cell}u_n^*\left(\vec{k},\vec{r}\right)\nabla u_l\left(\vec{q},\vec{r}\right)d^3r\right\}$$

$$=\left[\frac{(2\pi)^3}{\Omega}\right]\left\{\delta\left(\vec{q}+\vec{s}_0-\vec{k}\right)\int_{cell}u_n^*\left(\vec{k},\vec{r}\right)\nabla u_l\left(\vec{q},\vec{r}\right)d^3r\right\} \tag{2.1H}$$

where, Ω is the volume of the unit cell and $\int u_n^*(\vec{k},\vec{r})u_l(\vec{q},\vec{r})d^3r=\delta(\vec{q}-\vec{k})\delta_{nl}=0$, since $n\neq l$

The delta function expresses the conservation of wave vector in the absorption of microwaves and $\vec{s}_0$ is small compared to the dimension of a typical Brillouin zone and we set $\vec{q}=\vec{k}$.

From equations (2.1G) and (2.1H), we can write,

$$\hat{H}'_{nl}=\frac{|e|A_0}{2m}\vec{\varepsilon}_s\bullet\hat{p}_{nl}(\vec{k})\delta(\vec{q}-\vec{k})\cos(\omega t) \tag{2.1I}$$

where, $\hat{p}_{nl}(\vec{k})=-i\hbar\int u_n^*\nabla u_l d^3r=\int u_n^*(\vec{k},\vec{r})\hat{p}u_l(\vec{k},\vec{r})d^3r$

Therefore, we can write

$$\hat{H}'_{nl}=\frac{|e|A_0}{2m}\vec{\varepsilon}\bullet\hat{p}_{nl}(\vec{k}) \tag{2.1J}$$

where, $\vec{\varepsilon}=\vec{\varepsilon}_s\cos\omega t$.

When a photon interacts with a semiconductor, the carriers (i.e., electrons) are generated in the bands which are followed by the interband transitions. For example, when the carriers are generated in the valence band, the carriers then make interband transition to the conduction band. The transition of the electrons within the same band i.e., $\hat{H}'_{nn}=\left\langle n\vec{k}\,|\,\hat{H}'\,|\,n\vec{k}\right\rangle$ is neglected. Because, in such a case, i.e., when the carriers are generated within the same bands by photons, are lost by recombination within the aforementioned band resulting zero carriers.

Therefore, $<n\vec{k}\left|\hat{H}'\right|n\vec{k}>=0$ (2.1K)

With $n=c$ stands for conduction band and $l=v$ stand for valance band, the energy equation for the conduction electron can approximately be written as

$$\gamma(E)=\left(\frac{\hbar^2k^2}{2m^*}\right)+\frac{\left(\frac{|e|A_0}{2m}\right)^2\left\langle\left|\vec{\varepsilon}\bullet\hat{p}_{cv}(\vec{k})\right|^2\right\rangle_{av}}{E_c(\vec{k})-E_v(\vec{k})} \tag{2.1L}$$

where, $\gamma(E)\equiv E(aE+1)(bE+1)/(cE+1)$, $a\equiv 1/E_{g_0}$, E_{g_0} is the un-perturbed band-gap, $b\equiv 1/\left(E_{g_0}+\Delta\right)$, $c\equiv 1/\left(E_{g_0}+2\Delta/3\right)$, and $\left\langle\left|\vec{\varepsilon}\bullet\hat{p}_{cv}(\vec{k})\right|^2\right\rangle_{av}$ represents the average of the square of the optical matrix element (OME).

For the three-band model of Kane, we can write,

$$\xi_{1k}=E_c(\vec{k})-E_v(\vec{k})=(E_{g_0}^2+E_{g_0}\hbar^2k^2/m_r)^{1/2} \tag{2.1M}$$

where, m_r is the reduced mass and is given by $m_r^{-1}=\left(m^*\right)^{-1}+m_v^{-1}$, and m_v is the effective mass of the heavy hole at the top of the valance band in the absence of any field.

Thus, combining the appropriate equations we can write

$$\left(\frac{|e|A_0}{2m}\right)^2\frac{\left\langle\left|\vec{\varepsilon}\bullet\hat{p}_{cv}(\vec{k})\right|^2\right\rangle_{av}}{E_c(\vec{k})-E_v(\vec{k})}=\left(\frac{|e|A_0}{2m}\right)^2\frac{2\pi}{3}\left|\vec{k}\bullet\hat{p}_{cv}(0)\right|^2\frac{\beta^2}{4}\left(t+\frac{\rho}{\sqrt{2}}\right)^2$$
$$\frac{1}{\xi_{1k}}\left\{\left(1+\frac{E_{g_0}-\delta'}{\xi_{1k}+\delta'}\right)+(E_{g_0}-\delta')\left[\frac{1}{\xi_{1k}+\delta'}-\frac{1}{E_{g_0}+\delta'}\right]\right]^{1/2}\left[\frac{1}{\xi_{1k}+\delta'}-\frac{E_{g_0}+\delta'}{\left(E_{g_0}-\delta'\right)^2}\right]^{1/2}\right\}^2 \tag{2.1P}$$

It is well-known that [14]

$$A_0^2=\frac{I\lambda^2}{2\pi^2c^3\sqrt{\varepsilon_{sc}\varepsilon_0}} \tag{2.1Q}$$

where, I is the intensity of microwave of wavelength λ, ε_0 is the permittivity of free space and c is the velocity of light. Thus, the simplified electron energy spectrum in III-V, ternary and quaternary materials up to the second order in the presence of microwaves can approximately be written as

$$\frac{\hbar^2k^2}{2m^*}=\beta_0\left(E,\lambda\right) \tag{2.2}$$

where,

$$\beta_0(E,\lambda) \equiv \left[\gamma(E) - \theta_0(E,\lambda)\right],$$

$$\theta_0(E,\lambda) \equiv \frac{|e|^2}{96 m_r \pi c^3} \frac{I\lambda^2}{\sqrt{\varepsilon_{sc}\varepsilon_0}} \frac{E_{g_0}\left(E_{g_0}+\Delta\right)}{\left(E_{g_0}+\frac{2}{3}\Delta\right)} \frac{\beta^2}{4}\left(t+\frac{\rho}{\sqrt{2}}\right)^2 \frac{1}{\phi_0(E)}$$

$$\left\{\left(1+\frac{E_{g_0}-\delta'}{\phi_0(E)+\delta'}\right)+(E_{g_0}-\delta')\left[\frac{1}{\phi_0(E)+\delta'}-\frac{1}{E_{g_0}+\delta'}\right]^{1/2}\left[\frac{1}{\phi_0(E)+\delta'}-\frac{E_{g_0}+\delta'}{\left(E_{g_0}-\delta'\right)^2}\right]^{1/2}\right\}^2$$

$$\phi_0(E) \equiv E_{g_0}\left(1+2\left(1+\frac{m^*}{m_v}\right)\frac{\gamma(E)}{E_{g_0}}\right)^{1/2}$$

and

1) For the two-band model of Kane, we have $\Delta \to 0$. Under this condition, $\gamma(E) \to E(1+aE) = \frac{\hbar^2 k^2}{2m^*}$. Since, $\beta \to 1$, $t \to 1$, $\rho \to 0$, $\delta' \to 0$ for $\Delta \to 0$, from (2.2), we can write the energy spectrum of III-V, ternary and quaternary materials in the presence of external microwaves whose unperturbed conduction electrons obey the two band model of Kane as

$$\frac{\hbar^2 k^2}{2m^*} = \tau_0(E,\lambda) \tag{2.3}$$

$$\textit{where}\ \ \tau_0(E,\lambda) \equiv E(1+aE) - B_0(E,\lambda),$$

$$B_0(E,\lambda) \equiv \frac{|e|^2 I\lambda^2 E_{g_0}}{384\pi c^3 m_r \sqrt{\varepsilon_{sc}\varepsilon_0}} \frac{1}{\phi_1(E)}\left\{\left(1+\frac{E_{g_0}}{\phi_1(E)}\right)+E_{g_0}\left[\frac{1}{\phi_1(E)}-\frac{1}{E_{g_0}}\right]\right\}^2,$$

$$\phi_1(E) \equiv E_{g_0}\left\{1+\frac{2m^*}{m_r}aE(1+aE)\right\}^{1/2}.$$

For relatively wide band gap semiconductors, one can write, $a \to 0$, $b \to 0$, $c \to 0$

and

$\gamma(E) \to E$.

Thus, from (2.2), we get,

$$\frac{\hbar^2 k^2}{2m^*} = \rho_0(E,\lambda)$$

$$\rho_0(E,\lambda) \equiv E - \frac{|e|^2 I\lambda^2}{96\pi c^3 m_r \sqrt{\varepsilon_{sc}\varepsilon_0}}\left[1+\left(\frac{2m^*}{m_r}\right)aE\right]^{-3/2} \tag{2.4}$$

The dispersion relations of the electrons in QDs optoelectronic materials in the presence of microwaves can respectively be expressed from (2.2), (2.3) and (2.4) as

$$\frac{2m^*\beta_0(E_{Q1},\lambda)}{\hbar^2} = H_{71}(n_{x71}, n_{y71}, n_{z71}) \tag{2.5}$$

$$\frac{2m^*\tau_0(E_{Q2},\lambda)}{\hbar^2} = H_{72}(n_{x72}, n_{y72}, n_{z72}) \tag{2.6}$$

$$\frac{2m^*\rho_0(E_{Q3},\lambda)}{\hbar^2} = H_{73}(n_{x73}, n_{y73}, n_{z73}) \tag{2.7}$$

where E_{Qi} is the totally quantized energy and $H_{7i}(n_{x7i}, n_{y7i}, n_{z7i}) = \left(\frac{\pi n_{x7i}}{d_x}\right)^2 + \left(\frac{\pi n_{y7i}}{d_y}\right)^2 + \left(\frac{\pi n_{z7i}}{d_z}\right)^2$

The electron concentration can, in general, be written as

$$n_{0DL} = \left(\frac{2g_v}{d_x d_y d_z}\right)\sum_{n_{x7i}=1}^{n_{x7i_{\max}}}\sum_{n_{y7i}=1}^{n_{y7i_{\max}}}\sum_{n_{z7i}=1}^{n_{z7i_{\max}}} F_{-1}(\eta_{7i0D}) \tag{2.8}$$

where $\eta_{7i0D} = \frac{E_{F0DL} - E_{\theta i}}{k_B T}$ and E_{F0DL} is the Fermi energy in QDs in the presence of microwaves as measured from the edge of the conduction band in the vertically upward direction in the absence of any quantization,

The photo-emitted current densities in this case are given by the following equations

$$J_{0DL} = \frac{(\alpha_0 e g_v)}{d_x d_y d_z}\left(\frac{m^*}{2}\right)^{-1/2}\sum_{n_{x71}=1}^{n_{x71\max}}\sum_{n_{y71}=1}^{n_{y71\max}}\sum_{n_{z71\min}}^{n_{z71\max}}\left[\frac{\sqrt{\beta_0(E_{n_{z71}},\lambda)}}{\beta_0'(E_{n_{z71}},\lambda)}\right]F_{-1}(\eta_{710D}) \tag{2.9}$$

$$J_{0DL}=\frac{(\alpha_0 e g_v)}{d_x d_y d_z}\left(\frac{m^*}{2}\right)^{-1/2}\sum_{n_{x72}=1}^{n_{x72\max}}\sum_{n_{y72}=1}^{n_{y72\max}}\sum_{n_{z72\min}}^{n_{z72\max}}\left[\frac{\sqrt{\tau_0\left(E_{n_{z72}},\lambda\right)}}{\tau_0'\left(E_{n_{z72}},\lambda\right)}\right]F_{-1}(\eta_{720D}) \tag{2.10}$$

and $$J_{0DL}=\frac{(\alpha_0 e g_v)}{d_x d_y d_z}\left(\frac{m^*}{2}\right)^{-1/2}\sum_{n_{x73}=1}^{n_{x73\max}}\sum_{n_{y73}=1}^{n_{y73\max}}\sum_{n_{z73\min}}^{n_{z73\max}}\left[\frac{\sqrt{\rho_0\left(E_{n_{z73}},\lambda\right)}}{\rho_0'\left(E_{n_{z73}},\lambda\right)}\right]F_{-1}(\eta_{730D}) \tag{2.11}$$

2.2.2. Photoemission from Quantum dot Effective Mass Superlattices

The electron energy spectrum in this case can be expressed as

$$k_x^2=\left[\frac{1}{L_0^2}\left[\cos^{-1} f_{30}\left(E,k_y,k_z\right)\right]^2-k_\perp^2\right] \tag{2.12}$$

where

$$f_{30}\left(E,k_y,k_z\right)=\left[a_1\cos\left[a_0 g_{30}\left(E,k_\perp\right)+b_0 h_{30}\left(E,k_\perp\right)\right]-a_2\cos\left[a_0 g_{30}\left(E,k_\perp\right)-b_0 h_{30}\left(E,k_\perp\right)\right]\right]$$

$$g_{30}\left(E,k_\perp\right)=\left[\frac{2m_1^*}{\hbar^2}\beta_0\left(E,\lambda,E_{g_{01}},\Delta_1\right)-k_\perp{}^2\right]^{1/2},$$

$$\beta_0\left(E,\lambda,E_{g_{0i}},\Delta_i\right)=\left[\gamma_i\left(E,E_{g_{0i}},\Delta_i\right)-\theta_i\left(E,E_{g_{0i}},\Delta_i\right)\right],$$

$$i=1,2\quad \gamma_i\left(E,E_{g_{0i}},\Delta_i\right)=\frac{E\left(E+E_{g_{0i}}\right)\left(E+E_{g_{0i}}+\Delta_i\right)\left(E_{g_{0i}}+\frac{2}{3}\Delta_i\right)}{E_{g_{0i}}\left(E_{g_{0i}}+\Delta_i\right)\left(E+E_{g_{0i}}+\frac{2}{3}\Delta_i\right)},$$

$$\theta_i\left(E,E_{g_{0i}},\Delta_i\right)=\left[\frac{|e|^2}{96 m_{ri}\pi c^3}\cdot\frac{I_0\lambda^2.E_{g_{0i}}\left(E_{g_{0i}}+\Delta_i\right)}{\sqrt{\varepsilon_0\varepsilon_{sci}}\left(E_{g_{0i}}+\frac{2}{3}\Delta_i\right)}\times\right.$$

$$\left.\frac{\beta_i^2}{4}\left(t_i+\frac{\rho_i}{\sqrt{2}}\right)^2\frac{1}{\phi_{0i}(E)}\right]\left\{\left(1+\frac{\left(E_{g_{0i}}-\delta_i'\right)}{\left(\phi_{0i}(E)+\delta_i'\right)}\right)+\left(E_{g_{0i}}-\delta_i'\right)\left[\frac{1}{\phi_{0i}(E)+\delta_i'}-\frac{1}{E_{g_{0i}}+\delta_i'}\right]^{1/2}.\right.$$

$$\left.\left.\cdot\left[\frac{1}{\phi_{0i}(E)+\delta_i'}-\frac{E_{g_{0i}}+\delta_i'}{\left(E_{g_{0i}}-\delta_i'\right)^2}\right]^{1/2}\right\}^2\right\},$$

$$m_{ri}=\left[\left(\frac{1}{m_i^*}\right)+\left(\frac{1}{m_{\upsilon_i}}\right)\right]^{-1},\ \beta_i=\left[\frac{6\left(E_{g_{0i}}+\frac{2}{3}\Delta_i\right)\left(E_{g_{0i}}+\Delta_i\right)}{\chi_i}\right]^{1/2},$$

$$\chi_i=\left(6E_{g_{0i}}^{\ 2}+9E_{g_{0i}}\Delta_i+4\Delta_i^{\ 2}\right),t_i=\left[\frac{6\left(E_{g_{0i}}+\frac{2}{3}\Delta_i\right)^2}{\chi_i}\right]^{1/2},$$

$$\rho_i\equiv\left(\frac{4\Delta_i^2}{3\chi_i}\right)^{1/2},\phi_{0i}(E)=E_{g_{0i}}\left[1+2\left(1+\frac{m_i^*}{m_{\upsilon_i}}\right)\frac{\gamma_0\left(E,E_{g_{0i}},\Delta_i\right)}{E_{g_{0i}}}\right]^{1/2},\delta_i'=\left[\frac{E_{g_{0i}}^{\ 2}\Delta_i}{\chi_i}\right]$$

and $h_{30}(E,k_\perp)=\left[\frac{2m_2^*}{\hbar^2}\beta_0\left(E,\lambda,E_{g_{02}},\Delta_2\right)-k_\perp^2\right]^{1/2}$

When the unperturbed bulk dispersion law of the constituent materials is defined by the two band model of Kane, (2.12) remains as it is only $\beta_0\left(E,\lambda,E_{g_{0i}},\Delta_i\right)$ is being replaced by

$$\tau_0\left(E,\lambda,E_{g_{0i}}\right)\text{where}\ \tau_0\left(E,\lambda,E_{g_{0i}}\right)=E\left(1+\alpha_{0i}E\right)-\left[\frac{|e|^2}{384m_{ri}\pi c^3}\cdot\frac{I_0\lambda^2.E_{g_{0i}}}{\sqrt{\varepsilon_0\varepsilon_{sci}}}\right.$$

$$\left.\left[\frac{1}{U_i(E)}\left\{\left(1+\frac{E_{g_{0i}}}{U_i(E)}\right)+E_{g_{0i}}\left[\frac{1}{U_i(E)}-\frac{1}{E_{g_{0i}}}\right]\right\}^2\right],\ \alpha_{0i}=\frac{1}{E_{g_{0i}}}$$

and

$$U_i(E)=E_{g_{0i}}\left[1+\left(\frac{2m_i^*}{m_r}\right)\alpha_{0i}E\left[1+\alpha_{0i}E\right]\right]^{1/2}.$$

When the unperturbed bulk dispersion law of the constituent materials is defined by parabolic energy bands, the (2.12) remains as it is only $\beta_0\left(E,\lambda,E_{g_{0i}},\Delta_i\right)$ should be replaced by $\rho_0\left(E,\lambda,E_{g_{0i}}\right)$ where

$$\rho_0\left(E,\lambda,E_{g_{0i}}\right)=E-\left[\frac{|e|^2 I_0\lambda^2}{96m_{ri}\pi c^3\sqrt{\varepsilon_0\varepsilon_{sci}}}\right]\left[1+\left(\frac{2m_i^*}{m_{ri}}\right)\alpha_{0i}E\right]^{-3/2}$$

The electron energy spectrum in quantum dot effective mass super lattices this case is given by

$$\left(\frac{\pi n_x}{d_x}\right)^2=\left[\frac{1}{L_0^2}\left[\cos^{-1} f_{31}\left(E,n_y,n_z\right)\Big|_{E=E_{32}}\right]^2-\left(\frac{\pi n_y}{d_y}\right)^2-\left(\frac{\pi n_z}{d_z}\right)^2\right] \tag{2.13}$$

where E_{32} is the totally quantized electron energy in this case and

$$f_{31}\left(E,n_y,n_z\right)=\left[a_1\cos\left[a_0 g_{31}\left(E,n_y,n_z\right)+b_0 h_{31}\left(E,n_y,n_z\right)\right]\right.$$
$$\left.-a_2\cos\left[a_0 g_{31}\left(E,n_y,n_z\right)-b_0 h_{31}\left(E,n_y,n_z\right)\right]\right]$$

$$g_{31}\left(E,n_y,n_z\right)=\left[\frac{2m_1^*}{\hbar^2}\beta_0\left(E,\lambda,E_{g_{01}},\Delta_1\right)-\left[\left(\frac{\pi n_y}{d_y}\right)^2+\left(\frac{\pi n_z}{d_z}\right)^2\right]\right]^{1/2}$$

and $$h_{31}\left(E,n_y,n_z\right)=\left[\frac{2m_2^*}{\hbar^2}\beta_0\left(E,\lambda,E_{g_{02}},\Delta_2\right)-\left[\left(\frac{\pi n_y}{d_y}\right)^2+\left(\frac{\pi n_z}{d_z}\right)^2\right]\right]^{1/2}$$

The electron concentration is given by

$$N_{0L}=\frac{2g_v}{d_x d_y d_z}\sum_{n_x=1}^{n_{x_{\max}}}\sum_{n_y=1}^{n_{y_{\max}}}\sum_{n_z=1}^{n_{z_{\max}}}F_{-1}\left(\eta_{32}\right) \tag{2.14}$$

where $\eta_{32}=\dfrac{E_{FQDSLEM}-E_{32}}{k_B T}$, in which $E_{FQDSLEM}$ is the Fermi energy in the present case.

The photo-emitted current density is given by

$$J_{1L}=\left(\frac{2L_0 e g_v\alpha_0}{d_x d_y d_z\hbar}\right)\sum_{n_{x_{\min}}}^{n_{x_{\max}}}\sum_{n_y=1}^{n_{y_{\max}}}\sum_{n_z=1}^{n_{z_{\max}}}L_{30}\left(E_{n_x},0\right)F_{-1}\left(\eta_{32}\right) \tag{2.16}$$

where $L_{30}\left(E_{n_x},0\right)=I_{30}\left(E_{n_x},0\right)\left[1-f_{30}^{2}\left(E_{n_x},0\right)\right]^{-1/2}$,

$$I_{30}\left(E_{n_x},0\right)=\left[a_2\left[a_0 g_{30}'\left(E_{n_x},0\right)-b_0 h_{30}'\left(E_{n_x},0\right)\right]\right.$$
$$\left[\sin\left[a_0 g_{30}\left(E_{n_x},0\right)-b_0 h_{30}\left(E_{n_x},0\right)\right]\right]$$
$$\left.-a_1\left[a_0 g_{30}'\left(E_{n_x},0\right)+b_0 h_{30}'\left(E_{n_x},0\right)\right]\left[\sin\left[a_0 g_{30}\left(E_{n_x},0\right)+b_0 h_{30}\left(E_{n_x},0\right)\right]\right]\right],$$

$$g_{30}'\left(E_{n_x},0\right)=\left[g_{30}\left(E_{n_x},0\right)\right]^{-1}\left[\left(\frac{m_1^*}{\hbar^2}\right)\beta_0'\left(E_{n_x},\lambda,E_{g_{01}},\Delta_1\right)\right],$$

$$g_{30}\left(E_{n_x},0\right)=\left[\frac{2m_1^*}{\hbar^2}\beta_0\left(E_{n_x},\lambda,E_{g_{01}},\Delta_1\right)\right]^{1/2},$$

$$\beta_0'\left(E_{n_x},\lambda,E_{g_{0i}},\Delta_i\right)=\beta_0'\left(E,\lambda,E_{g_{0i}},\Delta_i\right)\Big|_{E=E_{n_x}},$$

$$\beta_0'\left(E,\lambda,E_{g_{0i}},\Delta_i\right)=\left[\gamma_i\left(E,E_{g_{0i}},\Delta_i\right)\right.$$

$$\left[\frac{1}{E}+\frac{1}{E+E_{g_{0i}}}+\frac{1}{E+E_{g_{0i}}+\Delta_i}-\frac{1}{E+E_{g_{0i}}+\frac{2}{3}\Delta_i}\right]+\theta_i\left(E,E_{g_{0i}},\Delta_i\right)$$

$$\left.\left[-\frac{\phi_{0i}'(E)}{\phi_{0i}(E)}+\frac{2\psi_i'\left(E,E_{g_{0i}},\Delta_i\right)}{\psi_i\left(E,E_{g_{0i}},\Delta_i\right)}\right]\right],\ \psi_i'\left(E,E_{g_{0i}},\Delta_i\right)=\frac{\phi_{0i}'(E)\left(\delta_i'-E_{g_{0i}}\right)}{2\left(\phi_{0i}(E)+\delta_i'\right)^2}$$

$$\left[2+\left[\left[\frac{1}{\phi_{0i}(E)+\delta_i'}-\frac{1}{E_{g_{0i}}+\delta_i'}\right]^{-1/2}\times\left[\frac{1}{\left(\phi_{0i}(E)+\delta_i'\right)}-\frac{\left(E_{g_{0i}}+\delta_i'\right)}{\left(E_{g_{0i}}-\delta_i'\right)^2}\right]^{1/2}\right.\right.$$

$$\left.\left.+\left[\frac{1}{\left(\phi_{0i}(E)+\delta_i'\right)}-\frac{\left(E_{g_{0i}}+\delta_i'\right)}{\left(E_{g_{0i}}-\delta_i'\right)^2}\right]^{-1/2}\left[\frac{1}{\phi_{0i}(E)+\delta_i'}-\frac{1}{\left(E_{g_{0i}}+\delta_i'\right)}\right]^{1/2}\right]\right],$$

$$\psi_i\left(E,E_{g_{0i}},\Delta_i\right)=\left\{\left(1+\frac{\left(E_{g_{0i}}-\delta_i'\right)}{\left(\phi_{0i}(E)+\delta_i'\right)}\right)+\left(E_{g_{0i}}-\delta_i'\right)\left[\frac{1}{\phi_{0i}(E)+\delta_i'}-\frac{1}{E_{g_{0i}}+\delta_i'}\right]^{1/2}\right.$$

$$\left.\left[\frac{1}{\phi_{0i}(E)+\delta_i'}-\frac{\left(E_{g_{0i}}+\delta_i'\right)}{\left(E_{g_{0i}}-\delta_i'\right)^2}\right]^{1/2}\right\},$$

$$h'_{30}\left(E_{n_x},0\right)=\left[h_{30}\left(E_{n_x},0\right)\right]^{-1}\left[\left(\frac{m_2^*}{\hbar^2}\right)\beta_0'\left(E_{n_x},\lambda,E_{g_{02}},\Delta_2\right)\right],$$

$$h_{30}\left(E_{n_x},0\right)=\left[\frac{2m_2^*}{\hbar^2}\beta_0\left(E_{n_x},\lambda,E_{g_{02}},\Delta_2\right)\right]^{1/2},$$

$$f_{30}\left(E,0\right)=\left[a_1\cos\left[a_0g_{30}\left(E,0\right)+b_0h_{30}\left(E,0\right)\right]-a_2\cos\left[a_0g_{30}\left(E,0\right)-b_0h_{30}\left(E,0\right)\right]\right]$$

$$g_{30}{}^2\left(E,0\right)=\frac{2m_1^*}{\hbar^2}\beta_0\left(E,\lambda,E_{g_{01}},\Delta_1\right)\text{and } h_{30}{}^2\left(E,0\right)=\frac{2m_2^*}{\hbar^2}\beta_0\left(E,\lambda,E_{g_{02}},\Delta_2\right)$$

$$\text{and } n_{x_{\min}}\geq\left(\frac{d_x}{\pi L_0}\right)\left[\cos^{-1}\left\{f_{30}\left(W-h\upsilon,0\right)\right\}\right] \tag{2.17}$$

E_{n_x} should be determined from the equation

$$f_{30}\left(E_{n_x},0\right)=\cos\left[\frac{L_0n_x\pi}{d_x}\right] \tag{2.18}$$

The sub-band energy E_{32} is given by

$$\frac{1}{L_0{}^2}\left[\cos^{-1}f_{31}\left(E,n_y,n_z\right)\Big|_{E=E_{32}}\right]^2=\left[\left(\frac{\pi n_y}{d_y}\right)^2+\left(\frac{\pi n_z}{d_z}\right)^2+\left(\frac{\pi n_x}{d_x}\right)^2\right] \tag{2.19}$$

2.2.3. Photo Emission from Quantum Dot Super Lattices with Graded Interfaces

The electron energy spectrum in this case can be expressed as

$$k_z^2=\frac{1}{L_0^2}\left[\cos^{-1}\left[\frac{1}{2}\phi_{91}\left(E,k_s\right)\right]\right]^2-k_s{}^2 \tag{2.20}$$

where, $L_0\left(=a_0+b_0\right)$ is the SL period, a_0 and b_0 are the widths of the barrier and the well respectively,

$$\phi_{91}\left(E,k_s\right)=\left[2\cosh\left\{\beta_{91}\left(E,k_s\right)\right\}\cos\left\{\gamma_{91}\left(E,k_s\right)\right\}+\varepsilon_{91}\left(E,k_s\right)\sinh\left\{\beta_{91}\left(E,k_s\right)\right\}\sin\left\{\gamma_{91}\left(E,k_s\right)\right\}\right.$$

$$+\Delta_0\left[\left(\frac{K_{91}^2\left(E,k_s\right)}{K_{92}\left(E,k_s\right)}-3K_{92}\left(E,k_s\right)\right)\cosh\left\{\beta_{91}\left(E,k_s\right)\right\}\sin\left\{\gamma_{91}\left(E,k_s\right)\right\}\right.$$

$$+\left(3K_{91}(E,k_s)-\frac{K_{92}^2(E,k_s)}{K_{91}(E,k_s)}\right)\sinh\{\beta_{91}(E,k_s)\}\cos\{\gamma_{91}(E,k_s)\}\Big]$$

$$+\Delta_0\Big[2\{K_{91}^2(E,k_s)-K_{92}^2(E,k_s)\}\cosh\{\beta_{91}(E,k_s)\}\cos\{\gamma_{91}(E,k_s)\}$$

$$+\frac{1}{12}\left[\frac{5K_{92}^2(E,k_s)}{K_{91}(E,k_s)}+\right.$$

$$\left.\frac{5K_{91}^2(E,k_s)}{K_{92}(E,k_s)}-34K_{91}(E,k_s)K_{92}(E,k_s)\right]\sinh\{\beta_{91}(E,k_s)\}\sin\{\gamma_{91}(E,k_s)\}\Big]\Big],$$

$$\beta_{91}(E,k_s)=K_{91}(E,k_s)(a_0-\Delta_0),K_{91}(E,k_s)=\left[k_s^2-\frac{2m_2^*}{\hbar^2}\beta_0\left(E-V_0,\lambda,E_{g_{02}},\Delta_2\right)\right]^{1/2}$$

$$,k_s^2=k_x^2+k_y^2,V_0=\left|E_{g_{02}}-E_{g_{01}}\right|,\gamma_{91}(E,k_s)=K_{92}(E,k_s)(b_0-\Delta_0),$$

$K_{92}(E,k_s)=\left[\frac{2m_1^*}{\hbar^2}\beta_0\left(E,\lambda,E_{g_{01}},\Delta_1\right)-k_s^2\right]^{1/2}$ and Δ_0 is the interface width.

The electron energy spectrum in this case is given by

$$\left(\frac{\pi n_z}{d_z}\right)^2=\left[\frac{1}{L_0^2}\left[\cos^{-1}\left[\frac{1}{2}\phi_{91}(E,n_x,n_y)\right]\Big|_{E=E_{93}}\right]^2-\left(\frac{\pi n_x}{d_x}\right)^2-\left(\frac{\pi n_y}{d_y}\right)^2\right] \tag{2.21}$$

where E_{93} is the totally quantized energy in this case and

$$\phi_{91}(E,n_x,n_y)=\Big[2\cosh\{\beta_{91}(E,n_x,n_y)\}\cos\{\gamma_{91}(E,n_x,n_y)\}+\varepsilon_{91}(E,n_x,n_y)\sinh\{\beta_{91}(E,n_x,n_y)\}$$

$$\sin\{\gamma_{91}(E,n_x,n_y)\}$$

$$+\Delta_0\left[\left(\frac{K_{91}^2(E,n_x,n_y)}{K_{92}(E,n_x,n_y)}-3K_{92}(E,n_x,n_y)\right)\cosh\{\beta_{91}(E,n_x,n_y)\}\sin\{\gamma_{91}(E,n_x,n_y)\}\right.$$

$$\left.+\left(3K_{91}(E,n_x,n_y)-\frac{K_{92}^2(E,n_x,n_y)}{K_{91}(E,n_x,n_y)}\right)\sinh\{\beta_{91}(E,n_x,n_y)\}\cos\{\gamma_{91}(E,n_x,n_y)\}\right]$$

$$+\Delta_0\Big[2\{K_{91}^2(E,n_x,n_y)$$

$$-K_{92}^2(E,n_x,n_y)\}\cosh\{\beta_{91}(E,n_x,n_y)\}\cos\{\gamma_{91}(E,n_x,n_y)\}+\frac{1}{12}\left[\frac{5K_{92}^2(E,n_x,n_y)}{K_{91}(E,n_x,n_y)}\right.$$

$$+\frac{5K_{91}^2(E,n_x,n_y)}{K_{92}(E,n_x,n_y)}-34K_{91}(E,n_x,n_y)K_{92}(E,n_x,n_y)\Bigg]$$

$$\sinh\{\beta_{91}(E,n_x,n_y)\}\sin\{\gamma_{91}(E,n_x,n_y)\}\Big]\Big],$$

$$\beta_{91}(E,n_x,n_y)=K_{91}(E,n_x,n_y)(a_0-\Delta_0),$$

$$K_{91}(E,n_x,n_y)=\left[\left(\frac{\pi n_x}{d_x}\right)^2+\left(\frac{\pi n_y}{d_y}\right)^2-\frac{2m_2^*}{\hbar^2}\beta_0(E-V_0,\lambda,E_{g_{02}},\Delta_2)\right]^{1/2},$$

$$\gamma_{91}(E,n_x,n_y)=K_{92}(E,n_x,n_y)(b_0-\Delta_0),\ K_{92}(E,n_x,n_y)=\left[-\left(\frac{\pi n_x}{d_x}\right)^2-\left(\frac{\pi n_y}{d_y}\right)^2+\frac{2m_1^*}{\hbar^2}\beta_0(E,\lambda,E_{g_{01}},\Delta_1)\right]^{1/2}$$

and $\varepsilon_{91}(E,n_x,n_y)=\left[\frac{K_{91}(E,n_x,n_y)}{K_{92}(E,n_x,n_y)}-\frac{K_{92}(E,n_x,n_y)}{K_{91}(E,n_x,n_y)}\right]$

The sub-band energy E_{93} is given by

$$\frac{1}{L_0^2}\left[\cos^{-1}\left[\frac{1}{2}\phi_{91}(E,n_x,n_y)\Big|_{E=E_{93}}\right]\right]^2=\left(\frac{\pi n_x}{d_x}\right)^2+\left(\frac{\pi n_y}{d_y}\right)^2+\left(\frac{\pi n_z}{d_z}\right)^2 \tag{2.22}$$

The electron concentration is given by $N_{0L}=\frac{2g_v}{d_xd_yd_z}\sum_{n_x=1}^{n_{x_{\max}}}\sum_{n_y=1}^{n_{y_{\max}}}\sum_{n_z=1}^{n_{z_{\max}}}F_{-1}(\eta_{93})$ (2.23)

where $\eta_{93}=\frac{E_{FQDSLGI}-E_{93}}{k_BT}$, in which $E_{FQDSLGI}$ is the Fermi energy in this case. The photoelectric current density is given by

$$J_{1L}=\frac{2eL_0g_v\alpha_0}{d_xd_yd_z\hbar}\sum_{n_x=1}^{n_{x_{\max}}}\sum_{n_y=1}^{n_{y_{\max}}}\sum_{n_{z_{\min}}}^{n_{z_{\max}}}F_{-1}(\eta_{93})\rho_{91}(E_{n_z}) \tag{2.24}$$

where $\rho_{91}(E_{n_z})=\left|\frac{\sin\left(\frac{L_0\pi n_z}{d_z}\right)}{\phi_{91}'(E_{n_z},0)}\right|$ in which $\phi_{91}'(E_{n_z},0)=\phi_{91}'(E,0)\Big|_{E=E_{n_z}}$

where

$$\phi'_{91}(E,0)=\Big[\big(2\beta_{91}(E,0)\sinh\{\beta_{91}(E,0)\}\cos\{\gamma_{91}(E,0)\}-\big(2\gamma'_{91}(E,0)\big)\cosh\{\beta_{91}(E,0)\}\sin\{\gamma_{91}(E,0)\}$$

$$+\omega_{91}(E,0)+\Delta_0\Bigg[\Bigg[\left(\frac{2K_{91}(E,0)K'_{91}(E,0)}{K_{92}(E,0)}-\frac{K_{91}^2(E,0)K'_{92}(E,0)}{K_{92}^2(E,0)}-3K'_{92}(E,0)\right)\cosh\{\beta_{91}(E,0)\}$$

$$\sin\{\gamma_{91}(E,0)\}+\left(\frac{K_{91}^2(E,0)}{K_{92}(E,0)}-3K_{92}(E,0)\right)$$

$$\{(\beta'_{91}(E,0))\sinh\{\beta_{91}(E,0)\}\sin\{\gamma_{91}(E,0)\}+\gamma'_{91}(E,0)\cos\{\gamma_{91}(E,0)\}\times$$

$$\cosh\{\beta_{91}(E,0)\}\}+\sinh\{\beta_{91}(E,0)\}\cos\{\gamma_{91}(E,0)\}$$

$$\left\{3K'_{91}(E,0)-\frac{2K_{92}(E,0)K'_{92}(E,0)}{K_{91}(E,0)}+\frac{K_{92}^2(E,0)K'_{91}(E,0)}{K_{91}^2(E,0)}\right\}$$

$$+\left(3K_{91}(E,0)-\frac{K_{92}^2(E,0)}{K_{91}(E,0)}\right)\{(\beta'_{91}(E,0))\cosh\{\beta_{91}(E,0)\}\cos\{\gamma_{91}(E,0)\}-$$

$$\gamma'_{91}(E,0)\sin\{\gamma_{91}(E,0)\}\sinh\{\beta_{91}(E,0)\}\}\Bigg]$$

$$+\Delta_0\Big[4\{K_{91}(E,0)K'_{91}(E,0)-K_{92}(E,0)K'_{92}(E,0)\}$$

$$\cosh\{\beta_{91}(E,0)\}\cos\{\gamma_{91}(E,0)\}+2\big(K_{91}^2(E,0)-K_{92}^2(E,0)\big)$$

$$\{\beta'_{91}(E,0)\sinh\{\beta_{91}(E,0)\}\cos\{\gamma_{91}(E,0)\}-$$

$$\gamma'_{91}(E,0)\sin\{\gamma_{91}(E,0)\}\cosh\{\beta_{91}(E,0)\}\}$$

$$+\frac{1}{12}\Bigg[\frac{10K_{92}(E,0)K'_{92}(E,0)}{K_{91}(E,0)}-\frac{5K_{92}^2(E,0)K'_{91}(E,0)}{K_{91}^2(E,0)}+$$

$$\frac{10K_{91}(E,0)K'_{91}(E,0)}{K_{92}(E,0)}-\frac{5K_{91}^2(E,0)K'_{92}(E,0)}{K_{92}^2(E,0)}-34K'_{91}(E,0)K_{92}(E,0)-34K_{91}(E,0)K'_{92}(E,0)\Bigg]$$

$$\big[\sinh\{\beta_{91}(E,0)\}\sin\{\gamma_{91}(E,0)\}\big]$$

$$+\frac{1}{12}\left[\frac{5K_{92}^2(E,0)}{K_{91}(E,0)}+\frac{5K_{91}^2(E,0)}{K_{92}(E,0)}-34K_{91}(E,0)K_{92}(E,0)\right]$$

$$\big[\beta'_{91}(E,0)\cosh\{\beta_{91}(E,0)\}\sin\{\gamma_{91}(E,0)\}+\gamma'_{91}(E,0)\cos\{\gamma_{91}(E,0)\}\sinh\{\beta_{91}(E,0)\}\big]\Big]\Big],$$

$$\beta'_{91}(E,0)=K'_{91}(E,0)(a_0-\Delta_0),\ K'_{91}(E,0)=\big[K_{91}(E,0)\big]^{-1}\left(\frac{m_2^*}{\hbar^2}\right)$$

$$\big[-\gamma_2(E-V_0,E_{g_{02}},\Delta_2)+(V_0-E)\gamma_2'(E-V_0,E_{g_{02}},\Delta_2)+\psi_2'(E-V_0,E_{g_{02}},\Delta_2)\big],$$

$$\gamma_2'\left(E-V_{0,}E_{g_{02}},\Delta_2\right)=\gamma_2'\left(E,E_{g_{02}},\Delta_2\right)\Big|_{E=E-V_0},$$

$$\gamma_2'\left(E,E_{g_{02}},\Delta_2\right)=\gamma_2\left(E,E_{g_{02}},\Delta_2\right)\left[\frac{1}{E+E_{g_{02}}}+\frac{1}{E+E_{g_{02}}+\Delta_2}-\frac{1}{E+E_{g_{02}}+\frac{2}{3}\Delta_2}\right],$$

$$\psi_2'\left(E-V_{0,}E_{g_{02}},\Delta_2\right)=\psi_2'\left(E,E_{g_{02}},\Delta_2\right)\Big|_{E=E-V_0}\left(\psi_i'\left(E,E_{g_{0i}},\Delta_i\right),\right.$$

$$\gamma_{91}'(E,0)=K_{92}'(E,0)(b_0-\Delta_0),\,K_{92}'(E,0)=\left[K_{92}(E,0)\right]^{-1}\left(\frac{m_1^*}{\hbar^2}\right)\beta_0'\left(E,\lambda,E_{g_{01}},\Delta_1\right),$$

$$\omega_{91}(E,0)=\left[\Delta_{91}(E,0)\right.$$
$$\left\{\beta_{91}(E,0)\right\}\sin\left\{\gamma_{91}(E,0)\right\}+\varepsilon_{91}(E,0)\left(\beta_{91}'(E,0)\right)\cosh\left\{\beta_{91}(E,0)\right\}\sin\left\{\gamma_{91}(E,0)\right\}+$$
$$\left.\gamma_{91}'(E,0)\,\varepsilon_{91}(E,0)\sinh\left\{\beta_{91}(E,0)\right\}\cos\left\{\gamma_{91}(E,0)\right\}\right],$$

$$\Delta_{91}(E,0)=\left[\frac{K_{91}'(E,0)}{K_{92}(E,0)}-\frac{K_{91}(E,0)K_{92}'(E,0)}{K_{92}^2(E,0)}\right.$$
$$\left.-\frac{K_{92}'(E,0)}{K_{91}(E,0)}+\frac{K_{92}(E,0)K_{91}'(E,0)}{K_{91}^2(E,0)}\right]$$

$n_{z_{\min}}$ and E_{n_z} are determined from

$$n_{z_{\min}}\geq\frac{d_z}{L_0\pi}\cos^{-1}\left[\frac{1}{2}\phi_{91}(W-h\upsilon,0)\right] \tag{2.25}$$

$$\phi_{91}\left(E_{n_z},0\right)=2\cos\left[\frac{L_0\pi n_z}{d_z}\right] \tag{2.26}$$

in which $\phi_{91}\left(E_{n_z},0\right)=\phi_{91}(E,0)\Big|_{E=E_{n_z}}$

where

$$\phi_{91}(E,0)=\left[2\cosh\left\{\beta_{91}(E,0)\right\}\cos\left\{\gamma_{91}(E,0)\right\}+\varepsilon_{91}(E,0)\sinh\left\{\beta_{91}(E,0)\right\}\sin\left\{\gamma_{91}(E,0)\right\}\right.$$
$$+\Delta_0\left[\left(\frac{K_{91}^2(E,0)}{K_{92}(E,0)}-3K_{92}(E,0)\right)\cosh\left\{\beta_{91}(E,0)\right\}\sin\left\{\gamma_{91}(E,0)\right\}+\right.$$
$$\left(3K_{91}(E,0)-\frac{K_{92}^2(E,0)}{K_{91}(E,0)}\right)$$

$$\times \sinh\{\beta_{91}(E,0)\}\cos\{\gamma_{91}(E,0)\}]$$

$$+\Delta_0\left[2\{K_{91}^2(E,0)-K_{92}^2(E,0)\}\cosh\{\beta_{91}(E,0)\}\cos\{\gamma_{91}(E,0)\}\right.$$

$$+\frac{1}{12}\left[\frac{5K_{92}^2(E,0)}{K_{91}(E,0)}+\right.$$

$$\left.\left.\frac{5K_{91}^2(E,0)}{K_{92}(E,0)}-34K_{91}(E,0)K_{92}(E,0)\right]\sinh\{\beta_{91}(E,0)\}\sin\{\gamma_{91}(E,0)\}\right]\Bigg],$$

$$\beta_{91}(E,0)=K_{91}(E,0)(a_0-\Delta_0),$$

$$K_{91}(E,0)=\left[\frac{2m_2^*}{\hbar^2}\{(V_0-E)\gamma_2(E-V_0,E_{g_{02}},\Delta_2)+\psi_2(E-V_0,E_{g_{02}},\Delta_2)\}\right]^{1/2},$$

$$\gamma_{91}(E,0)=K_{92}(E,0)(b_0-\Delta_0),\ K_{92}(E,0)=\left[\frac{2m_1^*}{\hbar^2}\beta_0(E,\lambda,E_{g_{01}},\Delta_1)\right]^{1/2}$$

and $\varepsilon_{91}(E,0)=\left[\frac{K_{91}(E,0)}{K_{92}(E,0)}-\frac{K_{92}(E,0)}{K_{91}(E,0)}\right]$

For the purpose of relative assessment we also formulate the expression of photoemission from bulk specimens of optoelectronic materials under microwaves.

Using (2.2), $v_z(E')$ and $N(E')$ for optoelectronic materials in the presence of microwaves whose unperturbed conduction electrons obey the three band model of Kane can be written as

$$v_z(E')=\frac{\sqrt{2}}{\sqrt{m^*}}\frac{[\beta_0(E',\lambda)]^{1/2}}{\beta_0'(E',\lambda)} \tag{2.27}$$

and

$$N(E')=\left[4\pi\left(\frac{2m^*}{h^2}\right)^{3/2}\cdot g_v\right]\sqrt{\beta_0(E',\lambda)}\beta_0'(E',\lambda) \tag{2.28}$$

where, $\beta_0'(E,\lambda)\equiv\frac{\partial}{\partial E}[\beta_0(E,\lambda)]$.

Using the appropriate equations the photo-emitted current density in this case can be written using the generalized Sommerfield's lemma as

$$J_L = \left[\frac{4\pi e m^* (k_B T)^2 \alpha_0 g_v}{h^3}\right]\left[\left[\frac{\left(1+\frac{2}{3}\alpha\Delta\right)}{(1+\alpha\Delta)}\right.\right.$$

$$\left[2\alpha k_B T F_2(\eta_L) + \left(1+2\alpha E_0 + \frac{1}{3}\alpha\Delta\right)F_1(\eta_L) + \frac{E_0 + \alpha E_0^2 + \frac{1}{3}\alpha\Delta E_0}{k_B T}F_{-1}(\eta_L)\right.$$

$$\left.\left. + a_0\left[\boldsymbol{ln}\left|\frac{a_0+\eta_L}{a_0}\right| + \phi_5(\eta_L)\right]\right] - \left\{\frac{B_{51}}{(k_B T)^2}\right\} I_2\right] \qquad (2.29)$$

$\eta_L \equiv \left(E_{F_L} - E_0\right)(k_B T)^{-1}$, E_{F_L} is the Fermi energy in the presence of microwaves as measured from the edge of the conduction band in the absence of any field,

$$a_0 \equiv \left(E_0 + E_{g_0} + \frac{2}{3}\Delta\right)(k_B T)^{-1}, \phi(\eta_L) \equiv \sum_{r=1}^{s_0} 2\left(1-2^{1-2r}\right)\xi(2r)\frac{(-1)^{2r-1}(2r-1)!}{(a_0+\eta_L)^{2r}},$$

$$B_{51} \equiv \frac{e^2}{48 m_r \pi c^3}\frac{I_0\lambda^2}{\sqrt{\varepsilon_{sc}\varepsilon_0}}\frac{E_{g_0}\left(E_{g_0}+\Delta\right)}{\left(E_{g_0}+\frac{2}{3}\Delta\right)}\frac{\beta^2}{4}\left(t+\frac{\rho}{\sqrt{2}}\right)^2, I_2 \equiv \int_{E_0}^{\infty} f_1(E') f(E)\, dE'$$

and

$$f_1(E) \equiv \frac{1}{\phi_0(E)}\left\{\left(1+\frac{E_{g_0}-\delta}{\phi_0(E)+\delta}\right) + (E_{g_0}-\delta)\left[\frac{1}{\phi_0(E)+\delta} - \frac{1}{E_{g_0}+\delta}\right]^{1/2}\left[\frac{1}{\phi_0(E)+\delta} - \frac{E_{g_0}+\delta}{\left(E_{g_0}-\delta\right)^2}\right]^{1/2}\right\}^2.$$

It appears than that the determination of J_L as a function of electron concentration requires an expression of electron statistics which can, in-turn, be written as

$$n_{0L} = \left(3\pi^2\right)^{-1}\left(\frac{2m^*}{\hbar^2}\right)^{3/2} g_v\left[M_{51}\left(E_{F_L},\lambda\right) + N_{51}\left(E_{F_L},\lambda\right)\right] \qquad (2.30)$$

where, $M_{51}(E_{F_L},\lambda) \equiv \left[\beta_L\left(E_{F_L},\lambda\right)\right]^{3/2}, N_{51}(E_{F_L},\lambda) \equiv \sum_{r=1}^{s_0} Z_{r_Y}\left[M_{51}(E_{F_L},\lambda)\right]$ and Y=L.

The expressions of J_L and n_{0L} for optoelectronic materials in the presence of microwaves whose unperturbed conduction electrons obey the two band model of Kane can be written following (2.29) as

$$J_L = \left[\frac{4\pi e m^* (k_B T)^2 \alpha_0 g_v}{h^3}\right]\left[F_1(\eta_L) + 2\alpha k_B T F_2(\eta_L) - \left(\frac{\overline{B}_{51}\alpha}{k_B T}\right)\right.$$

$$\left.\left. F_0(\eta_L) - \left(\frac{3\overline{C}_{51}}{2}\right) k_B T F_1(\eta_L) + 2C_{52}(k_B T)^2 F_2(\eta_L)\right]\right] \tag{2.31}$$

and $$n_{0L} = (3\pi^2)^{-1}\left(\frac{2m^*}{\hbar^2}\right)^{3/2} g_v\left[M_{52}(E_{F_L},\lambda) + N_{52}(E_{F_L},\lambda)\right] \tag{2.32}$$

where, $\overline{B}_{51} \equiv \frac{e^2}{192 m_r \pi c^3}\frac{I_0\lambda^2 E_{g_0}}{\sqrt{\varepsilon_{sc}\varepsilon_0}}$, $\overline{C}_{51} \equiv \frac{2m^*\alpha}{m_v}$, $C_{52} \equiv \left[\frac{15(\overline{C}_{51})^2}{8} - \left(\frac{3}{2}\alpha(\overline{C}_{51})\right)\right]$,

$M_{52}(E_{F_L},\lambda) \equiv \left[\omega_0(E_{F_L},\lambda)\right]^{3/2}$,

$N_{52}(E_{F_L},\lambda) \equiv \sum_{r=1}^{S_0} Z_{r_Y} M_{52}(E_{F_L},\lambda)$ and $Y = L$.

The expressions of J_L and n_{0L} for optoelectronic materials in the presence of microwaves whose unperturbed conduction electrons obey the parabolic energy bands can be expressed as

$$J_L = \left[\frac{4\pi e m^* (k_B T)^2 \alpha_0 g_v}{h^3}\right]\left[F_1(\eta_L) - \left(\frac{B_{53}}{k_B T}\right)\left[F_0(\eta_L) - 3\alpha k_B T\left(1 + \frac{m^*}{m_v}\right)F_1(\eta_L)\right]\right] \tag{2.33}$$

and $$n_{0L} = (3\pi^2)^{-1}\left(\frac{2m^*}{\hbar^2}\right)^{3/2} \cdot g_v\left[M_{53}(E_{F_L},\lambda) + N_{53}(E_{F_L},\lambda)\right] \tag{2.34}$$

where, $B_{53} \equiv \frac{e^2}{8 m_r \pi (m^*)^2 c^3}\frac{I_0\lambda^2}{\sqrt{\varepsilon_{sc}\varepsilon_0}}$, $M_{53}(E_{F_L},\lambda) \equiv \left[\rho_0(E_{F_L},\lambda)\right]^{3/2}$

and $N_{53}(E_{F_L},\lambda) \equiv \sum_{r=1}^{S_0} Z_{r_Y} M_{53}(E_{F_L},\lambda)$.

Special Case:

Formulation of electron concentration and current density for unperturbed three and two band models of Kane for optoelectronic materials:

(i) The expressions of J and n_0 in accordance with the unperturbed three band model of Kane assume the forms

$$J = \left[\frac{4\pi e m^* (k_B T)^2 \alpha_0 g_v}{h^3}\right]\left[\frac{\left(1+\frac{2}{3}\alpha\Delta\right)}{(1+\alpha\Delta)}\right.$$

$$\left[2\alpha k_B T F_2(\eta_0) + \left(1+2\alpha E_0+\frac{1}{3}\alpha\Delta\right)F_1(\eta_0) + \frac{E_0+\alpha E_0^2+\frac{1}{3}\alpha\Delta E_0}{k_B T}F_{-1}(\eta_0)\right.$$

$$\left.\left.+a_0\left[\boldsymbol{ln}\left|\frac{\overline{a_0}+\eta_0}{\overline{a_0}}\right|+\phi(\eta_0)\right]\right]\right] \tag{2.35}$$

$$\text{and } n_0 = (3\pi^2)^{-1}\left(\frac{2m^*}{\hbar^2}\right)^{3/2} g_v\left[M_{54}(E_F)+N_{54}(E_F)\right] \tag{2.36}$$

where $a_0 = \dfrac{E_0+E_{g_0}+\frac{2}{3}\Delta}{k_B T}$, $\overline{a_0} = \dfrac{2}{9}\dfrac{\Delta^2}{(k_B T)^2}\left(1+\dfrac{2}{3}\alpha\Delta\right)$,

$$\phi(\eta_0) = \sum_{r=1}^{S_0} 2(1-2^{1-2r})\zeta(2r)\frac{(-1)^{2r-1}(2r-1)!}{(\overline{a_0}+\eta_0)^{2r}},\ M_{54}(E_F) \equiv [I(E_F)]^{3/2},$$

$$N_{54}(E_F) \equiv \sum_{r=1}^{S_0} Z_r M_{54}(E_F) \text{ and } Z_r\left[\equiv 2(k_B T)^{2r}(1-2^{1-2r})\zeta(2r)\frac{\partial^{2r}}{\partial E_F^{\ 2r}}\right].$$

(ii) In accordance with the unperturbed two band model of Kane, the corresponding expressions of J and n_0 are given by

$$J = \left[\frac{4\pi e m^* (k_B T)^2 \alpha_0 g_v}{h^3}\right]\left[F_1(\eta_0)+2\alpha k_B T F_2(\eta_0)\right] \tag{2.37}$$

and

$$n_0 = \left(3\pi^2\right)^{-1}\left(\frac{2m^*}{\hbar^2}\right)^{3/2} g_v\left[M_{55}\left(E_F\right)+N_{55}\left(E_F\right)\right] \tag{2.38}$$

where, $M_{55}(E_F) \equiv \left[E_F\left(1+\alpha E_F\right)\right]^{3/2}$ and $N_{55}(E_F) \equiv \sum_{r=1}^{S_o} Z_r M_{55}(E_F)$.

For $\alpha \to 0$, (2.37) and (2.38) get simplified into the well-known forms as given in the literature.

The two different applications of the results of this chapter in the field of materials science in general:

1. The nonlinear response from the optical excitation of the free carriers is given by[15]

$$G_0 = \frac{-e^2}{\left(2\pi f_0\right)^2 \hbar^2}\int_0^\infty \left(k_x \frac{\partial k_x}{\partial E}\right)^{-1} N(E) f\ (E) dE \tag{2.39}$$

where f_0 is the optical frequency. From the band structure we can derive the term $\left(k_x \frac{\partial k_x}{\partial E}\right)$ as a function of E and by using the DOS function as formulated, we can study the G_0 for all types of optoelectronic and III-V materials.

2. The effective mass of the charge carriers in electronic materials, being inversely related with the mobility is known to be one of the central concepts in the whole field of materials science and related disciplines [16-17]. It is important to note that various definitions exist for the effective carrier mass. Among them, it is the momentum effective mass (EMM) that should be regarded as the basic quantity [18]. This is because of the fact that the EMM appears in the analysis of transport phenomena and all other properties of the carriers in a band with arbitrary band non-parabolicity.

It can be proved that it is this mass which enters in various transport coefficients and plays the most dominant role in explaining the experimental results of different mechanisms of scattering[19-20]. The doping in degenerate materials influences the effective mass when it is energy dependent. Under the condition of carrier degeneracy, only the carriers at the Fermi surface of electronic materials participate in the conduction process and hence, the EMM of the electrons corresponding to the Fermi level would be of interest in carrier transport under such conditions. The carrier dispersion laws are of prime importance in determining the EMM and in recent years various carrier energy spectra in one, two and three dimensions have been developed [21-32]. The Fermi energy is again determined by the carrier dispersion laws and the carrier concentration and therefore, these two important features would determine the dependence of the EMM in degenerate materials under the degree of carrier degeneracy [33-40].

The expression of the EMM is given by

$$m^*\left(E_{F_L}\right)=\left[\left(\hbar k\right)\Big/\left(\frac{1}{\hbar}\frac{\partial E}{\partial k}\right)\right]\Bigg|_{E=E_{F_L}}=\hbar^2 k\frac{\partial k}{\partial E}\Bigg|_{E=E_{F_L}} \tag{2.40}$$

Using (2.2), (2.3), (2.4) and (2.40) the expressions of the EMMs in the presence of microwaves for III-V, ternary and quaternary materials whose unperturbed conduction electrons obey the there and two band models of Kane together with parabolic energy bands can be written as

$$m^*\left(E_{F_L}\right)=m^*\left[I'\left(E_{F_L}\right)-\theta_0'\left(E_{F_L},\lambda\right)\right] \tag{2.41}$$

$$m^*\left(E_{F_L}\right)=m^*\left[\left(1+2\alpha E_{F_L}\right)-B_0'\left(E_{F_L},\lambda\right)\right] \tag{2.42}$$

$$m^*\left(E_{F_L}\right)=m^*\left[\rho_0'\left(E_{F_L},\lambda\right)\right] \tag{2.43}$$

where, the primes indicate the differentiation of the differentiable functions with respect to Fermi energy. In the absence of microwaves the corresponding well known expressions are, respectively, given by

$$m^*\left(E_{F_L}\right)=m^*\left[I'\left(E_F\right)\right] \tag{2.44}$$

$$m^*\left(E_{F_L}\right)=m^*\left[\left(1+2\alpha E_{F_L}\right)\right] \tag{2.45}$$

$$m^*\left(E_F\right)=m^* \tag{2.46}$$

Comparing (2.41) and (2.44) and again (2.42) and (2.45), we observe that the EMM, in addition to doping, changes with light intensity and wavelength when the material is being exposed to photo-excitation and it is important to note that in (2.44) to (2.46), the outcome of conventional analysis cannot explain this important phenomenon. The (2.46) exhibits the fact that the EMM is a constant quantity for electronic materials having unperturbed parabolic energy bands, whereas, (2.44) reflects the idea that the EMM of the same material in the presence of microwaves changes with concentration, light intensity and wavelength respectively in various manners.

2.3. Results and Discussions

The dependences of the normalized photo-emitted current density from QDs of n-$Hg_{1-x}Cd_xTe$ on the film thickness, normalized carrier degeneracy, normalized intensity and wavelength of the microwaves have been drawn in Figures 2.1 to 2.4 in accordance with perturbed three (using (2.8) and (2.9)) and two (using (2.8) and (2.10)) band models of Kane together with parabolic (using (2.8) and (2.11)) energy bands as shown by curves (a), (b) and (c) respectively.

The variations of normalized photo-current density from QDs of n-$In_{1-x}Ga_xAs_yP_{1-y}$ lattice matched to InP, have been drawn in Figures 2.5 to 2.8 as functions of film thickness, normalized carrier degeneracy, normalized intensity and wavelength of the microwaves respectively for all the cases of Figures 2.5.

From Figures 2.1 and 2.5, we observe that the photoemission from QDs of optoelectronic materials decreases with increasing film thickness exhibiting prominent trapezoidal variation for relatively large thickness bandwidth.

These types of variations are the special signatures of 3D confinement in QDs of optoelectronic materials respectively in the presence of light. From Figures 2.2 and 2.6, it appears that the normalized photo-emitted current density from QDs increases with increasing carrier degeneracy and for relatively large values of the same variable; it exhibits quantum jumps for all types of band models when the size quantum number changes from one fixed value to another.

From Figures 2.3 and 2.7, it appears that the photoemission increases with decreasing intensity. From Figs 2.4 and 2.8 we can conclude that the normalized photoemission decreases with increasing wavelength for QDs of optoelectronic materials. Finally, it is apparent from all the figures that the photoemission from quantum confined ternary materials is larger as compared with the quantum confined quaternary compounds. Using (2.14) and (2.15), the normalized photo-emitted current density from $HgTe/Hg_{1-x}Cd_xTe$ and $In_xGa_{1-x}As/InP$ effective mass QD SLs respectively has been plotted for all types of band models as a function of film thickness as shown in Figure 2.9. Figures 2.10 to 2.13 exhibit the plots of normalized photo-emitted current density from the said SLs as functions of normalized electron degeneracy, normalized intensity, wavelength and normalized incident photon energy respectively for all cases of Figure 2.9.

From Figure 2.9, it appears that photo-emitted current density from QD effective mass $HgTe/Hg_{1-x}Cd_xTe$ and $In_xGa_{1-x}As/InP$ SLs exhibits the oscillations with increase in thickness. From Figures 2.10 to 2.12, it appears that photoemission current density from QD effective mass $HgTe/Hg_{1-x}Cd_xTe$ and $In_xGa_{1-x}As/InP$ SLs increases with increasing carrier concentration, decreasing intensity and decreasing wavelength respectively in various manners. Figure 2.13 demonstrates the fact that the photoemission current density from QD effective mass $HgTe/Hg_{1-x}Cd_xTe$ and $In_xGa_{1-x}As/InP$ SLs exhibit quantum steps with increasing photon energy for both the cases.

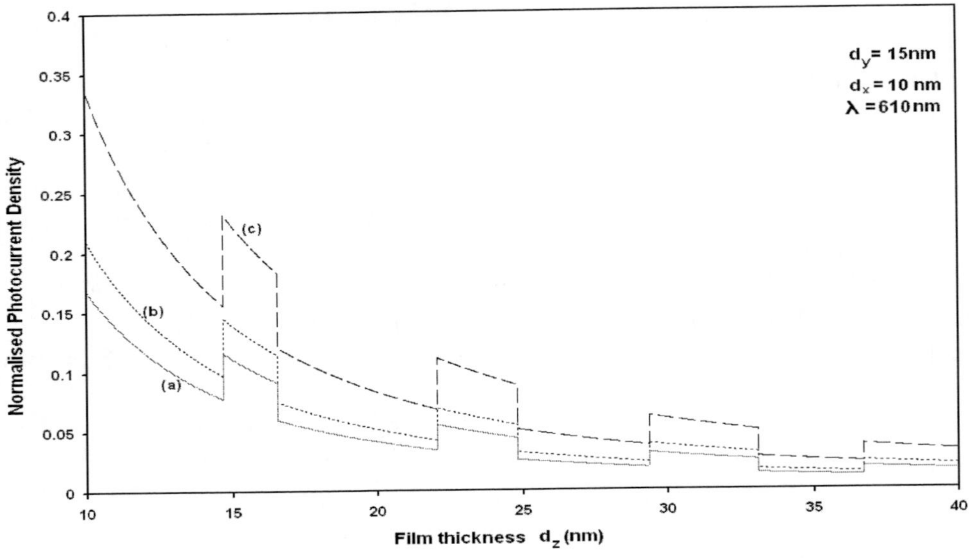

Figure 2.1. Plot of the normalized photocurrent density from QDs of n-$Hg_{1-x}Cd_xTe$ as a function of film thickness in which the curves (a), (b) and (c) represent the perturbed three and two band models of Kane together with parabolic energy bands respectively.

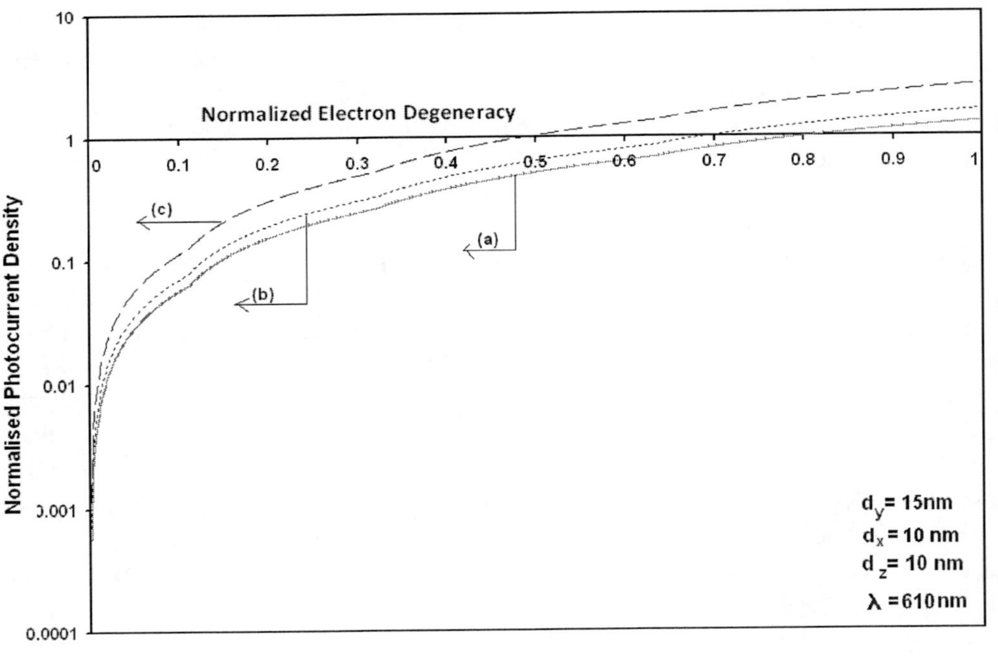

Figure 2.2. Plot of the normalized photocurrent density from QDs of n-$Hg_{1-x}Cd_xTe$ as a function of normalized electron degeneracy for all cases of Figure 2.1.

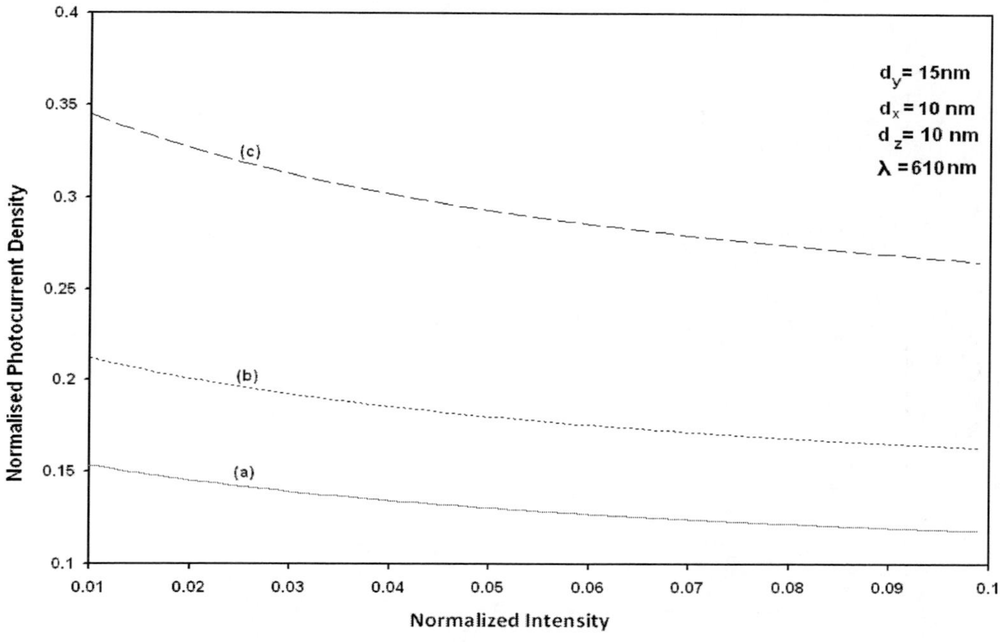

Figure 2.3. Plot of the normalized photocurrent density from QDs of n-$Hg_{1-x}Cd_xTe$ as a function of normalized intensity for all cases of Figure 2.1.

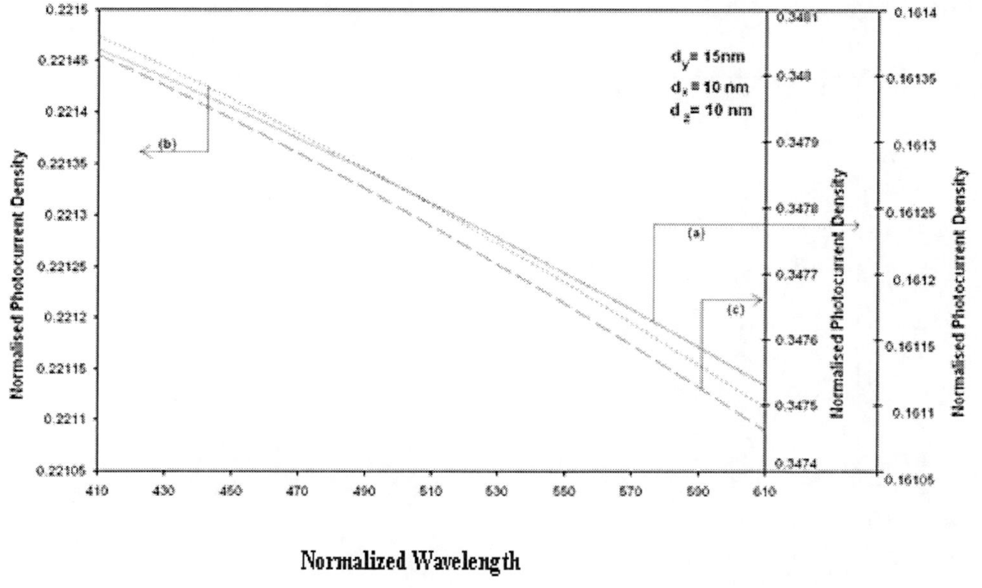

Figure 2.4. Plot of the normalized photocurrent density from QDs of n-$Hg_{1-x}Cd_xTe$ as a function of microwaves length for all cases of Figure 2.1.

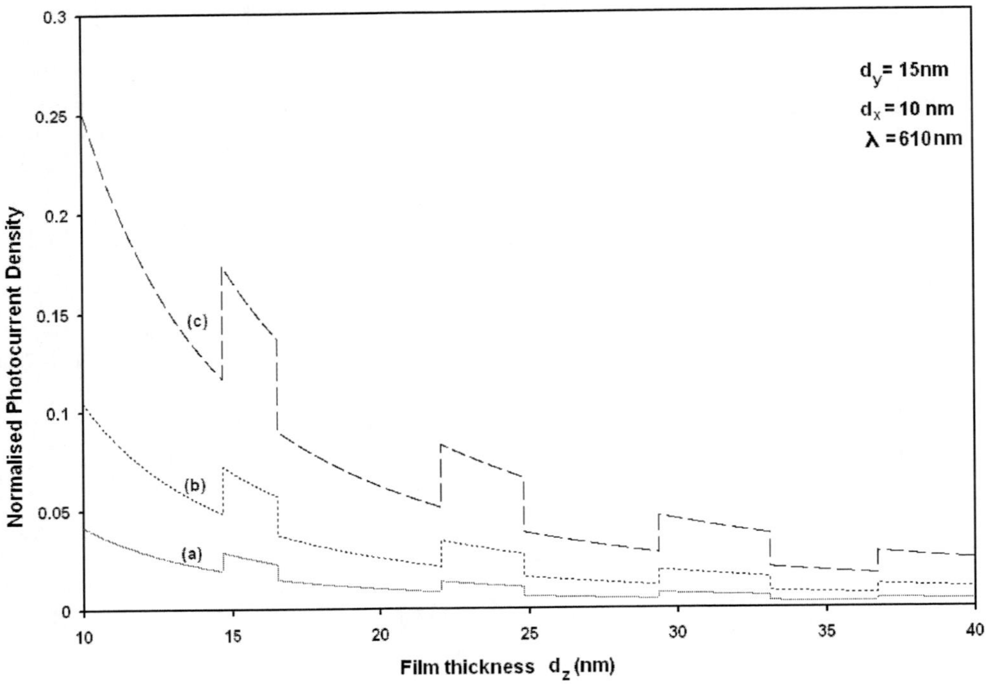

Figure 2.5. Plot of the normalized photocurrent density from QDs of n-$In_{1-x}Ga_xAs_yP_{1-y}$ lattice matched to InP as a function of film thickness in which the curves (a), (b) and (c) represent the perturbed three and two band models of Kane together with parabolic energy bands respectively.

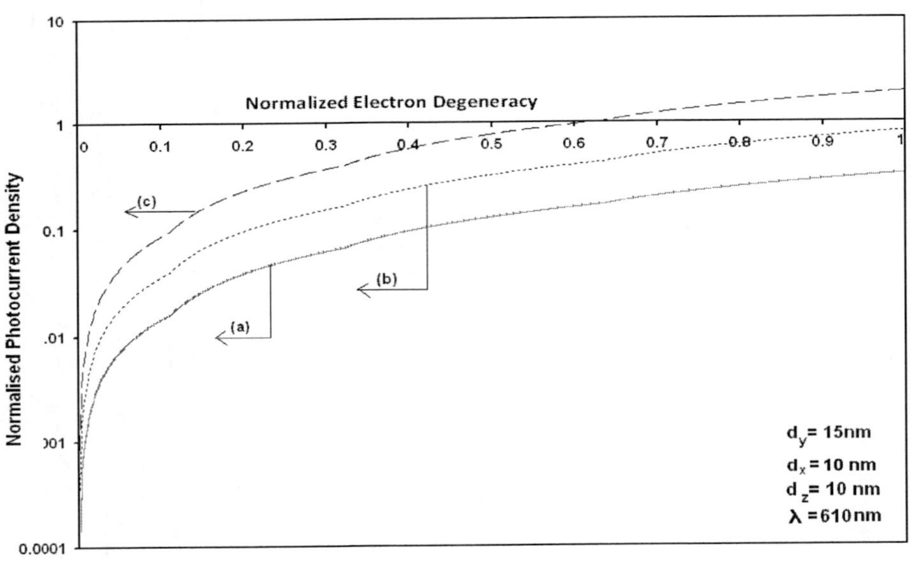

Figure 2.6. Plot of the normalized photocurrent density from QDs of n-$In_{1-x}Ga_xAs_yP_{1-y}$ lattice matched to InP as a function of normalized electron degeneracy for all cases of Figure 2.5.

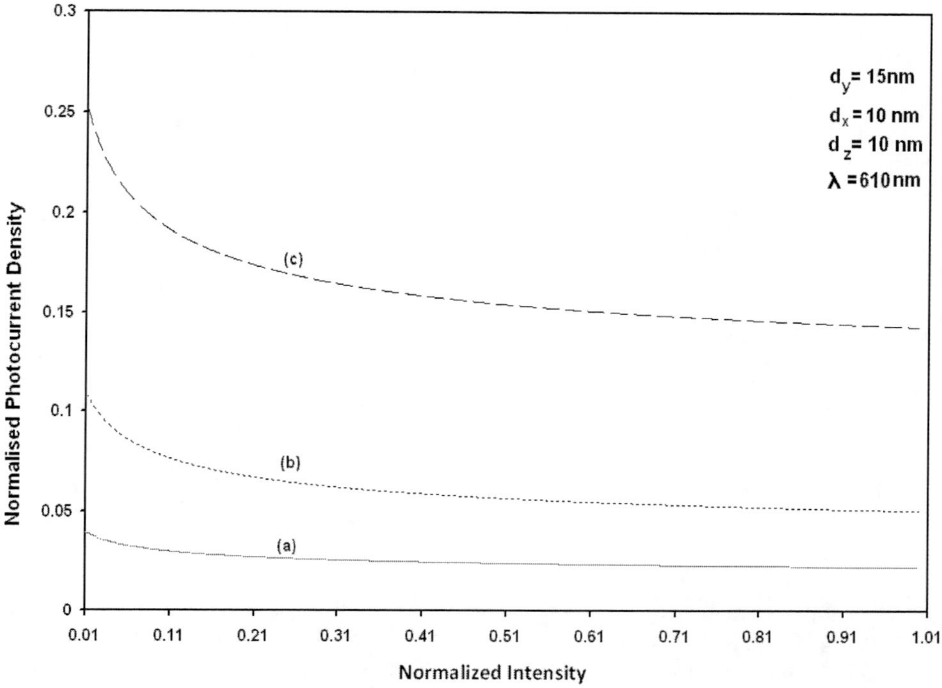

Figure 2.7. Plot of the normalized photocurrent density from QDs of n-$In_{1-x}Ga_xAs_yP_{1-y}$ lattice matched to InP as a function of normalized light intensity for all cases of Figure 2.5.

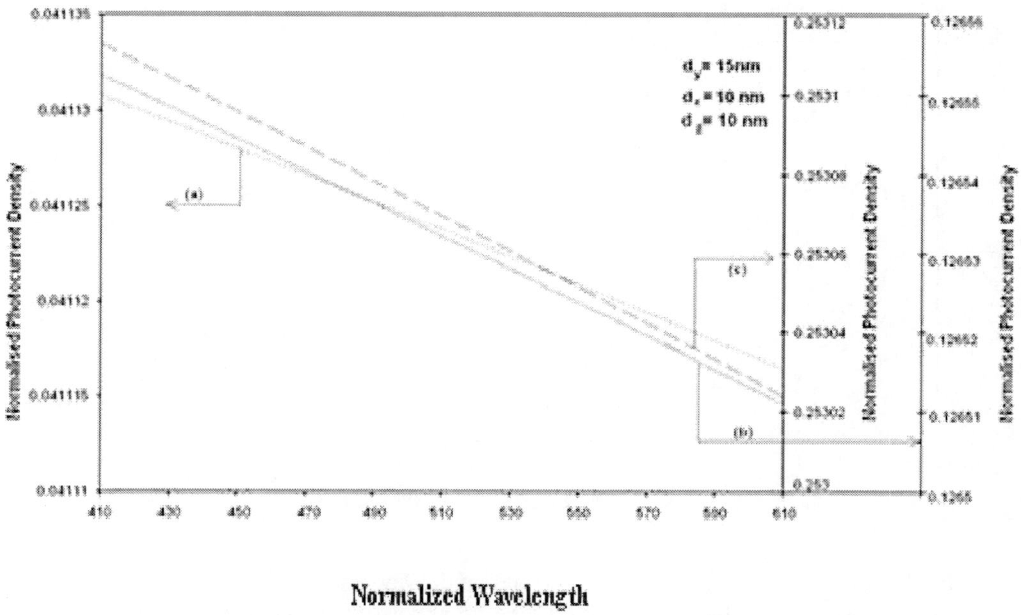

Figure 2.8. Plot of the normalized photo-emitted current density from QDs of n-$In_{1-x}Ga_xAs_yP_{1-y}$ lattice matched to InP as a function of normalized waveslength for all cases of Figure 2.5

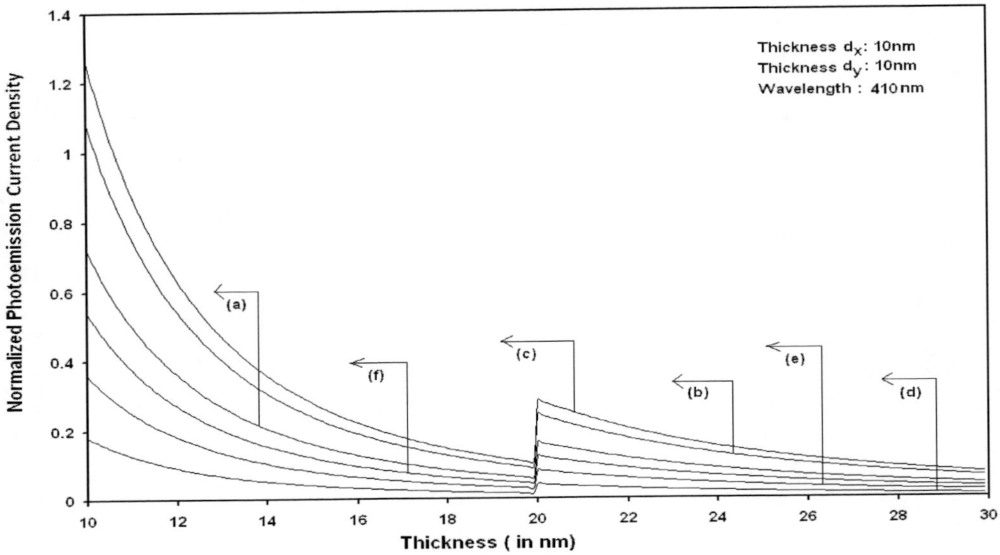

Figure 2.9. Plot of the normalized photoemission current density from quantum dot effective mass superlattices of $HgTe/Hg_{1-x}Cd_xTe$ as function of film thickness in which the curves (a), (b) and (c) represent the perturbed three and two band models of Kane together with parabolic energy bands respectively. The curves (d), (e) and (f) exhibit the corresponding plots of $In_xGa_{1-x}As/InP$.

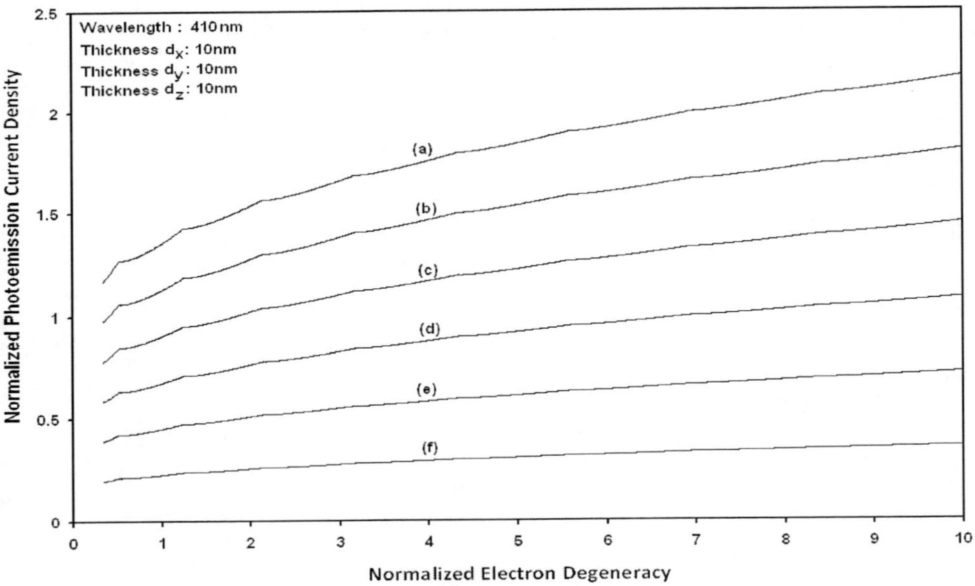

Figure 2.10. Plot of the normalized photoemission current density from quantum dot effective mass superlattices of $HgTe/Hg_{1-x}Cd_xTe$ and $In_xGa_{1-x}As/InP$ as function of normalized electron degeneracy for all cases of Figure 2.9.

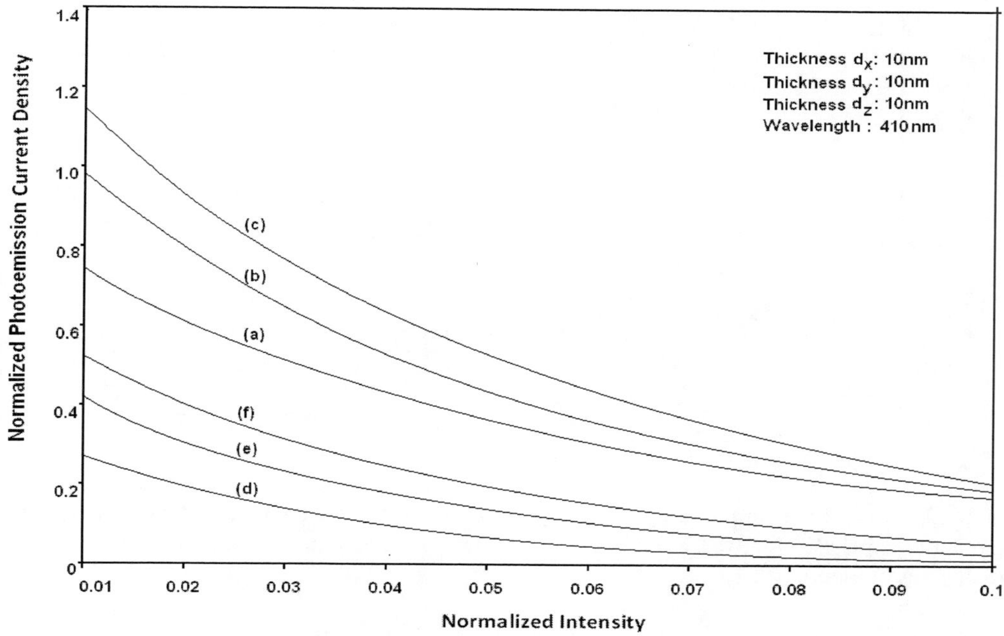

Figure 2.11. Plot of the normalized photoemission current density from quantum dot effective mass superlattices of $HgTe/Hg_{1-x}Cd_xTe$ and $In_xGa_{1-x}As/InP$ as function of normalized light intensity for all cases of Figure 2.9.

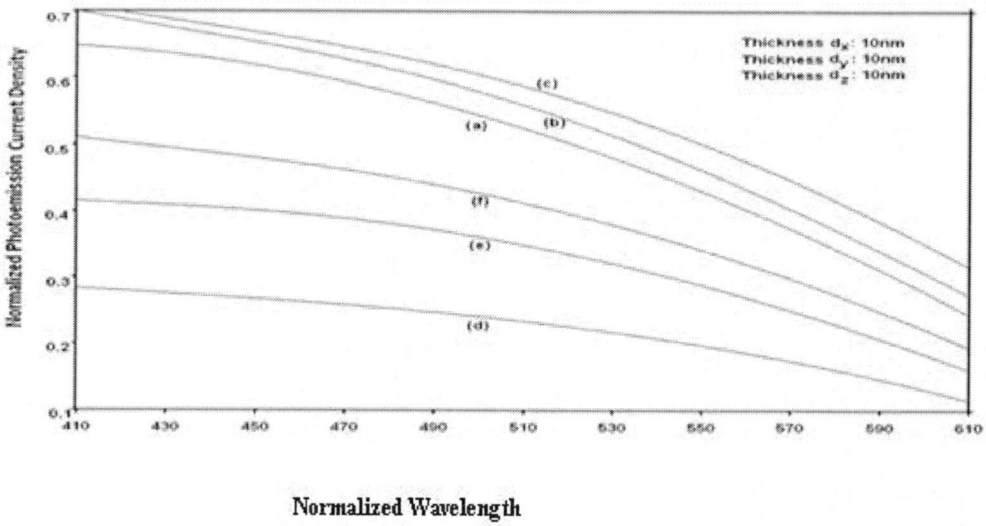

Figure 2.12. Plot of the normalized photoemission current density from quantum dot effective mass superlattices of $HgTe/Hg_{1-x}Cd_xTe$ and $In_xGa_{1-x}As/InP$ as function of normalized wavelength for all cases of Figure 2.9.

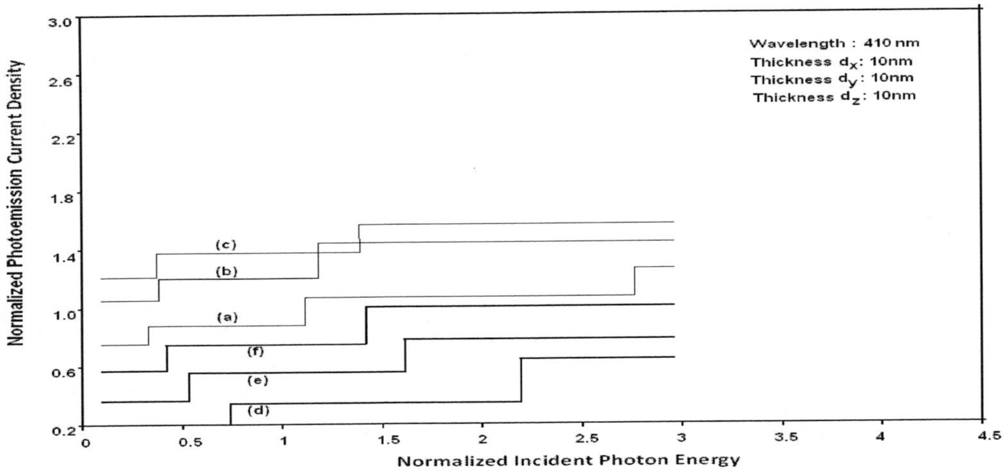

Figure 2.13. Plot of the normalized photoemission current density from quantum dot effective mass superlattices of $HgTe/Hg_{1-x}Cd_xTe$ and $In_xGa_{1-x}As/InP$ as function of normalized incident photon energy for all cases of Figure 2.9.

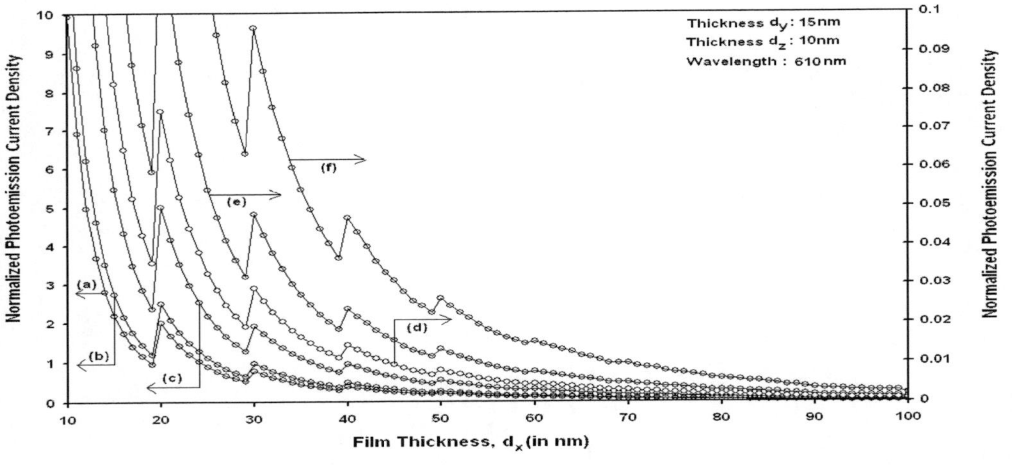

Figure 2.14. Plot of the normalized photocurrent density from quantum dot superlattices of graded interface of $HgTe/Hg_{1-x}Cd_xTe$ as function of film thickness in which the curves (a), (b) and (c) represent the perturbed three and two band models of Kane together with parabolic energy bands respectively. The curves (d), (e) and (f) exhibit the corresponding plots of $In_xGa_{1-x}As/InP$.

The normalized photo-emitted current density from $HgTe/Hg_{1-x}Cd_xTe$ and $In_xGa_{1-x}As/InP$ QD SLs with graded interfaces respectively has been shown in Figure 2.14 for all types of band models as a function of film thickness. Besides, the Figures 2.15 to 2.18 further exhibits the plots of normalized photo-emitted current density from the said SLs as functions of normalized carrier concentration, normalized intensity, wavelength and normalized incident photon energy respectively for all cases of Figure 2.14. From Figure 2.14 it appears that the photoemission from QD $HgTe/Hg_{1-x}Cd_xTe$ and $In_xGa_{1-x}As/InP$ SLs with graded interfaces increases with decreasing film thickness with larger oscillations again exhibiting the quantization effect. From Figures 2.15, it is observed that the photoemission for the present system increases with increasing carrier degeneracy in an oscillatory way. From Figures 2.16 and 2.17, it appears that the photoemission from the said systems increases with decreasing intensity and decreasing wavelength respectively in various manners. Figure 2.18 demonstrates the fact that the photocurrent density for the present system exhibit quantum steps with increasing photon energy for both the cases.

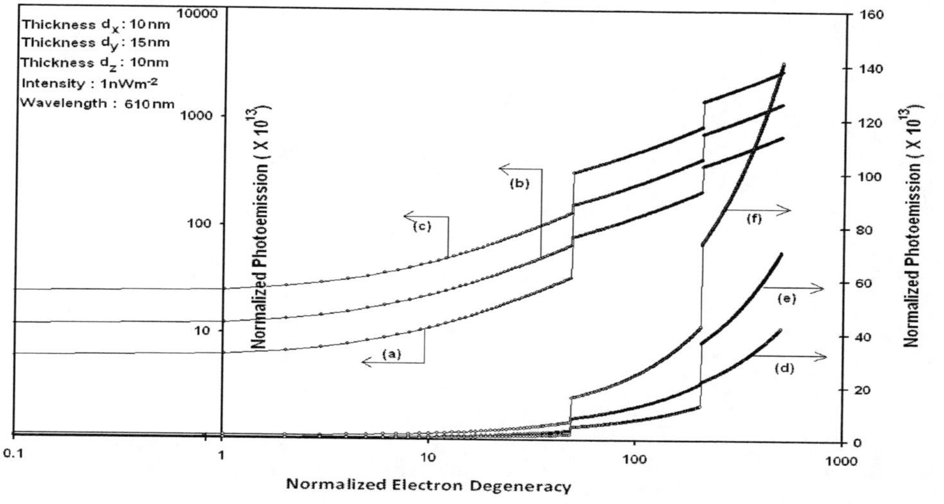

Figure 2.15. Plot of the normalized photocurrent density from quantum dot superlattices of graded interface of HgTe/$Hg_{1-x}Cd_xTe$ and $In_xGa_{1-x}As$/InP as function of normalized electron degeneracy for all cases of Figure 2.14.

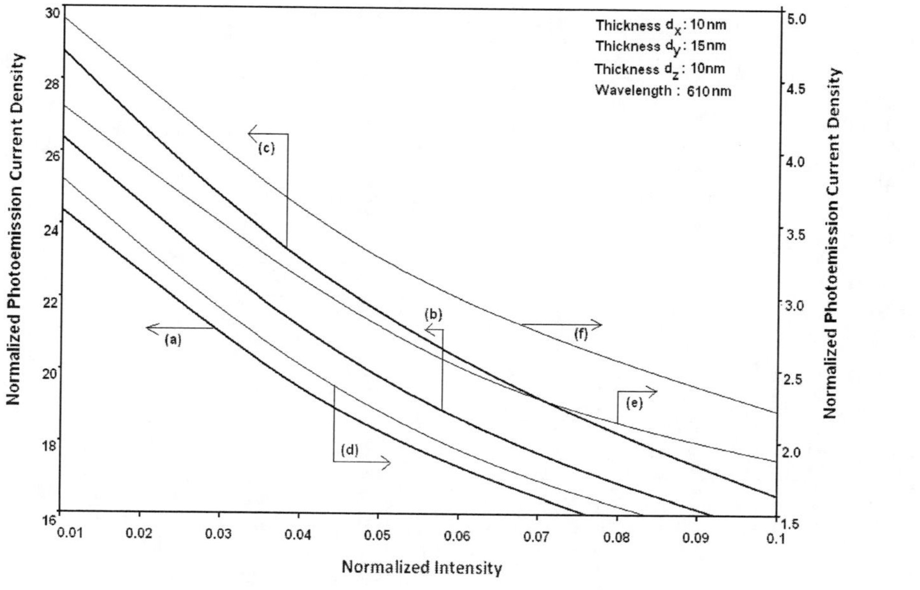

Figure 2.16. Plot of the normalized photocurrent density from quantum dot superlattices of graded interface of HgTe/$Hg_{1-x}Cd_xTe$ and $In_xGa_{1-x}As$/InP as function of normalized light intensity for all cases of Figure 2.14.

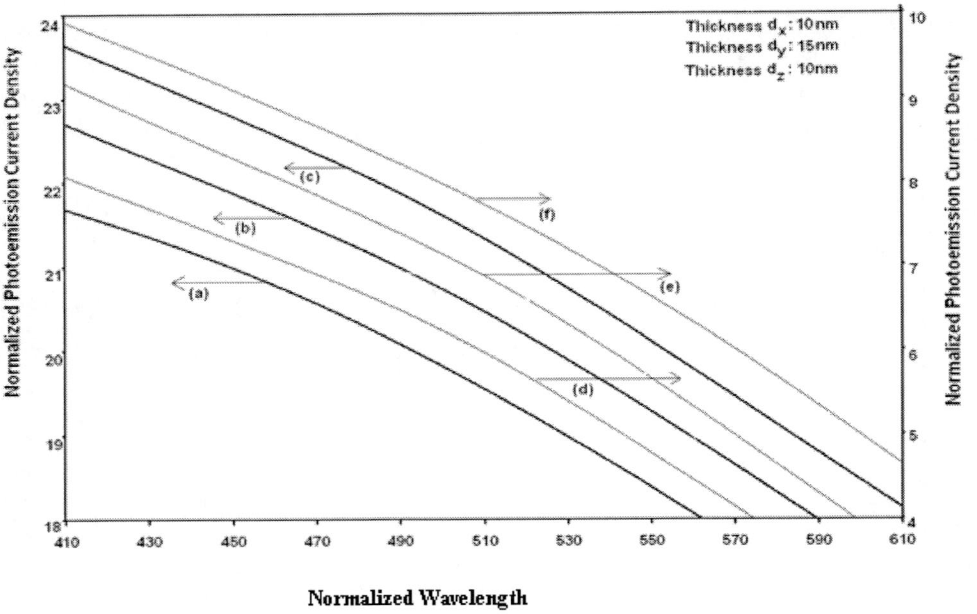

Figure 2.17. Plot of the normalized photocurrent density from quantum dot superlattices of graded interface of HgTe/$Hg_{1-x}Cd_xTe$ and $In_xGa_{1-x}As$/InP as function of microwaveslength for all cases of Figure 2.14.

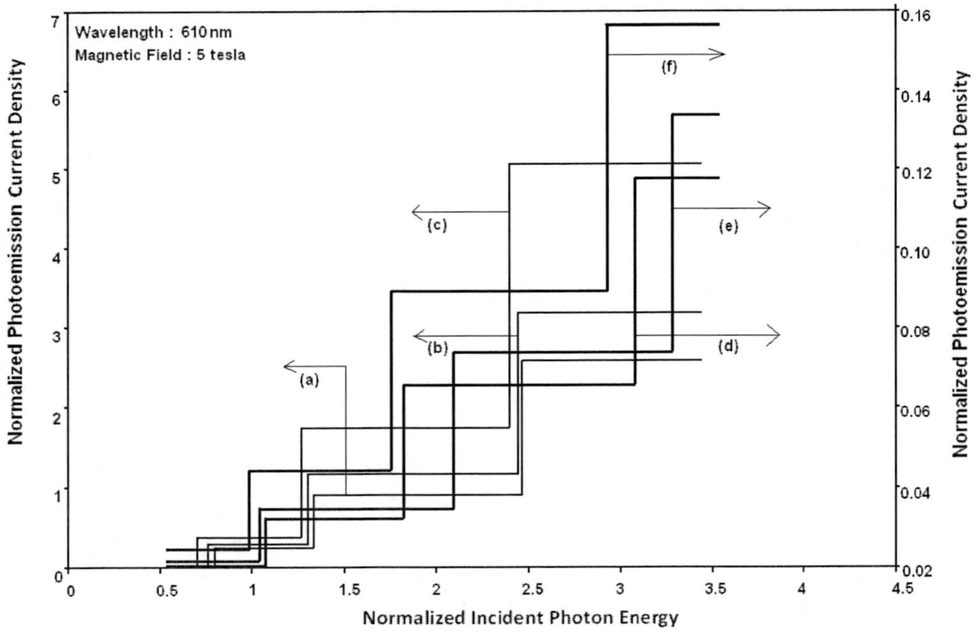

Figure 2.18. Plot of the normalized photocurrent density from quantum dot superlattices of graded interface of HgTe/$Hg_{1-x}Cd_xTe$ and $In_xGa_{1-x}As$/InP as function of normalized incident photon energy for all cases of Figure 2.14.

ACKNOWLEDGMENTS

This chapter is dedicated to the sweet memories of Late Professors P.N Butcher, J.S.Blakemore, P.T Landsberg and V.S Letokov respectively. K.P.Ghatak and D.De are grateful to UGC for sanctioning the research project having the No-F 40-469/2011(SR), Dt-13.07.2011 under which this chapter has been completed.

REFERENCES

[1] P. K. Basu, (1997) Theory of Optical Processes in Semiconductors, Bulk and Microstructures (Oxford University Press, Oxford.

[2] K. P. Ghatak, S. Bhattacharya, (2007) *J. Appl. Phys.* vol.102, 073704.

[3] K. P. Ghatak, S. Bhattacharya, S. K. Biswas, A. De, A. K. Dasgupta, (2007) *Phys. Scr.* vol.75, 820.

[4] K. P. Ghatak, S. Bhattacharya, D. De, S. Pahari, A. Dey, A. K. Dasgupta, S. N. Biswas, (2008) *J. Comput. Theo. Nanosci.vol.* 5, 1345.

[5] P. Y. Lu, C.H. Wung, C.M. Williams, S.N.G. Chu, C.M. Stiles, (1986),*Appl. Phys. Letts.* vol.49, 1372.

[6] N. R. Taskar, I. B. Bhat, K. K. Prat, D. Terry, H. Ehasani, S. K. Ghandhi, (1989) *J. Vac. Sci. Tech.* vol.7A, 281.

[7] F. Koch, (1984) *Springer Series in Solid States Sciences*, vol. 53, pp. 20 (Springer-Verlag, Germany).

[8] L. R. Tomasetta, H. D. Law, R. C. Eden, I. Reyhimy, K. Nakano, (1978) *IEEE, J. Quant. Electron.vol.* 14, 800.

[9] T. Yamato, K. Sakai, S. Akiba, Y. Suematsu, (1978). *IEEE J. Quant. Electron.* vol.14, 95.

[10] T. P. Pearsall, B. I. Miller, R. J. Capik, (1976) *Appl. Phys. Letts.* vol.28, 499.

[11] M. A. Washington, R. E. Nahory, M. A. Pollack, E. D. Beeke, (1978) *Appl. Phys. Letts.* vol.33, 854.

[12] M. I. Timmons, S. M. Bedair, R. J. Markunas, J. A. Hutchby, (1982) *Proceedings of the 16th IEEE Photovoltaic Specialist Conference* (IEEE, San Diego, California 666.

[13] E.Haga, H.Kimura, (1964) *J.Phys. Soc.* Japan, 18, 777, (1963); vol.19, 471.

[14] K. P. Ghatak, S. Bhattacharya, D. De, (2008) *Einstein Relation in Compound Semiconductors and Their Nanostructures*, Springer Series in Materials Science, vol. 116 (Springer-Verlag, Germany).

[15] A.S. Filipchenko, I.G. Lang, D.N. Nasledov, S.T. Pavlov, L.N. Radaikine, *Phys. Stat. Sol* (b) 66 (1974) 417.

[16] Adachi, S. *J. Appl. Phys.* 58, R11 (1985).

[17] Dornhaus, R and Nimtz, G., *"Springer Tracts in Modern Physics,"*, Vol. 78, (Berlin-Heidelberg, N., Springer, 1, 1976).

[18] Zawadzki, W., *"Handbook of Semiconductor Physics"*, (Paul, W., ed.). Vol. 1, (Amsterdam, North Holland, 719, 1982).

[19] Tsidilkovski, I.M., *Cand. Thesis* Leningrad University SSR (1955).

[20] Bass, F.G. and Tsidilkovski, I.M., *Ivz. Acad. Nauk Azerb SSR* 10, 3 (1966).

[21] Mitra, B. and Ghatak., K.P., *Physica Scripta* 40, 776 (1989); Biswas, S.K., Ghatak, A.R., Neogi, A., Sharma, A., Bhattacharya, S. and Ghatak, K.P., Physica E 36, 163 (2007).

[22] Mondal, M, Banik, S. and Ghatak, K.P., *J. Low Temp. Phys.* 74, 423 (1989).

[23] Ghatak, K.P, Bhattacharya, S., Saikia, H. and Sinha, A. *J. Comp. Theor. Nanoscience* 3, 1 (2006).

[24] Ghatak, K.P. and Biswas, S.N., J. Vac. Sc. Tech. 7B, 104 (1989); Chakraborty, P.K., Choudhury, S., and Ghatak, K.P. *Physica* B 387, 333 (2007).

[25] Bhattacharya, S., Pahari, S., Basu, D.K. and Ghatak, K.P., J. Comp. Theo. Nanoscience 3, 280 (2006); Sinha, A., Sharma, A.K., Barui, R., Ghatak, A.R., Bhattacharya, S. and Ghatak, K.P., *Physica* B 391, 141 (2007).

[26] Choudhury, S., Singh, L.J. and Ghatak, K.P., *Nanotechnology* 15, 180 (2004).

[27] Chowdhary, S., Singh, L.J. and Ghatak, Physica B B365, 5 (2005); Chakraborty, P.K., Sinha, A., Bhattacharya, S., Ghatak, K.P., Physica B 390, 325 (2007); Ghatak, K.P., Banerjee, J.P. and Bhattacharyya, D., *Nanotechnology* 7, 110 (1996).

[28] Chakraborty, P.K., Nag, B. and Ghatak. K.P., J. Phys. Chem. Solids 64, 2191 (2003).

[29] Ghatak, K.P., Banerjee, J.P. and Nag, B., *J. Appl. Phys.* 83, 1420 (1998).

[30] Nag, B. and Ghatak, K.P., J. Phys. Chem. Solids 59, 713 (1998); Chakraborty, P.K., Nag, B. and Ghatak. K.P., J. Phys. Chem. Sol. 64, 2191 (2003); Chakraborty, P.K., and Ghatak, K.P., Physics Letts. 288, 335 (2001); J. Phys. Chem. Sol. 62, 1061 (2001); J. Phys. D, Appl. Phys. 32, 2438 (1999); Nag, B., and Ghatak, K.P., *J. Phys. Chem. Sol.* 58, 427 (1997).

[31] Ghatak, K.P., Basu D.K. and Nag. B., J. Phys. Chem. Solids 58, 133 (1997); Nag, B. and Ghatak, K.P., Nonlinear Optics and Quantum Optics 19, 1 (1998); Ghatak, K.P. and Nag. B, *Nanostructured Materials* 10, 923 (1998).

[32] Ghatak, K.P., Nag, B. and Bhattacharyya, D., J. *Low Temp. Phys.* 14, 1 (1995); Mondal, M. and Ghatak, K.P., *Phys. Scripta* 30, 217 (1984); Mondal, M. and Ghatak, K.P., *Jour. of Mag. and Magnetic Mat.* 62, 115 (1986); Mondal, M. and Ghatak, K.P., *Thin Solid Films* 148, 219 (1987); Ghatak, K.P., Choudhury, A.K., Ghosh, S. and Chakravarti, A.N., *Appl. Phys.* 23, 241 (1980); Mondal, M., Chattopadhyay, N. and Ghatak, K.P., J. *of Low Temp. Phys*. 73, 321 (1988); Ghatak, K.P., Chattopadhyay. N. and Mondal, M., *Appl. Phys.* A 44, 305 (1987); Mitra, B. and Ghatak, K.P., *Physics Letts.* 135A, 397 (1989); Mitra, B.and Ghatak, K.P., *Solid State Electron*. 32, 810 (1989); Mitra, B., and Ghatak, K.P., *Phys. Scripta* 42, 103 (1990).

[33] K.P., Ghatak, M., Mondal, Z.F. Naturforschung 41A, 821 (1986); A.N., Chakravarti, A.K Choudhury,., K.P., Ghatak, S. Ghosh, A.Dhar, *Appl. Phys.* 25,105 (1981).

[34] K.P. Ghatak, M.Mondal, Z.F. Physik B B69, 471 (1988); A.N. Chakravarti, K.P.Ghatak, K.K.Ghosh, S. Ghosh, A. Dhar, Z.F, *Physik* B. 47, 149 (1982).

[35] H.A Lyden, Phys. Rev. 135, A514(1964); E.D Palik, G.B. Wright, in *Semiconductors and Semimetals,* edited by R.K Willardson and A.C Beer,3, p-421(1967,Academic Press, New York, USA).

[36] H.I. Zhang, Phys. Rev, 1B, 3450(1970); Y. Zhang, A. Mascarenhas, H. P. Xin, C. W. Tu, *Phys. Rev.* B 61, 7479 (2000); P. N. Hai, W. M. Chen, I. A. Buyanova, H. P. Xin, C. W. Tu,*Appl. Phys. Lett.* 77,1843 (2000); J. Wu, W. Shan, W. Walukiewicz, K. M. Yu, J. W. Ager, III, E. E.Haller, H. P. Xin, C. W. Tu, *Phys. Rev*. B 64, 085320(2001); D. L. Young, D. L. Geisz, T. J. Coutts, *Appl. Phys. Lett.* 82,1236 (2003); Y. J. Wang, X. Wei, Y. Zhang, A. Mascarenhas, H. P. Xin, Y. G.Hong, C. W. Tu, *Appl. Phys. Lett.* 82, 4453 (2003); F. Masia, A. Polimeni, G. Baldassarri Höger von Högersthal, M.Bissiri, M. Capizzi, P. J. Klar, W. Stolz, *Appl. Phys. Lett.* 82, 4474 (2003); A. Polimeni, G. Baldassarri Höger von Högersthal, F. Masia, A.Frova, M. Capizzi, S. Sanna, V. Fiorentini, P. J. Klar, W.Stolz, *Phys. Rev*. B 69, 041201(R) (2004); G. Baldassarri Höger von Högersthal, A. Polimeni, F. Masia, M.Bissiri, M. Capizzi, D. Gollub, M. Fischer, A. Forchel, *Phys. Rev.* B 67, 233304 (2003).

[37] M.Mondal, K.P.Ghatak, Phys. Letts. 131 A, 529 (1988); K.P. Ghatak, B. Mitra, *Int. J. Electron.* 72, 541 (1992). B.Mitra, A. Ghoshal, K.P.Ghatak, Nouvo Cimento D 12D, 891 (1990)., K.P.Ghatak,, S.N. Biswas, *Nonlinear Optics and Quantum Optics* 4, 347 (1993), 12, 83 (1995).,

[38] K.P.Ghatak, A. Ghoshal. B. Mitra., *Nouvo Cimento* 14D, 903 (1992); 13D, 867 (1991)., K.P Ghatak, A. Ghoshal, B. Mitra, Nouvo Cimento 13 D, 867 (1991)., B. Mitra, K.P. Ghatak, *Solid State Electron.* 32, 177 (1989)., M. Mondal,, N. Chattapadhyay, K.P. Ghatak, J. *Low Temp. Phys*. 66 131 (1987).

[39] P. N. Hai, W. M. Chen, I. A. Buyanova, H. P. Xin, C. W. Tu; *Appl. Phys. Lett.,* 77,1843(2000); D.P. DiVincenzo, E.J Mele; *Phys. Rev*. B, 29,1685 (1984); P. Perlin, E. Litwin-Staszewska, B. Suchanek, W. Knap, J. Camassel, T. Suski, R. Piotrzkowski, I. Grzegory, S. Porowski, E. Kaminska, J. C. Chervin; *Appl. Phys. Lett.*,68, 1114(1996); G.E. Smith; *Phys. Rev. Lett.,* 9, 487 (1962);

[40] D. Schneider, D. Rurup, A. Plichta, H.-U. Grubert, A. Schlachetzki, K. Hansen; *Z. Phys*. B, 95, 281(1994).

In: Quantum Dots and Quantum Cellular Automata
Editors: D. De, S. Bhattacharaya and K.P. Ghatak
ISBN: 978-1-62257-920-4

Chapter 3

SEEBECK COEFFICIENT IN QUANTUM DOTS AND QUANTUM DOT SUPER LATTICES OF HEAVILY DOPED SEMICONDUCTORS UNDER LARGE MAGNETIC FIELD

K. P. Ghatak[1*], S. Bhattacharya[2], D. De[3,4], S. M. Adhikari[1], N. Paitya[1,5], M. Mitra[6], B. Nag[6], P. K. Sarkar[1] and S.Ghosh[7]

[1]Department of Electronics and Communication Engineering, National Institute of Technology, Agartala Jirania, Tripura, India
[2]Nanoscale Device Research Laboratory, Center for Electronics Design and Technology, Indian Institute Science, Bangalore, India
[3]Department of Computer Science and Engineering, West Bengal University of Technology, Kolkata, India
[4]School of Physics, University of Western Australia, The University of Western Australia, Perth, Western Australia
[5]Department of Applied Electronics and Instrumentation Engineering, Sikkim Manipal Institute of Technology, Majitar, Rangpo, East Sikkim,India
[6]Department of Applied Physics, University of Calcutta, Kolkata, India
[7]Department of Electronics and Telecommunication Engineering, Bengal Engineering and Science University, Howrah, India

ABSTRACT

In this chapter, an attempt is made to study the Seebeck coefficient under large magnetic field (S) in quantum dots of heavily doped nonlinear optical, III-V, II-VI, GaP, Ge, Te, $PtSb_2$, II-V, GaSb, stressed materials, IV-VI, Lead Germanium Telluride, Zinc and Cadmium diphosphides and Bi_2Te_3 on the basis of newly formulated carrier dispersion laws respectively. We have also investigated the S in heavily doped III-V,II-VI,IV-VI, HgTe/CdTe and strained layer Quantum Dot Superlattices (QDSL) with graded interfaces together with the effective mass superlattices of the afore mentioned materials by formulating new carrier energy spectra. It has been found that the S for the said heavily doped quantum dots and QDSL oscillate with increasing thickness and changes with increasing electron concentration in various manners for all types of superlattices with two entirely different signatures of quantization as appropriate in respective cases of the aforementioned quantized structures.

3.1. INTRODUCTION

In recent years, with the advent of Quantum Hall Effect (QHE) [1], there has been considerable interest in studying the Seebeck coefficient under large magnetic field (S) in various types of nanostructured materials having quantum confinement of their charge carriers in one, two and three dimensions of the respective wave vector space leading to different carrier energy spectra [2-33]. The classical equation of S is valid only under the condition of carrier non-degeneracy, being independent of carrier concentration and reflects the fact that the signature of the band structure of any material is totally absent in the same.

Zawadzki [7] demonstrated that the S for electronic materials having degenerate electron concentration is essentially determined by their respective energy band structures. It has, therefore, different values in different materials and changes with the doping, magnitude of the reciprocal quantizing magnetic field under magnetic quantization, quantizing electric field as in inversion layers, nanothickness as in quantum wells, wires and dots, with superlattice period as in quantum confined semiconductor superlattices with graded interfaces having various carrier energy spectra and also in other types of field assisted nanostructured materials. Some of the significant features, which have emerged from these studies, are:

(a) The S decreases with the increase in carrier concentration;
(b) The S decreases with increasing doping in heavily doped semiconductors forming band tails;
(c) The nature of variations is significantly influenced by the spectrum constants of various materials having different band structures;
(d) The S exhibits oscillatory dependence with inverse quantizing magnetic field because of the to Shubnikov de Haas effect;
(e) The S decreases with the magnitude of the quantizing electric field in inversion layers;
(f) The S exhibits composite oscillations with significantly different values in superlattices and various other quantized field aided structures.

In this chapter, an attempt is made to study the S in quantum dots of heavily doped nonlinear optical, III-V, II-VI, GaP, Ge, Te, $PtSb_2$, II-V, GaSb, stressed materials, IV-VI, Lead Germanium Telluride, Zinc and Cadmium diphosphides and Bi_2Te_3 on the basis of newly formulated respective carrier dispersion laws from sections 3.2.1 to 3.2.14 of theoretical background respectively. We have also investigated the S in heavily doped III-V,II-VI,IV-VI, HgTe/CdTe and strained layer Quantum Dot Superlattices (QDSL) with graded interfaces together with the effective mass superlattices of the afore mentioned materials byin sections 3.2.15 to 3.2.24 by formulating new carrier energy spectra.

It is well known that the band tails are being formed in the forbidden zone of heavily doped semiconductors and can be explained by the overlapping of the impurity band with the conduction and valence bands [34]. Kane [35] and Bonch Bruevich [36] have independently derived the theory of band tailing for semiconductors having unperturbed parabolic energy bands. Kane's model [34] was used to explain the experimental results on tunneling [37] and the optical absorption edges [38, 39] in this context. Halperin and Lax [40] developed a model for band tailing applicable only to the deep tailing states. Although Kane's concept is often used in the literature for the investigation of band tailing [41,42], it may be noted that this model [35, 43] suffers from serious assumptions in the sense that the local impurity potential is assumed to be small and slowly varying in space coordinates [42]. In this respect, the local impurity potential may be assumed to be a constant. In order to avoid these approximations, we have developed in this chapter the electron energy spectra for heavily doped semiconductors for studying the S based on the concept of the variation of the kinetic energy [34, 42] of the electron with the local point in space coordinates. This kinetic energy is then averaged over the entire region of variation using a Gaussian type potential energy. On the basis of the *E–k* dispersion relation, we have obtained the electron statistics for different quantum dots of heavily doped materials and quantum dots of heavily doped superlattices for the purpose of numerical computation of the respective Ss. It may be noted that, a more general treatment of many-body theory for the density-of-states of heavily doped semi-conductors merges with one-electron theory under macroscopic conditions [34]. Also, the experimental results for the Fermi energy and others are the average effect of this macroscopic case. So, the present treatment of the one-electron system is more applicable to the experimental point of view and it is also easy to understand the overall effect in such a case [44]. In a heavily-doped semiconductors, each impurity atom is surrounded by the electrons, assuming a regular distribution of atoms, and it is screened independently [41, 43, 45]. The interaction energy between electrons and impurities is known as the impurity screening potential. This energy is determined by the inter-impurity distance and the screening radius, which is known as the screening length. The screening radius grows with the electron concentration and the effective mass. Furthermore, these entities are important for heavily doped materials in characterizing the semiconductor properties [46, 47] and the devices [41,48]. The works on Fermi energy and the screening length in an n-type GaAs have already been initiated in the literature [49,50,51], based on Kane's model. Incidentally, the limitations of Kane's model [43], as mentioned above, are also present in their studies.

At this point, it may be noted that many band tail models are proposed using the Gaussian distribution of the impurity potential variation [35, 42]. In this chapter, we have used the Gaussian band tails to study the S for heavily doped nonlinear optical, III-V, II-VI, GaP, Ge, Te, $PtSb_2$, II-V, GaSb, stressed materials, IV-VI, Lead Germanium Telluride, Zinc and Cadmium diphosphides, Bi_2Te_3 and quantum dots of heavily doped superlattices. Our method is not at al related with the density-of-states (DOS) technique as used in the aforementioned works. From the electron energy spectrum, one can obtain the DOS but the DOS technique, as used in the literature cannot provide the *E-k* dispersion relation. Therefore, our study is more fundamental than those in the existing literature, because the Boltzmann transport equation, which controls the study of the charge transport properties of the semiconductor devices, can be solved if and only if the *E-k* dispersion relation is known. We wish to note that the Gaussian function for the impurity potential distribution has been used by many authors. It has been widely used since 1963 when Kane first proposed it. We will also use the Gaussian distribution for the present study. The section 3.3 contains the results and discussion of this chapter.

3.2. THEORETICAL BACKGROUND

3.2.1. Study of the Seebeck Coefficient under Large Magnetic Field in Quantum Dots of Heavily Doped Nonlinear Optical Materials

The generalized unperturbed electron energy spectrum for the bulk specimens of the nonlinear optical materials in the absence of any doping can be expressed following equation (1.2) as

$$\frac{\hbar^2 k_z^2}{2m_{\|}^*}+\left(\frac{b_{\|}}{b_{\perp}}\frac{c_{\perp}}{c_{\|}}\right)\frac{\hbar^2 k_s^2}{2m_{\perp}^*}=\left\{\frac{E(\alpha E+1)(b_{\|}E+1)}{(c_{\|}E+1)}+\frac{\alpha b_{\|}}{c_{\|}}\right.$$
$$\left.\left[\delta E+\frac{2}{9}\left(\Delta_{\|}^2-\Delta_{\perp}^2\right)\right]-\left(\frac{2}{9}\right)\frac{\alpha b_{\|}}{c_{\|}}\frac{\left(\Delta_{\|}^2-\Delta_{\perp}^2\right)}{\left(c_{\|}E+1\right)}\right\}$$
$$-\left(\frac{\hbar^2 k_s^2}{2m_{\perp}^*}\right)\left\{\left(\frac{b_{\|}}{b_{\perp}}\frac{c_{\perp}}{c_{\|}}\right)\left[\left(\frac{\delta}{2}+\frac{\Delta_{\|}^2-\Delta_{\perp}^2}{6\Delta_{\|}}\right)\frac{\alpha_{\|}}{\alpha_{\|}E+1}\right.\right.$$
$$\left.\left.+\left(\frac{\delta}{2}-\left\{\frac{\Delta_{\|}^2-\Delta_{\perp}^2}{6\Delta_{\|}}\right\}\right)\frac{c_{\|}}{c_{\|}E+1}\right]\right\} \tag{3.1}$$

where

$b_{\|}\equiv 1/\left(E_g+\Delta_{\|}\right)$, $c_{\perp}\equiv 1/\left(E_g+\frac{2}{3}\Delta_{\perp}\right)$, $b_{\perp}\equiv 1/\left(E_g+\Delta_{\perp}\right)$, $c_{\|}\equiv 1/\left(E_g+\frac{2}{3}\Delta_{\|}\right)$

and $\alpha\equiv 1/E_g$.

The Gaussian distribution $F(V)$ of the impurity potential is given by [35]

$$F(V)=\left(\pi\eta_g^2\right)^{-1/2}\exp\left(-V^2/\eta_g^2\right) \tag{3.2}$$

where, η_g is the impurity scattering potential. It appears from (3.2) that the variance parameter η_g is not equal to zero, but the mean value is zero. Further, the impurities are assumed to be uncorrelated and the band mixing effect has been neglected in this simplified theoretical formalism.

We have to average the Kinetic energy in the order to obtain the E-k dispersion relation in nonlinear optical materials including the band tailing effect. Using the (3.1) and (3.2), we get

$$\left[\frac{\hbar^2 k_z^2}{2m_\parallel^*}\int_{-\infty}^{E} F(V)dV\right]+\left[\left(\frac{b_\parallel}{b_\perp}\frac{c_\perp}{c_\parallel}\right)\frac{\hbar^2 k_s^2}{2m_\perp^*}\int_{-\infty}^{E} F(V)dV\right]$$
$$=\left\{\int_{-\infty}^{E}\frac{(E-V)[\alpha(E-V)+1]\left[b_\parallel(E-V)+1\right]}{\left[c_\parallel(E-V)+1\right]}F(V)dV\right.$$
$$+\frac{\alpha b_\parallel}{c_\parallel}\left[\delta\int_{-\infty}^{E}(E-V)F(V)dV+\frac{2}{9}\left(\Delta_\parallel^2-\Delta_\perp^2\right)\int_{-\infty}^{E}F(V)dV\right]$$
$$\left.-\left(\frac{2}{9}\right)\frac{\alpha b_\parallel}{c_\parallel}\left(\Delta_\parallel^2-\Delta_\perp^2\right)\int_{-\infty}^{E}\frac{F(V)dV}{\left[c_\parallel(E-V)+1\right]}\right\}-$$
$$\left(\frac{\hbar^2 k_s^2}{2m_\perp^*}\right)\left\{\left(\frac{b_\parallel}{b_\perp}\frac{c_\perp}{c_\parallel}\right)\left[\left(\frac{\delta}{2}+\frac{\Delta_\parallel^2-\Delta_\perp^2}{6\Delta_\parallel}\right)\alpha\int_{-\infty}^{E}\frac{F(V)dV}{[\alpha(E-V)+1]}\right.\right.$$
$$\left.\left.+\left(\frac{\delta}{2}-\frac{\Delta_\parallel^2-\Delta_\perp^2}{6\Delta_\parallel}\right)c_\parallel\int_{-\infty}^{E}\frac{F(V)dV}{\left[c_\parallel(E-V)+1\right]}\right]\right\} \tag{3.3}$$

Equation (3.3) can be rewritten as

$$\frac{\hbar^2 k_z^2}{2m_\parallel^*}I(1)+\left(\frac{b_\parallel}{b_\perp}\frac{c_\perp}{c_\parallel}\right)\frac{\hbar^2 k_s^2}{2m_\perp^*}I(1)=\left\{I_3\left(c_\parallel\right)+\frac{\alpha b_\parallel}{c_\parallel}\left[\delta I(4)\right.\right.$$
$$\left.\left.+\frac{2}{9}(\Delta_\parallel^2-\Delta_\perp^2)I(1)\right]-\left(\frac{2}{9}\right)\frac{\alpha b_\parallel}{c_\parallel}(\Delta_\parallel^2-\Delta_\perp^2)I_6\left(c_\parallel\right)\right\}-$$
$$\left(\frac{\hbar^2 k_s^2}{2m_\perp^*}\right)\left\{\left(\frac{b_\parallel}{b_\perp}\frac{c_\perp}{c_\parallel}\right)\left[\left(\frac{\delta}{2}+\frac{\Delta_\parallel^2-\Delta_\perp^2}{6\Delta_\parallel}\right)\alpha I(\alpha)+\left(\frac{\delta}{2}-\left\{\frac{\Delta_\parallel^2-\Delta_\perp^2}{6\Delta_\parallel}\right\}\right)c_\parallel I\left(c_\parallel\right)\right]\right\} \tag{3.4}$$

where,

$$I(1)\equiv\int_{-\infty}^{E}F(V)dV \tag{3.5}$$

$$I_3\left(c_\parallel\right)\equiv\int_{-\infty}^{E}\frac{(E-V)[\alpha(E-V)+1]\left[b_\parallel(E-V)+1\right]}{\left[c_\parallel(E-V)+1\right]}F(V)dV \tag{3.6}$$

$$I(4)\equiv\int_{-\infty}^{E}(E-V)F(V)dV \tag{3.7}$$

$$I(\alpha) \equiv \int_{-\infty}^{E} \frac{F(V)dV}{[\alpha(E-V)+1]} \tag{3.8}$$

Let us substitute $E-V \equiv x$ and $x/\eta_g \equiv t_0$, we get from (3.5)

$$I(1) = \left(\exp(-E^2/\eta_g^2)/\sqrt{\pi}\right)\int_0^{\infty}\exp\left[-t_0^2+\left(2Et_0/\eta_g\right)\right]dt_0$$

Thus,

$$I(1) = \left[\frac{1+Erf\left(E/\eta_g\right)}{2}\right] \tag{3.9}$$

From (3.7), one can write

$$I(4) = \left(1/\eta_g\sqrt{\pi}\right)\int_{-\infty}^{E}(E-V)\exp(-V^2/\eta_g^2)dV$$

$$= \frac{E}{2}\left[1+Erf(E/\eta_g)\right] - \left\{\frac{1}{\sqrt{\pi\eta_g^2}}\int_{-\infty}^{E} V\exp\left(-E^2/\eta_g^2\right)dV\right\}$$

After computing this simple integration, one obtains

Thus, $I(4) = \eta_g \exp\left(-E^2/\eta_g^2\right)\left(2\sqrt{\pi}\right)^{-1} + \frac{E}{2}\left(1+Erf(E/\eta_g)\right) = \gamma_0\left(E,\eta_g\right)$ (3.10)

From (3.8), we can write

$$I(\alpha) = \frac{1}{\sqrt{\pi\eta_g^2}}\int_{-\infty}^{E}\frac{\exp\left(-V^2/\eta_g^2\right)dV}{[\alpha(E-V)+1]} \tag{3.11}$$

when, $V \to \pm\infty$, $\frac{1}{[\alpha(E-V)+1]} \to 0$ and $\exp\left(-V^2/\eta_g^2\right) \to 0$; therefore, using (3.11) one can write

$$I(\alpha) = \frac{1}{\sqrt{\pi\eta_g^2}}\int_{-\infty}^{+\infty}\frac{\exp\left(-V^2/\eta_g^2\right)dV}{[\alpha E+1-\alpha V]} \tag{3.12}$$

The (3.12) can be expressed as

$$I(\alpha)=\left(1/\alpha\eta_g\sqrt{\pi}\right)\int_{-\infty}^{\infty}\exp\left(-t^2\right)(u-t)^{-1}\,dt \tag{3.13}$$

where, $\dfrac{V}{\eta_g}\equiv t$ and $u\equiv\left(\dfrac{1+\alpha E}{\alpha\eta}\right)$.

It is well known that [53-54]

$$W(Z)=(i/\pi)\int_{-\infty}^{\infty}(Z-t)^{-1}\exp(-t^2)dt \tag{3.14}$$

in which $i=\sqrt{-1}$ and Z is, in general, a complex number.

We also know [53-54],

$$W(Z)=\exp\left(-Z^2\right)Erfc(-iZ) \tag{3.15}$$

where, $Erfc(Z)\equiv 1-Erf(Z)$.

Thus, $Erfc(-iu)=1-Erf(-iu)$

Since, $Erf(-iu)=-Erf(iu)$

Therefore, $Erfc(-iu)=1+Erf(iu)$.

Thus,

$$I(\alpha)=\left[-i\sqrt{\pi}/\alpha\eta_g\right]\exp\left(-u^2\right)[1+Erf(iu)] \tag{3.16}$$

We also know that [53-54]

$$Erf(x+iy)=Erf(x)+\left(\frac{e^{-x^2}}{2\pi x}\right)\left[(1-\cos(2xy))+i\sin(2xy)+\frac{2}{\pi}e^{-x^2}\sum_{p=1}^{\infty}\frac{\exp\left(-p^2/4\right)}{\left(p^2+4x^2\right)}\right]$$

$$\left[f_p(x,y)+ig_p(x,y)+\varepsilon(x,y)\right] \tag{3.17}$$

where,

$$f_p(x,y) \equiv \left[2x - 2x\cosh(py)\cos(2xy) + p\sinh(py)\sin(2xy)\right],$$
$$g_p(x,y) \equiv \left[2x\cosh(py)\sin(2xy) + p\sinh(py)\cos(2xy)\right],$$
$$\left|\varepsilon(x,y)\right| \approx 10^{-16}\left|Erf(x+iy)\right|.$$

Substituting $x=0$ and $y=u$ in (3.17), one obtains,

$$Erf(iu) = \left(\frac{2i}{\pi}\right)\sum_{p=1}^{\infty}\left\{\frac{\exp(-p^2/4)}{p}\sinh(pu)\right\} \quad (3.18)$$

Therefore, one can write

$$I(\alpha) = C_{21}(\alpha, E, \eta_g) - iD_{21}(\alpha, E, \eta_g) \quad (3.19)$$

where,

$$C_{21}(\alpha, E, \eta_g) \equiv \left[\frac{2}{\alpha\eta_g\sqrt{\pi}}\right]\exp(-u^2)\left[\sum_{p=1}^{\infty}\left\{\frac{\exp(-p^2/4)}{p}\sinh(pu)\right\}\right]$$

and $D_{21}(\alpha, E, \eta_g) \equiv \left[\frac{\sqrt{\pi}}{\alpha\eta_g}\exp(-u^2)\right]$.

The equation (3.19) has both real and imaginary parts and therefore, $I(\alpha)$ is complex, which can also be prove by using the method of analytic continuation.

The integral $I_3(c_{\|})$ in (3.6) can be written as

$$I_3(c_{\|}) = \left(\frac{\alpha b_{\|}}{c_{\|}}\right)I(5) + \left(\frac{\alpha c_{\|} + b_{\|}c_{\|} - \alpha b_{\|}}{c_{\|}^2}\right)I(4) + \frac{1}{c_{\|}}\left(1 - \frac{\alpha}{c_{\|}}\right)\left(1 - \frac{b_{\|}}{c_{\|}}\right)I(1) - \left\{\frac{1}{c_{\|}}\left(1 - \frac{\alpha}{c_{\|}}\right)\left(1 - \frac{b_{\|}}{c_{\|}}\right)I(c_{\|})\right\} \quad (3.20)$$

where

$$I(5) \equiv \int_{-\infty}^{E}(E-V)^2 F(V)\,dV \quad (3.21)$$

From (3.21) one can write

$$I(5)=\frac{1}{\sqrt{\pi\eta_g^2}}\left[E^2\int_{-\infty}^{E}\exp\left(\frac{-V^2}{\eta_g^2}\right)dV-2E\int_{-\infty}^{E}V\exp\left(\frac{-V^2}{\eta_g^2}\right)dV+\int_{-\infty}^{E}V^2\exp\left(\frac{-V^2}{\eta_g^2}\right)dV\right]$$

The evaluations of the component integrals lead us to write

$$I(5)=\frac{\eta_g E}{2\sqrt{\pi}}\exp\left(\frac{-E^2}{\eta_g^2}\right)+\frac{1}{4}\left(\eta_g^2+2E^2\right)\left[1+Erf\left(\frac{E}{\eta_g}\right)\right]=\theta_0\left(E,\eta_g\right) \tag{3.22}$$

Thus combining the aforementioned equations, $I_3\left(c_\parallel\right)$ can be expressed as

$$I_3\left(c_\parallel\right)=A_{21}\left(E,\eta_g\right)+iB_{21}\left(E,\eta_g\right) \tag{3.23}$$

where,

$$A_{21}\left(E,\eta\right)\equiv\left[\frac{\alpha b_\parallel}{c_\parallel}\left[\frac{\eta_g E}{2\sqrt{\pi}}\exp\left(\frac{-E^2}{\eta_g^2}\right)+\frac{1}{4}\left(\eta_g^2+2E^2\right)\left\{1+Erf\left(\frac{E}{\eta_g}\right)\right\}\right]\right]$$
$$+\left[\frac{\alpha c_\parallel+b_\parallel c_\parallel-\alpha b_\parallel}{c_\parallel^2}\right]\left\{\frac{E}{2}\left[1+Erf\left(E/\eta\right)\right]+\frac{\eta_g\exp\left(-E^2/\eta_g^2\right)}{2\sqrt{\pi}}\right\}$$
$$+\frac{1}{c_\parallel}\left(1-\frac{\alpha}{c_\parallel}\right)\left(1-\frac{b_\parallel}{c_\parallel}\right)\frac{1}{2}\left[1+Erf(E/\eta)\right]-\left\{\frac{2}{c_\parallel^2\eta_g\sqrt{\pi}}\left(1-\frac{\alpha}{c_\parallel}\right)\left(1-\frac{b_\parallel}{c_\parallel}\right)\exp(-u_1^2)\right\}\left[\sum_{p=1}^{\infty}\left\{\frac{\exp(-p^2/4)}{p}\sinh(pu_1)\right\}\right],$$

$$u_1\equiv\left[\frac{1+c_\parallel E}{c_\parallel\eta_g}\right]\text{ and } B_{21}\left(E,\eta_g\right)\equiv\frac{\sqrt{\pi}}{c_\parallel^2\eta_g}\left(1-\frac{\alpha}{c_\parallel}\right)\left(1-\frac{b_\parallel}{c_\parallel}\right)\exp(-u_1^2).$$

Therefore, the combination of all the appropriate equations together with the algebraic manipulations lead to the dispersion relation of the conduction electrons of heavily doped nonlinear optical materials forming Gaussian band tails as

$$\frac{\hbar^2k_z^2}{2m_\parallel^*T_{21}\left(E,\eta_g\right)}+\frac{\hbar^2k_s^2}{2m_\perp^*T_{22}\left(E,\eta_g\right)}=1 \tag{3.24}$$

where $T_{21}\left(E,\eta_g\right)$ and $T_{22}\left(E,\eta_g\right)$ have both real and complex parts and they are given by

$$T_{21}\left(E,\eta_g\right) \equiv \left[T_{27}\left(E,\eta_g\right) + iT_{28}\left(E,\eta_g\right)\right],\ T_{27}\left(E,\eta_g\right) \equiv \left[\frac{T_{23}\left(E,\eta_g\right)}{T_5\left(E,\eta_g\right)}\right],$$

$$T_{23}\left(E,\eta_g\right) \equiv \left[A_{21}\left(E,\eta_g\right) + \frac{\alpha b_{\|}}{c_{\|}}\left[\delta\gamma_0\left(E,\eta_g\right) + \frac{1}{9}\left(\Delta_{\|}^2 - \Delta_{\perp}^2\right)\left[1 + Erf\left(E/\eta_g\right)\right]\right]\right.$$
$$\left. - \left\{\frac{2}{9}\left(\frac{\alpha b_{\|}}{c_{\|}}\right)\left(\Delta_{\|}^2 - \Delta_{\perp}^2\right)G_{21}\left(c_{\|}, E, \eta_g\right)\right\}\right],$$

$$G_{21}\left(E,\eta_g\right) \equiv \frac{2}{c_{\|}\eta_g\sqrt{\pi}}\exp\left(-u_1^2\right)\sum_{p=1}^{\infty}\left\{\frac{\exp\left(-p^2/4\right)}{p}sinh\left(pu_1\right)\right\},$$

$$T_5\left(E,\eta_g\right) \equiv \frac{1}{2}\left[1 + Erf\left(E/\eta_g\right)\right],$$

$$T_{28}\left(E,\eta_g\right) \equiv \left[\frac{T_{24}\left(E,\eta_g\right)}{T_5\left(E,\eta_g\right)}\right],$$

$$T_{24}\left(E,\eta_g\right) \equiv \left[B_{21}\left(E,\eta_g\right) + \frac{2}{9}\frac{\alpha b_{\|}}{c_{\|}}\left(\Delta_{\|}^2 - \Delta_{\perp}^2\right)H_{21}\left(c_{\|}, E, \eta_g\right)\right],$$

$$H_{21}\left(c_{\|}, E, \eta_g\right) \equiv \left[\frac{\sqrt{\pi}}{\eta_g c_{\|}}\exp\left(-u_1^2\right)\right],$$

$$T_{22}\left(E,\eta_g\right) \equiv \left[T_{29}\left(E,\eta_g\right) + iT_{30}\left(E,\eta_g\right)\right],$$

$$T_{29}\left(E,\eta_g\right) \equiv \frac{T_{23}\left(E,\eta_g\right)T_{25}\left(E,\eta_g\right) - T_{24}\left(E,\eta_g\right)T_{26}\left(E,\eta_g\right)}{\left[\left(T_{25}\left(E,\eta_g\right)\right)^2 + \left(T_{26}\left(E,\eta_g\right)\right)^2\right]},$$

$$T_{25}\left(E,\eta_g\right) \equiv \left[\left(\frac{b_{\|}}{b_{\perp}}\frac{c_{\perp}}{c_{\|}}\right)\frac{1}{2}\left[1 + Erf\left(\frac{E}{\eta_g}\right)\right] + \left(\frac{b_{\|}}{b_{\perp}}\frac{c_{\perp}}{c_{\|}}\right)\left(\frac{\delta}{2} + \left[\frac{\Delta_{\|}^2 - \Delta_{\perp}^2}{6\Delta_{\|}}\right]\right)\alpha_{\|}C_{21}\left(\alpha_{\|}, E, \eta_g\right) + \left(\frac{b_{\|}c_{\perp}}{b_{\perp}}\right)\left(\frac{\delta}{2} - \left[\frac{\Delta_{\|}^2 - \Delta_{\perp}^2}{6\Delta_{\|}}\right]\right)G_{21}\left(\alpha_{\|}, E, \eta_g\right)\right],$$

$$C_{21}\left(\alpha, E, \eta_g\right) \equiv \left[\frac{2}{\alpha\sqrt{\pi}\eta_g}\exp\left(-u^2\right)\left[\sum_{p=1}^{\infty}\frac{\exp\left(-p^2/4\right)}{p}sinh(pu)\right]\right],$$

$$T_{26}\left(E,\eta_g\right) \equiv \left(\frac{b_{\|}}{b_{\perp}}\frac{c_{\perp}}{c_{\|}}\right)\left(\frac{\delta}{2} - \frac{\Delta_{\|}^2 - \Delta_{\perp}^2}{6\Delta_{\|}}\right)\alpha D_{21}\left(\alpha, E, \eta_g\right) + \frac{b_{\|}c_{\perp}}{b_{\perp}}\left(\frac{\delta}{2} - \frac{\Delta_{\|}^2 - \Delta_{\perp}^2}{6\Delta_{\|}}\right)H_{21}\left(c_{\|}, E, \eta_g\right),$$

and $T_{30}\left(E,\eta_g\right) \equiv \frac{T_{24}\left(E,\eta_g\right)T_{25}\left(E,\eta_g\right)+T_{23}\left(E,\eta_g\right)T_{26}\left(E,\eta_g\right)}{\left[\left(T_{25}\left(E,\eta_g\right)\right)^2+\left(T_{26}\left(E,\eta_g\right)\right)^2\right]}$.

From (3.24), it appears that the energy spectrum in heavily doped nonlinear optical semiconductors is complex. The complex nature of the electron dispersion law in heavily doped semiconductors occurs from the existence of the essential poles in the corresponding undoped electron energy spectrum. It may be noted that the complex band structures have already been studied for bulk semiconductors and superlattices without heavy doping [55] and bears no relationship with the complex electron dispersion law as indicated by (3.24). The physical picture behind the formulation of the complex energy spectrum in heavily doped nonlinear optical semiconductors is the interaction of the impurity atoms in the tails with the splitting constants of the valance bands. More is the interaction causes more prominence of the complex part than the other case. When there is no heavy doping, $\eta_g \to 0$, and there is no interaction of the impurity atoms in the tails with the spin orbit constants. As a result, there exist no complex energy spectrum and (3.24) gets converted into (3.2) when $\eta_g \to 0$. Besides, the complex spectra are not related to same evanescent modes in the band tails and the conduction bands.

The simplified dispersion relation of the conduction electrons in quantum dots of heavily doped nonlinear optical semiconductors can be written as

$$\frac{(\hbar n_z\pi/d_z)^2}{2m_{\parallel}^{*}T_{21}\left(E',\eta_g\right)}+\frac{(\hbar n_x\pi/d_x)^2+(\hbar n_y\pi/d_y)^2}{2m_{\perp}^{*}T_{22}\left(E',\eta_g\right)}=1 \tag{3.25}$$

where E' is the real part of the totally quantized energy.

The electron concentration is given by

$$n_0=\left(\frac{2g_v}{d_xd_yd_z}\right)\sum_{n_x=1}^{n_{x_{\max}}}\sum_{n_y=1}^{n_{y_{\max}}}\sum_{n_z=1}^{n_{z_{\max}}}F_{-1}(\eta_{3D}) \tag{3.26}$$

where $\eta_{3D}=((E_{FD}-E')/k_BT)$, E_{FD} is the Fermi energy in the present case and E' is the lowest positive real root of (3.25).

The Seebeck coefficient under large magnetic field (S) in the present case can be written as

$$S=\left(\pi^2k_B^2T/3en_0\right)\left(\frac{\partial n_0}{\partial(E_{FD}-E')}\right) \tag{3.27}$$

Thus using (3.26) and (3.27) one can obtain the Seebeck coefficient in the presence of strong magnetic field for quantum dots of heavily doped nonlinear optical materials.

3.2.2. Study of the Seebeck Coefficient under Large Magnetic Field in Quantum Dots of Heavily Doped III-V Materials

As known from chapter 1 that the dispersion relation of the conduction electrons of III-V compounds are described by the models of Kane (both three and two bands), Stillman et. al. Newson and Kurobe, Rossler, Palik et. al. Johnson and Dickey and Agafonov et.al. respectively. For the purpose of complete and coherent presentation, the Seebeck coefficient under large magnetic field in QDs of heavily doped III-V compounds have also been investigated in accordance with the aforementioned different dispersion relations for the purpose of relative comparison as follows:

(a)

Under the conditions, $\delta = 0$, $m_{\parallel}^* = m_{\perp}^* = m^*$ and $\Delta_{\parallel} = \Delta_{\perp} = \Delta$ in (3.24), the electron dispersion law in this case assumes the form

$$\frac{\hbar^2 k^2}{2m^*} = T_{31}(E,\eta_g) + iT_{32}(E,\eta_g) \tag{3.28}$$

where,

$$T_{31}(E,\eta_g) \equiv \left(\frac{2}{1+Erf(E/\eta_g)}\right)\left[\frac{\alpha b}{c}\theta_0(E,\eta_g) + \left[\frac{\alpha c + bc - \alpha b}{c^2}\right]\gamma_0(E,\eta_g) + \frac{1}{c}\left(1-\frac{\alpha}{c}\right)\left(1-\frac{b}{c}\right)\frac{1}{2}\left[1+Erf\left(\frac{E}{\eta_g}\right)\right]\right.$$

$$\left. -\frac{1}{c}\left(1-\frac{\alpha}{c}\right)\left(1-\frac{b}{c}\right)\frac{2}{c\eta_g\sqrt{\pi}}\exp(-u_2^2)\left[\sum_{p=1}^{\infty}\frac{\exp(-p^2/4)}{p}\sinh(pu_2)\right]\right],$$

$$b \equiv \left(\frac{1}{E_g+\Delta}\right),\ c \equiv \left(\frac{1}{E_g+\frac{2}{3}\Delta}\right),$$

$$u_2 \equiv \frac{1+cE}{c\eta_g} \text{ and } T_{32}(E,\eta_g) \equiv \left(\frac{2}{1+Erf(E/\eta_g)}\right)\frac{1}{c}\left(1-\frac{\alpha}{c}\right)\left(1-\frac{b}{c}\right)\frac{\sqrt{\pi}}{c\eta_g}\exp(-u_2^2).$$

Thus, the complex energy spectrum occurs due to the term $T_{32}(E,\eta_g)$ and this imaginary band is quite different from the forbidden energy band.

The simplified dispersion relation of the conduction electrons in quantum dots of heavily doped III-V ternary and quaternary materials can be written as

$$((\hbar n_z \pi / d_z)^2 + (\hbar n_x \pi / d_x)^2 + (\hbar n_y \pi / d_y)^2) / 2m^* = T_{31}\left(E_1', \eta_g\right) + iT_{32}\left(E_1', \eta_g\right) \tag{3.29}$$

where E_1' is the real part of the totally quantized energy I this case.

The electron concentration in this case is given by

$$n_0 = \left(\frac{2g_v}{d_x d_y d_z}\right) \sum_{n_x=1}^{n_{x_{\max}}} \sum_{n_y=1}^{n_{y_{\max}}} \sum_{n_z=1}^{n_{z_{\max}}} F_{-1}(\eta_{3D1}) \tag{3.30}$$

where $\eta_{3D1} = ((E_{FD1} - E_1') / k_B T)$ and E_{FD1} is the Fermi energy in the present case .

Thus using (3.30) and (3.27) one can obtain the Seebeck coefficient in the presence of large magnetic field for quantum dots of heavily doped III-V, ternary and quaternary materials whose undoped energy spectrum obeys the three band model of Kane.

(b)

The dispersion relation in heavily doped III-V, ternary and quaternary materials whose undoped energy spectrum obeys the two band model of Kane is given by

$$(\hbar^2 k^2 / 2m^*) = \gamma_2\left(E, \eta_g\right) \tag{3.31}$$

where, $\gamma_2\left(E, \eta_g\right) \equiv \left[\dfrac{2}{1 + Erf\left(E/\eta_g\right)}\right]\left[\gamma_0\left(E, \eta_g\right) + \alpha\theta_0\left(E, \eta_g\right)\right]$.

Since, the original two band Kane model is an all zero and no pole function, therefore, the heavily doped counterpart will be totally real and the complex band vanishes.

Therefore the electron energy spectrum in quantum dots of heavily doped III-V, ternary and quaternary materials whose bulk dispersion relation follows (3.31) can be expressed as

$$\cdot\ \gamma_2\left(E_2', \eta_g\right) = ((\hbar n_z \pi / d_z)^2 + (\hbar n_x \pi / d_x)^2 + (\hbar n_y \pi / d_y)^2) / 2m^* \tag{3.32}$$

where E_2' is the totally quantized energy in this case.

The electron concentration in this case is given by

$$n_0 = \left(\frac{2g_v}{d_x d_y d_z}\right) \sum_{n_x=1}^{n_{x_{\max}}} \sum_{n_y=1}^{n_{y_{\max}}} \sum_{n_z=1}^{n_{z_{\max}}} F_{-1}(\eta_{3D2}) \tag{3.33}$$

where $\eta_{3D2} = ((E_{FD2} - E_2') / k_B T)$ and E_{FD2} is the Fermi energy in the present case .

Thus using (3.33) and (3.27) one can obtain the Seebeck coefficient in the presence of large magnetic field for quantum dots of heavily doped III-V, ternary and quaternary materials whose undoped energy spectrum obeys the two band model of Kane.

(c) The dispersion relation in heavily doped semiconductors whose unperturbed conduction electrons obeys parabolic energy bands is given by

$$\frac{\hbar^2 k^2}{2m^*} = \gamma_3\left(E,\eta_g\right) \tag{3.34}$$

where, $\gamma_3\left(E,\eta_g\right) \equiv \left[\frac{2}{\left(1+Erf\left(E/\eta_g\right)\right)}\right]\gamma_0\left(E,\eta_g\right)$.

Since, the original parabolic energy band is no pole function, therefore, the heavily doped counterpart will be totally real, which is also apparent form the expression (3.34).

Therefore the electron energy spectrum in quantum dots of heavily doped III-V, ternary and quaternary materials whose bulk dispersion relation follows (3.34) can be expressed as

$$\gamma_3\left(E_3',\eta_g\right) = ((\hbar n_z \pi / d_z)^2 + (\hbar n_x \pi / d_x)^2 + (\hbar n_y \pi / d_y)^2) / 2m^* \tag{3.35}$$

where E_3' is the totally quantized energy in this case.

The electron concentration in this case is given by

$$n_0 = \left(\frac{2g_v}{d_x d_y d_z}\right)\sum_{n_x=1}^{n_{x_{\max}}}\sum_{n_y=1}^{n_{y_{\max}}}\sum_{n_z=1}^{n_{z_{\max}}} F_{-1}\left(\eta_{3D3}\right) \tag{3.36}$$

where $\eta_{3D3} = ((E_{FD3} - E_3') / k_B T)$ and E_{FD3} is the Fermi energy in the present case .

Thus using (3.36) and (3.27) one can obtain the Seebeck coefficient in the presence of strong magnetic field for quantum dots of heavily doped materials whose undoped energy spectrum obeys the parabolic dispersion relation.

(d) The Model of Stillman et. al.

In accordance with the model of Stillman et. al., the electron dispersion law of heavily doped III-V materials assumes the form

$$\gamma_3(E,\eta_g) = t_{11}k^2 - t_{12}k^4 \tag{3.37}$$

where the notations are defined in (1.33) of chapter 1.

The simplified dispersion relation in heavily doped QDs of III-V materials in accordance with the model of Stillman et. al can be written as

$$I_{11}\left(E_4'\right)=\frac{\hbar^2\pi^2}{2m^*}\left[\left(\frac{n_x}{d_x}\right)^2+\left(\frac{n_y}{d_y}\right)^2+\left(\frac{n_z}{d_z}\right)^2\right] \tag{3.38}$$

where, $I_{11}\left(E_4'\right)\equiv a_{11}\left[1-\left(1-a_{12}\gamma_3(E_4',\eta_g)\right)^{1/2}\right]$, $a_{11}\equiv\left(\frac{\hbar^2 t_{11}}{4m^* t_{12}}\right), a_{12}\equiv\frac{4t_{12}}{t_{11}{}^2}$

and E_4' is the totally quantized energy in this case.

The electron concentration in this case is given by

$$n_0=\left(\frac{2g_v}{d_x d_y d_z}\right)\sum_{n_x=1}^{n_{x_{\max}}}\sum_{n_y=1}^{n_{y_{\max}}}\sum_{n_z=1}^{n_{z_{\max}}}F_{-1}\left(\eta_{3D4}\right) \tag{3.39}$$

where $\eta_{3D4}=((E_{FD4}-E_4')/k_BT)$ and E_{FD4} is the Fermi energy in the present case .

Thus using (3.39) and (3.27) one can obtain the Seebeck coefficient in the presence of large magnetic field for quantum dots of heavily doped III-V materials whose undoped energy spectrum obeys the model of Stillman et.al.

(e) The Model of Newson and Kurobe

In accordance with the model of Newson and Kurobe, the electron dispersion law in this case for heavily doped III-V semiconductors assumes the form

$$\gamma_3(E,\eta_g)=a_{13}k_z{}^4+\left[\frac{\hbar^2}{2m^*}+a_{14}k_s{}^2\right]k_z{}^2+\frac{\hbar^2}{2m^*}k_s{}^2+a_{14}k_x{}^2k_y{}^2+a_{13}\left(k_x{}^4+k_y{}^4\right) \tag{3.40}$$

where the notations are defined in (1.38) of chapter 1.

The simplified dispersion relation for heavily doped QDs in this case can be written following (3.40) as

$$\gamma_3(E_5',\eta_g)=a_{13}\left(\frac{\pi n_z}{d_z}\right)^4+\left[\frac{\hbar^2}{2m^*}+a_{14}\left(\left(\frac{\pi n_x}{d_x}\right)^2+\left(\frac{\pi n_y}{d_y}\right)^2\right)\right]\cdot\left(\frac{\pi n_z}{d_z}\right)^2+\frac{\hbar^2}{2m^*}\left[\left(\frac{\pi n_x}{d_x}\right)^2+\left(\frac{\pi n_y}{d_y}\right)^2\right]+a_{14}\pi^4\left(\frac{n_x n_y}{d_x d_y}\right)^2+a_{13}\pi^4\left[\left(\frac{n_x}{d_x}\right)^4+\left(\frac{n_y}{d_y}\right)^4\right] \tag{3.41}$$

where E_5' is the totally quantized energy in this case.

The electron concentration in this case is given by

$$n_0 = \left(\frac{2g_v}{d_x d_y d_z}\right) \sum_{n_x=1}^{n_{x_{\max}}} \sum_{n_y=1}^{n_{y_{\max}}} \sum_{n_z=1}^{n_{z_{\max}}} F_{-1}(\eta_{3D5}) \tag{3.42}$$

where $\eta_{3D5} = ((E_{FD5} - E_5') / k_B T)$ and E_{FD5} is the Fermi energy in the present case .

Thus using (3.42) and (3.27) one can obtain the Seebeck coefficient in the presence of large magnetic field for quantum dots of heavily doped III-V materials whose undoped energy spectrum obeys the model of Newson and Kurobe.

(f) The Model of Rossler

The dispersion relation of the conduction electrons in heavily doped III-V semiconductors in accordance with the model of Rossler can be written as

$$\gamma_3(E,\eta_g) = \frac{\hbar^2 k^2}{2m^*} + [\alpha_{11} + \alpha_{12}k]k^4 + (\beta_{11} + \beta_{12}k)\left[k_x^2 k_y^2 + k_y^2 k_z^2 + k_z^2 k_x^2\right] \pm [\gamma_{11} + \gamma_{12}k]$$
$$\left[k^2\left(k_x^2 k_y^2 + k_y^2 k_z^2 + k_z^2 k_x^2\right) - 9k_x^2 k_y^2 k_z^2\right]^{1/2} \tag{3.43}$$

where the notations are defined in (1.43) of chapter 1.

The simplified dispersion relation for heavily doped QDs in this case can be written following (3.43) as

$$\gamma_3(E_{6,\pm}',\eta_g) \equiv \frac{\hbar^2 \pi^2}{2m^*}\left[\left(\frac{n_x}{d_x}\right)^2 + \left(\frac{n_y}{d_y}\right)^2 + \left(\frac{n_z}{d_z}\right)^2\right] + \left[\alpha_{11} + \alpha_{12}\left[\left(\frac{\pi n_x}{d_x}\right)^2 + \left(\frac{\pi n_y}{d_y}\right)^2 + \left(\frac{\pi n_z}{d_z}\right)^2\right]^{1/2}\right].$$
$$\left[\left(\frac{\pi n_x}{d_x}\right)^2 + \left(\frac{\pi n_y}{d_y}\right)^2 + \left(\frac{\pi n_z}{d_z}\right)^2\right]^2 + \left[\beta_{11} + \beta_{12}\left[\left(\frac{\pi n_x}{d_x}\right)^2 + \left(\frac{\pi n_y}{d_y}\right)^2 + \left(\frac{\pi n_z}{d_z}\right)^2\right]^{1/2}\right].$$
$$\left[\pi^4\left(\frac{n_x n_y}{d_x d_y}\right)^2 + \pi^4\left(\frac{n_y n_z}{d_y d_z}\right)^2 + \pi^4\left(\frac{n_z n_x}{d_z d_x}\right)^2\right] \pm \left[\gamma_{11} + \gamma_{12}\left[\left(\frac{\pi n_x}{d_x}\right)^2 + \left(\frac{\pi n_y}{d_y}\right)^2 + \left(\frac{\pi n_z}{d_z}\right)^2\right]^{1/2}\right].$$

$$\left[\left[\left(\frac{\pi n_x}{d_x}\right)^2+\left(\frac{\pi n_y}{d_y}\right)^2+\left(\frac{\pi n_z}{d_z}\right)^2\right]\left[\pi^4\left(\frac{n_x n_y}{d_x d_y}\right)^2+\pi^4\left(\frac{n_y n_z}{d_y d_z}\right)^2+\pi^4\left(\frac{n_z n_x}{d_z d_x}\right)^2\right]-9\pi^6\left(n_x n_y n_z / d_x d_y d_z\right)^6\right] \tag{3.44}$$

where $E'_{6,\pm}$ is the totally quantized energy in this case.

The electron concentration in this case is given by

$$n_0=\left(\frac{g_v}{d_x d_y d_z}\right)\sum_{n_x=1}^{n_{x_{\max}}}\sum_{n_y=1}^{n_{y_{\max}}}\sum_{n_z=1}^{n_{z_{\max}}}F_{-1}\left(\eta_{3D6,\pm}\right) \tag{3.45}$$

where $\eta_{3D6,\pm}=((E_{FD6}-E'_{6,\pm})/k_B T)$ and E_{FD6} is the Fermi energy in the present case .Thus using (3.45) and (3.27) one can obtain the Seebeck coefficient in the presence of large magnetic field for quantum dots of heavily doped III-V materials whose undoped energy spectrum obeys the model of Rossler.

(g) The Model of Palik et al.

In accordance with the model of Palik et. al., the electron dispersion law of heavily doped III-V materials assumes the form

$$\gamma_3(E,\eta_g)=\frac{\hbar^2 k^2}{2m^*}-b_{11}k^4 \tag{3.46}$$

where the notations are defined in (1.50) of chapter 1.

The simplified dispersion relation in heavily doped QDs of III-V materials in accordance with (3.46)

$$I_{11}\left(E'_7\right)=\frac{\hbar^2\pi^2}{2m^*}\left[\left(\frac{n_x}{d_x}\right)^2+\left(\frac{n_y}{d_y}\right)^2+\left(\frac{n_z}{d_z}\right)^2\right] \tag{3.47}$$

where $I_{12}\left(E'_7\right)\equiv b_{12}[a_{12}-\left(a_{12}^2-4b_{11}\gamma_3(E'_7,\eta_g)\right)^{1/2}], b_{12}\equiv\frac{a_{12}}{2b_{11}}, a_{12}\equiv$

$\frac{\hbar^2}{2m^*}$ *and* E'_7 is the totally quantized energy in this case.

The electron concentration in this case is given by

$$n_0 = \left(\frac{2g_v}{d_x d_y d_z}\right) \sum_{n_x=1}^{n_{x_{\max}}} \sum_{n_y=1}^{n_{y_{\max}}} \sum_{n_z=1}^{n_{z_{\max}}} F_{-1}(\eta_{3D7}) \tag{3.48}$$

where $\eta_{3D7} = ((E_{FD7} - E'_7)/k_B T)$ and E_{FD7} is the Fermi energy in the present case.

Thus using (3.48) and (3.27) one can obtain the Seebeck coefficient in the presence of strong magnetic field for quantum dots of heavily doped III-V materials whose undoped energy spectrum obeys the model of Palik et.al.

(h) The Model of Agafonov et. al.

In accordance with the model of Agafonov et. al., the electron dispersion law in heavily doped III-V materials can be written as

$$\gamma_3(E,\eta_g) = \frac{\bar{y} - E_{g_0}}{2}\left[1 - T_5\left(\frac{k_x^{\ 4} + k_y^{\ 4} + k_z^{\ 4}}{\bar{y}k^2}\right)\right] \tag{3.49}$$

where the notations are defined in (1.60) of chapter 1.

The simplified dispersion relation in heavily doped quantum dots of III-V materials can be expressed following (3.49) as

$$\gamma_3(E_9,\eta_g) = \left(\frac{\psi_{30} - E_{g_0}}{2}\right)\left[1 - \left[T_5\left[\left(\frac{\pi n_x}{d_x}\right)^4 + \left(\frac{\pi n_y}{d_y}\right)^4 + \left(\frac{\pi n_z}{d_z}\right)^4\right]\right]\left[\psi_{30}\left[\left(\frac{\pi n_x}{d_x}\right)^2 + \left(\frac{\pi n_y}{d_y}\right)^2 + \left(\frac{\pi n_z}{d_z}\right)^2\right]\right]^{-1}\right] \tag{3.50}$$

where $\psi_{30} = \left[E_{g_0}^{\ 2} + \frac{8}{3}P_0^2\left[\left(\frac{\pi n_x}{d_x}\right)^2 + \left(\frac{\pi n_y}{d_y}\right)^2 + \left(\frac{\pi n_z}{d_z}\right)^2\right]\right]^{1/2}$

and E_9 is the totally quantized energy in this case.

The electron concentration in this case is given by

$$n_0 = \left(\frac{2g_v}{d_x d_y d_z}\right) \sum_{n_x=1}^{n_{x_{\max}}} \sum_{n_y=1}^{n_{y_{\max}}} \sum_{n_z=1}^{n_{z_{\max}}} F_{-1}(\eta_{3D9}) \tag{3.51}$$

where $\eta_{3D9} = ((E_{FD9} - E'_9)/k_B T)$ and E_{FD9} is the Fermi energy in the present case .

Thus using (3.51) and (3.27) one can obtain the Seebeck coefficient in the presence of large magnetic field for quantum dots of heavily doped III-V materials whose undoped energy spectrum obeys the model of Agafonov et.al.

3.2.3. Study of the Seebeck Coefficient under Large Magnetic Field in Quantum Dots of Heavily Doped II-VI Materials

Using (1.66) and (3.2), the dispersion relation of the carriers in heavily doped II-VI materials in the presence of Gaussian band tails can be expressed as

$$\gamma_3\left(E,\eta_g\right)=A_0k_s^{\ 2}+B_0k_z^{\ 2}\pm C_0k_s \tag{3.52}$$

Thus, the energy spectrum in this case is real since the corresponding undoped case as given by (1.66) is a no pole function.

Therefore the electron energy spectrum in quantum dots of heavily doped III-V, ternary and quaternary materials whose bulk dispersion relation follows (1.66) can be expressed as

$$\gamma_3\left(E_{10}',\eta_g\right)=A_0((n_x\pi/d_x)^2+(n_y\pi/d_y)^2)+B_0(n_z\pi/d_z)^2\pm C_0((n_x\pi/d_x)^2+(n_y\pi/d_y)^2)^{1/2} \tag{3.53}$$

where E_4' is the totally quantized energy in this case.

The electron concentration in this case is given by

$$n_0=\left(\frac{2g_v}{d_xd_yd_z}\right)\sum_{n_x=1}^{n_{x_{\max}}}\sum_{n_y=1}^{n_{y_{\max}}}\sum_{n_z=1}^{n_{z_{\max}}}F_{-1}\left(\eta_{3D10}\right) \tag{3.54}$$

where $\eta_{3D10}=((E_{FD10}-E_{10}')/k_BT)$ and E_{FD10} is the Fermi energy in the present case .

Thus using (3.54) and (3.27) one can obtain the Seebeck coefficient in the presence of large magnetic field for quantum dots of heavily doped II-IV materials.

3.2.4. Study of the Seebeck Coefficient under Large Magnetic Field in Quantum Dots of Heavily Doped n-Gallium Phosphide

The dispersion relation of the conduction electrons in bulk specimens of heavily doped n-GaP can be written following (1.73) of chapter 1 as

$$\gamma_3\left(E,\eta_g\right)=\frac{\hbar^2k_s^2}{2m_\perp^*}+\frac{\hbar^2}{2m_\parallel^*}\left[k_s^2+k_z^2\right]-\left[\frac{\hbar^4k_0^2}{m_\parallel^{*2}}\left(k_s^2+k_z^2\right)+\left|V_G\right|^2\right]^{1/2}+\left|V_G\right| \tag{3.55}$$

where the symbols for energy band structures are defined in (1.73).

The simplified dispersion relation in quantum dots of n-GaP assumes the form

$$\gamma_3\left(E'_{11},\eta_g\right)=\frac{\hbar^2}{2m_\perp^*}\left[\left(\frac{\pi n_x}{d_x}\right)^2+\left(\frac{\pi n_y}{d_y}\right)^2\right]+\frac{\hbar^2}{2m_\parallel^*}\left[\left(\frac{\pi n_x}{d_x}\right)^2+\left(\frac{\pi n_y}{d_y}\right)^2+\left(\frac{\pi n_z}{d_z}\right)^2\right]$$

$$-\left[\frac{\hbar^4 k_0^2}{m_\parallel^{*2}}\left[\left(\frac{\pi n_x}{d_x}\right)^2+\left(\frac{\pi n_y}{d_y}\right)^2+\left(\frac{\pi n_z}{d_z}\right)^2\right]+|V_G|^2\right]^{1/2}+|V_G| \quad (3.56)$$

where E'_{11} is the totally quantized energy in this case.

The electron concentration in this case is given by

$$n_0=\left(\frac{2g_v}{d_x d_y d_z}\right)\sum_{n_x=1}^{n_{x_{\max}}}\sum_{n_y=1}^{n_{y_{\max}}}\sum_{n_z=1}^{n_{z_{\max}}}F_{-1}\left(\eta_{3D11}\right) \quad (3.57)$$

where $\eta_{3D11}=((E_{FD11}-E'_{11})/k_BT)$ and E_{FD11} is the Fermi energy in the present case.

Thus using (3.57) and (3.27) one can obtain the Seebeck coefficient in the presence of large magnetic field for quantum dots of heavily doped n-GaP.

3.2.5 Study of the Seebeck Coefficient under Large Magnetic Field in Quantum Dots of heavily Doped n-Germanium

The energy spectrum of the conduction electrons in bulk specimens of heavily doped n-Ge can be written extending the model of Cardona et. al. (as given in (1.80) of chapter 1) as

$$\gamma_3\left(E,\eta_g\right)=-\frac{E_{g_0}}{2}+\frac{\hbar^2 k_z^2}{2m_\parallel^*}+\left[\frac{E_{g_0}^2}{4}+E_{g_0}k_s^2\left(\frac{\hbar^2}{2m_\perp^*}\right)\right]^{1/2} \quad (3.58)$$

The simplified electron dispersion law in quantum dots of heavily doped n-Ge can be expressed following (3.58) as

$$\gamma_3\left(E'_{12},\eta_g\right)=-\frac{E_{g_0}}{2}+\frac{\hbar^2}{2m_\parallel^*}\left(\frac{\pi n_z}{d_z}\right)^2+\left[\frac{E_{g_0}^2}{4}+E_{g_0}\left(\frac{\hbar^2}{2m_\perp^*}\right)\left\{\left(\frac{\pi n_x}{d_x}\right)^2+\left(\frac{\pi n_y}{d_y}\right)^2\right\}\right]^{1/2} \quad (3.59)$$

where E'_{12} is the totally quantized energy in this case

a) The electron concentration in this case is given by

$$n_0 = \left(\frac{2g_v}{d_x d_y d_z}\right) \sum_{n_x=1}^{n_{x_{\max}}} \sum_{n_y=1}^{n_{y_{\max}}} \sum_{n_z=1}^{n_{z_{\max}}} F_{-1}(\eta_{3D12}) \tag{3.60}$$

where $\eta_{3D12} = ((E_{FD12} - E'_{12}) / k_B T)$ and E_{FD12} is the Fermi energy in the present case.

Thus using (3.60) and (3.27) one can obtain the Seebeck coefficient in the presence of large magnetic field for quantum dots of heavily doped n-Ge.

(b) The dispersion relation of the conduction electron in bulk specimens of heavily doped n-Ge can be expressed extending the model of Wang and Ressler (as given in (1.87) of chapter 1) as

$$\gamma_3(E,\eta_g) = \frac{\hbar^2 k_z^2}{2m_\parallel^*} + \frac{\hbar^2 k_s^2}{2m_\perp^*} - \overline{c}_1\left(\frac{\hbar^2 k_s^2}{2m_\perp^*}\right)^2 - \overline{d}_1\left(\frac{\hbar^2 k_s^2}{2m_\perp^*}\right)\left(\frac{\hbar^2 k_z^2}{2m_\parallel^*}\right) - \overline{e}_1\left(\frac{\hbar^2 k_z^2}{2m_\parallel^*}\right)^2 \tag{3.61}$$

where the symbols have already been defined.

The simplified electron energy spectrum in quantum dots of n-Ge can be expressed following (3.61) as

$$\gamma_3(E'_{13},\eta_g) = \frac{\hbar^2}{2m_\parallel^*}\left(\frac{\pi n_z}{d_z}\right)^2 + \frac{\hbar^2}{2m_\perp^*}\left\{\left(\frac{\pi n_x}{d_x}\right)^2 + \left(\frac{\pi n_y}{d_y}\right)^2\right\} - \overline{c}_1\left(\frac{\hbar^2}{2m_\perp^*}\right)^2\left\{\left(\frac{\pi n_x}{d_x}\right)^2 + \left(\frac{\pi n_y}{d_y}\right)^2\right\}^2$$
$$-\overline{d}_1\left(\frac{\hbar^2}{2m_\perp^*}\right)\left\{\left(\frac{\pi n_x}{d_x}\right)^2 + \left(\frac{\pi n_y}{d_y}\right)^2\right\}\left(\frac{\hbar^2}{2m_\parallel^*}\right)\left(\frac{\pi n_z}{d_z}\right)^2 - \overline{e}_1\left(\frac{\hbar^2}{2m_\parallel^*}\right)^2\left(\frac{\pi n_z}{d_z}\right)^4 \tag{3.62}$$

where E'_{13} is the totally quantized energy in this case

The electron concentration in this case is given by

$$n_0 = \left(\frac{2g_v}{d_x d_y d_z}\right) \sum_{n_x=1}^{n_{x_{\max}}} \sum_{n_y=1}^{n_{y_{\max}}} \sum_{n_z=1}^{n_{z_{\max}}} F_{-1}(\eta_{3D13}) \tag{3.63}$$

where $\eta_{3D13} = ((E_{FD13} - E'_{13}) / k_B T)$ and E_{FD13} is the Fermi energy in the present case.

Thus using (3.63) and (3.27) one can obtain the Seebeck coefficient in the presence of large magnetic field for quantum dots of heavily doped n-Ge.

3.2.6. Study of the Seebeck Coefficient under Large Magnetic Field in Quantum Dots of Heavily Doped Tellurium

The dispersion relation of the conduction electrons of heavily doped Tellurium can be written in extending the model of Bouat et. al. (as given in (1.94) of chapter 1) as

$$\gamma_3\left(E,\eta_g\right)=A_6k_z^2+A_7k_s^2\pm\left[A_8k_z^2+A_9k_s^2\right]^{1/2} \tag{3.64}$$

The simplified electron energy spectrum in quantum dots of heavily doped Te can be expressed following (3.64) as

$$\gamma_3\left(E'_{14,\pm},\eta_g\right)=A_6\left(\frac{\pi n_z}{d_z}\right)^2+A_7\left[\left(\frac{\pi n_x}{d_x}\right)^2+\left(\frac{\pi n_y}{d_y}\right)^2\right]\pm \left[A_8\left(\frac{\pi n_z}{d_z}\right)^2+A_9\left[\left(\frac{\pi n_x}{d_x}\right)^2+\left(\frac{\pi n_y}{d_y}\right)^2\right]\right]^{1/2} \tag{3.65}$$

where $E'_{14,\pm}$ is the totally quantized energy in this case

a) The electron concentration in this case is given by

$$n_0=\left(\frac{g_v}{d_xd_yd_z}\right)\sum_{n_x=1}^{n_{x_{\max}}}\sum_{n_y=1}^{n_{y_{\max}}}\sum_{n_z=1}^{n_{z_{\max}}}F_{-1}\left(\eta_{3D14,\pm}\right) \tag{3.66}$$

where $\eta_{3D14,\pm}=((E_{FD14}-E'_{14,\pm})/k_BT)$ and E_{FD14} is the Fermi energy in the present case.

Thus using (3.66) and (3.27) one can obtain the Seebeck coefficient in the presence of large magnetic field for quantum dots of heavily doped Te.

The dispersion relation of the conduction electrons of heavily doped Tellurium can be written in extending the model of Ortenberg and Button (as given in (1.100) of chapter 1) as

$$\gamma_3\left(E,\eta_g\right)=t_1+t_2k_z^2+t_3k_s^2+t_4k_s^4+t_5k_s^2k_z^2\pm\left[\left(t_1+t_6k_s^2\right)^2+t_7k_z^2\right]^{1/2} \tag{3.67}$$

b) The simplified electron energy spectrum in quantum dots of heavily doped Te can be expressed following (3.67) as

$$\gamma_3\left(E'_{15,\pm},\eta_g\right)=t_1+t_2\left(\frac{\pi n_z}{d_z}\right)^2+t_3\left[\left(\frac{\pi n_x}{d_x}\right)^2+\left(\frac{\pi n_y}{d_y}\right)^2\right]+t_4\left[\left(\frac{\pi n_x}{d_x}\right)^2+\left(\frac{\pi n_y}{d_y}\right)^2\right]^2+$$

$$t_5\left[\left(\frac{\pi n_x}{d_x}\right)^2+\left(\frac{\pi n_y}{d_y}\right)^2\right]\left(\frac{\pi n_z}{d_z}\right)^2\pm\left[\left[t_1+t_6\left\{\left(\frac{\pi n_x}{d_x}\right)^2+\left(\frac{\pi n_y}{d_y}\right)^2\right\}\right]^2+t_7\left(\frac{\pi n_z}{d_z}\right)^2\right]^{1/2}$$

(3.68)

where $E'_{15,\pm}$ is the totally quantized energy in this case

The electron concentration in this case is given by

$$n_0=\left(\frac{g_v}{d_x d_y d_z}\right)\sum_{n_x=1}^{n_{x_{max}}}\sum_{n_y=1}^{n_{y_{max}}}\sum_{n_z=1}^{n_{z_{max}}}F_{-1}\left(\eta_{3D15,\pm}\right)$$

(3.69)

where $\eta_{3D15,\pm}=((E_{FD15}-E'_{15,\pm})/k_BT)$ and E_{FD15} is the Fermi energy in the present case.Thus using (3.69) and (3.27) one can obtain the Seebeck coefficient in the presence of large magnetic field for quantum dots of heavily doped Te.

3.2.7. Study of the Seebeck Coefficient under Large Magnetic Field in Quantum Dots of Heavily Doped Platinum Antimonide

The dispersion relation of the carriers in heavily doped Platinum Antimonide can be written as

$$\theta_0\left(E,\eta_g\right)+\gamma_0\left(E,\eta_g\right)\left[\left(\omega_1k_s^2+\omega_2k_z^2+\delta_0\right)-\left(\omega_3k_s^2+\omega_4k_z^2\right)\right]$$
$$=\left[\frac{1+Erf\left(E/\eta_g\right)}{2}\right]\left[I_1\left(k_s^2+k_z^2\right)^2+\left(\omega_1k_s^2+\omega_2k_z^2\right)\left(\omega_3k_s^2+\omega_4k_z^2\right)-\delta_0\left(\omega_1k_s^2+\omega_2k_z^2\right)\right]$$

where the notations have been defined in (1.112) of chapter 1.

(3.70)

The simplified electron energy spectrum in quantum dots of heavily doped Platinum Antimonide can be expressed as

$$\theta_0\left(E'_{16},\eta_g\right)+\gamma_0\left(E'_{16},\eta_g\right)\left[\left(\omega_1\left\{\left(\frac{\pi n_x}{d_x}\right)^2+\left(\frac{\pi n_y}{d_y}\right)^2\right\}+\omega_2\left\{\left(\frac{\pi n_z}{d_z}\right)^2\right\}+\delta_0\right)-\left(\omega_3\left\{\left(\frac{\pi n_x}{d_x}\right)^2+\left(\frac{\pi n_y}{d_y}\right)^2\right\}\right.\right.$$
$$\left.\left.+\omega_4\left\{\left(\frac{\pi n_z}{d_z}\right)^2\right\}\right)\right]=\left[\frac{1+Erf\left(E'_{16}/\eta_g\right)}{2}\right]\left[I_1\left(\left\{\left(\frac{\pi n_x}{d_x}\right)^2+\left(\frac{\pi n_y}{d_y}\right)^2\right\}+\left\{\left(\frac{\pi n_z}{d_z}\right)^2\right\}\right)^2+\left(\omega_1\left\{\left(\frac{\pi n_x}{d_x}\right)^2+\left(\frac{\pi n_y}{d_y}\right)^2\right\}\right.\right.$$
$$\left.\left.+\omega_2\left\{\left(\frac{\pi n_z}{d_z}\right)^2\right\}\right)\left(\omega_3\left\{\left(\frac{\pi n_x}{d_x}\right)^2+\left(\frac{\pi n_y}{d_y}\right)^2\right\}+\omega_4\left\{\left(\frac{\pi n_z}{d_z}\right)^2\right\}\right)-\delta_0\left(\omega_1\left\{\left(\frac{\pi n_x}{d_x}\right)^2+\left(\frac{\pi n_y}{d_y}\right)^2\right\}+\omega_2\left\{\left(\frac{\pi n_z}{d_z}\right)^2\right\}\right)\right] \quad (3.71)$$

where E'_{16} is the totally quantized energy in this case

The electron concentration in this case is given by

$$n_0=\left(\frac{2g_v}{d_x d_y d_z}\right)\sum_{n_x=1}^{n_{x_{max}}}\sum_{n_y=1}^{n_{y_{max}}}\sum_{n_z=1}^{n_{z_{max}}}F_{-1}\left(\eta_{3D16}\right) \quad (3.72)$$

where $\eta_{3D16}=((E_{FD16}-E'_{16})/k_BT)$ and E_{FD16} is the Fermi energy in the present case.Thus using (3.72) and (3.27) one can obtain the Seebeck coefficient in the presence of large magnetic field for quantum dots of heavily doped $PtSb_2$.

3.2.8. Study of the Seebeck Coefficient under Large Magnetic Field in Quantum Dots of Heavily Doped II-V Materials

The dispersion relation of the holes in heavily doped II-V compounds in accordance with Yamada can be expressed as

$$\gamma_3\left(E,\eta_g\right)=A_{10}k_x^2+A_{11}k_y^2+A_{12}k_z^2+A_{13}k_x\pm\left[\left(A_{14}k_x^2+A_{15}k_y^2+A_{16}k_z^2+A_{17}k_x\right)^2+A_{18}k_y^2+A_{19}^2\right]^{1/2} \quad (3.73)$$

where the notations have already been defined in (1.126) chapter1.

The hole-energy spectrum in heavily doped II-V compounds can be written following (3.73) as

$$\gamma_3\left(E'_{17,\pm},\eta_g\right)=A_{10}\left(\frac{\pi n_x}{d_x}\right)^2+A_{11}\left(\frac{\pi n_y}{d_y}\right)^2+A_{12}\left(\frac{\pi n_z}{d_z}\right)^2+A_{13}\left(\frac{\pi n_x}{d_x}\right)$$
$$\pm\left[\left(A_{14}\left(\frac{\pi n_x}{d_x}\right)^2+A_{15}\left(\frac{\pi n_y}{d_y}\right)^2+A_{16}\left(\frac{\pi n_z}{d_z}\right)^2+A_{17}\left(\frac{\pi n_x}{d_x}\right)\right)^2+A_{18}\left(\frac{\pi n_y}{d_y}\right)^2+A_{19}{}^2\right]^{1/2}$$

(3.74)

where $E'_{17,\pm}$ is the totally quantized energy in this case

The electron concentration in this case is given by

$$n_0 = \left(\frac{g_v}{d_x d_y d_z}\right) \sum_{n_x=1}^{n_{x_{\max}}} \sum_{n_y=1}^{n_{y_{\max}}} \sum_{n_z=1}^{n_{z_{\max}}} F_{-1}\left(\eta_{3D17,\pm}\right) \tag{3.75}$$

where $\eta_{3D17,\pm} = ((E_{FD17} - E'_{17,\pm}) / k_B T)$ and E_{FD17} is the Fermi energy in the present case. Thus using (3.75) and (3.27) one can obtain the Seebeck coefficient in the presence of large magnetic field for quantum dots of heavily doped II-V materials.

3.2.9. Study of the Seebeck Coefficient under Large Magnetic Field in Quantum Dots of Heavily Doped Gallium Antimonide

The conduction electrons of undoped n-GaSb obey the three dispersion relations as provided by Seiler et. al., Mathur et. al. and Zhang respectively as stated in chapter 1. The Seebeck coefficient in the presence of strong magnetic field in QDs of heavily doped GaSb is being presented in accordance with the aforementioned models for the joint purpose of coherent presentation and relative assessment.

The dispersion relation of the conduction electrons in heavily doped n-GaSb whose unperturbed electron energy spectrum follows the model of Seiler et. al. assume the form

$$\gamma_3\left(E,\eta_g\right) = \left[-\frac{E_{g_0}}{2} + \frac{E_{g_0}}{2}\left[1+\alpha_4 k^2\right]^{1/2} + \frac{\overline{\zeta_0}\hbar^2 k^2}{2m_0} + \frac{\overline{v_0}\theta_1(k)\hbar^2}{2m_0} \pm \frac{\overline{w_0}\theta_2(k)\hbar^2}{2m_0}\right] \tag{3.76}$$

where the notations have been defined in (1.133) chapter 1

The simplified electron energy spectrum in QDs of heavily doped n-GaSb can be written following (3.76) as

$$\gamma_3\left(E'_{18\pm},\eta_g\right) = \left[-\frac{E_{g_0}}{2} + \frac{E_{g_0}}{2}\left[1+\alpha_4\left[\left(\frac{\pi n_x}{d_x}\right)^2 + \left(\frac{\pi n_y}{d_y}\right)^2 + \left(\frac{\pi n_z}{d_z}\right)^2\right]\right]^{1/2} + \left(\frac{\overline{\zeta_0}\hbar^2}{2m_0}\right)\right.$$
$$\left.\left[\left(\frac{\pi n_x}{d_x}\right)^2 + \left(\frac{\pi n_y}{d_y}\right)^2 + \left(\frac{\pi n_z}{d_z}\right)^2\right] + \left(\frac{\overline{v_0}\hbar^2}{2m_0}\right)S_6 \pm \left(\frac{\overline{w_0}S_7\hbar^2}{2m_0}\right)\right] \tag{3.77}$$

where $E'_{18,\pm}$ is the totally quantized energy in this case.

a) The electron concentration in this case is given by

$$n_0 = \left(\frac{g_v}{d_x d_y d_z}\right) \sum_{n_x=1}^{n_{x_{\max}}} \sum_{n_y=1}^{n_{y_{\max}}} \sum_{n_z=1}^{n_{z_{\max}}} F_{-1}\left(\eta_{3D18,\pm}\right) \tag{3.78}$$

where $\eta_{3D18,\pm} = ((E_{FD18} - E'_{18,\pm}) / k_B T)$ and E_{FD18} is the Fermi energy in the present case.

Thus using (3.78) and (3.27) one can obtain the Seebeck coefficient in the presence of large magnetic field for quantum dots of heavily doped GaSb materials.

The dispersion relation of the conduction electrons in heavily doped n-GaSb whose unperturbed electron energy spectrum follows the model of Mathur et. al. assume the form

$$\gamma_3\left(E,\eta_g\right) = \alpha_9 k^2 + \frac{E_{g1}}{2}\left[\sqrt{1+\alpha_{10}k^2} - 1\right] \tag{3.79}$$

where the notations of the right hand side have already been defined in (1.140) chapter 1.

The simplified electron energy spectrum in QDs of heavily doped n-GaSb can be written following (3.79) as

$$\gamma_3\left(E'_{19},\eta_g\right) = \alpha_9\left[\left(\frac{\pi n_x}{d_x}\right)^2 + \left(\frac{\pi n_y}{d_y}\right)^2 + \left(\frac{\pi n_z}{d_z}\right)^2\right] + \frac{E_{g1}}{2}\left[\sqrt{1+\alpha_{10}\left[\left(\frac{\pi n_x}{d_x}\right)^2 + \left(\frac{\pi n_y}{d_y}\right)^2 + \left(\frac{\pi n_z}{d_z}\right)^2\right]} - 1\right] \tag{3.80}$$

where E'_{19} is the totally contized energy in this case.

b) The electron concentration in this case is given by

$$n_0 = \left(\frac{g_v}{d_x d_y d_z}\right) \sum_{n_x=1}^{n_{x_{\max}}} \sum_{n_y=1}^{n_{y_{\max}}} \sum_{n_z=1}^{n_{z_{\max}}} F_{-1}\left(\eta_{3D19}\right) \tag{3.81}$$

where $\eta_{3D19} = ((E_{FD19} - E'_{19}) / k_B T)$ and E_{FD19} is the Fermi energy in the present case.

Thus using (3.81) and (3.27) one can obtain the Seebeck coefficient in the presence of large magnetic field for quantum dots of heavily doped GaSb materials.

The dispersion relation of the conduction electrons in heavily doped n-GaSb whose unperturbed electron energy spectrum follows the model of Zhang assume the form as

$$\gamma_3\left(E,\eta_g\right) = \left[\overline{E}_1 + \overline{E}_2 Z_1(k)\right]k^2 + \left[\overline{E}_3 + \overline{E}_4 Z_1(k)\right]k^4 + k^6\left[\overline{E}_5 + \overline{E}_6 Z_1(k) + \overline{E}_7 Z_2(k)\right] \tag{3.82}$$

where the notations of the right hand side have already been defined in (1.148) chapter 1.

The simplified electron energy spectrum in QDs of heavily doped n-GaSb can be written following (3.82) as

$$\gamma_3\left(E'_{20},\eta_g\right)=\left[\overline{E}_1+\overline{E}_2 Z_3\right]\left[\left(\frac{\pi n_x}{d_x}\right)^2+\left(\frac{\pi n_y}{d_y}\right)^2+\left(\frac{\pi n_z}{d_z}\right)^2\right]+\left[\left(\frac{\pi n_x}{d_x}\right)^2+\left(\frac{\pi n_y}{d_y}\right)^2+\left(\frac{\pi n_z}{d_z}\right)^2\right]^2.$$

$$+\left[\left(\frac{\pi n_x}{d_x}\right)^2+\left(\frac{\pi n_y}{d_y}\right)^2+\left(\frac{\pi n_z}{d_z}\right)^2\right]^3.\left[\overline{E}_5+\overline{E}_6 Z_3+\overline{E}_7 Z_4\right] \tag{3.83}$$

where E'_{20} is the totally contized energy in this case.

c) The electron concentration in this case is given by

$$n_0=\left(\frac{2g_v}{d_x d_y d_z}\right)\sum_{n_x=1}^{n_{x_{\max}}}\sum_{n_y=1}^{n_{y_{\max}}}\sum_{n_z=1}^{n_{z_{\max}}}F_{-1}\left(\eta_{3D20}\right) \tag{3.84}$$

where $\eta_{3D20}=((E_{FD20}-E'_{20})/k_B T)$ and E_{FD20} is the Fermi energy in the present case.

Thus using (3.84) and (3.27) one can obtain the Seebeck coefficient in the presence of large magnetic field for quantum dots of heavily doped GaSb materials.

3.2.10. The Seebeck Coefficient in the Presence of Large Magnetic Field in Quantum Dots of Heavily Doped Stressed Materials

The (1.155) can be expressed as

$$Ek^2-T_{17}k_x^2-T_{27}k_y^2-T_{37}k_z^2=\left[q_{67}E^3-R_{67}E^2+V_{67}E+\rho_{67}\right] \tag{3.85}$$

where

$$T_{17}=[E_g-C_1\varepsilon-(\overline{a_0}+C_1)\varepsilon+\frac{3}{2}\overline{b_0}\varepsilon_{xx}-\frac{\overline{b_0}}{2}\varepsilon+\frac{\sqrt{3}}{2}\overline{d_0}\varepsilon_{xy}],$$

$$T_{27}=[E_g-C_1\varepsilon-(\overline{a_0}+C_1)\varepsilon+\frac{3}{2}\overline{b_0}\varepsilon_{xx}-\frac{\overline{b_0}}{2}\varepsilon-\frac{\sqrt{3}}{2}\overline{d_0}\varepsilon_{xy}],$$

$$T_{37} = [E_g - C_1\varepsilon + (\overline{a_0} + C_1)\varepsilon + \frac{3}{2}\overline{b_0}\varepsilon_{zz} - \frac{\overline{b_0}}{2}\varepsilon], q_{67} = \frac{3}{2B_2^2}, R_{67} = q_{67}[2A_3 + C_1\varepsilon],$$

$$A_3 = (E_g - C_1\varepsilon), V_{67} = q_{67}[A_3^2 - \frac{2C_2^2\varepsilon_{xy}}{3} + 2A_3C_1\varepsilon] and\, \rho_{67} = q_{67}[(2C_2^2\varepsilon_{xy}/3) - C_1\varepsilon A_3^2)$$

Using (3.85) and (3.2) we can write

$$I(4)k^2 - T_{17}I(1)k_x^2 - T_{27}I(1)k_y^2 - T_{37}k_z^2 I(1) = \left[q_{67}I(6) - R_{67}I(5) + V_{67}I(4) + \rho_{67}I(1)\right] \tag{3.86}$$

where $I(6) = \int_{-\infty}^{E} (E-V)^3 F(V)\,dV$ (3.87)

Equation ((3.87)) can be written as

$$I(6) = E^3 I(1) - 3E^2 I(7) + 3EI(8) - I(9) \tag{3.88}$$

In which,

$$I(7) = \int_{-\infty}^{E} VF(V)\,dV \tag{3.89}$$

$$I(8) = \int_{-\infty}^{E} V^2 F(V)\,dV \tag{3.90}$$

$$I(9) = \int_{-\infty}^{E} V^3 F(V)\,dV \tag{3.91}$$

Using (3.2) and the appropriate equations together with simple algebraic manipulations, one obtains

$$I(7) = \frac{-\eta_g}{2\sqrt{\pi}} \exp\left(\frac{-E^2}{\eta_g^2}\right) \tag{3.92}$$

$$I(8) = \frac{\eta_g^2}{4}\left[1 + Erf\left(\frac{E}{\eta_g}\right)\right] \tag{3.93}$$

and

$$I(9)=\frac{-\eta_g^3}{2\sqrt{\pi}}\exp\left(\frac{-E^2}{\eta_g^2}\right)\left[1+\frac{E^2}{\eta_g^2}\right] \tag{3.94}$$

Thus, (3.88) can be written as

$$I(6)=\left[\frac{E}{2}\left[1+Erf\left(\frac{E}{\eta_g}\right)\right]\left[E^2+\frac{3}{2}\eta_g^2\right]+\frac{\eta_g}{2\sqrt{\pi}}\exp\left(\frac{-E^2}{\eta_g^2}\right)\left[4E^2+\eta_g^2\right]\right] \tag{3.95}$$

Thus, combining the appropriate equations, the dispersion relations of the conduction electrons in heavily doped stressed materials can be expressed as

$$P_{11}(E,\eta_g)k_x^2+Q_{11}(E,\eta_g)k_y^2+S_{11}(E,\eta_g)k_z^2=1 \tag{3.96}$$

where, $P_{11}(E,\eta_g)\equiv\left[\frac{\gamma_0(E,\eta_g)-(T_{17}/2)\left[1+Erf(E/\eta_g)\right]}{\Delta_{14}(E,\eta_g)}\right]$,

$$\Delta_{14}(E,\eta_g)\equiv\left[q_{67}\left\{\frac{E}{2}\left[1+Erf\left(\frac{E}{\eta_g}\right)\right]\left[E^2+\frac{3}{2}\eta_g^2\right]+\frac{\eta_g}{2\sqrt{\pi}}\exp\left(\frac{-E^2}{\eta_g^2}\right)\left[4E^2+\eta_g^2\right]\right\}\right.$$
$$\left.-R_{67}\theta_0(E,\eta_g)+V_{67}\gamma_0(E,\eta_g)+\frac{\rho_{67}}{2}\left[1+Erf(E/\eta_g)\right]\right],$$

$$Q_{11}(E,\eta_g)\equiv\left[\frac{\gamma_0(E,\eta_g)-(T_{27}/2)\left[1+Erf(E/\eta_g)\right]}{\Delta_{14}(E,\eta_g)}\right]$$

and

$$S_{11}(E,\eta_g)\equiv\left[\frac{\gamma_0(E,\eta_g)-(T_{37}/2)\left[1+Erf(E/\eta_g)\right]}{\Delta_{14}(E,\eta_g)}\right].$$

Thus, the energy spectrum in this case is real since the corresponding undoped material as given by (3.85) is a pole-less function.

The simplified dispersion relation in quantum dots of stressed Kane type semiconductors can be written following (3.96) as

$$P_{11}(E_{21}',\eta_g)\left(\frac{n_x\pi}{d_x}\right)^2+Q_{11}(E_{21}',\eta_g)\left(\frac{n_y\pi}{d_y}\right)^2+S_{11}(E_{21}',\eta_g)\left(\frac{n_z\pi}{d_z}\right)^2=1 \tag{3.97}$$

where E_{21}' is the totally quantized energy in this case.

The electron concentration in this case is given by

$$n_0 = \left(\frac{2g_v}{d_x d_y d_z}\right) \sum_{n_x=1}^{n_{x_{\max}}} \sum_{n_y=1}^{n_{y_{\max}}} \sum_{n_z=1}^{n_{z_{\max}}} F_{-1}(\eta_{3D21}) \tag{3.98}$$

where $\eta_{3D21} = ((E_{FD21} - E'_{21}) / k_B T)$, E_{FD21} is the Fermi energy in the present case.

Thus using (3.98) and (3.27) one can obtain for quantum dots of the Seebeck coefficient in the presence of large magnetic field for heavily doped stressed Kane type materials.

3.2.11. Study of the Seebeck Coefficient under Large Magnetic Field in Quantum Dots of Heavily Doped IV-VI Materials

The carriers of IV-VI materials are also described by the five types of dispersion relations as provided by Cohen,Lax,Dimmock, Bangert and Kastner and Foley et. al. respectively. The formulation of S in QDs of heavily doped IV-VI materials in accordance with the said models have been discussed for the purpose of complete presentation:

(a) The Cohen Model

The dispersion relation in heavily doped IV-VI materials hose unpertuebed carrier energy spectrum is given by the Cohen model can be written extending (1.176) as

$$\gamma_3(E,\eta_g)\left(1+\frac{\alpha p_y^2}{2m_2'}(1-\frac{m_2'}{m_2})\right)+(\alpha.2.\theta_0(E,\eta_g))(1+Erf(E/\eta_g))^{-1} = \frac{p_x^2}{2m_1}+\frac{p_z^2}{2m_3}+\frac{p_y^2}{2m_2}+\frac{\alpha p_y^4}{4m_2 m_2'} \tag{3.99}$$

The simplified dispersion relation in quantum dots of heavily doped IV-VI semiconductors can be written following (3.99) as

$$\gamma_3(E'_{22},\eta_g)(1+\frac{\alpha\hbar^2}{2m_2'}(\frac{\pi n_y}{d_y})^2(1-\frac{m_2'}{m_2}))+(\alpha.2.\theta_0(E'_{22},\eta_g))(1+Erf(E'_{22}/\eta_g))^{-1}$$

$$= \frac{\hbar^2}{2m_1}\left(\frac{\pi n_x}{d_x}\right)^2 + \frac{\hbar^2}{2m_3}\left(\frac{\pi n_z}{d_z}\right)^2 + \frac{\hbar^2}{2m_2}\left(\frac{\pi n_y}{d_y}\right)^2 + \frac{\alpha\hbar^4}{4m_2 m_2'}\left(\frac{\pi n_y}{d_y}\right)^4 \tag{3.100}$$

where E'_{22} is the totally quantized energy in this case.

The electron concentration in this case is given by

$$n_0 = \left(\frac{2g_v}{d_x d_y d_z}\right) \sum_{n_x=1}^{n_{x_{max}}} \sum_{n_y=1}^{n_{y_{max}}} \sum_{n_z=1}^{n_{z_{max}}} F_{-1}(\eta_{3D22}) \tag{3.101}$$

where $\eta_{3D22} = ((E_{FD22} - E'_{22}) / k_B T)$, E_{FD22} is the Fermi energy in the present case.

Thus using (3.101) and (3.27) one can obtain the Seebeck coefficient in the presence of large magnetic field in quantum dots of heavily doped IV-VI materials.

(b) The Lax Model

The dispersion relation in heavily doped IV-VI materials in accordance with Lax model can be written as

$$\gamma_2(E, \eta_g) = \frac{p_x^{\ 2}}{2m_1} + \frac{p_y^{\ 2}}{2m_2} + \frac{p_z^{\ 2}}{2m_3} \tag{3.102}$$

The simplified dispersion relation in quantum dots of IV-VI compounds can be written following (3.102) as

$$\gamma_2(E'_{23}, \eta_g) = \frac{\hbar^2}{2m_1}\left(\frac{\pi n_x}{d_x}\right)^2 + \frac{\hbar^2}{2m_2}\left(\frac{\pi n_y}{d_y}\right)^2 + \frac{\hbar^2}{2m_3}\left(\frac{\pi n_z}{d_z}\right)^2 \tag{3.103}$$

where E'_{23} is the totally quantized energy in this case.

The electron concentration in this case is given by

$$n_0 = \left(\frac{2g_v}{d_x d_y d_z}\right) \sum_{n_x=1}^{n_{x_{max}}} \sum_{n_y=1}^{n_{y_{max}}} \sum_{n_z=1}^{n_{z_{max}}} F_{-1}(\eta_{3D23}) \tag{3.104}$$

where $\eta_{3D23} = ((E_{FD23} - E'_{23}) / k_B T)$, E_{FD23} is the Fermi energy in the present case.

Thus using (3.104) and (3.27) one can obtain the Seebeck coefficient in the presence of large magnetic field in quantum dots of heavily doped IV-VI materials.

(c) The Dimmock Model

The dispersion relation of the conduction electrons in heavily doped IV-VI materials can be expressed extending Dimmock Model as

$$\gamma_2(E, \eta_g) + \alpha\gamma_3(E, \eta_g)\left[\frac{\hbar^2 k_s^{\ 2}}{2m_t^{\ +}} + \frac{\hbar^2 k_z^{\ 2}}{2m_l^{\ +}} - \frac{\hbar^2 k_s^{\ 2}}{2m_t^{\ -}} - \frac{\hbar^2 k_z^{\ 2}}{2m_l^{\ -}}\right]$$

$$=\left[\frac{\hbar^2 k_s^{\ 2}}{2m_t^*}+\frac{\hbar^2 k_z^{\ 2}}{2m_l^*}+(\frac{\hbar^2 k_s^{\ 2}}{2m_t^-}+\frac{\hbar^2 k_z^{\ 2}}{2m_l^-})[1+\alpha\{(\frac{\hbar^2 k_s^{\ 2}}{2m_t^+})+(\frac{\hbar^2 k_z^{\ 2}}{2m_l^+})\}]\right] \quad (3.105)$$

where the symbols of (3.105) have been defined in (1.190)

The simplified dispersion relation in quantum dots of heavily doped IV-VI compounds can be written following (3.105) as

$$\gamma_2(E'_{24},\eta_g)+\alpha\gamma_3(E'_{24},\eta_g)[\frac{\hbar^2}{2m_t^+}((\pi n_x/d_x)^2+(\pi n_y/d_y)^2)$$
$$+\frac{\hbar^2}{2m_l^+}(\pi n_z/d_z)^2-\frac{\hbar^2}{2m_t^-}((\pi n_x/d_x)^2+(\pi n_y/d_y)^2)-\frac{\hbar^2}{2m_l^-}(\pi n_z/d_z)^2]$$

$$=[\frac{\hbar^2}{2m_t^*}((\pi n_x/d_x)^2+(\pi n_y/d_y)^2)+\frac{\hbar^2}{2m_l^*}(\pi n_z/d_z)^2+(\frac{\hbar^2}{2m_t^-}((\pi n_x/d_x)^2+(\pi n_y/d_y)^2)+\frac{\hbar^2 k_z^{\ 2}}{2m_l^-})$$
$$[1+\alpha\{(\frac{\hbar^2}{2m_t^+}((\pi n_x/d_x)^2+(\pi n_y/d_y)^2))+(\frac{\hbar^2}{2m_l^+}(\pi n_z/d_z)^2)\}]] \quad (3.106)$$

where E'_{24} is the totally quantized energy in this case.

The electron concentration in this case is given by

$$n_0=\left(\frac{2g_v}{d_x d_y d_z}\right)\sum_{n_x=1}^{n_{x_{\max}}}\sum_{n_y=1}^{n_{y_{\max}}}\sum_{n_z=1}^{n_{z_{\max}}}F_{-1}(\eta_{3D24}) \quad (3.107)$$

where $\eta_{3D24}=((E_{FD24}-E'_{24})/k_BT)$ and E_{FD24} is the Fermi energy in the present case.

Thus using (3.107) and (3.27) one can obtain the Seebeck coefficient in the presence of large magnetic field for quantum dots of heavily doped IV-VI materials.

(d) The Bangert and Kastner Model

The electron dispersion law in heavily doped IV-VI compounds can be written in accordance with Bangert and Kastner as

$$S_1(E,\eta_g)k_s^2+S_2(E,\eta_g)k_z^2=1 \quad (3.108)$$

where, $S_1(E,\eta_g)=[2\gamma_0(E,\eta_g)]^{-1}[((\bar{R})^2/E_{g0})\{C_1(\alpha_1,E,\eta_g)-iD_1(\alpha_1,E,\eta_g)\}$
$+((\bar{S})^2/\Delta'_c)\{C_2(\alpha_2,E,\eta_g)-iD_2(\alpha_2,E,\eta_g)\}+((\bar{Q})^2/\Delta''_c)\{C_3(\alpha_3,E,\eta_g)-iD_3(\alpha_3,E,\eta_g)\},$

$$S_2(E,\eta_g) = [2\gamma_0(E,\eta_g)]^{-1}[((2\bar{A})^2 / E_{g0})\{C_1(\alpha_1,E,\eta_g) - iD_1(\alpha_1,E,\eta_g)\}$$
$$+((\bar{S}+\bar{Q})^2 / \Delta_c'')\{C_3(\alpha_3,E,\eta_g) - iD_3(\alpha_3,E,\eta_g)\},$$

$$C_i(\alpha_i,E,\eta_g) \equiv \left[\frac{2}{\alpha_i\eta_g\sqrt{\pi}}\right]\exp(-u_i^2)\left[\sum_{p=1}^{\infty}\left\{\frac{\exp(-p^2/4)}{p}\sinh(pu_i)\right\}\right],$$

$u_i = ((1+\alpha_i E)/\alpha_i\eta_g)$, $i = 1, 2 and 3$, $\alpha_1 = (1/E_{g0})$, $\alpha_2 = (1/\Delta_c')$, $\alpha_3 = (1/\Delta_c'')$ and

$$D_i(\alpha_i,E,\eta_g) \equiv \left[\frac{\sqrt{\pi}}{\alpha_i\eta_g}\exp(-u_i^2)\right]$$

Since $S_1(E,\eta_g)$ and $S_2(E,\eta_g)$ are complex, the energy spectrum of heavily doped IV-VI materials is also complex in accordance with the model of Bangert and Kastner.

The simplified dispersion relation in quantum dots of heavily doped IV-VI semiconductors can be written following (3.108) as

$$S_1(E_{25}',\eta_g)((\pi n_x/d_x)^2 + (\pi n_y/d_y)^2) + S_2(E_{25}',\eta_g)(\pi n_z/d_z)^2 = 1 \qquad (3.109)$$

where E_{25}' is the totally quantized energy in this case.

The electron concentration in this case is given by

$$n_0 = \left(\frac{2g_v}{d_x d_y d_z}\right)\sum_{n_x=1}^{n_{x_{\max}}}\sum_{n_y=1}^{n_{y_{\max}}}\sum_{n_z=1}^{n_{z_{\max}}} F_{-1}(\eta_{3D25}) \qquad (3.110)$$

where $\eta_{3D25} = ((E_{FD25} - E_{25}')/k_B T)$, E_{FD25} is the Fermi energy in the present case.

Thus using (3.110) and (3.27) one can obtain the Seebeck coefficient in the presence of large magnetic field for quantum dots of heavily doped IV-VI materials.

(e) The Model of Foley et. al

The electron dispersion law in heavily doped IV-VI compounds can be written in accordance with the model of Foley et.al. as

$$\gamma_3(E,\eta_g) = \frac{\hbar^2 k_s^2}{2m_\perp^-} + \frac{\hbar^2 k_z^2}{2m_\parallel^-} + \left[\left[\frac{\hbar^2 k_s^2}{2m_\perp^+} + \frac{\hbar^2 k_z^2}{2m_\parallel^+} + \frac{E_{g_0}}{2}\right]^2 + P_\parallel^2 k_z^2 + P_\perp^2 k_s^2\right]^{1/2} - \frac{E_{g_0}}{2} \qquad (3.111)$$

The simplified dispersion relation in quantum dots of heavily doped IV-VI semiconductors can be written following (3.111) as

$$\gamma_3(E'_{26},\eta_g) = -\frac{E_{g_0}}{2} + \frac{\hbar^2}{2m_\perp^-}\left[\left(\frac{n_x\pi}{d_x}\right)^2 + \left(\frac{n_y\pi}{d_y}\right)^2\right] + \frac{\hbar^2}{2m_\parallel^-}\left(\frac{n_z\pi}{d_z}\right)^2 +$$

$$\left[\left[\frac{E_{g_0}}{2} + \frac{\hbar}{2m_\parallel^+}\left(\frac{n_z\pi}{d_z}\right)^2 + \frac{\hbar^2}{2m_\perp^+}\left\{\left(\frac{n_x\pi}{d_x}\right)^2 + \left(\frac{n_y\pi}{d_y}\right)^2\right\}\right]^2\right.$$

$$\left. + P_\parallel^2\left(\frac{n_z\pi}{d_z}\right)^2 + P_\perp^2\left\{\left(\frac{n_x\pi}{d_x}\right)^2 + \left(\frac{n_y\pi}{d_y}\right)^2\right\}\right]^{1/2} \tag{3.112}$$

where E'_{26} is the totally quantized energy in this case.

The electron concentration in this case is given by

$$n_0 = \left(\frac{2g_v}{d_x d_y d_z}\right)\sum_{n_x=1}^{n_{x_{\max}}}\sum_{n_y=1}^{n_{y_{\max}}}\sum_{n_z=1}^{n_{z_{\max}}} F_{-1}(\eta_{3D26}) \tag{3.113}$$

where $\eta_{3D26} = ((E_{FD26} - E'_{26})/k_B T)$ and E_{FD26} is the Fermi energy in the present case.

Thus using (3.113) and (3.27) one can obtain the Seebeck coefficient in the presence of large magnetic field for quantum dots of heavily doped IV-VI materials.

3.2.12. Study of the Seebeck Coefficient under Large Magnetic Field in Heavily Doped Quantum Dots of Heavily Doped Lead Germanium Telluride

The dispersion law of n-type heavily doped $Pb_{1-x}Ge_xTe$ with x = 0.01 can be expressed extending (1.212) as

$$[2/(Erf(E/\eta_g))]\theta_0(E,\eta_g) + \gamma_3(E,\eta_g)[\bar{E}_{g0} - .195k_s^2 - .345k_z^2]$$

$$= 0.23k_s^2 + 0.02k_z^2 \pm \left[0.06\bar{E}_{g_0} + 0.061k_s^2 + 0.0066k_z^2\right]k_s + \left[\bar{E}_{g_0} + 0.411k_s^2 + 0.377k_z^2\right]$$

$$\left[0.606k_s^2 + 0.722k_z^2\right] \tag{3.114}$$

The simplified dispersion relation in quantum dots of heavily doped IV-VI semiconductors can be written following (3.114) as

$$[2/(Erf(E'_{27}/\eta_g))]\theta_0(E'_{27},\eta_g)+\gamma_3(E'_{27},\eta_g)[\bar{E}_{g0}-.195(\left(\frac{n_x\pi}{d_x}\right)^2+\left(\frac{n_y\pi}{d_y}\right)^2)-.345(\frac{n_x\pi}{d_x})^2]$$

$$=0.23((\pi n_x/d_x)^2+(\pi n_y/d_y)^2)+0.02(\pi n_z/d_Z)^2$$

$$\pm\left[0.06\bar{E}_{g_0}+0.061((\pi n_x/d_x)^2+(\pi n_y/d_y)^2)+0.0066(\pi n_z/d_Z)^2\right]$$

$$((\pi n_x/d_x)^2+(\pi n_y/d_y)^2)^{1/2}+\left[\bar{E}_{g_0}+0.411((\pi n_x/d_x)^2+(\pi n_y/d_y)^2)+0.377(\pi n_z/d_Z)^2\right]$$

$$\left[0.606((\pi n_x/d_x)^2+(\pi n_y/d_y)^2)+0.722(\pi n_z/d_Z)^2\right] \tag{3.115}$$

where E'_{27} is the totally quantized energy in this case.

The electron concentration in this case is given by

$$n_0=\left(\frac{g_v}{d_x d_y d_z}\right)\sum_{n_x=1}^{n_{x_{max}}}\sum_{n_y=1}^{n_{y_{max}}}\sum_{n_z=1}^{n_{z_{max}}}F_{-1}(\eta_{3D27,\pm}) \tag{3.116}$$

where $\eta_{3D27,\pm}=((E_{FD27}-E'_{27})/k_BT)$, E_{FD27} is the Fermi energy in the present case.

Thus using (3.116) and (3.27) one can obtain the Seebeck coefficient in the presence of large magnetic field for quantum dots of heavily doped IV-VI materials.

3.2.13. Study of the Seebeck Coefficient under Large Magnetic Field in Quantum Dots of Heavily Doped Zinc and Cadmium Diphosphides

The dispersion relation of the holes of heavily doped Cadmium and Zinc diphosphides can approximately be written extending (1.219) as

$$\gamma_3(E,\eta_g)=\left[\beta_1+\frac{\beta_2\beta_3(k)}{8\beta_4}\right]k^2\pm\left\{\left[\beta_4\beta_3(k)\left(\beta_5-\frac{\beta_2\beta_3(k)}{8\beta_4}\right)k^2\right]+8\beta_4^2\left(1-\frac{\beta_3^2(k)}{4}\right)-\beta_2\left(1-\frac{\beta_3^2(k)}{4}\right)k^2\right\}^{1/2} \tag{3.117}$$

The simplified dispersion relation in quantum dots of heavily doped Cadmium and Zinc diphosphides can be written following (3.117) as

$$\gamma_3(E'_{28,\pm},\eta_g)=\left[\beta_1+\frac{\beta_2\beta_6}{8\beta_4}\right]\beta_7^2\pm\left\{\left[\beta_4\beta_6\left(\beta_5-\frac{\beta_2\beta_6}{8\beta_4}\right)\beta_7^2\right]+8\beta_4^2\left(1-\frac{\beta_6^2}{4}\right)-\beta_2\left(1-\frac{\beta_6^2}{4}\right)\beta_7^2\right\}^{1/2} \tag{3.118}$$

where $E'_{28,\pm}$ is the totally quantized energy in this case.

The electron concentration in this case is given by

$$n_0 = \left(\frac{g_v}{d_x d_y d_z}\right) \sum_{n_x=1}^{n_{x_{\max}}} \sum_{n_y=1}^{n_{y_{\max}}} \sum_{n_z=1}^{n_{z_{\max}}} F_{-1}\left(\eta_{3D28,\pm}\right) \tag{3.119}$$

where $\eta_{3D28,\pm} = ((E_{FD28} - E'_{28\pm}) / k_B T)$, E_{FD28} is the Fermi energy in the present case.

Thus using (3.119) and (3.27) one can obtain the Seebeck coefficient in the presence of large magnetic field for quantum dots of heavily doped IV-VI materials.

3.2.14. Study of the Seebeck Coefficient under Large Magnetic Field in Quantum Dots of Heavily Doped Bismuth Telluride

The dispersion relation of the holes in heavily doped Bi_2Te_3 can be expressed extending (1.226) as

$$\gamma_2(E,\eta_g) = \frac{\hbar^2}{2m_0}\left(\left(\overline{\overline{\alpha_{11}}}\right)k_x^{\,2} + \alpha_{22}k_y^{\,2} + \alpha_{33}k_z^{\,2} + 2\alpha_{23}k_y k_z\right) \tag{3.120}$$

where $\overline{\overline{\alpha_{11}}}, \alpha_{22}, \alpha_{33}$ and α_{23} are spectrum constants.

The simplified dispersion relation in quantum dots of heavily doped Bismuth Teluride can be written following (3.120) as

$$\gamma_2(E'_{29},\eta_g) = \frac{\hbar^2}{2m_0}\left[\left(\overline{\overline{\alpha_{11}}}\right)\left(\frac{n_x\pi}{d_x}\right)^2 + \alpha_{22}\left(\frac{n_y\pi}{d_y}\right)^2 + \alpha_{33}\left(\frac{n_z\pi}{d_z}\right)^2 + 2\alpha_{23}\left(\frac{\pi^2 n_y n_z}{d_y d_z}\right)\right] \tag{3.121}$$

where $E'_{29,\pm}$ is the totally quantized energy in this case.

The electron concentration in this case is given by

$$n_0 = \left(\frac{2g_v}{d_x d_y d_z}\right) \sum_{n_x=1}^{n_{x_{\max}}} \sum_{n_y=1}^{n_{y_{\max}}} \sum_{n_z=1}^{n_{z_{\max}}} F_{-1}\left(\eta_{3D29}\right) \tag{3.122}$$

where $\eta_{3D29} = ((E_{FD29} - E'_{29}) / k_B T)$ and E_{FD29} is the Fermi energy in the present case.

Thus using (3.122) and (3.27) one can obtain the Seebeck coefficient in the presence of large magnetic field for quantum dots of heavily doped Bismuth Teluride

3.2.15. Study of the Seebeck Coefficient under Large Magnetic Field in Heavily Doped III-V Quantum Dot Superlattices with Graded Interfaces

The electron dispersion law in bulk specimens of the heavily doped constituent materials of III-V SLs whose energy band structures are defined by three band model of Kane can be expressed following (3.28) as

$$\frac{\hbar^2 k^2}{2m_{cj}^*} = T_{1j}\left(E,\Delta_j,E_{gj},\eta_{gj}\right) + iT_{2j}\left(E,\Delta_j,E_{gj},\eta_{gj}\right) \tag{3.123}$$

where j=1,2, $T_{ij}\left(E,\Delta_j,E_{gj},\eta_{gj}\right) =$

$(2/(1+Erf(E/\eta_{gj}))[(\alpha_j b_j/c_j).\theta_0\left(E,\eta_{gj}\right)+[(\alpha_j c_j+b_j c_j-\alpha_j b_j)/c_j^2].$

$\gamma_0\left(E,\eta_{gj}\right)+[(1/c_j)(1-(\alpha_j/c_j))(1-(b_j/c_j))\frac{1}{2}[1+Erf(E/\eta_{gj})]-$

$(1/c_j)(1-(\alpha_j/c_j))(1-(b_j/c_j))(2/(c_j\eta_{gj}\sqrt{\pi})\exp\left(-u_j^2\right)[\sum_{p=1}^{\infty}(\exp\left(-p^2/4\right)/p)sinh(pu_j)]],$

$b_j \equiv (E_{gj}+\Delta_j)^{-1}$, $c_j \equiv (E_{gj}+\frac{2}{3}\Delta_j)^{-1}$, $u_j \equiv \frac{1+c_j E}{c_j\eta_{gj}}$ and

$$T_{2j}\left(E,\Delta_j,E_{gj},\eta_{gj}\right) \equiv \left(\frac{2}{1+Erf\left(E/\eta_{gj}\right)}\right)\frac{1}{c_j}\left(1-\frac{\alpha_j}{c_j}\right)\left(1-\frac{b_j}{c_j}\right)\frac{\sqrt{\pi}}{c_j\eta_{gj}}\exp\left(-u_j^2\right).$$

Therefore, the dispersion law of the electrons of heavily doped III-V SLs with graded interfaces can be expressed as

$$k_z^2 = G_8 + iH_8 \tag{3.124}$$

where $G_8 = \left[\frac{C_7^2 - D_7^2}{L_0^2} - k_s^2\right]$, $C_7 = \cos^{-1}(\overline{\omega_7})$, $\overline{\omega_7} = (2)^{\frac{-1}{2}}[(1-G_7^2-H_7^2)-\sqrt{(1-G_7^2-H_7^2)^2+4G_7^2}]^{\frac{1}{2}}$

$$G_7 = [G_1 + (\rho_5 G_2/2) - (\rho_6 H_2/2) + (\Delta_0/2)\{\rho_6 H_2 - \rho_8 H_3 + \rho_9 H_4 - \rho_{10} H_4 + \rho_{11} H_5 - \rho_{12} H_5 + (1/12)(\rho_{12} G_6 - \rho_{14} H_6)\}],\ G_1 = [(\cos(\mathrm{h}_1))(\cosh(\mathrm{h}_2))(\cosh(g_1))(\cos(g_2)) + (\sin(\mathrm{h}_1))(\sinh(\mathrm{h}_2))(\sinh(g_1))(\sin(g_2))],\ \mathrm{h}_1 = e_1(b_0 - \Delta_0),\ e_1 = 2^{\frac{-1}{2}}(\sqrt{t_1^2 + t_2^2} + t_1)^{\frac{1}{2}},$$

$$t_1 = [(2m_{c1}^*/\hbar^2).T_{11}(E, E_{g1}, \Delta_1, \eta_{g1}) - k_s^2],\ t_2 = [(2m_{c1}^*/\hbar^2)T_{21}(E, E_{g1}, \Delta_1, \eta_{g1})],$$

$$\mathrm{h}_2 = e_2(b_0 - \Delta_0),\ e_2 = 2^{\frac{-1}{2}}(\sqrt{t_1^2 + t_2^2} - t_1)^{\frac{1}{2}},\ \mathrm{g}_1 = d_1(a_0 - \Delta_0),\ \mathrm{d}_1 = 2^{\frac{-1}{2}}(\sqrt{x_1^2 + y_1^2} + x_1)^{\frac{1}{2}},$$

$$x_1 = [-(2m_{c2}^*/\hbar^2).T_{11}(E - V_0, E_{g2}, \Delta_2, \eta_{g2}) + k_s^2],\ \mathrm{y}_1 = [(2m_{c2}^*/\hbar^2)T_{22}(E - V_0, E_{g2}, \Delta_2, \eta_{g2})],$$

$$\mathrm{g}_2 = d_2(a_0 - \Delta_0), \mathrm{d}_2 = 2^{\frac{-1}{2}}(\sqrt{x_1^2 + y_1^2} - x_1)^{\frac{1}{2}},\ \rho_5 = (\rho_3^2 + \rho_4^2)^{-1}[\rho_1\rho_3 - \rho_2\rho_4],$$

$$\rho_1 = [d_1^2 + e_2^2 - d_2^2 - e_1^2], \rho_3 = [\mathrm{d}_1 e_1 + \mathrm{d}_2 e_2], \rho_2 = 2[d_1 d_2 + e_1 e_2], \rho_4 = [d_1 e_2 - e_1 d_2],$$

$$G_2 = [(\sin(\mathrm{h}_1))(\cosh(\mathrm{h}_2))(\sinh(g_1))(\cos(g_2)) + (\cos(\mathrm{h}_1))(\sinh(\mathrm{h}_2))(\cosh(g_1))(\sin(g_2))],$$

$$\rho_6 = (\rho_3^2 + \rho_4^2)^{-1}[\rho_1\rho_4 + \rho_2\rho_3],$$

$$H_2 = [(\sin(\mathrm{h}_1))(\cosh(\mathrm{h}_2))(\sin(g_2))(\cosh(g_1)) - (\cos(\mathrm{h}_1))(\sinh(\mathrm{h}_2))(\sinh(g_1))(\cos(g_2))],$$

$$\rho_7 = [(e_1^2 + e_2^2)^{-1}[e_1(d_1^2 - d_2^2) - 2d_1 d_2 e_2] - 3e_1],$$

$$G_3 = [(\sin(\mathrm{h}_1))(\cosh(\mathrm{h}_2))(\cosh(g_1))(\cos(g_2)) + (\cos(\mathrm{h}_1))(\sinh(\mathrm{h}_2))(\sinh(g_1))(\sin(g_2))],$$

$$\rho_8 = [(e_1^2 + e_2^2)^{-1}[e_2(d_1^2 - d_2^2) + 2d_1 d_2 e_1] + 3e_2],$$

$$H_3 = [(\sin(\mathrm{h}_1))(\cosh(\mathrm{h}_2))(\sin(g_2))(\sinh(g_1)) - (\cos(\mathrm{h}_1))(\sinh(\mathrm{h}_2))(\cosh(g_1))(\cos(g_2))],$$

$$\rho_9 = [(d_1^2 + d_2^2)^{-1}[d_1(e_2^2 - e_1^2) + 2e_2 d_2 e_1] + 3d_1],$$

$$G_4 = [(\cos(\mathrm{h}_1))(\cosh(\mathrm{h}_2))(\cos(g_2))(\sinh(g_1)) - (\sin(\mathrm{h}_1))(\sinh(\mathrm{h}_2))(\cosh(g_1))(\sin(g_2))],$$

$$\rho_{10} = [-(d_1^2 + d_2^2)^{-1}[d_2(-e_2^2 + e_1^2) + 2e_2 d_2 e_1] + 3d_2],$$

$$H_4 = [(cos(\mathrm{h}_1))(\cosh(\mathrm{h}_2))(\cosh(g_1))(\sin(g_2)) + (\sin(\mathrm{h}_1))(\sinh(\mathrm{h}_2))(\sinh(g_1))(\cos(g_2))],$$

$$\rho_{11} = 2[d_1^2 + e_2^2 - d_2^2 - e_1^2],$$

$$G_5 = [(\cos(\mathrm{h}_1))(\cosh(\mathrm{h}_2))(\cos(g_2))(\cosh(g_1)) - (\sin(\mathrm{h}_1))(\sinh(\mathrm{h}_2))(\sinh(g_1))(\sin(g_2))],$$

$$\rho_{12} = 4[d_1 d_2 + e_1 e_2],$$

$$H_5 = [(cos(\mathrm{h}_1))(\cosh(\mathrm{h}_2))(\sinh(g_1))(\sin(g_2)) + (\sin(\mathrm{h}_1))(\sinh(\mathrm{h}_2))(\cosh(g_1))(\cos(g_2))],$$

$$\rho_{13} = [\{5(d_1 e_1^3 - 3e_1 e_2^2 d_1) + 5d_2(e_1^3 - 3e_1^2 e_2)\}(d_1^2 + d_2^2)^{-1} + (e_1^2 + e_2^2)^{-1}\{5(e_1 d_1^3 - 3d_2 e_1^2 d_1) + 5(d_2^3 e_2 - 3d_1^2 d_2 e_2)\} - 34(d_1 e_1 + d_2 e_2)],$$

$G_6 = [(\sin(h_1))(\cosh(h_2))(\sinh(g_1))(\cos(g_2)) + (\cos(h_1))(\sinh(h_2))(\cosh(g_1))(\sin(g_2))]$,
$\rho_{14} = [\{5(d_1e_2^3 - 3e_2e_1^2d_1) + 5d_2(-e_1^3 + 3e_2^2e_1)\}(d_1^2 + d_2^2)^{-1} + (e_1^2 + e_2^2)^{-1}\{5(-e_1d_2^3 + 3d_1^2d_2e_1) + 5(-d_1^3e_2 + 3d_2^2d_1e_2)\} + 34(d_1e_2 - d_2e_1)]$,
$H_6 = [(\sin(h_1))(\cosh(h_2))(\cosh(g_1))(\sin(g_2)) - (\cos(h_1))(\sinh(h_2))(\sinh(g_1))(\cos(g_2))]$,
$H_7 = [H_1 + (\rho_5H_2/2) + (\rho_6G_2/2) + (\Delta_0/2)\{\rho_8G_3 + \rho_7H_3 + \rho_{10}G_4 + \rho_9H_4 + \rho_{12}G_5 + \rho_{11}H_5 + (1/12)(\rho_{14}G_6 + \rho_{13}H_6)\}]$,
$H_1 = [(\sin(h_1))(\sinh(h_2))(\cosh(g_1))(\cos(g_2)) + (\cos(h_1))(\cosh(h_2))(\sinh(g_1))(\sin(g_2))]$,
$D_7 = \sinh^{-1}(\overline{\omega_7})$, $\mathrm{H}_8 = (2C_7D_7/L_0^2)$

The simplified dispersion relation in quantum dots of heavily doped III-V superllatices with graded interfaces can be expressed as

$$\left(\frac{\pi n_z}{d_z}\right)^2 = G_{80D} + iH_{80D} \tag{3.125}$$

where $\mathrm{G}_{80D} = \left[\frac{C_{70D}^2 - D_{70D}^2}{L_0^2} - \{(\frac{n_x\pi}{d_x})^2 + (\frac{n_y\pi}{d_y})^2\}\right]$,

$C_{70D} = \cos^{-1}(\overline{\omega_{70D}})$, $\overline{\omega_{70D}} = (2)^{\frac{-1}{2}}[(1 - G_{70D}^2 - H_{70D}^2) - \sqrt{(1 - G_{70D}^2 - H_{70D}^2)^2 + 4G_{70D}^2}]^{\frac{1}{2}}$
$G_{70D} = [G_{10D} + (\rho_{50D}G_{20D}/2) - (\rho_{60D}H_{20D}/2) + (\Delta_0/2)\{\rho_{60D}H_{20D} - \rho_{80D}H_{30D} + \rho_{90D}H_{40D} - \rho_{100D}H_{40D} + \rho_{110D}H_{50D} - \rho_{120D}H_{50D} + (1/12)(\rho_{120D}G_{60D} - \rho_{140D}H_{60D})\}]$, $G_{10D} = [(\cos(\mathrm{h}_{10D}))(\cosh(\mathrm{h}_{20D}))(\cosh(g_{10D}))(\cos(g_{20D})) + (\sin(\mathrm{h}_{10D}))(\sinh(\mathrm{h}_{20D}))(\sinh(g_{10D}))(\sin(g_{20D}))]$, $\mathrm{h}_{10D} = e_{10D}(b_0 - \Delta_0)$, $e_{10D} = 2^{\frac{-1}{2}}(\sqrt{t_{10D}^2 + t_{20D}^2} + t_{10D})^{\frac{1}{2}}$,

$t_{10D} = [(2m_{c1}^*/\hbar^2)T_{11}(E, E_{g1}, \Delta_1, \eta_{g1}) - \{(\frac{n_x\pi}{d_x})^2 + (\frac{n_y\pi}{d_y})^2\}]$, $t_{20D} = [(2m_{c1}^*/\hbar^2)T_{21}(E, E_{g1}, \Delta_1, \eta_{g1})]$,

$\mathrm{h}_{20D} = e_{20D}(b_0 - \Delta_0)$, $e_{20D} = 2^{\frac{-1}{2}}(\sqrt{t_{10D}^2 + t_{20D}^2} - t_{10D})^{\frac{1}{2}}$, $g_{10D} = d_{10D}(a_0 - \Delta_0)$, $\mathrm{d}_{10D} = 2^{\frac{-1}{2}}(\sqrt{x_{10D}^2 + y_{10D}^2} + x_{10D})^{\frac{1}{2}}$,

$x_{10D} = [-(2m_{c2}^*/\hbar^2)T_{11}(E - V_0, E_{g2}, \Delta_2, \eta_{g2}) + \{(\frac{n_x\pi}{d_x})^2 + (\frac{n_y\pi}{d_y})^2\}]$, $\mathrm{y}_{10D} = [(2m_{c2}^*/\hbar^2)T_{220D}(E - V_0, E_{g2}, \Delta_2, \eta_{g2})]$,

$g_{20D} = d_{20D}(a_0 - \Delta_0)$, $\mathrm{d}_{20D} = 2^{\frac{-1}{2}}(\sqrt{x_{10D}^2 + y_{10D}^2} - x_{10D})^{\frac{1}{2}}$, $\rho_{50D} = (\rho_{30D}^2 + \rho_{40D}^2)^{-1}[\rho_{10D}\rho_{30D} - \rho_{20D}\rho_{40D}]$,
$\rho_{10D} = [d_{10D}^2 + e_{20D}^2 - d_{20D}^2 - e_{10D}^2]$, $\rho_{30D} = [\mathrm{d}_{10D}e_{10D} + \mathrm{d}_{20D}e_{20D}]$,
$\rho_{20D} = 2[d_{10D}d_{20D} + e_{10D}e_{20D}]$, $\rho_{40D} = [d_{10D}e_{20D} - e_{10D}d_{20D}]$,

$$G_{20D} = [(\sin(h_{10D}))(\cosh(h_{20D}))(\sinh(g_{10D}))(\cos(g_{20D})) + (\cos(h_{10D}))(\sinh(h_{20D}))(\cosh(g_{10D}))(\sin(g_{20D}))],$$

$$\rho_{60D} = (\rho_{30D}^2 + \rho_{40D}^2)^{-1}[\rho_{10D}\rho_{40D} + \rho_{20D}\rho_{30D}],$$

$$H_{20D} = [(\sin(h_{10D}))(\cosh(h_{20D}))(\sin(g_{20D}))(\cosh(g_{10D})) - (\cos(h_{10D}))(\sinh(h_{20D}))(\sinh(g_{10D}))(\cos(g_{20D}))],$$

$$\rho_{70D} = [(e_{10D}^2 + e_{20D}^2)^{-1}[e_{10D}(d_{10D}^2 - d_{20D}^2) - 2d_{10D}d_{20D}e_{20D}] - 3e_{10D}],$$

$$G_{30D} = [(\sin(h_{10D}))(\cosh(h_{20D}))(\cosh(g_{10D}))(\cos(g_{20D})) + (\cos(h_{10D}))(\sinh(h_{20D}))(\sinh(g_{10D}))(\sin(g_{20D}))],$$

$$\rho_{80D} = [(e_{10D}^2 + e_{20D}^2)^{-1}[e_{20D}(d_{10D}^2 - d_{20D}^2) + 2d_{10D}d_{20D}e_{10D}] + 3e_{20D}],$$

$$H_{30D} = [(\sin(h_{10D}))(\cosh(h_{20D}))(\sin(g_{20D}))(\sinh(g_{10D})) - (\cos(h_{10D}))(\sinh(h_{20D}))(\cosh(g_{10D}))(\cos(g_{20D}))],$$

$$\rho_{90D} = [(d_{10D}^2 + d_{20D}^2)^{-1}[d_{10D}(e_{20D}^2 - e_{10D}^2) + 2e_{20D}d_{20D}e_{10D}] + 3d_{10D}],$$

$$G_{40D} = [(\cos(h_{10D}))(\cosh(h_{20D}))(\cos(g_{20D}))(\sinh(g_{10D})) - (\sin(h_{10D}))(\sinh(h_{20D}))(\cosh(g_{10D}))(\sin(g_{20D}))],$$

$$\rho_{100D} = [-(d_{10D}^2 + d_{20D}^2)^{-1}[d_{20D}(-e_{20D}^2 + e_{10D}^2) + 2e_{20D}d_{20D}e_{10D}] + 3d_{20D}],$$

$$H_{40D} = [(\cos(h_{10D}))(\cosh(h_{20D}))(\cosh(g_{10D}))(\sin(g_{20D})) + (\sin(h_{10D}))(\sinh(h_{20D}))(\sinh(g_{10D}))(\cos(g_{20D}))],$$

$$\rho_{110D} = 2[d_{10D}^2 + e_{20D}^2 - d_{20D}^2 - e_{10D}^2],$$

$$G_{50D} = [(\cos(h_{10D}))(\cosh(h_{20D}))(\cos(g_{20D}))(\cosh(g_{10D})) - (\sin(h_{10D}))(\sinh(h_{20D}))(\sinh(g_1))(\sin(g_{20D}))],$$

$$\rho_{120D} = 4[d_{10D}d_{20D} + e_{10D}e_{20D}],$$

$$H_{50D} = [(\cos(h_{10D}))(\cosh(h_{20D}))(\sinh(g_{10D}))(\sin(g_{20D})) + (\sin(h_{10D}))(\sinh(h_{20D}))(\cosh(g_{10D}))(\cos(g_{20D}))],$$

$$\rho_{130D} = [\{5(d_{10D}e_{10D}^3 - 3e_{10D}e_{20D}^2 d_{10D}) + 5d_{20D}(e_{10D}^3 - 3e_{10D}^2 e_{20D})\}(d_{10D}^2 + d_{20D}^2)^{-1} +$$
$$(e_{10D}^2 + e_{20D}^2)^{-1}\{5(e_{10D}d_{10D}^3 - 3d_{20D}e_{10D}^2 d_{10D}) + 5(d_{20D}^3 e_{20D} - 3d_{10D}^2 d_{20D}e_{20D})\} - 34(d_{10D}e_{10D} + d_{20D}e_{20D})],$$

$$G_{60D} = [(\sin(h_{10D}))(\cosh(h_{20D}))(\sinh(g_{10D}))(\cos(g_{20D})) + (\cos(h_{10D}))(\sinh(h_{20D}))(\cosh(g_{10D}))(\sin(g_{20D}))],$$

$$\rho_{140D} = [\{5(d_{10D}e_{20D}^3 - 3e_{20D}e_{10D}^2 d_{10D}) + 5d_{20D}(-e_{10D}^3 + 3e_{20D}^2 e_{10D})\}(d_{10D}^2 + d_{20D}^2)^{-1} +$$
$$(e_{10D}^2 + e_{20D}^2)^{-1}\{5(-e_{10D}d_{20D}^3 + 3d_{10D}^2 d_{20D}e_{10D}) + 5(-d_{10D}^3 e_{20D} + 3d_{20D}^2 d_{10D}e_{20D})\} + 34(d_{10D}e_{20D} - d_{20D}e_{10D})],$$

$$H_{60D} = [(\sin(h_{10D}))(\cosh(h_{20D}))(\cosh(g_{10D}))(\sin(g_{20D})) - (\cos(h_{10D}))(\sinh(h_{20D}))(\sinh(g_{10D}))(\cos(g_{20D}))],$$

$$H_{70D} = [H_{10D} + (\rho_{50D}H_{20D}/2) + (\rho_{60D}G_{20D}/2) + (\Delta_0/2)\{\rho_{80D}G_{30D} + \rho_{70D}H_{30D} + \rho_{100D}G_{40D} + \rho_{90D}H_{40D} +$$
$$\rho_{120D}G_{50D} + \rho_{110D}H_{50D} + (1/12)(\rho_{140D}G_{60D} + \rho_{130D}H_{60D})\}],$$

$$H_{10D} = [(\sin(h_{10D}))(\sinh(h_{20D}))(\cosh(g_{10D}))(\cos(g_{20D})) + (\cos(h_{10D}))(\cosh(h_{20D}))(\sinh(g_{10D}))(\sin(g_{20D}))],$$

$$D_{70D} = \sinh^{-1}(\overline{\omega_{70D}}) \text{ and } H_{80D} = (2C_{70D}D_{70D}/L_0^2)$$

The electron concentration in this case is given by

$$n_0 = \left(\frac{2g_v}{d_x d_y d_z}\right)\sum_{n_x=1}^{n_{x_{\max}}}\sum_{n_y=1}^{n_{y_{\max}}}\sum_{n_z=1}^{n_{z_{\max}}} F_{-1}(\eta_{3D30}) \tag{3.126}$$

where $\eta_{3D30} = ((E_{FD30} - E'_{30}) / k_B T)$, E_{FD30} is the Fermi energy in the present case and E'_{30} is the real part of (3.125).Thus using (3.126) and (3.27) one can obtain the Seebeck coefficient in the presence of large magnetic field for heavily doped III-V quantum dot superlattices with graded interfaces

3.2.16. Study of the Seebeck Coefficient under Large Magnetic Field in Heavily Doped II-VI Quantum Dot Superlattices with Graded Interfaces

The electron energy spectra of the heavily doped constituent materials of II-VI SLs are given by

$$\gamma_3(E,\eta_{g1}) = \frac{\hbar^2 k_s^2}{2m_{\perp,1}^*} + \frac{\hbar^2 k_z^2}{2m_{\|,1}^*} \pm C_0 k_s \tag{3.127}$$

and $$\frac{\hbar^2 k^2}{2m_{c2}^*} = T_{12}\left(E,\Delta_2,E_{g2},\eta_{g2}\right) + iT_{22}\left(E,\Delta_2,E_{g2},\eta_{g2}\right) \tag{3.128}$$

where $m_{\perp,1}^*$ and $m_{\|,1}^*$ are the transverse and longitudinal effective electron masses respectively at the edge of the conduction band for the first material. The energy-wave vector dispersion relation of the conduction electrons in heavily doped II-VI SLs with graded interfaces can be expressed as

$$k_z^2 = G_{19} + iH_{19} \tag{3.129}$$

where $\mathrm{G}_{19} = \left[\frac{C_{18}^2 - D_{18}^2}{L_0^2} - k_s^2\right]$,

$C_{18} = \cos^{-1}(\omega_{18}), \omega_{18} = (2)^{\frac{-1}{2}}[(1 - G_{18}^2 - H_{18}^2) - \sqrt{(1 - G_{18}^2 - H_{18}^2)^2 + 4G_{18}^2}]^{\frac{1}{2}}$,

$G_{18} = \frac{1}{2}[G_{11} + G_{12} + \Delta_0(G_{13} + G_{14}) + \Delta_0(G_{15} + G_{16})]$,

$G_{11} = 2(\cos(g_1))(\cos(g_2))(\cos\gamma_{11}(E,k_s))$

$\gamma_{11}(E,k_s) = k_{21}(E,k_s)(b_0 - \Delta_0), k_{21}(E,k_s) = \{[\gamma_3(E,\eta_{g1}) - \frac{\hbar^2 k_s^2}{2m_{\perp,1}^*} \pm C_0 k_s]\frac{2m_{\|,1}^*}{\hbar^2}\}^{1/2}$

$, G_{12} = ([\Omega_1(E,k_s)(\sinh g_1)(\cos g_2) - \Omega_2(E,k_s)(\sin g_2)(\cosh g_1)](\sin\gamma_{11}(E,k_s)))$

$\Omega_1(E,k_s) = [\frac{d_1}{k_{21}(E,k_s)} - \frac{k_{21}(E,k_s)d_1}{d_1^2 + d_2^2}]$

and $\Omega_2(E,k_s)=[\frac{d_2}{k_{21}(E,k_s)}+\frac{k_{21}(E,k_s)d_2}{d_1^2+d_2^2}]$

$$G_{13}=([\Omega_3(E,k_s)(\cosh g_1)(\cos g_2)-\Omega_4(E,k_s)(\sinh g_1)(\sin g_2)](\sin\gamma_{11}(E,k_s)))$$

$$\Omega_3(E,k_s)=[\frac{d_1^2-d_2^2}{k_{21}(E,k_s)}-3k_{21}(E,k_s)],\Omega_4(E,k_s)=[\frac{2d_1d_2}{k_{21}(E,k_s)}]$$

$$G_{14}=([\Omega_5(E,k_s)(\sinh g_1)(\cos g_2)-\Omega_6(E,k_s)(\sin g_1)(\cosh g_2)](\cos\gamma_{11}(E,k_s))).$$

$$\Omega_5(E,k_s)=[3d_1-\frac{d_1}{d_1^2+d_2^2}k_{21}^2(E,k_s)],\ \Omega_6(E,k_s)=[3d_2+\frac{d_2}{d_1^2+d_2^2}k_{21}^2(E,k_s)]$$

$$G_{15}=([\Omega_9(E,k_s)(\cosh g_1)(\cos g_2)-\Omega_{10}(E,k_s)(\sinh g_1)(\sin g_2)](\cos\gamma_{11}(E,k_s)))$$

$$\Omega_9(E,k_s)=[2d_1^2-2d_2^2-k_{21}^2(E,k_s)],\Omega_{10}(E,k_s)=[2d_1d_2]$$

$$G_{16}=([\Omega_7(E,k_s)(\sinh g_1)(\cos g_2)-\Omega_8(E,k_s)(\sin g_1)(\cosh g_2)](\sin\gamma_{11}(E,k_s)/12)),$$

$$\Omega_7(E,k_s)=[\frac{5d_1}{d_1^2+d_2^2}k_{21}^3(E,k_s)+\frac{5(d_1^3-3d_2^2d_1)}{k_{21}(E,k_s)}-34k_{21}(E,k_s)d_1],$$

$$\Omega_8(E,k_s)=[\frac{5d_2}{d_1^2+d_2^2}k_{21}^3(E,k_s)+\frac{5(d_2^3-3d_2^2d_1)}{k_{21}(E,k_s)}+34k_{21}(E,k_s)d_2]$$

$$H_{18}=\frac{1}{2}[H_{11}+H_{12}+\Delta_0(H_{13}+H_{14})+\Delta_0(H_{15}+H_{16})],$$

$$H_{11}=2(\sinh g_1\sin g_2\cos\gamma_{11}(E,k_s)),$$

$$H_{12}=([\Omega_2(E,k_s)(\sinh g_1)(\cos g_2)+\Omega_1(E,k_s)(\sin g_2)(\cosh g_1)](\sin\gamma_{11}(E,k_s))),$$

$$H_{13}=([\Omega_4(E,k_s)(\cosh g_1)(\cos g_2)+\Omega_3(E,k_s)(\sinh g_1)(\sin g_2)](\sin\gamma_{11}(E,k_s))),$$

$$H_{14}=([\Omega_6(E,k_s)(\sinh g_1)(\cos g_2)+\Omega_5(E,k_s)(\sin g_1)(\cosh g_2)](\cos\gamma_{11}(E,k_s))),$$

$$H_{15}=([\Omega_{10}(E,k_s)(\cosh g_1)(\cos g_2)+\Omega_9(E,k_s)(\sinh g_1)(\sin g_2)](\cos\gamma_{11}(E,k_s))),$$

$$H_{16}=([\Omega_8(E,k_s)(\sinh g_1)(\cos g_2)+\Omega_7(E,k_s)(\sin g_1)(\cosh g_2)](\sin\gamma_{11}(E,k_s)/12)),$$

$$\mathrm{H}_{19}=\left[\frac{2C_{18}D_{18}}{L_0^2}\right]$$

and

$$D_{18}=\sinh^{-1}(\omega_{18})$$

The simplified dispersion relation in quantum dots of heavily doped II-VI superllatices with graded interfaces can be expressed as

$$(\frac{n_z\pi}{d_z})^2=G_{190D}+iH_{190D} \tag{3.130}$$

where $G_{190D} = \left[\frac{C_{180D}^2 - D_{180D}^2}{L_0^2} - ((\frac{n_x \pi}{d_x})^2 + (\frac{n_y \pi}{d_y})^2) \right]$,

$$C_{180D} = \cos^{-1}(\omega_{180D}), \omega_{180D} = (2)^{\frac{-1}{2}}[(1 - G_{180D}^2 - H_{180D}^2) - \sqrt{(1 - G_{180D}^2 - H_{180D}^2)^2 + 4G_{180D}^2}]^{\frac{1}{2}},$$

$$G_{180D} = \frac{1}{2}[G_{110D} + G_{120D} + \Delta_0(G_{130D} + G_{140D}) + \Delta_0(G_{150D} + G_{160D})],$$

$$G_{110D} = 2(\cos(g_1))(\cos(g_2))(\cos\gamma_{110D}(E)), \gamma_{110D}(E) = k_{210D}(E)(b_0 - \Delta_0),$$

$$k_{210D}(E) = \{[\gamma_3(E, \eta_{g1}) - \frac{\hbar^2}{2m_{\perp,1}^*}((\frac{n_x \pi}{d_x})^2 + (\frac{n_y \pi}{d_y})^2) \pm C_0((\frac{n_x \pi}{d_x})^2 + (\frac{n_y \pi}{d_y})^2)^{1/2}]\frac{2m_{\parallel,1}^*}{\hbar^2}\}^{1/2},$$

$$G_{120D} = ([\Omega_{10D}(E)(\sinh g_{10D})(\cos g_{20D}) - \Omega_{20D}(E)(\sin g_{20D})(\cosh g_{10D})](\sin\gamma_{110D}(E)))$$

$$\Omega_{10D}(E) = [\frac{d_{10D}}{k_{210D}(E)} - \frac{k_{210D}(E)d_{10D}}{d_{10D}^2 + d_{20D}^2}], \ \Omega_{20D}(E) = [\frac{d_{20D}}{k_{210D}(E)} + \frac{k_{210D}(E)d_{20D}}{d_{10D}^2 + d_{20D}^2}]$$

$$G_{130D} = ([\Omega_{30D}(E)(\cosh g_{10D})(\cos g_{20D}) - \Omega_{40D}(E)(\sinh g_{10D})(\sin g_{20D})](\sin\gamma_{110D}(E)))$$

$$\Omega_{30D}(E) = [\frac{d_{10D}^2 - d_{20D}^2}{k_{210D}(E)} - 3k_{210D}(E)], \Omega_{40D}(E) = [\frac{2d_{10D}d_{20D}}{k_{210D}(E)}]$$

$$G_{140D} = ([\Omega_{50D}(E)(\sinh g_{10D})(\cos g_{20D}) - \Omega_{60D}(E)(\sin g_{10D})(\cosh g_{20D})](\cos\gamma_{110D}(E))).$$

$$\Omega_{50D}(E) = [3d_{10D} - \frac{d_{10D}}{d_{10D}^2 + d_{20D}^2}k_{210D}^2(E)],$$

$$\Omega_{60D}(E) = [3d_{20D} + \frac{d_{20D}}{d_{10D}^2 + d_{20D}^2}k_{210D}^2(E)]$$

$$G_{150D} = ([\Omega_{90D}(E)(\cosh g_{10D})(\cos g_{20D}) - \Omega_{100D}(E)(\sinh g_{10D})(\sin g_{20D})](\cos\gamma_{110D}(E)))$$

$$\Omega_{90D}(E) = [2d_{10D}^2 - 2d_{20D}^2 - k_{210D}^2(E)], \Omega_{100D}(E) = [2d_{10D}d_{20D}]$$

$$G_{160D} = ([\Omega_{70D}(E)(\sinh g_{10D})(\cos g_{20D}) - \Omega_{80D}(E)(\sin g_{10D})(\cosh g_{20D})](\sin\gamma_{110D}(E)/12)),$$

$$\Omega_{70D}(E) = [\frac{5d_{10D}}{d_{10D}^2 + d_{20D}^2}k_{210D}^3(E) + \frac{5(d_{10D}^3 - 3d_{20D}^2 d_{10D})}{k_{210D}(E)} - 34k_{210D}(E)d_{10D}],$$

$$\Omega_{80D}(E) = [\frac{5d_{20D}}{d_{10D}^2 + d_{20D}^2}k_{210D}^3(E) + \frac{5(d_{20D}^3 - 3d_{20D}^2 d_{10D})}{k_{210D}(E)} + 34k_{210D}(E)d_{20D}]$$

$$H_{180D} = \frac{1}{2}[H_{110D} + H_{120D} + \Delta_0(H_{130D} + H_{140D}) + \Delta_0(H_{150D} + H_{160D})],$$

$$H_{110D} = 2(\sinh g_{10D})(\sin g_{20D})(\cos\gamma_{10D}(E))),$$

$$H_{120D} = ([\Omega_{20D}(E)(\sinh g_{10D})(\cos g_{20D}) + \Omega_{10D}(E)(\sin g_{20D})(\cosh g_{10D})](\sin\gamma_{110D}(E))),$$

$$H_{130D} = ([\Omega_{40D}(E)(\cosh g_{10D})(\cos g_{20D}) + \Omega_{30D}(E)(\sinh g_{10D})(\sin g_{20D})](\sin\gamma_{110D}(E))),$$

$$H_{140D} = ([\Omega_{60D}(E)(\sinh g_{10D})(\cos g_{20D}) + \Omega_{50D}(E)(\sin g_{10D})(\cosh g_{20D})](\cos\gamma_{110D}(E))),$$

$$H_{150D} = ([\Omega_{100D}(E)(\cosh g_{10D})(\cos g_{20D}) + \Omega_{90D}(E)(\sinh g_{10D})(\sin g_{20D})](\cos\gamma_{110D}(E))),$$
$$H_{160D} = ([\Omega_{80D}(E)(\sinh g_{10D})(\cos g_{20D}) + \Omega_{70D}(E)(\sin g_{10D})(\cosh g_{20D})](\sin\gamma_{110D}(E)/2)),$$
$$\mathrm{H}_{190D} = \left[\frac{2C_{180D}D_{180D}}{L_0^2}\right] \text{ and } D_{180D} = \sinh^{-1}(\omega_{180D})$$

The electron concentration in this case is given by

$$n_0 = \left(\frac{2g_v}{d_x d_y d_z}\right)\sum_{n_x=1}^{n_{x_{\max}}}\sum_{n_y=1}^{n_{y_{\max}}}\sum_{n_z=1}^{n_{z_{\max}}} F_{-1}(\eta_{3D31}) \tag{3.131a}$$

where $\eta_{3D31} = ((E_{FD31} - E'_{31})/k_B T)$, E_{FD31} is the Fermi energy in the present case and E'_{31} is the real part of (3.130).Thus using (3.131) and (3.27) one can obtain the Seebeck coefficient in the presence of large magnetic field for heavily doped II-VI quantum dot superlattices with graded interfaces

3.2.17. Study of the Seebeck Coefficient under Large Magnetic Field in Heavily Doped IV-VI Quantum dot Superlattices with Graded Interfaces

The **E-k** dispersion relation of the conduction electrons of the heavily doped constituent materials of the IV-VI SLs can be expressed as

$$k_z^2 = [2\bar{p}_{9,i}]^{-1}[-\bar{q}_{9,i}(E,k_s,\eta_{gi}) + [[\bar{q}_{9,i}(E,k_s,\eta_{gi})]^2 + 4\bar{p}_{9,i}\bar{R}_{9,i}(E,k_s,\eta_{gi})]^{\frac{1}{2}}] \tag{3.131b}$$

where, $\bar{p}_{9,i} = (\alpha_i\hbar^4)/(4m_{l,i}^- m_{li}^+)$, i=1,2 , $\bar{q}_{9,i}(E,k_s,\eta_{gi}) = [(\hbar^2/2)((1/m_{li}^*) + (1/m_{li}^-)) + \alpha_i(\hbar^4/4)k_s^2((1/m_{li}^+ m_{ti}^-) + (1/m_{ti}^+ m_{li}^-)) - \alpha_i\gamma_3(E,\eta_{gi})((1/m_{li}^+) - (1/m_{li}^-))]$ and $\bar{R}_{9,i}(E,k_s,\eta_{gi}) = [\gamma_2(E,\eta_{gi}) + \gamma_3(E,\eta_{gi})][(\hbar^2/2)\alpha_i k_s^2((1/m_{ti}^*) - (1/m_{ti}^-))] - [(\hbar^2/2)k_s^2((1/m_{ti}^*) + (1/m_{ti}^-))] - \alpha_i(\hbar^6/4)k_s^4((1/m_{ti}^+ m_{ti}^-))]$,

The electron dispersion law in heavily doped IV-VI SLs with graded interfaces can be expressed as

$$\cos(L_o k) = \frac{1}{2}\Phi_2(E,k_s) \tag{3.131c}$$

where

$$\Phi_2(E,k_s) \equiv \Big[2\cosh\{\beta_2(E,k_s)\}\cos\{\gamma_{22}(E,k_s)\} + \varepsilon_2(E,k_s)\sinh\{\beta_2(E,k_s)\}\sin\{\gamma_{22}(E,k_s)\}$$

$$+\Delta_0\left[\left(\frac{\{K_{112}(E,k_s)\}^2}{K_{212}(E,k_s)} - 3K_{212}(E,k_s)\right)\cosh\{\beta_2(E,k_s)\}\sin\{\gamma_{22}(E,k_s)\}\right.$$

$$+\left.\left(3K_{112}(E,k_s) - \frac{\{K_{212}(E,k_s)\}^2}{K_{112}(E,k_s)}\right)\sinh\{\beta_2(E,k_s)\}\cos\{\gamma_{22}(E,k_s)\}\right]$$

$$+\Delta_0\Big[2\left(\{K_{112}(E,k_s)\}^2 - \{K_{212}(E,k_s)\}^2\right)\cosh\{\beta_2(E,k_s)\}\cos\{\gamma_{22}(E,k_s)\}$$

$$+\frac{1}{12}\left[\frac{5\{K_{112}(E,k_s)\}^3}{K_{212}(E,k_s)} + \frac{5\{K_{212}(E,k_s)\}^3}{K_{112}(E,k_s)} - 34K_{212}(E,k_s)K_{112}(E,k_s)\right]\sinh\{\beta_2(E,k_s)\}\sin\{\gamma_{22}(E,k_s)\}\Big]\Big],$$

$$\beta_2(E,k_s) \equiv K_{112}(E,k_s)[a_0 - \Delta_0],$$

$$k_{112}^2(E,k_s) = [2\bar{p}_{9,2}]^{-1}[-\bar{q}_{9,2}(E-V_0,k_{s,}\eta_{g2}) - [[\bar{q}_{9,2}(E-V_0,k_{s,}\eta_{g2})]^2 + 4\bar{p}_{9,2}\bar{R}_{9,2}(E-V_0,k_{s,}\eta_{g2})]^{\frac{1}{2}}],$$

$$\gamma_{22}(E,k_s) = K_{212}(E,k_s)[b_0 - \Delta_0],$$

$$k_{212}^2(E,k_s) = [2\bar{p}_{9,1}]^{-1}[-\bar{q}_{9,1}(E,k_{s,}\eta_{g1}) + [[\bar{q}_{9,1}(E,k_{s,}\eta_{g1})]^2 + 4\bar{p}_{9,1}\bar{R}_{9,1}(E,k_{s,}\eta_{g1})]^{\frac{1}{2}}]$$

and $\varepsilon_2(E,k_s) \equiv \left[\frac{K_{112}(E,k_s)}{K_{212}(E,k_s)} - \frac{K_{212}(E,k_s)}{K_{112}(E,k_s)}\right]$.

The simplified dispersion relation in quantum dots of heavily doped IV-VI superllatices with graded interfaces can be expressed as

$$\cos(L_o k_0) = \frac{1}{2}\Phi_{20D}(E,k_s) \tag{3.131d}$$

where $k_0 = ((\pi n_x / d_x)^2 + (\pi n_y / d_y)^2 + (\pi n_z / d_z)^2)^{\frac{1}{2}}$,

$$\Phi_{20D}(E) \equiv \Big[2\cosh\{\beta_2(E)\}\cos\{\gamma_{22}(E,\eta_{g1})\} + \varepsilon_2(E)\sinh\{\beta_2(E)\}\sin\{\gamma_{22}(E,\eta_{g1})\}$$

$$+\Delta_0[((\{K_{1120D}(E)\}^2 / K_{2120D}(E)) - 3K_{2120D}(E))\cosh\{\beta_2(E)\}\sin\{\gamma_{22}(E,\eta_{g1})\}$$

$$+\left(3K_{1120D}(E) - \frac{\{K_{2120D}(E)\}^2}{K_{1120D}(E)}\right)\sinh\{\beta_2(E)\}\cos\{\gamma_{22}(E,\eta_{g1})\}]$$

$$+\Delta_0\Big[2\left(\{K_{1120D}(E)\}^2 - \{K_{2120D}(E)\}^2\right)\cosh\{\beta_2(E)\}\cos\{\gamma_{22}(E,\eta_{g1})\}$$

$$+\frac{1}{12}\left[\frac{5\{K_{1120D}(E)\}^3}{K_{2120D}(E)}+\frac{5\{K_{2120D}(E)\}^3}{K_{1120D}(E)}-34K_{2120D}(E)K_{1120D}(E)\right]\sinh\{\beta_2(E)\}\sin\{\gamma_{22}(E,\eta_{g1})\}\Bigg]\Bigg],$$

$$\beta_2(E)\equiv K_{1120D}(E)[a_0-\Delta_0],$$

$$k_{1120D}^2(E)=[2\bar{p}_{9,2}]^{-1}[-\bar{q}_{9,2}(E-V_0,\eta_{g2})-[[\bar{q}_{9,2}(E-V_0,\eta_{g2})]^2+4\bar{p}_{9,2}\bar{R}_{9,2}(E-V_0,\eta_{g2})]^{\frac{1}{2}}],$$

$$\bar{q}_{9,2}(E-V_0,\eta_{g2})=[(\hbar^2/2)((1/m_{t2}^*)+(1/m_{t2}^-))+\alpha_2(\hbar^4/4)k_{s0}^2((1/m_{t2}^+m_{t2}^-)+$$

$$(1/m_{t2}^+m_{t2}^-))-\alpha_2\gamma_3(E-V_0,\eta_{g2})((1/m_{t2}^+)-(1/m_{t2}^-)],\ k_{s0}=((\pi n_x/d_x)^2+(\pi n_y/d_y)^2)^{\frac{1}{2}},$$

$$\bar{R}_{9,2}(E,\eta_{g2})=[\gamma_{22}(E-V_0,\eta_{g2})+\gamma_3(E-V_0,\eta_{g2})[(\hbar^2/2)\alpha_2k_{s0}^2((1/m_{t2}^*)-$$

$$(1/m_{t2}^-))]-[(\hbar^2/2)k_{s0}^2((1/m_{t2}^*)+(1/m_{t2}^-))]-\alpha_2(\hbar^6/4)k_{s0}^4((1/m_{t2}^+m_{t2}^-))],$$

$$\gamma_{22}(E,\eta_{g1})=K_{2120D}(E)[b_0-\Delta_0],k_{2120D}^2(E)=[2\bar{p}_{9,1}]^{-1}[-\bar{q}_{9,1}(E,\eta_{g1})+[[\bar{q}_{9,1}(E,\eta_{g1})]^2+4\bar{p}_{9,1}\bar{R}_{9,1}(E,\eta_{g1})]^{\frac{1}{2}}]$$

$$\bar{R}_{9,1}(E,\eta_{g1})=[\gamma_{22}(E,\eta_{g1})+\gamma_3(E,\eta_{g1})[(\hbar^2/2)\alpha_1k_{s0}^2((1/m_{t1}^*)-$$

$$(1/m_{t1}^-))]-[(\hbar^2/2)k_{s0}^2((1/m_{t1}^*)+(1/m_{t1}^-))]-\alpha_1(\hbar^6/4)k_{s0}^4((1/m_{t1}^+m_{t1}^-))]\ \text{and}$$

$$\varepsilon_{20D}(E)\equiv\left[\frac{K_{1120D}(E)}{K_{2120D}(E)}-\frac{K_{2120D}(E)}{K_{1120D}(E)}\right].$$

The electron concentration in this case is given by

$$n_0=\left(\frac{2g_v}{d_xd_yd_z}\right)\sum_{n_x=1}^{n_{x_{\max}}}\sum_{n_y=1}^{n_{y_{\max}}}\sum_{n_z=1}^{n_{z_{\max}}}F_{-1}(\eta_{3D32}) \tag{3.131e}$$

where $\eta_{3D32}=((E_{FD32}-E'_{32})/k_BT)$, E_{FD32} is the Fermi energy in the present case and E'_{32} is the lowest positive root of (3.131b).Thus using (3.131e) and (3.27) one can obtain the Seebeck coefficient in the presence of large magnetic field for heavily doped II-VI quantum dot superlattices with graded interfaces

3.2.18. Study of the Seebeck Coefficient under Large Magnetic Field in Heavily Doped HgTe/CdTe Quantum Dot Superlattices with Graded Interfaces

The electron energy spectra of the constituent materials of HgTe/CdTe SLs are given by

$$k^2 = \left[\frac{B_{01}^2 + 4A_1E - B_{01}\sqrt{B_{01}^2 + 4A_1E}}{2A_1^2}\right] \quad (3.132)$$

and
$$\frac{\hbar^2 k^2}{2m_{c2}^*} = T_{12}\left(E,\Delta_2,E_{g2},\eta_{g2}\right) + iT_{22}\left(E,\Delta_2,E_{g2},\eta_{g2}\right) \quad (3.133)$$

where $B_{01} = \left(3|e|^2/128\varepsilon_{sc1}\right)$, $A_1 = \left(\hbar^2/2m_{c1}^*\right)$. ε_{sc1} is the semiconductor permittivity of the first material. The energy-wave vector dispersion relation of the conduction electrons in heavily doped HgTe/CdTe SLs with graded interfaces can be expressed as

$$k_z^2 = G_{192} + iH_{192} \quad (3.134)$$

where $G_{192} = [((C_{182}^2 - D_{182}^2)/L_0^2) - k_s^2]$,

$C_{182} = \cos^{-1}(\omega_{182}), \omega_{182} = (2)^{\frac{-1}{2}}[(1-G_{182}^2 - H_{182}^2) - \sqrt{(1-G_{182}^2 - H_{182}^2)^2 + 4G_{182}^2}]^{\frac{1}{2}}$,

$G_{182} = \frac{1}{2}[G_{112} + G_{122} + \Delta_0(G_{132} + G_{142}) + \Delta_0(G_{152} + G_{162})]$,

$G_{112} = 2(\cos(g_{12}))(\cos(g_{22}))(\cos\gamma_8(E,k_s))$

$\gamma_8(E,k_s) = k_8(E,k_s)(b_0 - \Delta_0)$,

$$k_8(E,k_s) = \left[\frac{B_{01}^2 + 4A_1E - B_{01}\sqrt{B_{01}^2 + 4A_1E}}{2A_1^2} - k_s^2\right]^{1/2},$$

$G_{122} = ([\Omega_{12}(E,k_s)(\sinh g_{12})(\cos g_{22}) - \Omega_{22}(E,k_s)(\sin g_{22})(\cosh g_{12})](\sin\gamma_8(E,k_s)))$

$\Omega_{12}(E,k_s) = [\frac{d_{12}}{k_8(E,k_s)} - \frac{k_8(E,k_s)d_{12}}{d_{12}^2 + d_{22}^2}]$, $\Omega_{22}(E,k_s) = [\frac{d_{22}}{k_8(E,k_s)} + \frac{k_8(E,k_s)d_{22}}{d_{12}^2 + d_{22}^2}]$,

$G_{132} = ([\Omega_{32}(E,k_s)(\cosh g_{12})(\cos g_{22}) - \Omega_{42}(E,k_s)(\sinh g_{12})(\sin g_{22})](\sin\gamma_8(E,k_s)))$,

$\Omega_{32}(E,k_s) = [\frac{d_{12}^2 - d_{22}^2}{k_8(E,k_s)} - 3k_8(E,k_s)], \Omega_{42}(E,k_s) = [\frac{2d_{12}d_{22}}{k_8(E,k_s)}]$,

$G_{142} = ([\Omega_{52}(E,k_s)(\sinh g_{12})(\cos g_{22}) - \Omega_{62}(E,k_s)(\sin g_{12})(\cosh g_{22})](\cos\gamma_8(E,k_s)))$,

$\Omega_{52}(E,k_s) = [3d_{12} - \frac{d_{12}}{d_{12}^2 + d_{22}^2}k_8^2(E,k_s)]$,

$\Omega_{62}(E,k_s) = [3d_{22} + \frac{d_{22}}{d_{12}^2 + d_{22}^2}k_8^2(E,k_s)]$,

$G_{152} = ([\Omega_{92}(E,k_s)(\cosh g_{12})(\cos g_{22}) - \Omega_{102}(E,k_s)(\sinh g_{12})(\sin g_{22})](\cos\gamma_8(E,k_s)))$,

$\Omega_{92}(E,k_s) = [2d_{12}^2 - 2d_{22}^2 - k_8^2(E,k_s)], , \Omega_{102}(E,k_s) = [2d_{12}d_{22}]$,

$$G_{162} = ([\Omega_{72}(E,k_s)(\sinh g_{12})(\cos g_{22}) - \Omega_{82}(E,k_s)(\sin g_{12})(\cosh g_{22})](\sin \gamma_8(E,k_s)/12)),$$

$$\Omega_{72}(E,k_s) = [\frac{5d_{12}}{d_{12}^2 + d_{22}^2} k_8^3(E,k_s) + \frac{5(d_{12}^3 - 3d_{22}^2 d_{12})}{k_8(E,k_s)} - 34k_8(E,k_s)d_{12}],$$

$$\Omega_{82}(E,k_s) = [\frac{5d_{22}}{d_{12}^2 + d_{22}^2} k_8^3(E,k_s) + \frac{5(d_{22}^3 - 3d_{22}^2 d_{12})}{k_8(E,k_s)} + 34k_8(E,k_s)d_{22}]$$

$$H_{182} = \frac{1}{2}[H_{112} + H_{122} + \Delta_0(H_{132} + H_{142}) + \Delta_0(H_{152} + H_{162})],$$

$$H_{112} = 2(\sinh g_{12} \sin g_{22} \cos \gamma_8(E,k_s)),$$

$$H_{122} = ([\Omega_{22}(E,k_s)(\sinh g_{12})(\cos g_{22}) + \Omega_{12}(E,k_s)(\sin g_{22})(\cosh g_{12})](\sin \gamma_8(E,k_s))),$$

$$H_{132} = ([\Omega_{42}(E,k_s)(\cosh g_{12})(\cos g_{22}) + \Omega_{32}(E,k_s)(\sinh g_{12})(\sin g_{22})](\sin \gamma_8(E,k_s))),$$

$$H_{142} = ([\Omega_{62}(E,k_s)(\sinh g_{12})(\cos g_{22}) + \Omega_{52}(E,k_s)(\sin g_{12})(\cosh g_{22})](\cos \gamma_8(E,k_s))),$$

$$H_{152} = ([\Omega_{102}(E,k_s)(\cosh g_{12})(\cos g_{22}) + \Omega_{92}(E,k_s)(\sinh g_{12})(\sin g_{22})](\cos \gamma_8(E,k_s))),$$

$$H_{162} = ([\Omega_{82}(E,k_s)(\sinh g_{12})(\cos g_{22}) + \Omega_{72}(E,k_s)(\sin g_{12})(\cosh g_{22})](\sin \gamma_8(E,k_s)/12)),$$

$$\mathrm{H}_{192} = [((2C_{182}D_{182})/L_0^2)]$$

and $D_{182} = \sinh^{-1}(\omega_{182})$

The simplified dispersion relation in quantum dots of heavily doped HgTe/CdTe superllatices with graded interfaces can be expressed as

$$(n_z\pi/d_z)^2 = G_{1920D} + iH_{1920D} \tag{3.135}$$

where $\mathrm{G}_{1920D} = \left[\frac{C_{1820D}^2 - D_{1820D}^2}{L_0^2} - ((\frac{n_x\pi}{d_x})^2 + (\frac{n_y\pi}{d_y})^2)\right]$,

$$C_{1820D} = \cos^{-1}(\omega_{1820D}), \omega_{1820D} = (2)^{\frac{-1}{2}}[(1 - G_{1820D}^2 - H_{1820D}^2) - \sqrt{(1 - G_{1820D}^2 - H_{1820D}^2)^2 + 4G_{1820D}^2}]^{\frac{1}{2}},$$

$$G_{1820D} = \frac{1}{2}[G_{1120D} + G_{1220D} + \Delta_0(G_{1320D} + G_{1420D}) + \Delta_0(G_{1520D} + G_{1620D})],$$

$$(n_z\pi/d_z)^2 = G_{1920D} + iH_{1920D}$$

$$G_{1120D} = 2(\cos(g_{12}))(\cos(g_{22}))(\cos \gamma_{80D}(E)), \gamma_{80D}(E) = k_{80D}(E)(b_0 - \Delta_0),$$

$$k_{80D}(E) = \left[\frac{B_{01}^2 + 4A_1E - B_{01}\sqrt{B_{01}^2 + 4A_1E}}{2A_1^2} - ((\frac{n_x\pi}{d_x})^2 + (\frac{n_y\pi}{d_y})^2)\right]^{1/2},$$

$$G_{120D} = ([\Omega_{120D}(E)(\sinh g_{120D})(\cos g_{220D}) - \Omega_{220D}(E)(\sin g_{220D})(\cosh g_{120D})](\sin \gamma_{80D}(E))),$$

$$\Omega_{120D}(E)=[\frac{d_{120D}}{k_{80D}(E)}-\frac{k_{80D}(E)d_{120D}}{d_{120D}^2+d_{220D}^2}],\ \Omega_{220D}(E)=[\frac{d_{220D}}{k_{80D}(E)}+\frac{k_{80D}(E)d_{220D}}{d_{120D}^2+d_{220D}^2}],$$

$$G_{1320D}=([\Omega_{320D}(E)(\cosh g_{120D})(\cos g_{220D})-\Omega_{420D}(E)(\sinh g_{120D})(\sin g_{220D})](\sin\gamma_{80D}(E))),$$

$$\Omega_{320D}(E)=[\frac{d_{120D}^2-d_{20D}^2}{k_{80D}(E)}-3k_{80D}(E)],\ \Omega_{420D}(E)=[\frac{2d_{120D}d_{220D}}{k_{80D}(E)}],$$

$$G_{1420D}=([\Omega_{520D}(E)(\sinh g_{120D})(\cos g_{220D})-\Omega_{620D}(E)(\sin g_{120D})(\cosh g_{220D})](\cos\gamma_{80D}(E))),$$

$$\Omega_{520D}(E)=[3d_{120D}-\frac{d_{120D}}{d_{120D}^2+d_{220D}^2}k_{80D}^2(E)],$$

$$\Omega_{620D}(E)=[3d_{220D}+\frac{d_{220D}}{d_{120D}^2+d_{220D}^2}k_{80D}^2(E)],$$

$$G_{1520D}=([\Omega_{920D}(E)(\cosh g_{120D})(\cos g_{220D})-\Omega_{1020D}(E)(\sinh g_{120D})(\sin g_{220D})](\cos\gamma_{80D}(E))),$$

$$\Omega_{920D}(E)=[2d_{120D}^2-2d_{220D}^2-k_{80D}^2(E)],\ \Omega_{1020D}(E)=[2d_{120D}d_{220D}],$$

$$G_{1620D}=([\Omega_{720D}(E)(\sinh g_{120D})(\cos g_{220D})-\Omega_{820D}(E)(\sin g_{120D})(\cosh g_{220D})](\sin\gamma_{80D}(E)/12)),$$

$$\Omega_{720D}(E)=[\frac{5d_{120D}}{d_{120D}^2+d_{220D}^2}k_{80D}^3(E)+\frac{5(d_{120D}^3-3d_{220D}^2d_{120D})}{k_{80D}(E)}-34k_{80D}(E)d_{120D}],$$

$$\Omega_{820D}(E)=[\frac{5d_{220D}}{d_{120D}^2+d_{220D}^2}k_{80D}^3(E)+\frac{5(d_{220D}^3-3d_{220D}^2d_{120D})}{k_{80D}(E)}+34k_{80D}(E)d_{220D}],$$

$$H_{1820D}=\frac{1}{2}[H_{1120D}+H_{1220D}+\Delta_0(H_{1320D}+H_{1420D})+\Delta_0(H_{1520D}+H_{1620D})],$$

$$H_{1120D}=2(\sinh g_{120D})(\sin g_{220D})(\cos\gamma_{80D}(E))),$$

$$H_{1220D}=([\Omega_{220D}(E)(\sinh g_{120D})(\cos g_{220D})+\Omega_{120D}(E)(\sin g_{220D})(\cosh g_{120D})](\sin\gamma_{80D}(E))),$$

$$H_{1320D}=([\Omega_{420D}(E)(\cosh g_{120D})(\cos g_{220D})+\Omega_{320D}(E)(\sinh g_{120D})(\sin g_{220D})](\sin\gamma_{80D}(E))),$$

$$H_{1420D}=([\Omega_{620D}(E)(\sinh g_{120D})(\cos g_{220D})+\Omega_{520D}(E)(\sin g_{120D})(\cosh g_{220D})](\cos\gamma_{80D}(E))),$$

$$H_{1520D}=([\Omega_{1020D}(E)(\cosh g_{120D})(\cos g_{220D})+\Omega_{920D}(E)(\sinh g_{120D})(\sin g_{220D})](\cos\gamma_{80D}(E))),$$

$$H_{1620D}=([\Omega_{820D}(E)(\sinh g_{120D})(\cos g_{220D})+\Omega_{720D}(E)(\sin g_{120D})(\cosh g_{220D})](\sin\gamma_{80D}(E)/2)),$$

$$\mathrm{H}_{1920D}=[((2C_{1820D}D_{1820D})/L_0^2)]\ \text{and}\ D_{1820D}=\sinh^{-1}(\omega_{1820D})$$

The electron concentration in this case is given by

$$n_0=\left(\frac{2g_v}{d_xd_yd_z}\right)\sum_{n_x=1}^{n_{x_{\max}}}\sum_{n_y=1}^{n_{y_{\max}}}\sum_{n_z=1}^{n_{z_{\max}}}F_{-1}(\eta_{3D33}) \tag{3.136}$$

where $\eta_{3D33} = ((E_{FD33} - E'_{33}) / k_B T)$, E_{FD33} is the Fermi energy in the present case and E'_{33} is the lowest positive root of (3.135).Thus using (3.136) and (3.27) one can obtain the Seebeck coefficient in the presence of large magnetic field for heavily doped HgTe/CdTe quantum dot superlattices with graded interfaces.

3.2.19. Study of the Seebeck Coefficient under Large Magnetic Field in Heavily Doped Strained Layer Quantum Dot Superlattices with Graded Interfaces

The dispersion relation of the conduction electrons of the constituent materials of the starined layer super lattices can be expreesed as

$$[E - T_{1i}]k_x^2 + [E - T_{2i}]k_y^2 + [E - T_{3i}]k_z^2 = q_i E^3 - R_i E^2 + V_i E + \zeta_i \tag{3.137}$$

where

$$T_{1i} = \overline{\theta}_i,\ \overline{\theta}_i = \left[E_{gi} - C_{1i}^c \varepsilon_i - \left(a_i + C_{1i}^c\right)\varepsilon_i + \frac{3}{2} b_i \varepsilon_{xxi} - \frac{b_i \varepsilon_i}{2} + \frac{\sqrt{3} d_i \varepsilon_{xyi}}{2}\right],$$

$$T_{2i} = \omega_i,\ \omega_i = \left[E_{gi} - C_{1i}^c \varepsilon_i - \left(a_i + C_{1i}^c\right)\varepsilon_i + \frac{3}{2} b_i \varepsilon_{xxi} - \frac{b_i \varepsilon_i}{2} - \frac{\sqrt{3} d_i \varepsilon_{xyi}}{2}\right],$$

$$\mathrm{T}_{3i} = \delta_i,\ \delta_i = \left[E_{gi} - C_{1i}^c \varepsilon_i + \left(a_i + C_{1i}^c\right)\varepsilon_i + \frac{3}{2} b_i \varepsilon_{zzi} - \frac{b_i \varepsilon_i}{2}\right]$$

$$R_i = q_i\left[2A_i + C_{1i}^c \varepsilon_i\right],\ \mathrm{q}_i = \frac{3}{2B_{2i}^2},\ \mathrm{A}_i = E_{gi} - C_{1i}^c \varepsilon_i,$$

$$V_i = q_i\left[A_i^2 - \frac{2C_{2i}^2 \varepsilon_{xyi}}{3} + 2A_i C_{1i}^c \varepsilon_i\right],\ \zeta_i = q_i\left[\frac{2C_{2i}^2 \varepsilon_{xyi}}{3} - C_{1i}^c \varepsilon_i A_i^2\right]$$

Therefore the electron energy spectrum in heavily doped stressed materials can be written as

$$\overline{P}_i(E, \eta_{gi})k_x^2 + \overline{Q}_i(E, \eta_{gi})k_y^2 + \overline{S}_i(E, \eta_{gi})k_z^2 = 1 \tag{3.138}$$

$$where\ \overline{P}_i(E,\eta_{gi})=\frac{\left[\gamma_0(E,\eta_{gi})-I_0T_{1i}\right]}{\overline{\Delta}_i(E,\eta_{gi})},$$

$$\overline{\Delta}_i(E,\eta_{gi})=\left[\frac{-q_i\eta_{gi}^3}{2\sqrt{\pi}}\exp\left(\frac{-E^2}{\eta_{gi}^2}\right)\left[1+\frac{E^2}{\eta_{gi}^2}\right]-R_i\theta_0(E,\eta_{gi})+V_i\gamma_0(E,\eta_{gi})+\frac{\zeta_i}{2}\left[1+Erf\left(\frac{E}{\eta_{gi}}\right)\right]\right],$$

$$I_0=\frac{1}{2}\left[1+Erf(E/\eta_{gi})\right],\ \overline{Q}_i(E,\eta_{gi})=\frac{\left[\gamma_0(E,\eta_{gi})-I_0T_{2i}\right]}{\overline{\Delta}_i(E,\eta_{gi})}\ \text{and}\ \overline{S}_i(E,\eta_{gi})=\frac{\left[\gamma_0(E,\eta_{gi})-I_0T_{3i}\right]}{\overline{\Delta}_i(E,\eta_{gi})}$$

The energy-wave vector dispersion relation of the conduction electrons in heavily doped strained layer SLs with graded interfaces can be expressed as

$$\cos(L_0k)=\frac{1}{2}\overline{\phi}_6(E,k_s) \tag{3.139}$$

where

$$\begin{aligned}\overline{\phi}_6(E,k_s)=&\left[2\cosh\left[T_4(E,\eta_{g2})\right]\cos\left[T_5(E,\eta_{g1})\right]\right]+\left[T_6(E,k_s)\right]\sinh\left[T_4(E,\eta_{g2})\right]\sin\left[T_5(E,\eta_{g1})\right]\\&+\Delta_0\left[\left(\frac{k_0^2(E,\eta_{g2})}{k'(E,\eta_{g1})}-3k'(E,\eta_{g1})\right)\cosh\left[T_4(E,\eta_{g2})\right]\sin\left[T_5(E,\eta_{g1})\right]\right.\\&+\left.\left(3k_0(E,\eta_{g2})-\frac{k'^2(E,\eta_{g1})}{k_0(E,\eta_{g2})}\right)\sinh\left[T_4(E,\eta_{g2})\right]\cos\left[T_5(E,\eta_{g1})\right]\right]\\&+\Delta_0\left[2\left(k_0^2(E,\eta_{g2})-k'^2(E,\eta_{g1})\right)\cosh\left[T_4(E,\eta_{g2})\right]\cos\left[T_5(E,\eta_{g1})\right]\right.\\&+\left.\frac{1}{12}\left(\frac{5k_0^3(E,\eta_{g2})}{k'(E,\eta_{g1})}+\frac{5k'^3(E,\eta_{g1})}{k_0(E,\eta_{g2})}-34k_0(E,\eta_{g2})k'(E,\eta_{g1})\right)\sinh\left[T_4(E,\eta_{g2})\right]\sin\left[T_5(E,\eta_{g1})\right]\right]\end{aligned}$$

$$\left[T_4(E,\eta_{g2})\right]=k_0(E,\eta_{g2})\left[a_0-\Delta_0\right],$$

$$k_0(E,\eta_{g2})=\left[\overline{S}_2(E-V_0,\eta_{g2})\right]^{-1/2}\left[\overline{P}_2(E-V_0,\eta_{g2})k_x^2+\overline{Q}_2(E-V_0,\eta_{g2})k_y^2-1\right]^{1/2},$$

$$T_5(E,\eta_{g1})=k'(E,\eta_{g1})\left[b_0-\Delta_0\right],$$

$$k'(E,\eta_{g1})=\left[\overline{S}_1(E,\eta_{g1})\right]^{-1/2}\left[1-\overline{P}_1(E,\eta_{g1})k_x^2-\overline{Q}_1(E,\eta_{g1})k_y^2\right]^{1/2}\ and$$

$$T_6(E,k_s)=\left[\frac{k_0(E,\eta_{g2})}{k'(E,\eta_{g1})}-\frac{k'(E,\eta_{g1})}{k_0(E,\eta_{g2})}\right]$$

Therefore the dispersion relation of the conduction electrons in heavily doped strained layer QDSLs with graded interfaces can be expressed as

Therefore the dispersion relation of the conduction electrons in heavily doped strained layer QDSLs with graded interfaces can be expressed as

$$\cos\left(L_0 k_0\right)=\frac{1}{2}\overline{\phi_{60D}}\left(E\right) \tag{3.140}$$

where

$$\overline{\phi_{60D}}(E)=\Big[2\cosh\left[T_{40D}\left(E,\eta_{g2}\right)\right]\cos\left[T_{50D}\left(E,\eta_{g1}\right)\right]\Big]+\left[T_{60D}(E)\right]\sinh\left[T_{40D}\left(E,\eta_{g2}\right)\right]\sin\left[T_{50D}\left(E,\eta_{g1}\right)\right]$$
$$+\Delta_0\left[\left(\frac{k_{00D}^2\left(E,\eta_{g2}\right)}{k_{00D}{}'\left(E,\eta_{g1}\right)}-3k_{00D}{}'\left(E,\eta_{g1}\right)\right)\cosh\left[T_{40D}\left(E,\eta_{g2}\right)\right]\sin\left[T_{50D}\left(E,\eta_{g1}\right)\right]\right.$$
$$\left.+\left(3k_{00D}\left(E,\eta_{g2}\right)-\frac{k_{00D}{}'^2\left(E,\eta_{g1}\right)}{k_{00D}\left(E,\eta_{g2}\right)}\right)\sinh\left[T_{40D}\left(E,\eta_{g2}\right)\right]\cos\left[T_{50D}\left(E,\eta_{g1}\right)\right]\right]$$
$$+\Delta_0\left[2\left(k_{00D}^2\left(E,\eta_{g2}\right)-k_{00D}'^2\left(E,\eta_{g1}\right)\right)\cosh\left[T_{40D}\left(E,\eta_{g2}\right)\right]\cos\left[T_{50D}\left(E,\eta_{g1}\right)\right]\right]$$
$$\left.+\frac{1}{12}\left(\frac{5k_{00D}^3\left(E,\eta_{g2}\right)}{k_{00D}{}'\left(E,\eta_{g1}\right)}+\frac{5k_{00D}{}'^3\left(E,\eta_{g1}\right)}{k_{00D}\left(E,\eta_{g2}\right)}-34k_{00D}\left(E,\eta_{g2}\right)k_{00D}{}'\left(E,\eta_{g1}\right)\right)\sinh\left[T_{40D}\left(E,\eta_{g2}\right)\right]\sin\left[T_{50D}\left(E,\eta_{g1}\right)\right]\right]$$

$$\left[T_{40D}\left(E,\eta_{g2}\right)\right]=k_{00D}\left(E,\eta_{g2}\right)\left[a_0-\Delta_0\right],$$

$$k_{00D}\left(E,\eta_{g2}\right)=\left[\overline{S_2}\left(E-V_0,\eta_{g2}\right)\right]^{-1/2}\left[\overline{P_2}\left(E-V_0,\eta_{g2}\right)\left(\frac{\pi n_x}{d_x}\right)^2+\overline{Q_2}\left(E-V_0,\eta_{g2}\right)\left(\frac{\pi n_y}{d_y}\right)^2-1\right]^{1/2},$$

$$T_{50D}\left(E,\eta_{g1}\right)=k_{0D}{}'\left(E,\eta_{g1}\right)\left[b_0-\Delta_0\right],$$

$$k_{00D}{}'\left(E,\eta_{g1}\right)=\left[\overline{S_1}\left(E,\eta_{g1}\right)\right]^{-1/2}\left[1-\overline{P_1}\left(E,\eta_{g1}\right)\left(\frac{\pi n_x}{d_x}\right)^2-\overline{Q_1}\left(E,\eta_{g1}\right)\left(\frac{\pi n_y}{d_y}\right)^2\right]^{1/2} \text{ and}$$

$$T_{60D}\left(E\right)=\left[\frac{k_{00D}\left(E,\eta_{g2}\right)}{k_{00D}{}'\left(E,\eta_{g1}\right)}-\frac{k_{00D}{}'\left(E,\eta_{g1}\right)}{k_{00D}\left(E,\eta_{g2}\right)}\right]$$

The electron concentration in this case is given by

$$n_0=\left(\frac{2g_v}{d_x d_y d_z}\right)\sum_{n_x=1}^{n_{x_{\max}}}\sum_{n_y=1}^{n_{y_{\max}}}\sum_{n_z=1}^{n_{z_{\max}}}F_{-1}\left(\eta_{3D34}\right) \tag{3.141}$$

where $\eta_{3D34}=((E_{FD34}-E'_{34})/k_BT)$, E_{FD34} is the Fermi energy in the present case and E'_{34} is the lowest positive root of (3.140). Thus using (3.141) and (3.27) one can obtain the Seebeck coefficient in the presence of large magnetic field for heavily doped strained layer quantum dot superlattices with graded interfaces.

3.2.20. Study of the Seebeck Coefficient under Large Magnetic Field in Heavily Doped III-V Quantum Dot Effective Mass Superlattices

Following Sasaki [24], the electron dispersion law in III-V heavily doped effective mass superlattices (EMSLs) can be written as

$$k_x^2 = \left[\frac{1}{L_0^2} \left\{ \cos^{-1}\left(f_{21}\left(E, k_y, k_z\right)\right)\right\}^2 - k_\perp^2 \right] \tag{3.142}$$

in which $f_{21}\left(E,k_y,k_z\right) = a_1 \cos\left[a_0 C_{21}\left(E,k_\perp,\eta_{g1}\right) + b_0 D_{21}\left(E,k_\perp,\eta_{g2}\right)\right]$
$-a_2 \cos\left[a_0 C_{21}\left(E,k_\perp,\eta_{g1}\right) - b_0 D_{21}\left(E,k_\perp,\eta_{g2}\right)\right]$, $k_\perp^2 = k_y^2 + k_z^2$,

$$a_1 = \left[\sqrt{\frac{M_2\left(0,\eta_{g2}\right)}{M_1\left(0,\eta_{g1}\right)}} + 1\right]^2 \left[4\left(\frac{M_2\left(0,\eta_{g2}\right)}{M_1\left(0,\eta_{g1}\right)}\right)^{1/2}\right]^{-1},$$

$$a_2 = \left[\sqrt{\frac{M_2\left(0,\eta_{g2}\right)}{M_1\left(0,\eta_{g1}\right)}} - 1\right]^2 \left[4\left(\frac{M_2\left(0,\eta_{g2}\right)}{M_1\left(0,\eta_{g1}\right)}\right)^{1/2}\right]^{-1},$$

$$M_i\left(0,\eta_{gi}\right) = m_{ci}^*\left[\frac{-2}{\sqrt{\pi}} T\left(0,\eta_{gi}\right) + 2\left[\frac{\alpha_i b_i}{c_i}\frac{\eta_{gi}}{\sqrt{\pi}} + \frac{1}{2}\left(\frac{\alpha_i c_i + c_i b_i - \alpha_i b_i}{c_i^2}\right) + \frac{1}{\sqrt{\pi} c_i}\left(1 - \frac{\alpha_i}{c_i}\right)\left(1 - \frac{b_i}{c_i}\right)\right.\right.$$
$$-\frac{1}{c_i}\left(1-\frac{\alpha_i}{c_i}\right)\left(1-\frac{b_i}{c_i}\right)\frac{2}{c_i\eta_{gi}\sqrt{\pi}}\left\{\frac{-2}{c_i\eta_{gi}^2}\exp\left(\frac{-1}{c_i^2\eta_{gi}^2}\right)\left(\sum_{p=1}^{\alpha}\exp\left(\frac{-p^2}{4}\right)\frac{1}{p}\sinh\left(\frac{p}{c_i\eta_{gi}}\right)\right)\right.$$
$$\left.\left.\left.+\exp\left(\frac{-1}{c_i^2\eta_{gi}^2}\right)\left(\sum_{p=1}^{\alpha}\exp\left(\frac{-p^2}{4}\right)\frac{1}{\eta_{gi}}\cosh\left(\frac{p}{c_i\eta_{gi}}\right)\right)\right\}\right]\right],$$

$$\mathrm{T}\left(0,\eta_{gi}\right) = 2\left[\frac{\alpha_i b_i}{c_i}\frac{\eta_{gi}^2}{4} + \left(\frac{\alpha_i c_i + b_i c_i - \alpha_i b_i}{c_i^2}\right)\frac{\eta_{gi}}{2\sqrt{\pi}} + \frac{1}{2c_i}\left(1-\frac{\alpha_i}{c_i}\right)\left(1-\frac{b_i}{c_i}\right)\right.$$
$$\left.-\frac{1}{c_i}\left(1-\frac{\alpha_i}{c_i}\right)\left(1-\frac{b_i}{c_i}\right)\frac{2}{c_i\eta_{gi}\sqrt{\pi}}\exp\left(\frac{-1}{c_i^2\eta_{gi}^2}\right)\sum_{p=1}^{\alpha}\frac{\exp\left(-p^2/4\right)}{p}\sinh\left(\frac{p}{c_i\eta_{gi}}\right)\right],$$

$$C_{21}\left(E,k_\perp,\eta_{g1}\right) = e_1 + ie_2,\ D_{21}\left(E,k_\perp,\eta_{g2}\right) = e_3 + ie_4,$$

$$e_1 = [((\sqrt{t_1^2 + t_2^2} + t_1)/2)]^{\frac{1}{2}},\ \mathrm{e}_2 = [((\sqrt{t_1^2 + t_2^2} - t_1)/2)]^{\frac{1}{2}},$$

$$t_1 = \left[\frac{2m_{c1}^*}{\hbar^2} T_{11}\left(E, \Delta_1, \eta_{g1}, E_{g1}\right) - k_\perp^2 \right],\ t_2 = \frac{2m_{c1}^*}{\hbar^2} T_{21}\left(E, \Delta_1, \eta_{g1}, E_{g1}\right),$$

$$e_3 = \left[\frac{\sqrt{t_3^2 + t_4^2} + t_3}{2} \right]^{1/2},\ e_4 = \left[\frac{\sqrt{t_3^2 + t_4^2} - t_3}{2} \right]^{1/2}$$

$$t_3 = \left[\frac{2m_{c2}^*}{\hbar^2} T_{12}\left(E, \Delta_2, \eta_{g2}, E_{g2}\right) - k_\perp^2 \right],\ t_4 = \frac{2m_{c2}^*}{\hbar^2} T_{22}\left(E, \Delta_2, \eta_{g2}, E_{g2}\right),$$

Therefore the equ (3.142) can be expressed as

$$k_x^2 = \delta_7 + i\delta_8 \tag{3.143}$$

$$\text{where } \delta_7 = \left[\frac{1}{L_0^2}\left(\delta_5^2 - \delta_6^2\right) - k_\perp^2 \right],\ \delta_5 = \cos^{-1} p_5,$$

$$p_5 = \left[\frac{1 - \delta_3^2 - \delta_4^2 - \sqrt{\left(1 - \delta_3^2 - \delta_4^2\right)^2 + 4\delta_4^2}}{2} \right]^{1/2},$$

$$\delta_3 = \left(a_1 \cos\Delta_1 \cosh\Delta_2 - a_2 \cos\Delta_3 \cosh\Delta_4\right),$$
$$\delta_4 = \left(a_1 \sin\Delta_1 \sinh\Delta_2 - a_2 \sin\Delta_3 \sinh\Delta_4\right),$$
$$\Delta_1 = \left(a_0 e_1 + b_0 e_3\right),\ \Delta_2 = \left(a_0 e_2 + b_0 e_4\right),\ \Delta_3 = \left(a_0 e_1 - b_0 e_3\right),\ \Delta_4 = \left(a_0 e_2 - b_0 e_4\right),$$
$$\delta_6 = \sinh^{-1} p_5 \text{ and } \delta_8 = \left[2\delta_5\delta_6 / L_0^2\right]$$

Therefore the electron dispersion relation in heavily doped III-V QDSL assumes the form

$$\left(\frac{\pi n_x}{d_x}\right)^2 = \delta_{7oD} + i\delta_{80D} \tag{3.144}$$

$$\text{where } \delta_{70D} = \left[\frac{1}{L_0^2}\left(\delta_{50D}^2 - \delta_{60D}^2\right) - \left\{ \left(\frac{\pi n_y}{d_y}\right)^2 + \left(\frac{\pi n_z}{d_z}\right)^2 \right\} \right],\ \delta_{50D} = \cos^{-1} p_{50D},$$

$$p_{50D} = \left[\frac{1 - \delta_{30D}^2 - \delta_{40D}^2 - \sqrt{\left(1 - \delta_{30D}^2 - \delta_{40D}^2\right)^2 + 4\delta_{40D}^2}}{2} \right]^{1/2},$$

$$\delta_{30D} = (a_1 \cos\Delta_{10D} \cosh\Delta_{20D} - a_2 \cos\Delta_{30D} \cosh\Delta_{40D}),$$
$$\delta_{40D} = (a_1 \sin\Delta_{10D} \sinh\Delta_{20D} - a_2 \sin\Delta_{30D} \sinh\Delta_{40D}),$$
$$\Delta_{10D} = (a_0 e_{10D} + b_0 e_{30D}),\ \Delta_{20D} = (a_0 e_{20D} + b_0 e_{40D}),\ \Delta_{30D} = (a_0 e_{10D} - b_0 e_{30D}),\ \Delta_{40D} = (a_0 e_{20D} - b_0 e_{40D}),$$
$$\delta_{60D} = \sinh^{-1} \mathrm{p}_{50D} \text{ and } \delta_{80D} = \left[2\delta_{50D}\delta_{60D} / L_0^2\right],$$

The electron concentration in this case is given by

$$n_0 = \left(\frac{2g_v}{d_x d_y d_z}\right) \sum_{n_x=1}^{n_{x_{\max}}} \sum_{n_y=1}^{n_{y_{\max}}} \sum_{n_z=1}^{n_{z_{\max}}} F_{-1}(\eta_{3D340}) \tag{3.145}$$

where $\eta_{3D340} = ((E_{FD340} - E'_{340}) / k_B T)$, E_{FD340} is the Fermi energy in the present case and E'_{340} is the real part of the lowest positive root of (3.144).Thus using (3.145) and (3.27)

3.2.21. Study of the Seebeck Coefficient under Large Magnetic Field in Heavily Doped II-VI quantum Dot Effective Mass Superlattices

Following Sasaki [24], the electron dispersion law in heavily doped II-VI EMSLs can be written as

$$k_z^2 = \Delta_{13} + i\Delta_{14}, \tag{3.146}$$

$$where\ \Delta_{13} = \left[\frac{1}{L_0^2}\left(\Delta_{11}^2 - \Delta_{12}^2\right) - k_s^2\right]$$

$$\Delta_{11} = \cos^{-1} p_6,\ \mathrm{p}_6 = \left[\frac{1 - \Delta_9^2 - \Delta_{10}^2 - \sqrt{\left(1 - \Delta_9^2 - \Delta_{10}^2\right)^2 + 4\Delta_{10}^2}}{2}\right]^{1/2},$$

$\Delta_9 = \left(\overline{a_1}\cos\Delta_6\cosh\Delta_7 - \overline{a_2}\cos\Delta_8\cosh\Delta_7\right)$,

$\Delta_{10} = \left(\overline{a_1}\sin\Delta_6\sinh\Delta_7 + \overline{a_2}\sin\Delta_8\sinh\Delta_7\right)$,

$\Delta_6 = \left[a_0 C_{22}\left(E, k_s, \eta_{g1}\right) + b_0 e_3\right]$, $\Delta_7 = b_0 e_4$, $\Delta_8 = \left[a_0 C_{22}\left(E, k_s, \eta_{g1}\right) - b_0 e_3\right]$,

$$C_{22}\left(E, k_s, \eta_{g1}\right) = \left[\frac{2m_{\parallel,1}^*}{\hbar^2}\left\{\gamma_3\left(E, \eta_{g1}\right) - \frac{\hbar^2 k_s^2}{2m_{\perp,1}^*} \mp C_0 k_s\right\}\right]^{1/2},$$

$$\overline{a_1} = \left[\sqrt{\frac{M_2\left(0,\eta_{g2}\right)}{\overline{M_1}\left(0,\eta_{g1}\right)}} + 1\right]^2\left[4\left(\frac{M_2\left(0,\eta_{g2}\right)}{\overline{M_1}\left(0,\eta_{g1}\right)}\right)^{1/2}\right]^{-1}, \quad \overline{M_1}\left(0,\eta_{g1}\right) = m_{c1}^*\left(1 - \frac{2}{\pi}\right),$$

$$\overline{a_2} = \left[\sqrt{\frac{M_2\left(0,\eta_{g2}\right)}{\overline{M_1}\left(0,\eta_{g1}\right)}} - 1\right]^2\left[4\left(\frac{M_2\left(0,\eta_{g2}\right)}{\overline{M_1}\left(0,\eta_{g1}\right)}\right)^{1/2}\right]^{-1}$$

$\Delta_{12} = \cos^{-1} p_6$, $\Delta_{14} = \dfrac{2\Delta_{11}\Delta_{12}}{L_0^2}$

The electron dispersion law in heavily doped II-VI QDSL can be written as

$$\left(\frac{\pi n_z}{d_z}\right)^2 = \Delta_{130D} + i\Delta_{140D}, \tag{3.147}$$

$$\text{where } \Delta_{130D} = \left[\frac{1}{L_0^2}\left(\Delta_{110D}^2 - \Delta_{120D}^2\right) - \left\{\left(\frac{\pi n_y}{d_y}\right)^2 + \left(\frac{\pi n_x}{d_x}\right)^2\right\}\right]$$

$$\Delta_{110D} = \cos^{-1} p_{60D}, \quad p_{60D} = \left[\frac{1 - \Delta_{90D}^2 - \Delta_{100D}^2 - \sqrt{\left(1 - \Delta_{90D}^2 - \Delta_{100D}^2\right)^2 + 4\Delta_{100D}^2}}{2}\right]^{1/2},$$

$\Delta_{90D} = \left(\overline{a_1}\cos\Delta_{60D}\cosh\Delta_{70D} - \overline{a_2}\cos\Delta_{80D}\cosh\Delta_{70D}\right)$,

$\Delta_{100D} = \left(\overline{a_1}\sin\Delta_{60D}\sinh\Delta_{70D} + \overline{a_2}\sin\Delta_{80D}\sinh\Delta_{70D}\right)$,

$\Delta_{60D} = \left[a_0 C_{220D}\left(E_{0D}, \eta_{g1}\right) + b_0 e_{30D}\right]$, $\Delta_{70D} = b_0 e_{40D}$, $\Delta_{80D} = \left[a_0 C_{220D}\left(E_{0D}, \eta_{g1}\right) - b_0 e_{30D}\right]$,

$$C_{220D}\left(E_{0D}, \eta_{g1}\right) = \left[\frac{2m_{\parallel,1}^*}{\hbar^2}\left\{\gamma_3\left(E_{0D}, \eta_{g1}\right) - \frac{\hbar^2}{2m_{\perp,1}^*}\left\{\left(\frac{\pi n_y}{d_y}\right)^2 + \left(\frac{\pi n_x}{d_x}\right)^2\right\} \mp C_0\left\{\left(\frac{\pi n_y}{d_y}\right)^2 + \left(\frac{\pi n_z}{d_z}\right)^2\right\}^{1/2}\right\}\right]^{1/2},$$

$\Delta_{120D} = \cos^{-1} p_{60D}$, $\Delta_{140D} = \dfrac{2\Delta_{110D}\Delta_{120D}}{L_0^2}$,

The electron concentration in this case is given by

$$n_0 = \left(\frac{2g_v}{d_x d_y d_z}\right) \sum_{n_x=1}^{n_{x_{\max}}} \sum_{n_y=1}^{n_{y_{\max}}} \sum_{n_z=1}^{n_{z_{\max}}} F_{-1}(\eta_{3D35}) \tag{3.148}$$

where $\eta_{3D35} = ((E_{FD35} - E'_{35}) / k_B T)$, E_{FD35} is the Fermi energy in the present case and E'_{35} is the real part of the lowest positive root of (3.147).Thus using (3.148) and (3.27) one can obtain the Seebeck coefficient in this case.

3.2.22. Study of the Seebeck Coefficient under Large Magnetic Field in IV-VI Quantum Dot Effective Mass Superlattices

Following Sasaki [24], the electron dispersion law in IV-VI, EMSLs can be written as

$$k_z^2 = \left[\frac{1}{L_0^2}\left\{\cos^{-1}\left(f_{23}\left(E, k_x, k_y\right)\right)\right\}^2 - k_s^2\right] \tag{3.149}$$

$$\text{where } f_{23}\left(E, k_x, k_y\right) = a_3 \cos\left[a_0 C_{23}\left(E, k_x, k_y, \eta_{g1}\right) + b_0 D_{23}\left(E, k_x, k_y, \eta_{g1}\right)\right]$$
$$-a_4 \cos\left[a_0 C_{23}\left(E, k_x, k_y, \eta_{g2}\right) - b_0 D_{23}\left(E, k_x, k_y, \eta_{g2}\right)\right],$$

$$a_3 = \left[\sqrt{\frac{M_3(0,\eta_{g2})}{M_3(0,\eta_{g1})}} + 1\right]^2 \left[4\left(\frac{M_3(0,\eta_{g2})}{M_3(0,\eta_{g1})}\right)^{1/2}\right]^{-1},$$

$$a_4 = \left[\sqrt{\frac{M_3(0,\eta_{g2})}{M_3(0,\eta_{g1})}} - 1\right]^2 \left[4\left(\frac{M_3(0,\eta_{g2})}{M_3(0,\eta_{g1})}\right)^{1/2}\right]^{-1}$$

$$M_3(0,\eta_{gi})=\left(4\overline{p_{9,i}}\right)^{-1}\left[\left\{\alpha_i\left(1-\frac{2}{\pi}\right)\left(\frac{1}{m_{l,i}^+}-\frac{1}{m_{l,i}^-}\right)\right\}+\left[\left[\overline{q_{9,i}}(0,\eta_{gi})\right]^2\right.\right.$$

$$\left.\left.+\left(4\overline{p_{9,i}}\right)\overline{R_{9,i}}(0,\eta_{gi})\right]^{-1/2}\left[\alpha_i\left(1-\frac{2}{\pi}\right)\left(\frac{1}{m_{l,i}^+}-\frac{1}{m_{l,i}^-}\right)\overline{q_{9,i}}(0,\eta_{gi})+2\overline{p_{9,i}}\left(1-\frac{2}{\pi}+\frac{\alpha_i\eta_{gi}}{\sqrt{\pi}}\right)\right]\right],$$

$$\overline{p_{9,i}}=\frac{\alpha_i\hbar^4}{4m_{l,i}^+m_{l,i}^-},\ \overline{q_{9,i}}(0,\eta_{gi})=\left[\frac{\hbar^2}{2}\left(\frac{1}{m_{l,i}^+}+\frac{1}{m_{l,i}^-}\right)-\frac{\alpha_i\eta_{gi}}{\sqrt{\pi}}\left(\frac{1}{m_{l,i}^+}-\frac{1}{m_{l,i}^-}\right)\right],\ \overline{R_{9,i}}(0,\eta_{gi})=\left[\frac{\eta_{gi}}{\sqrt{\pi}}+\frac{\alpha_i\eta_{gi}^2}{2}\right],$$

$$C_{23}(E,k_x,k_y,\eta_{g1})=\left[\left[2\overline{p_{9,1}}\right]^{-1}\left[-\overline{q_{9,1}}(E,k_x,k_y,\eta_{g1})+\left[\left\{\overline{q_{9,1}}(E,k_x,k_y,\eta_{g1})\right\}^2\right.\right.\right.$$

$$\left.\left.\left.+\left(4\overline{p_{9,1}}\right)\overline{R_{9,1}}(E,k_x,k_y,\eta_{g1})\right]^{1/2}\right]\right]^{1/2},$$

$$D_{23}(E,k_x,k_y,\eta_{g2})=\left[\left[2\overline{p_{9,2}}\right]^{-1}\left[-\overline{q_{9,2}}(E,k_x,k_y,\eta_{g2})+\left[\left\{\overline{q_{9,2}}(E,k_x,k_y,\eta_{g2})\right\}^2\right.\right.\right.$$

$$\left.\left.\left.+\left(4\overline{p_{9,2}}\right)\overline{R_{9,2}}(E,k_x,k_y,\eta_{g2})\right]^{1/2}\right]\right]^{1/2},$$

$$\overline{q_{9,i}}(E,k_x,k_y,\eta_{gi})=\left[\frac{\hbar^2}{2}\left(\frac{1}{m_{l,i}^*}+\frac{1}{m_{l,i}^-}\right)+\alpha_i\frac{\hbar^4}{4}k_s^2\left(\frac{1}{m_{l,i}^+m_{t,i}^-}+\frac{1}{m_{t,i}^+m_{l,i}^-}\right)\right.$$

$$\left.-\alpha_i\gamma_3(E,\eta_{gi})\left(\frac{1}{m_{l,i}^+}-\frac{1}{m_{l,i}^-}\right)\right],$$

$$\overline{R_{9,i}}(E,k_x,k_y,\eta_{gi})=\left[\gamma_2(E,\eta_{gi})+\gamma_3(E,\eta_{gi})\alpha_i\frac{\hbar^2}{2}k_s^2\left(\frac{1}{m_{t,i}^+}-\frac{1}{m_{t,i}^-}\right)\right.$$

$$\left.-\frac{\hbar^2}{2}k_s^2\left(\frac{1}{m_{t,i}^*}-\frac{1}{m_{t,i}^-}\right)-\frac{\alpha\hbar^6}{4}\frac{k_s^4}{m_{t,i}^-m_{t,i}^+}\right],\quad a_5=\left[\sqrt{\frac{m_2^*}{m_1^*}}+1\right]^2\left[4\left(\frac{m_2^*}{m_1^*}\right)^{1/2}\right]^{-1}$$

Therefore the electron dispersion law in heavily doped IV-VI, EMSLs can be written as

$$(\pi n_z/d_z)^2=[[1/L_0^2]\{\cos^{-1}(f_{23}(E_{0D}))\}^2-((\pi n_x/d_x)^2+(\pi n_y/d_y)^2)] \quad (3.150)$$

where $f_{230D}(E_{0D})=a_3\cos\left[a_0C_{230D}(E_{0D},\eta_{g1})+b_0D_{230D}(E_{0D},\eta_{g1})\right]-a_4\cos\left[a_0C_{230D}(E_{0D},\eta_{g2})-b_0D_{23}(E_{0D},\eta_{g2})\right],$

$$C_{230D}\left(E_{0D},\eta_{g1}\right)=\left[\left[2\overline{p_{9,1}}\right]^{-1}\left[-\overline{q_{9,10D}}\left(E_{0D},\eta_{g1}\right)+\left[\left\{\overline{q_{9,10D}}\left(E_{0D},\eta_{g1}\right)\right\}^{2}\right.\right.\right.$$

$$\left.\left.\left.+\left(4\overline{p_{9,1}}\right)\overline{R_{9,10D}}\left(E_{0D},\eta_{g1}\right)\right]^{1/2}\right]\right]^{1/2},$$

$$D_{230D}\left(E_{0D},\eta_{g2}\right)=\left[\left[2\overline{p_{9,2}}\right]^{-1}\left[-\overline{q_{9,20D}}\left(E_{0D},\eta_{g2}\right)+\left[\left\{\overline{q_{9,20D}}\left(E_{0D},\eta_{g2}\right)\right\}^{2}\right.\right.\right.$$

$$\left.\left.\left.+\left(4\overline{p_{9,2}}\right)\overline{R_{9,20D}}\left(E_{0D},\eta_{g2}\right)\right]^{1/2}\right]\right]^{1/2},$$

$$\overline{q_{9,i0D}}\left(E_{0D},\eta_{gi}\right)=\left[\frac{\hbar^{2}}{2}\left(\frac{1}{m_{l,i}^{*}}+\frac{1}{m_{l,i}^{-}}\right)+\alpha_{i}\frac{\hbar^{4}}{4}((\pi n_{x}/d_{x})^{2}+(\pi n_{y}/d_{y})^{2})\left(\frac{1}{m_{l,i}^{+}m_{t,i}^{-}}+\frac{1}{m_{t,i}^{+}m_{l,i}^{-}}\right)\right.$$

$$\left.-\alpha_{i}\gamma_{3}\left(E_{0D},\eta_{gi}\right)\left(\frac{1}{m_{l,i}^{+}}-\frac{1}{m_{l,i}^{-}}\right)\right],$$

$$\overline{R_{9,i0D}}\left(E_{0D},\eta_{gi}\right)=\left[\gamma_{2}\left(E_{0D},\eta_{gi}\right)+\gamma_{3}\left(E_{0D},\eta_{gi}\right)\alpha_{i}\frac{\hbar^{2}}{2}((\pi n_{x}/d_{x})^{2}+(\pi n_{y}/d_{y})^{2})\left(\frac{1}{m_{t,i}^{+}}-\frac{1}{m_{t,i}^{-}}\right)\right.$$

$$\left.-\frac{\hbar^{2}}{2}((\pi n_{x}/d_{x})^{2}+(\pi n_{y}/d_{y})^{2})\left(\frac{1}{m_{t,i}^{*}}-\frac{1}{m_{t,i}^{-}}\right)-\frac{\alpha\hbar^{6}}{4}\frac{((\pi n_{x}/d_{x})^{2}+(\pi n_{y}/d_{y})^{2})^{2}}{m_{t,i}^{-}m_{t,i}^{+}}\right],$$

The electron concentration in this case is given by

$$n_{0}=\left(\frac{2g_{v}}{d_{x}d_{y}d_{z}}\right)\sum_{n_{x}=1}^{n_{x_{\max}}}\sum_{n_{y}=1}^{n_{y_{\max}}}\sum_{n_{z}=1}^{n_{z_{\max}}}F_{-1}\left(\eta_{3D36}\right) \tag{3.151}$$

where $\eta_{3D36}=((E_{FD36}-E'_{36})/k_{B}T)$, E_{FD36} is the Fermi energy in the present case and E'_{36} is the lowest positive root of (3.150).Thus using (3.151) and (3.27) one can obtain the Seebeck coefficient in this case.

3.2.23. Study of the Seebeck Coefficient under Large Magnetic Field in Heavily Doped HgTe/CdTe Quantum Dot Effective Mass Superlattices

Following Sasaki [24], the electron dispersion law in heavily doped HgTe/CdTe EMSLs can be written as

$$k_{z}^{2}=\Delta_{13H}+i\Delta_{14H} \tag{3.152}$$

$$where\ \Delta_{13H}=\left[\frac{1}{L_0^2}\left(\Delta_{11H}^2-\Delta_{12H}^2\right)-k_s^2\right]$$

$$\Delta_{11H}=\cos^{-1}p_{6H},\ p_{6H}=\left[\frac{1-\Delta_{9H}^2-\Delta_{10H}^2-\sqrt{\left(1-\Delta_{9H}^2-\Delta_{10H}^2\right)^2+4\Delta_{10H}^2}}{2}\right]^{1/2},$$

$$\Delta_{9H}=\left(\overline{a_{1H}}\cos\Delta_{5H}\cosh\Delta_{6H}-\overline{a_{2H}}\cos\Delta_{7H}\cosh\Delta_{6H}\right),$$

$$\Delta_{10H}=\left(\overline{a_{1H}}\sin\Delta_{5H}\sinh\Delta_{6H}+\overline{a_{2H}}\sin\Delta_{7H}\sinh\Delta_{6H}\right),$$

$$\Delta_{5H}=\left[a_0C_{22H}\left(E,k_s,\eta_{g1}\right)+b_0e_3\right],\ \Delta_{6H}=b_0e_4,\ \Delta_{7H}=\left[a_0C_{22H}\left(E,k_s,\eta_{g1}\right)-b_0e_3\right],$$

$$C_{22H}\left(E,k_s,\eta_{g1}\right)=\left[\frac{B_{01}^2+2A_1E-B_{01}(B_{01}^2+4A_1E)}{2A_1^2}-k_s^2\right]^{1/2},$$

$$\overline{a_{1H}}=\left[\sqrt{\frac{M_2\left(0,\eta_{g2}\right)}{m_{c1}^*}}+1\right]^2\left[4\left(\frac{M_2\left(0,\eta_{g2}\right)}{m_{c1}^*}\right)^{1/2}\right]^{-1},\overline{a_{2H}}=\left[\sqrt{\frac{M_2\left(0,\eta_{g2}\right)}{m_{c1}^*}}-1\right]^2\left[4\left(\frac{M_2\left(0,\eta_{g2}\right)}{m_{c1}^*}\right)^{1/2}\right]^{-1}$$

$$\Delta_{12H}=\cos^{-1}p_{6H},\ \Delta_{14H}=\frac{2\Delta_{11H}\Delta_{12H}}{L_0^2}$$

The electron dispersion law in heavily doped HgTe/CdTe **QD**EMSLs can be written as

$$(\pi n_z/d_z)^2=\Delta_{13H0D}+i\Delta_{14H0D} \tag{3.153}$$

$$where\ \Delta_{13H0D}=[(1/L_0^2)\left(\Delta_{11H0D}^2-\Delta_{12H0D}^2\right)-[(\pi n_x/d_x)^2+(\pi n_y/d_y)^2]]$$

$$\Delta_{11H0D}=\cos^{-1}p_{6H0D},\ p_{6H0D}=[((1-\Delta_{9H}^2-\Delta_{10H}^2-\sqrt{\left(1-\Delta_{9H}^2-\Delta_{10H}^2\right)^2+4\Delta_{10H}^2}\ )/2)]^{\frac{1}{2}}$$

$$\Delta_{9H0D}=\left(\overline{a_{1H}}\cos\Delta_{5H0D}\cosh\Delta_{6H0D}-\overline{a_{2H}}\cos\Delta_{7H0D}\cosh\Delta_{6H0D}\right),$$

$$\Delta_{10H0D}=\left(\overline{a_{1H}}\sin\Delta_{5H0D}\sinh\Delta_{6H0D}+\overline{a_{2H}}\sin\Delta_{7H0D}\sinh\Delta_{6H0D}\right),$$

$$\Delta_{5H0D}=\left[a_0C_{22H0D}\left(E_{0D},\eta_{g1}\right)+b_0e_3\right],\ \Delta_{6H0D}=b_0e_4,\ \Delta_{7H0D}=\left[a_0C_{22H0D}\left(E_{0D},\eta_{g1}\right)-b_0e_3\right],$$

$$C_{22H0D}\left(E_{0D},\eta_{g1}\right)=\left[\frac{B_{01}^2+2A_1E_{0D}-B_{01}(B_{01}^2+4A_1E_{0D})}{2A_1^2}-[(\pi n_x/d_x)^2+(\pi n_y/d_y)^2]\right]^{1/2},$$

$$\Delta_{12H0D}=\cos^{-1}p_{6H0D},\ \Delta_{14H0D}=\frac{2\Delta_{11H0D}\Delta_{12H0D}}{L_0^2}$$

The electron concentration in this case is given by

$$n_0 = \left(\frac{2g_v}{d_x d_y d_z}\right) \sum_{n_x=1}^{n_{x_{\max}}} \sum_{n_y=1}^{n_{y_{\max}}} \sum_{n_z=1}^{n_{z_{\max}}} F_{-1}\left(\eta_{3D3400}\right) \tag{3.154}$$

where $\eta_{3D3400} = ((E_{FD3400} - E'_{3400}) / k_B T)$, E_{FD3400} is the Fermi energy in the present case and E'_{3400} is the real part of the lowest positive root of (3.153).Thus using (3.154) and (3.27)

3.2.24. Study of the Seebeck Coefficient under Large Magnetic Field in Heavily doPed Strained Layer Quantum Dot Effective Mass Superlattices

The dispersion relation of the constituent materials of heavily doped III-V super lattices can be written as

$$\bar{P}_i(E,\eta_{gi})k_x^2 + \bar{Q}_i(E,\eta_{gi})k_y^2 + \bar{S}_i(E,\eta_{gi})k_z^2 = 1 \tag{3.155}$$

where $\bar{P}_i(E,\eta_{gi}) = (\gamma_0(E,\eta_{gi}) - I_0 T_{1i})(\bar{\Delta}_i(E,\eta_{gi})^{-1}$,$I_0 = (1/2)[1 + Erf(E/\eta_{gi})]$,

$T_{1i} = [E_{gi} - C_{1i}^c \varepsilon_i - (a_i + C_{1i}^c)\varepsilon_i + (3/2)b_i \varepsilon_{xxi} - (b_i \varepsilon_i / 2) + (\sqrt{3} d_i \varepsilon_{xyi} / 2)]$,

$\bar{\Delta}_i(E,\eta_{gi}) = [(-q_i \eta_{gi}{}^3 / 2\sqrt{\pi})\exp(-(E^2/\eta_{gi}{}^2))[1 + (E^2/\eta_{gi}{}^2)] - R_i \theta_0(E,\eta_{gi}) + V_i \gamma_0(E,\eta_{gi})$

$+(\zeta_i / 2)[1 + Erf(E/\eta_{gi})]], q_i = (3/2B_{2i}^2), R_i = q_i[2A_i + C_{1i}^c \varepsilon_i], A_i = E_{gi} - C_{1i}^c \varepsilon_i$,

$V_i = q_i[A_i^2 - (2C_{2i}^2 \varepsilon_{xyi}/3) + 2A_i C_{1i}^c \varepsilon_i], \zeta_i = q_i[(2C_{2i}^2 \varepsilon_{xyi}/3) - C_{1i}^c \varepsilon_i A_i^2]$,

$\bar{Q}_i(E,\eta_{gi}) = (\gamma_0(E,\eta_{gi}) - I_0 T_{2i})(\bar{\Delta}_i(E,\eta_{gi}))^{-1}, T_{2i} = [E_{gi} - C_{1i}^c \varepsilon_i - (a_i + C_{1i}^c)\varepsilon_i + (3/2)b_i \varepsilon_{xxi} - (b_i \varepsilon_i / 2)$

$-(\sqrt{3} d_i \varepsilon_{xyi} / 2)], \bar{S}_i(E,\eta_{gi}) = (\gamma_0(E,\eta_{gi}) - I_0 T_{3i})(\bar{\Delta}_i(E,\eta_{gi}))^{-1}$,

$T_{3i} = [E_{gi} - C_{1i}^c \varepsilon_i + (a_i + C_{1i}^c)\varepsilon_i + (3/2)b_i \varepsilon_{zzi} - (b_i \varepsilon_i / 2)]$,

The electron energy spectrum in heavily doped strained layer effective mass superlattices can be written as

$$k_z^2 = \left[\frac{1}{L_0^2}\left\{\cos^{-1}\left(f_{40}\left(E,k_x,k_y\right)\right)\right\}^2 - k_s^2\right] \tag{3.156}$$

where $f_{40}\left(E,k_x,k_y\right) = a_{20}\cos\left[a_0 C_{40}\left(E,k_x,k_y,\eta_{g1}\right) + b_0 D_{40}\left(E,k_x,k_y,\eta_{g1}\right)\right]$

$-a_{21}\cos\left[a_0 C_{40}\left(E,k_x,k_y,\eta_{g2}\right) - b_0 D_{40}\left(E,k_x,k_y,\eta_{g2}\right)\right]$,

$$a_{20} = \left[\sqrt{\frac{M_{s2}\left(0,\eta_{g2}\right)}{M_{s1}\left(0,\eta_{g1}\right)}} + 1\right]^2 \left[4\left(\frac{M_{s2}\left(0,\eta_{g2}\right)}{M_{s1}\left(0,\eta_{g1}\right)}\right)^{1/2}\right]^{-1}, \quad M_{si}\left(0,\eta_{gi}\right) = (\hbar/2)\rho_i(\eta_{gi})$$

$$\rho_i(\eta_{gi})=[(\eta_{gi}/2\sqrt{\pi})-(T_{3i}/2)]^{-2}\times[\{(\eta_{gi}/2\sqrt{\pi})-(T_{3i}/2)\}\{(V_i/2)-(R_i\eta_{gi}/\sqrt{\pi})+(\zeta_i/\eta_{gi}\sqrt{\pi})\}$$
$$-((1/2)-(T_{3i}/\eta_{gi}\sqrt{\pi}))\{(\zeta_i/2)+(V_i\eta_{gi}/2\sqrt{\pi})-(R_i\eta^2_{gi}/4)-(q_i\eta^3_{gi}/2\sqrt{\pi})\}]$$

$$a_{20}=\left[\sqrt{\frac{M_{s2}(0,\eta_{g2})}{M_{s1}(0,\eta_{g1})}}+1\right]^2\left[4\left(\frac{M_{s2}(0,\eta_{g2})}{M_{s1}(0,\eta_{g1})}\right)^{1/2}\right]^{-1},a_{21}=\left[\sqrt{\frac{M_{s2}(0,\eta_{g2})}{M_{s1}(0,\eta_{g1})}}-1\right]^2\left[4\left(\frac{M_{s2}(0,\eta_{g2})}{M_{s1}(0,\eta_{g1})}\right)^{1/2}\right]^{-1}$$

$$C_{40}(E,k_x,k_y,\eta_{g1})=[1-\bar{P}_1(E,\eta_{g1})k_x^2-\bar{Q}_1(E,\eta_{g1})k_y^2]^{1/2}[\bar{S}_1(E,\eta_{g1})]^{-1/2}$$

$$D_{40}(E,k_x,k_y,\eta_{g2})=[1-\bar{P}_2(E,\eta_{g2})k_x^2-\bar{Q}_2(E,\eta_{g2})k_y^2]^{1/2}[\bar{S}_2(E,\eta_{g2})]^{-1/2}$$

Therefore, the electron dispersion law in heavily doped strained layer effective mass quantum dot superlattices can be expressed as

$$(\pi n_z/d_z)^2=\left[\frac{1}{L_0^2}\left\{\cos^{-1}\left(f_{400D}\left(E_{0D}\right)\right)\right\}^2-((\pi n_x/d_x)^2+(\pi n_y/d_y)^2)\right] \tag{3.157}$$

where $f_{400D}\left(E_{0D}\right)=a_{20}\cos\left[a_0C_{400D}\left(E_{0D},\eta_{g1}\right)+b_0D_{400D}\left(E_{0D},\eta_{g1}\right)\right]$
$-a_{21}\cos\left[a_0C_{400D}\left(E_{0D},\eta_{g2}\right)-b_0D_{400D}\left(E,\eta_{g2}\right)\right]$,

$$C_{400D}(E_{0D},\eta_{g1})=[1-\bar{P}_1(E_{0D},\eta_{g1})(\pi n_x/d_x)^2-\bar{Q}_1(E_{0D},\eta_{g1})(\pi n_y/d_y)^2]^{1/2}[\bar{S}_1(E_{0D},\eta_{g1})]^{-1/2}$$

$$D_{400D}(E_{0D},\eta_{g2})=[1-\bar{P}_2(E_{0D},\eta_{g2})(\pi n_x/d_x)^2-\bar{Q}_2(E_{0D},\eta_{g2})(\pi n_y/d_y)^2]^{1/2}[\bar{S}_2(E_{0D},\eta_{g2})]^{-1/2}$$

The electron concentration in this case is given by

$$n_0=\left(\frac{2g_v}{d_xd_yd_z}\right)\sum_{n_x=1}^{n_{x_{\max}}}\sum_{n_y=1}^{n_{y_{\max}}}\sum_{n_z=1}^{n_{z_{\max}}}F_{-1}(\eta_{3D40}) \tag{3.158}$$

where $\eta_{3D40}=((E_{FD40}-E'_{40})/k_BT)$, E_{FD40} is the Fermi energy in the present case and E'_{40} is the lowest positive root of (3.157).Thus using (3.158) and (3.27) one can obtain the Seebeck coefficient in this case.

3.3. Result and Discussions

Using the appropriate eqations and the band constants from Table 1.1, the normalized S in QDs of heavily doped non-linear optical materials has been plotted as a function of film thickness in Fig 3.1 for the generalized band model as shown by curve (a) where the curves (b), (c) and (d) are valid for three band models of Kane together with parabolic energy bands respectively. The case for $\delta=0$ has been plotted in the same figure and is represented by curve (e) for the purpose of assessing the influence of crystal field splitting on the S in this case. It appears from the figure that the S in the present case oscillates with film thickness and for specific values of the film thickness as determined by the energy band constants and the transition of Fermi energy from one allowed set of quantum numbers to another allowed set and the band structure of the material, the oscillatory dependence shows spikes. The influence of crystal field splitting effectively reduces the value of the S for relatively large values of the thickness in the whole range of thickness as considered. From the numerical values for the three and two band models of Kane, it appears that the influence of spin orbit splitting constant lessens the value of the S for relatively large values of the nano thickness.

The Figure 3.2 exhibits the variations of normalized S in this case as a function of electron concentration for all the cases mentioned as above for Figure 3.1. From Figure 3.2, we observe that the S for the range of concentrations below the concentration zone 10^{23} m^{-3}, the influence of said band structures of the constituent materials on the S is not prominent, and they exhibit converging tendencies, whereas for higher values of the carrier degeneracy, the S decreases with increasing carrier concentration. For generalized band model for quantum dots of heavily doped nonlinear optical materials, the numerical values of the S are least, whereas, for the most approximate parabolic energy bands of the same, the S exhibits the highest values with respect to concentration in this case. It may be noted that that the QDs lead to the discrete energy levels, somewhat like atomic energy levels, which produce very large changes. This is in accordance to the inherent nature of the quantum confinement of the carrier gas as dealt with here. In QDs, there remain no free carrier states in between any two allowed sets of size-quantized levels. Consequently, the crossing of the Fermi level by the size-quantized levels in QDs would have much greater impact on the redistribution of the carriers among the allowed levels, as compared to that found for 2D and 1D quantum confinements.

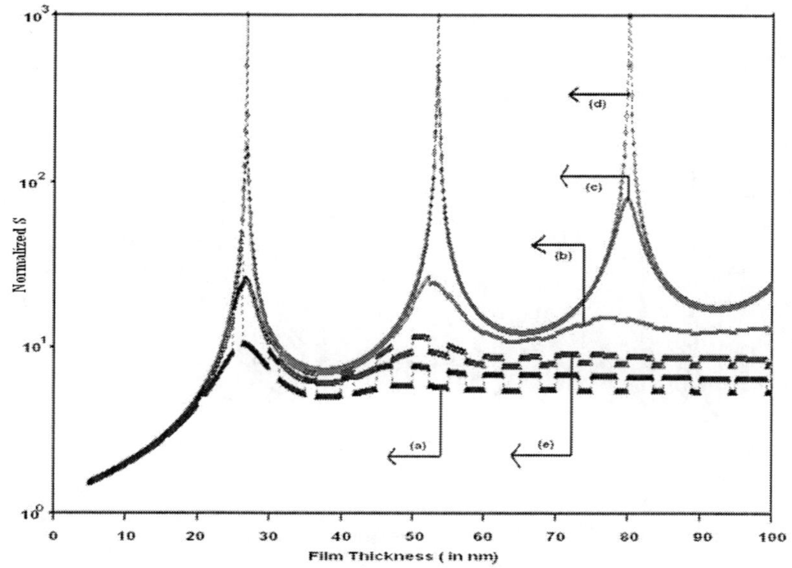

Figure 3.1. Plot of the normalized S in quantum dots of heavily doped n-$CdGeAs_2$ as a function of film thickness has been shown in accordance with the (a) generalized band model ($\delta \neq 0$), (b) three and (c) two band models of Kane together with (d) parabolic energy bands. The special case for $\delta = 0$ (e) has also been shown to assess the influence of crystal field splitting.

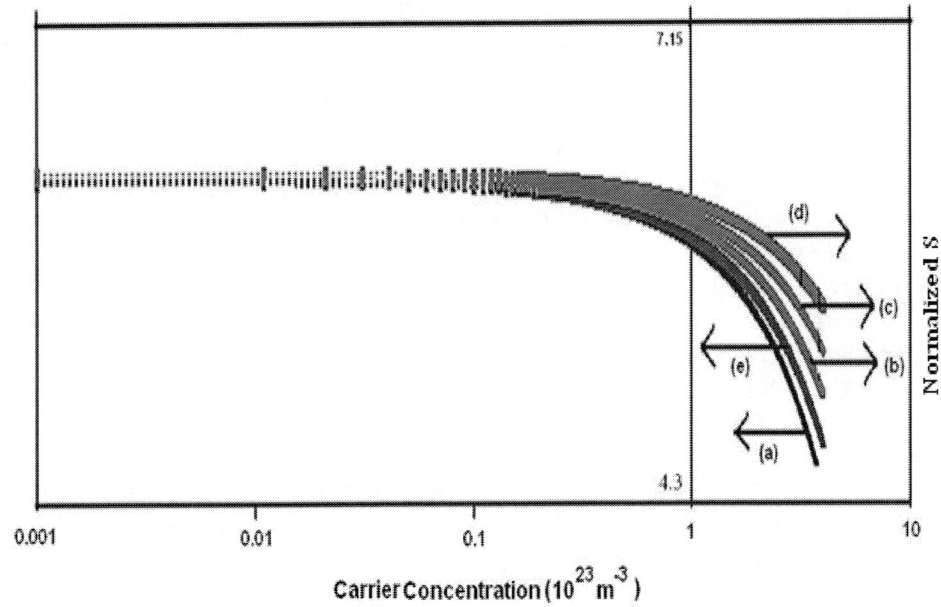

Figure 3.2. Plot of the normalized S in quantum dots of heavily doped n-$CdGeAs_2$ as a function of carrier concentration for all cases of Figure 3.1.

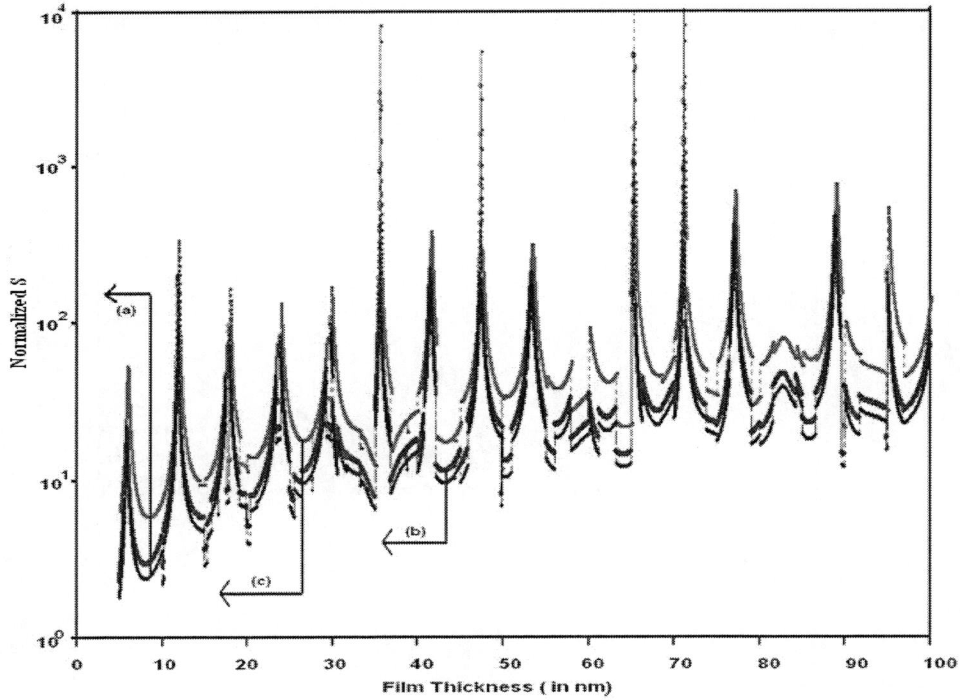

Figure 3.3. Plot of the normalized S in quantum dots of heavily doped InAs as a function of film thickness in accordance with the (a) three and (b) two band models of Kane together with (c) parabolic energy bands.

In Figure 3.3, the S has been plotted for QDs of heavily doped InAs as a function of film thickness in accordance with the three and two band models of Kane together with parabolic energy bands, it is observed from Figure 3.3 that the S shows oscillatory dependence with respect to film thickness and the influence of the heavily doped three band model of Kane is to change the numerical values of the S as compared with the heavily doped two band model of Kane and parabolic energy bands. The Figure 3.4 exhibits the variation of S with carrier concentration in QDs of InAs for all the cases of Figure 3.3. It appears that the S in accordance with all the band models exhibit wide separation from one another and for three band model of Kane, although the S is less as compared with other two types of band models is very prominent and decreases rapidly with concentration after a particular value of carrier degeneracy.

In Figure 3.5, the plot of the normalized S in QDs of heavily doped GaAs as a function of film thickness in accordance with the (a) three and (b) two band models of Kane together with (c) the parabolic energy bands has been shown. The influence of energy band models on the S in QDs of heavily doped GaAs is apparent from the plots of Figure 3.5. The Figure 3.6 exhibits the dependence of the S in QDs of n-GaAs and in this case the shape of the curves for two band model of Kane and that of parabolic energy bands exhibit wide difference although their nature are same. From the same figure, it is apparent that the S in QDs of heavily doped n-GaAs in accordance with three band Kane model decreases rapidly with increasing electron concentration after the value of 10^{23} m^{-3}. The influence of the spin orbit splitting constant is rather very large after a fixed value of carrier concentration. In Figures 3.7 and 3.8, we have shown the normalized S in QDs of heavily doped InSb as a function of film thickness and concentration respectively in accordance with the heavily doped three and two band models of Kane together with parabolic energy bands as shown by curves (a), (b) and (c) in the respective figures. The nature of variation of S in QDs of heavily doped InSb with respect to thickness and concentration remains same as compared with the S in heavily doped InAs and GaAs, although the numerical value changes precisely which is the manifestation of the energy band constants. The Figures 3.9, 3.10 and 3.12 illustrate the film thickness dependence of the normalized S in QDs of InAs, GaAs and InSb in accordance with the models of Stillman et. al., Newson et. al. and Rossler et. al. as represented by curves (a), (b) and (c) respectively.

From Figure 3.9, it appears that the S in QDs of heavily doped InAs exhibits greatest value for the model of Stillman et. al. and the lowest for the model of Rossler et. al. The model of Newson et. al. generates the values of S which lies in between. In Figures 3.10 and 3.12, the S has been plotted for QDs of heavily doped GaAs and InSb as a function of film thickness for all the cases of Figure 3.9 and the same observation of the aforementioned figure also holds true for Figures 3.10 and 3.12 respectively. In Figure 3.11, the S in QDs of heavily doped GaAs has been plotted with respect to carrier concentration in accordance with all the models of Figure 3.9. It appears that the model of Stillman et. al. exhibits the highest value, although for the whole range of carrier concentration, the S in accordance with the models of Newson et. al. and Rossler et. al. maintain the same separation and the same shape. Besides, the S in this case for the model of Rossler et. al. decreases sharply with increase in carrier concentration for relatively large values of concentration, whereas for small values of electron degeneracy, the S in accordance with the models of Newson et. al. and Rossler et. al. exhibit converging tendency.

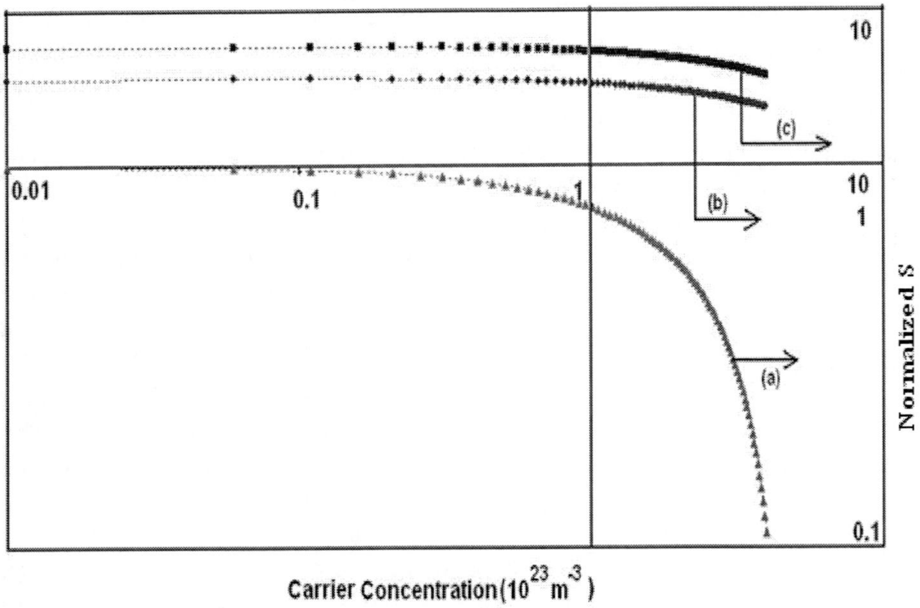

Figure 3.4. Plot of the normalized S in QDs of heavily doped InAs as a function of carrier concentration for all the cases of Figure 3.3.

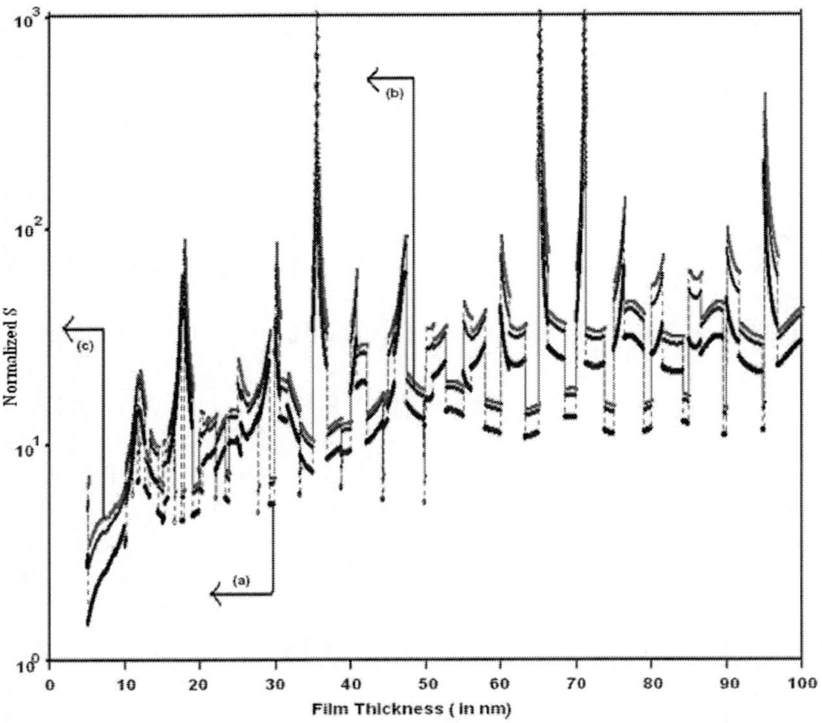

Figure 3.5. Plot of the normalized S in QDs of heavily doped GaAs as a function of film thickness in accordance with the (a) three and (b) two band models of Kane together with (c) the parabolic energy bands.

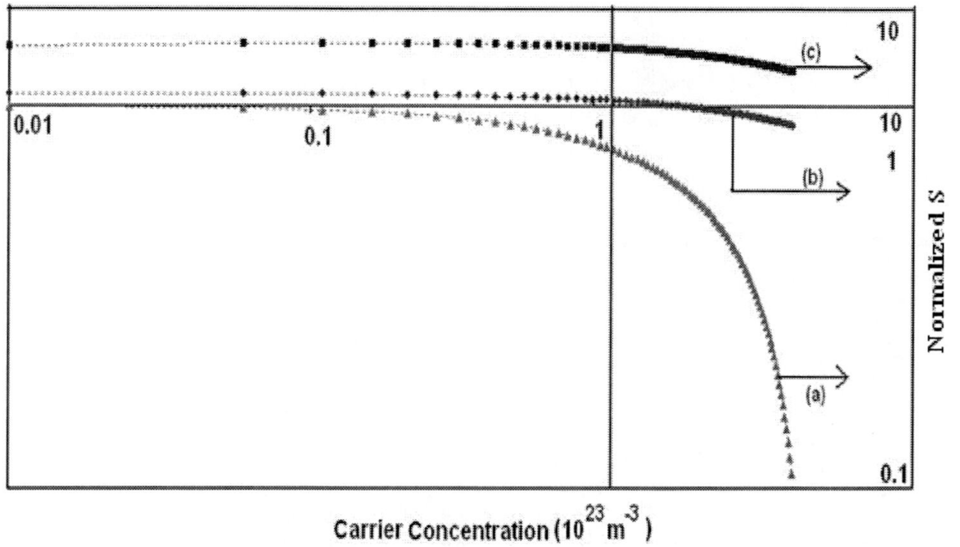

Figure 3.6. Plot of the normalized S in QDs of heavily doped GaAs as a function of carrier concentration for all the cases of Figure 3.5.

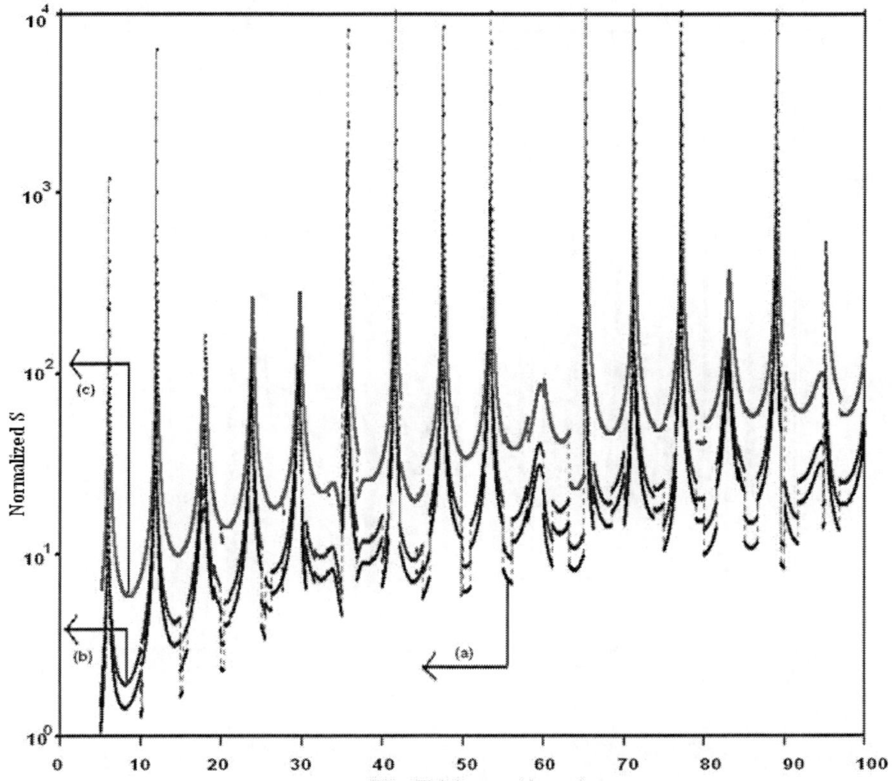

Figure 3.7. Plot of the normalized S in QDs of heavily doped InSb as a function of film thickness for all the cases of Figure 3.5.

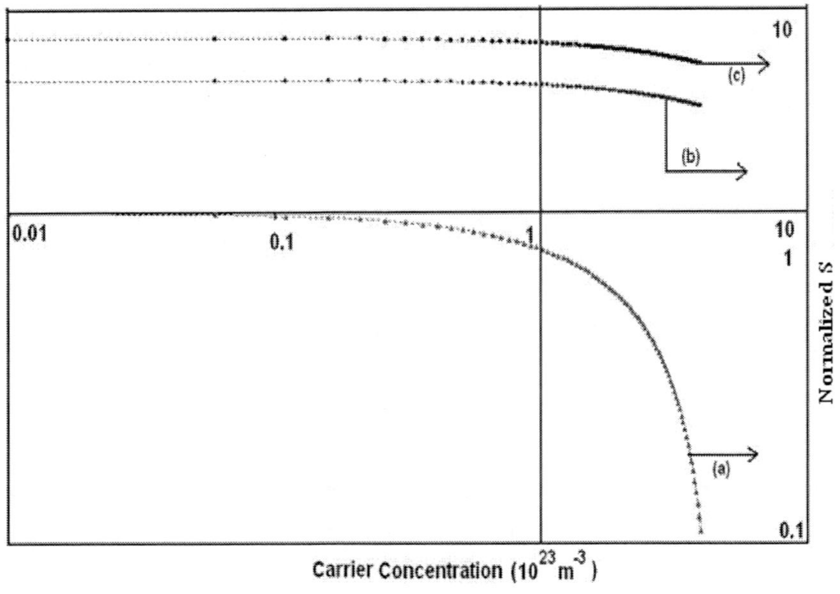

Figure 3.8. Plot of the normalized S in QDs of heavily doped InSb as a function of carrier concentration for all the cases of Figure 3.4.

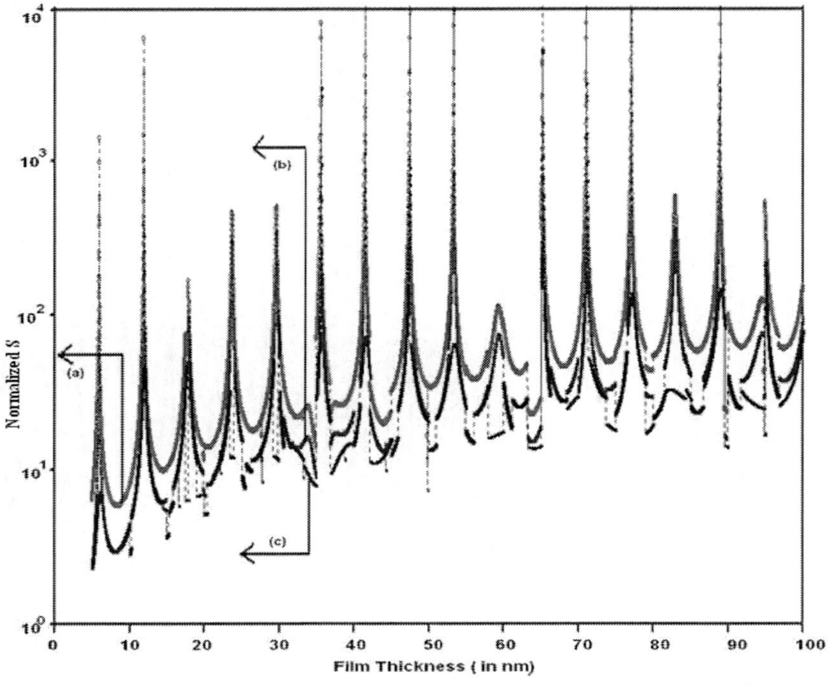

Figure 3.9. Plot of the normalized S in QDs of heavily doped InAs as a function of film thickness in accordance with the models of (a) Stillman et. al., (b) Newson et. al. and (c) Rossler et. al. respectively.

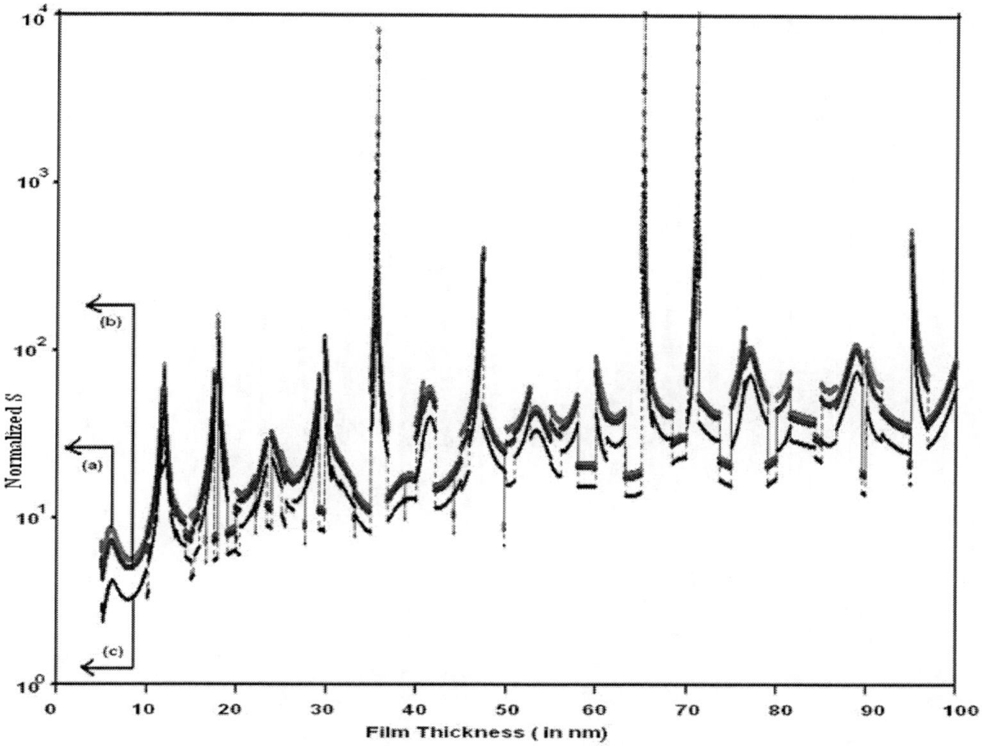

Figure 3.10. Plot of the normalized S in QDs of heavily doped GaAs as a function of film thickness for all the cases of Figure 3.9.

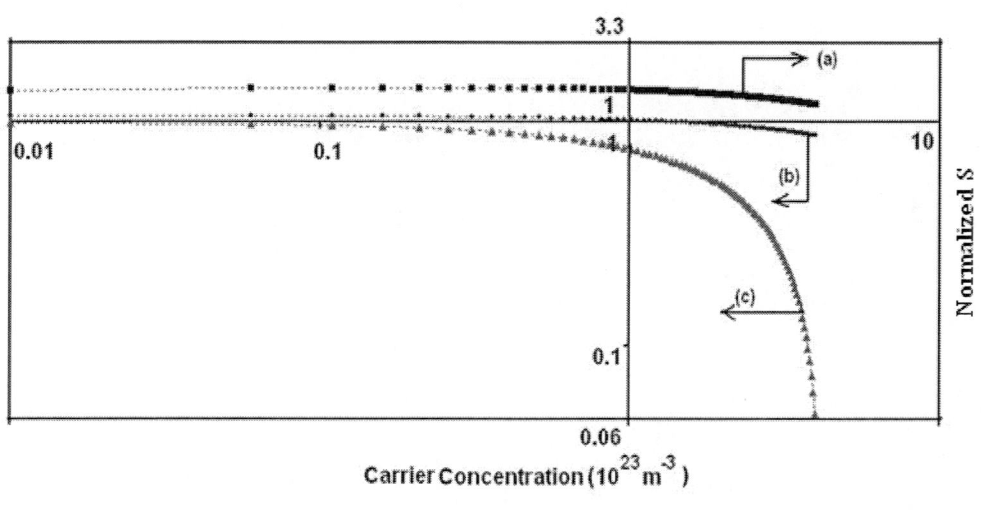

Figure 3.11 Plot of the normalized S in QDs of heavily doped GaAs as a function of carrier concentration for all the cases of Figure 3.9.

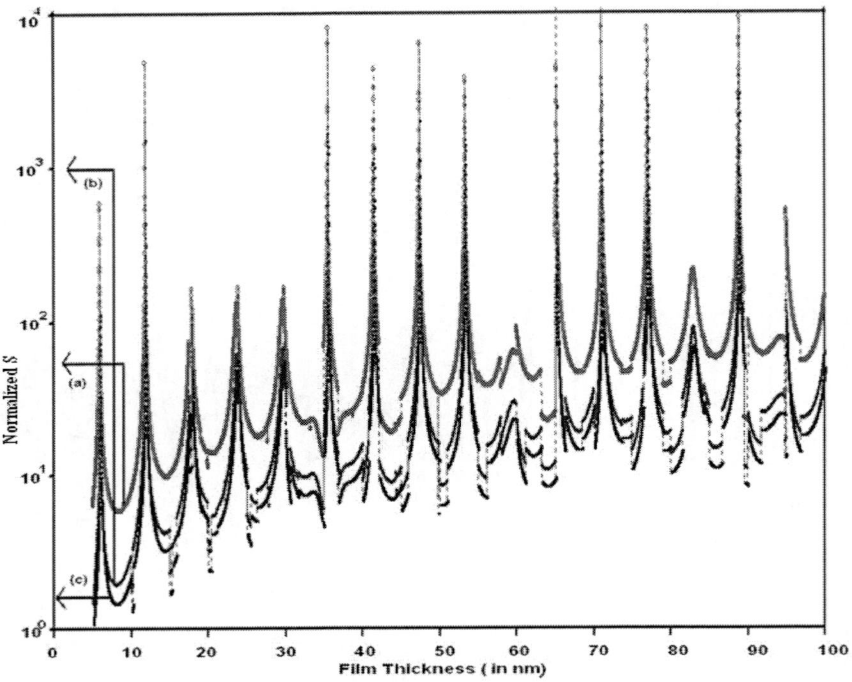

Figure 3.12. Plot of the normalized S in QDs of heavily doped InSb as a function of film thickness for all the cases of Figure 3.9.

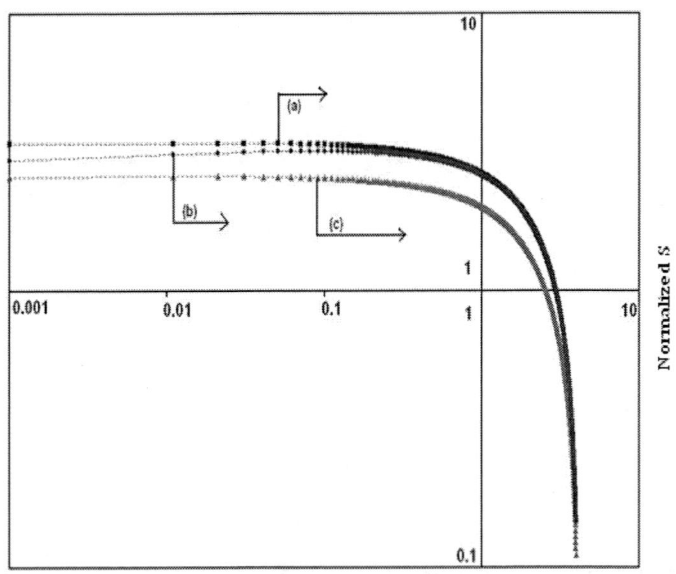

Figure 3.13. Plot of the normalized S in QDs of heavily doped InSb as a function of carrier concentration for all the cases of Figure 3.9.

The Figure 3.13 exhibits the influence of carrier concentration on the normalized S in QDs of InSb for all the cases of Figure 3.9. For small values of carrier concentration, the S in QDs of InSb exhibit relative difference for all the models and for higher values of carrier degeneracy, the decreasing nature of S with increasing concentration is very prominent for all the models and they exhibit converging nature.

The Figure 3.14 exhibits the film thickness dependence of the normalized S for QDs of heavily doped InSb and InAs in accordance with the model of Agafonov et. al. as represented by the curves (a) and (b) respectively. The numerical values of the S of heavily doped InSb are greater than heavily doped InAs in accordance with the said figure which is the direct influence of the energy band constants. The Figure 3.15 illustrates the concentration dependence of normalized S for all the cases of Figure 3.14. For small values of the electron concentration, the S in both cases of heavily doped InSb and InAs show wide difference and with increasing values of the carrier degeneracy, the S decreases and ultimately converges. The Figure 3.16 demonstrates the film thickness dependence of the normalized S for QDs of heavily doped InSb and InAs in accordance with the model of Johnson et. al. as represented by the numerical values of the S of heavily doped InSb are greater than heavily doped InAs in accordance with Figure 3.16 which is the consequence of constants of the energy spectra. The Figure 3.17 shows the concentration dependence of normalized S for all the cases of Figure 3.16. For small values of the electron concentration, the S in both cases of InSb and InAs show converging tendency and with increasing values of the carrier degeneracy, the S decreases and show appreciable difference with one another. The Figure 3.18 depicts the film thickness dependence of the normalized S for heavily doped QDs of InSb and InAs in accordance with the model of Palik et. al. as represented by the curves (a) and (b) in this context. The numerical values of the S of InSb are greater than InAs in accordance with Figure 3.18 which is the direct influence of energy band constants. The Figure 3.19 models the concentration dependence of normalized S for all the cases of Figure 3.18. For small values of the electron concentration, the influence of energy band constants of InSb and InAs on the S becomes very small whereas, and with increasing values of the carrier degeneracy, the effect of the constants of the carrier energy spectrum of the two said materials on the S is rather large. From Figure 3.19, we infer that the S decreases with increasing carrier concentration after the concentration value 10^{22} m^{-3} and the S for both the materials show substantial difference with one another. The Figure 3.20 exhibits the variation of the normalized S with the film thickness in QDs of II-VI materials in accordance with Hopfield model taking p-CdS as an example and considering both the cases $C_0 = 0$ and $C_0 \neq 0$) and GaP as shown by curves (a), (b) and (c) respectively. The S is higher for QDs of heavily doped CdS with $C_0 \neq 0$ as compared to the case for $C_0 = 0$. The S is lowest for heavily doped GaP as compared to the aforementioned two cases of heavily doped CdS. The Figure 3.21 demonstrates the concentration variation of the normalized S for all cases of Figure 3.20. For small values of the electron concentration, the S in Figure 3.21 for all cases of Figure 3.20 preserve difference for small values of carrier concentration which is the direct signature of the energy band constants of CdS and GaP respectively and with increasing values of the carrier degeneracy, the S decreases and show significant difference with one another. The effect of spin splitting of the carriers in QDs of heavily doped p-CdS on S can be numerically investigated from Figures 3.20 and 3.21. It appears that the absence of the spin splitting constant decreases the numerical value of the S in CdS for a particular range of film thickness. In Figure 3.22, the normalized S has been plotted as a function of film thickness for QDs of heavily doped Germanium for both the models of Wang et. al. and Cardona et. al. as shown by curves (a) and (b) respectively. The S for the model of Cardona et. al. shows sharp oscillatory peaks as compared to the model of Wang et. al. The Figure 3.23 exhibits the normalized S as a function of film thickness in QDs of Tellurium (by using the models of Bouat et. al., Ortenberg et. al. and stressed Kane type materials (taking n-InSb as an example) in accordance with the models of Seiler et. al. as shown by curves (a), (b) and (c) respectively. It appears from the figure that the S for QDs of heavily doped Te oscillates with increasing thickness, exhibiting oscillatory spikes in accordance with both the models of Te and the S in QDs of heavily doped stressed InSb in accordance with the model of Seiler et. al. exhibits the spikes in the whole range of thickness as considered. The Figure 3.24 shows the dependence of the normalized S on the carrier concentration for all the cases of Figure 3.23. From Figure 3.24, it appears that the S in heavily doped Te in accordance with both the aforementioned band models exhibits wide separation for relatively low values of carrier degeneracy, whereas for the relatively large values of the carrier concentration, both of them decrease and cut at a point. For stressed InSb, the model of Seiler et. al. predicts sharp fall of S with increasing electron statistics. It appears that at extremely low and high film thicknesses, the S in QDs of heavily doped Te dominates over that of the corresponding stressed heavily doped InSb, although, at mid zone thickness, the S in stressed InSb exhibits a high peak together with the fact that the periods of oscillations of the S are comparatively higher in heavily doped Te than that of stressed compounds. In Figure 3.25, the normalized S has been plotted as a function of carrier concentration for QDs of heavily doped Platimun Antimonide and heavily doped Lead Germanium Telluride in accordance with the models of Ushio et. al. Emtage and Vassilev as shown by curves (a), (b), (c) (this three graphs are valid for heavily doped Lead Germanium Telluride for three different temperetures) and (d) respectively. The S in all the cases decreases with increasing carrier degeneracy and is highest for QDs of heavily doped $Pb_{1-x}Ge_xTe$ The Figure 3.26 depicts the plot of normalized S for QDs of heavily doped GaSb as a function of carrier concentration in accordance with the models of Seiler et. al., Mathur et. al. and Zhang as shown by curves (a), (b) and (c) respectively. It appears from the figure that the S for QDs of heavily doped Gallium Antimonide decreases with increasing carrier concentration after the concentration zone 10^{22} m^{-3} exhibiting highest numerical value for the model of Zhang and the lowest for the model of Seiler et. al.

Besides, in the concentration regime from 0.001×10^{23} m^{-3} to less than 0.1×10^{23} m^{-3}, the influence of band structure on the S in this case is insignificant and they exhibit the converging tendency towards a fixed value. The Figure 3.27 exhibits the thickness dependence of the normalized S for all the cases of Figure 3.26 and the Figure 3.27 shows the oscillatory variation with respect to thickness as usual. The numerical value of the same is greatest in accordance with the model of Zhang and lowest for the band structure of GaSb as defined by Seiler et. al. The numerical value of the S as presented by the model of Mathur et. al. appears to fall in the mid-zone.

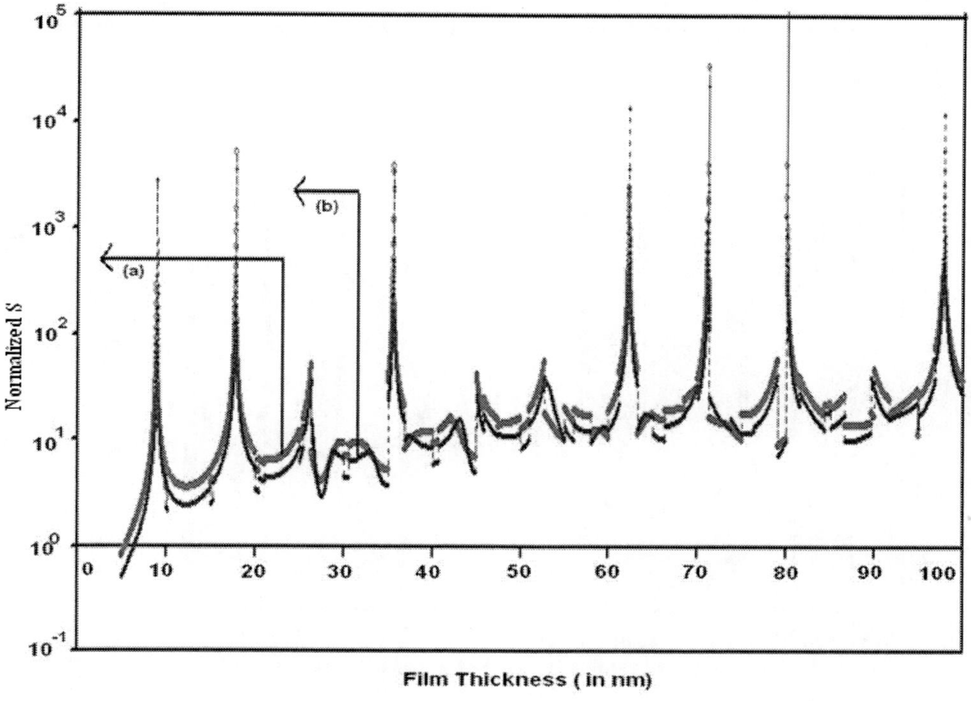

Figure 3.14. Plot of the normalized S in QDs of heavily doped (a) InSb and (b) InAs as a function of film thickness in accordance with model of Agafonov et. al.

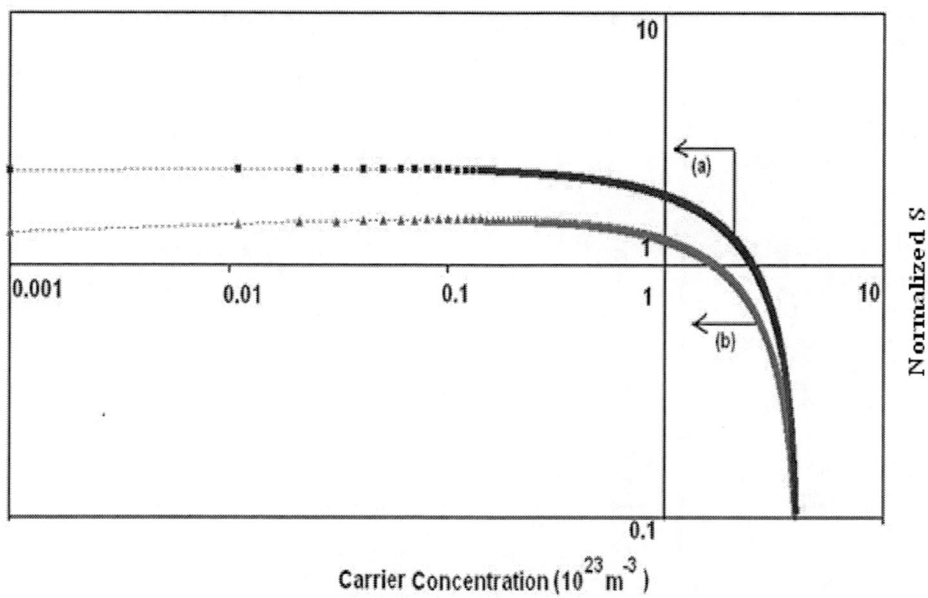

Figure 3.15. Plot of the normalized S in QDs of heavily doped (a) InSb and (b) InAs as a function of carrier concentration for the case of Figure 3.14.

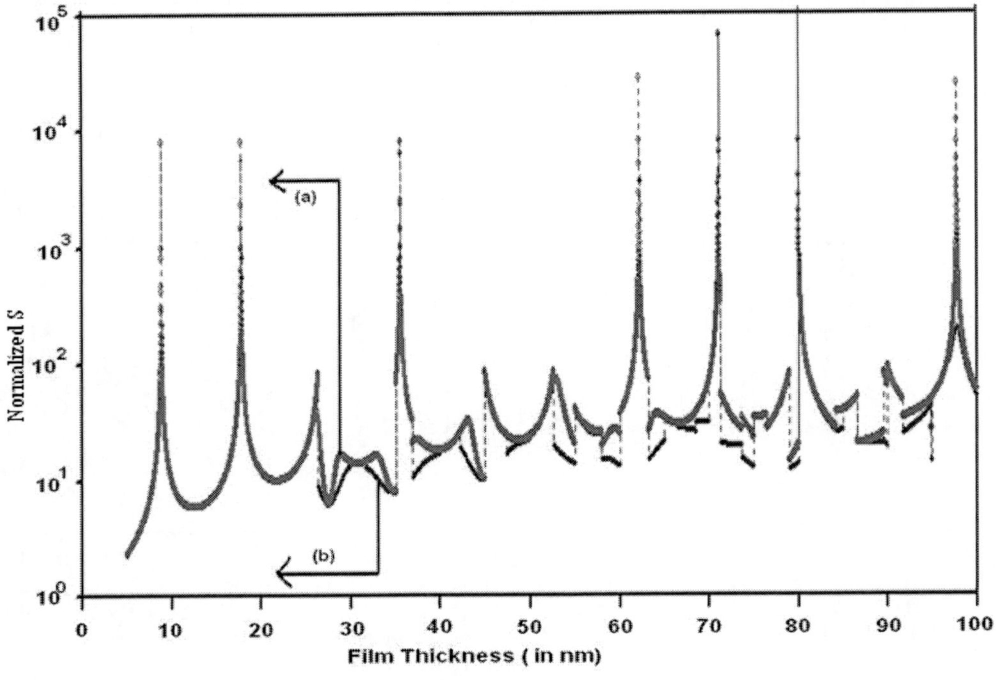

Figure 3.16. Plot of the normalized S in QDs of heavily doped (a) InSb and (b) InAs as a function of film thickness in accordance with the model of Johnson et. al.

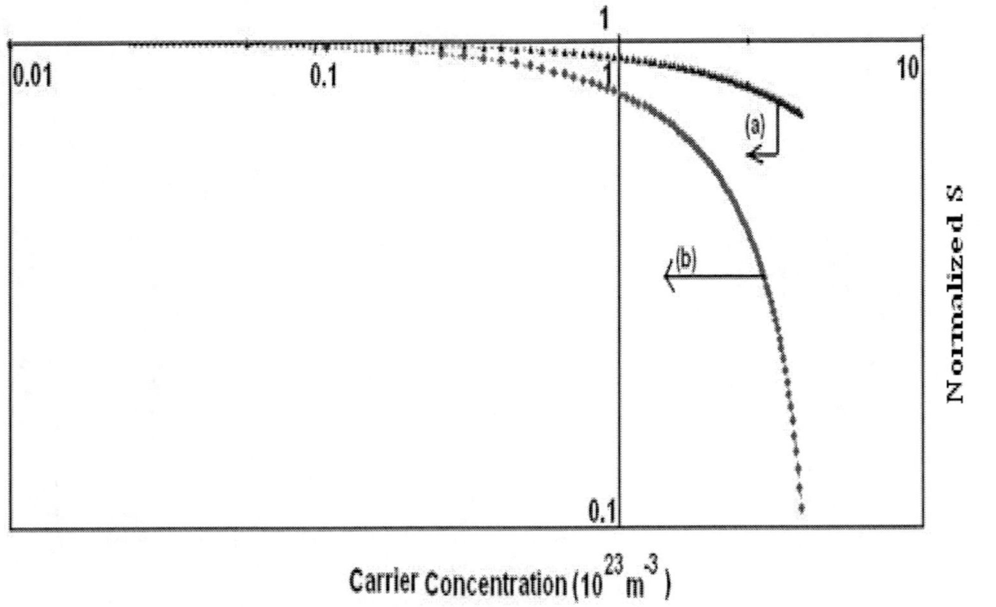

Figure 3.17. Plot of the normalized S in QDs of heavily doped (a) InSb and (b) InAs as a function of carrier concentration for the case of Figure 3.16.

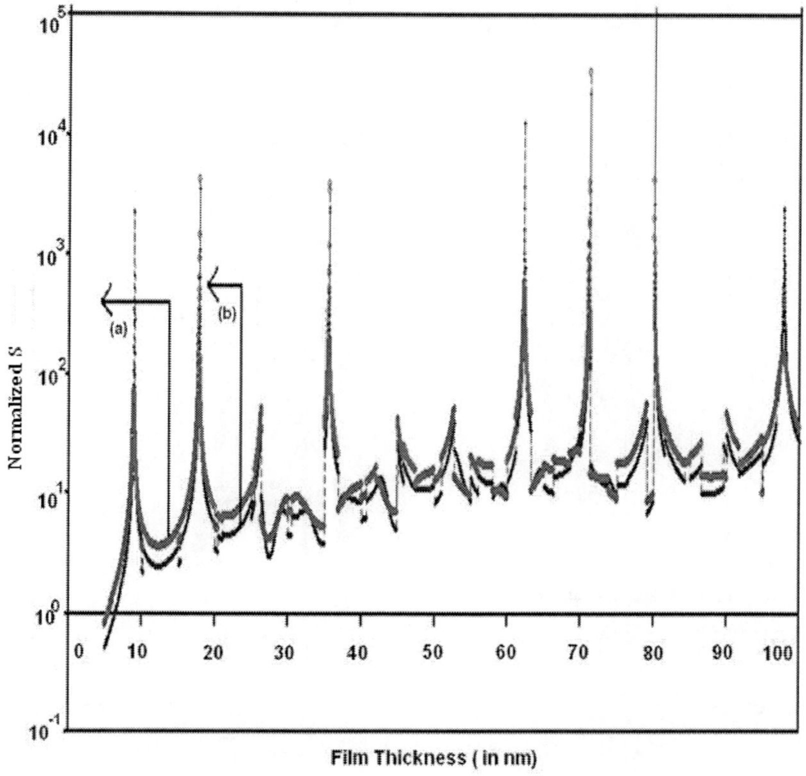

Figure 3.18. Plot of the normalized S in QDs of heavily doped (a) InSb and (b) InAs as a function of film thickness in accordance with the model of Palik et. al.

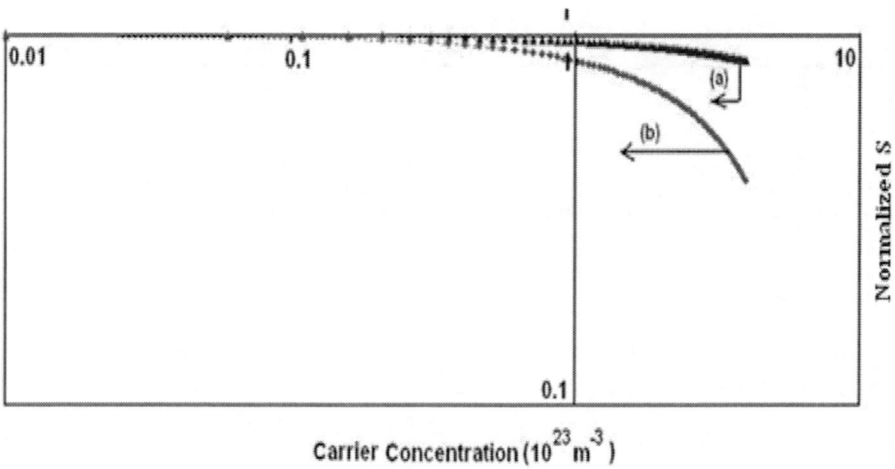

Figure 3.19. Plot of the normalized S in QDs of heavily doped (a) InSb and (b) InAs as a function of carrier concentration for the case of Figure 3.18.

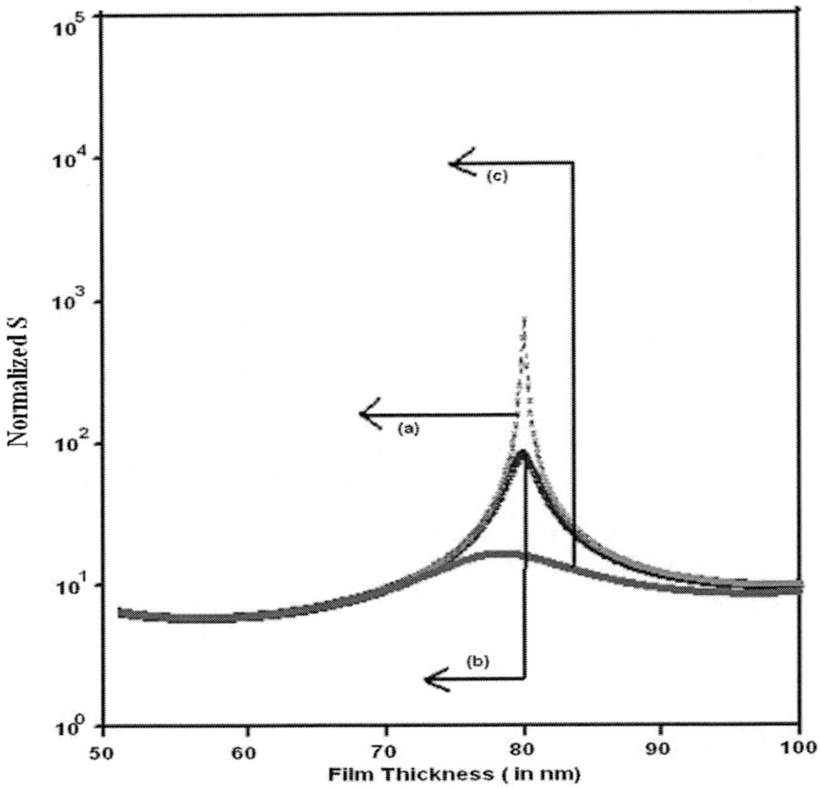

Figure 3.20. Plot of the normalized S in QDs of of heavily doped CdS with (a) $C_0 \neq 0$, (b) $C_0 = 0$ and (c) GaP as a function of film thickness in accordance with the models of Hopfield and Rees respectively.

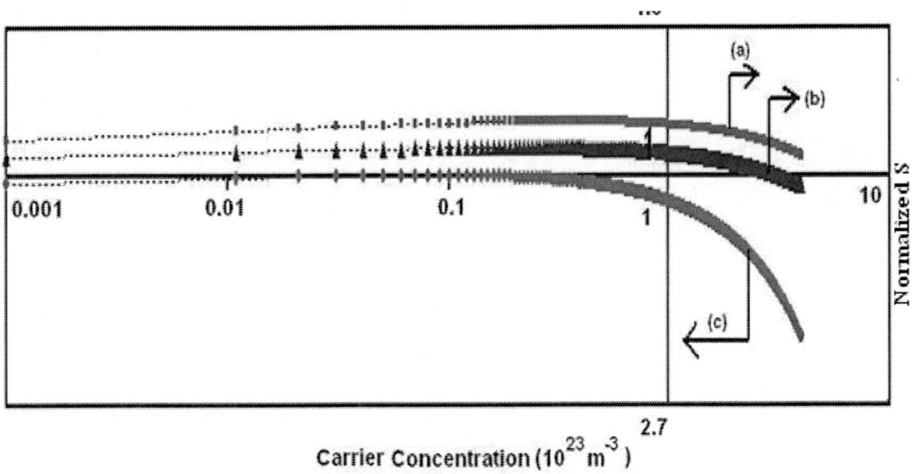

Figure 3.21. Plot of the normalized S in QDs of heavily doped CdS with (a) $C_0 \neq 0$, (b) $C_0 = 0$ and (c) GaP as a function of carrier concentration for all the cases of Figure 3.20.

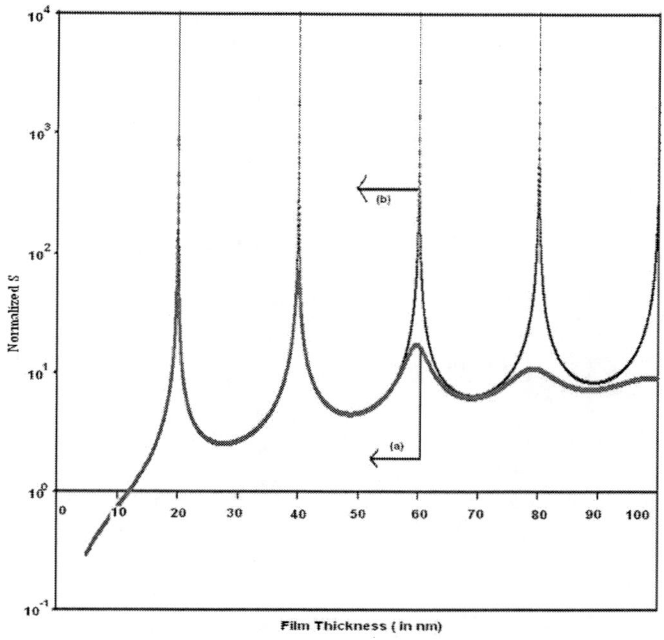

Figure 3.22. Plot of the normalized S in QDs of heavily doped Germanium as a function of film thickness in accordance with the models of (a) Wang et. al. and (b) Cardona et. al. respectively.

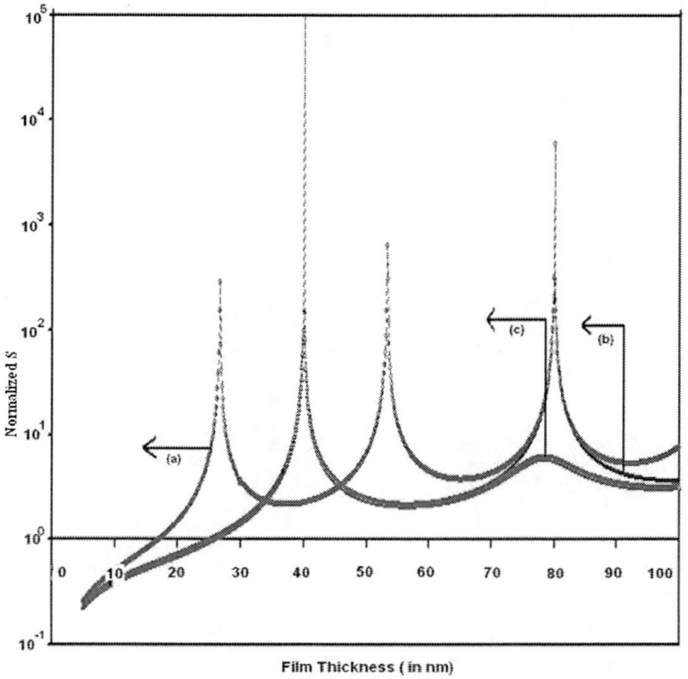

Figure 3.23. Plot of the normalized S in QDs of heavily doped Tellurium and heavily doped stressed Kane type material (n-InSb) as a function of film thickness in accordance with the undoped models of (a) Bouat et. al., (b) Ortenberg et. al. and (c) Seiler et. al. respectively.

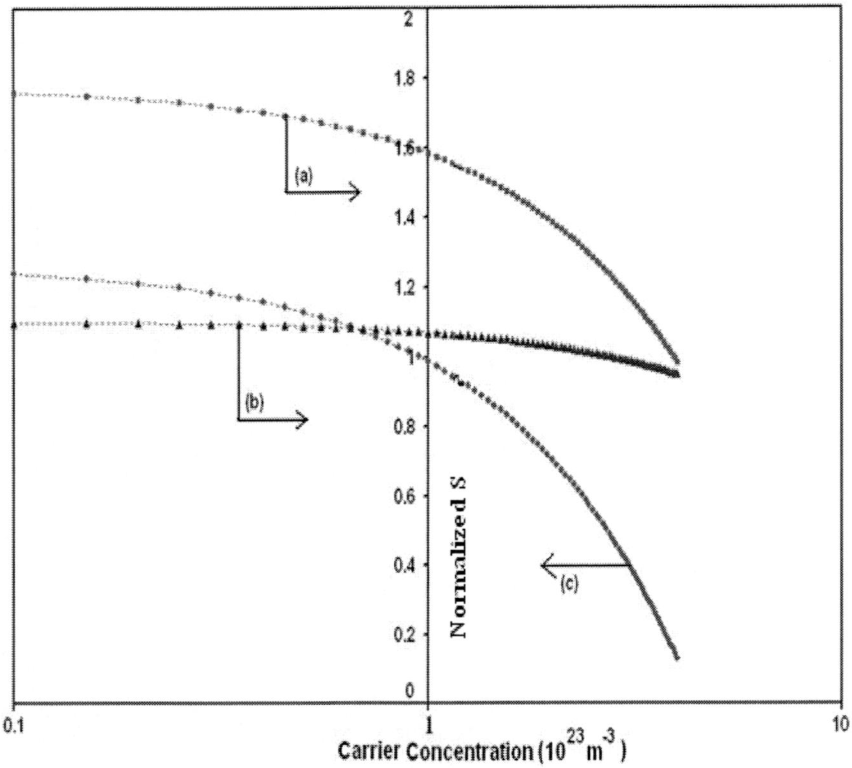

Figure 3.24. Plot of the normalized S in QDs of heavily doped Tellurium and heavily doped stressed Kane type material (n-InSb) as a function of carrier concentration for all the cases of Figure 3.23.

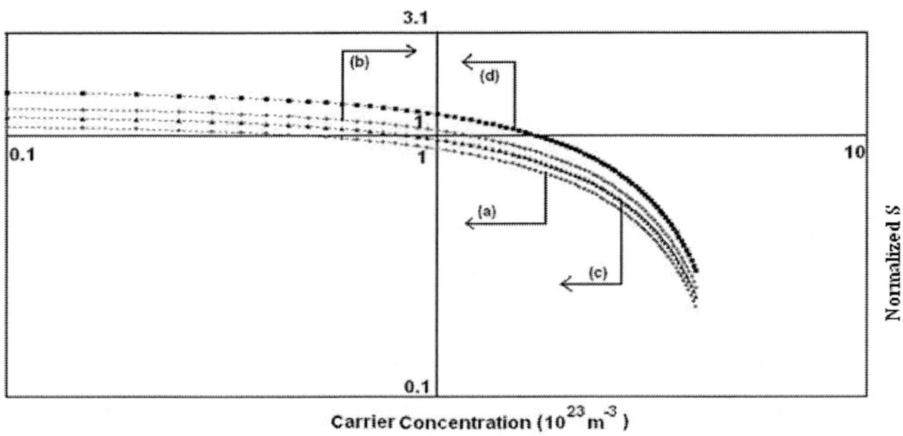

Figure 3.25. Plot of the normalized S in QDs of heavily doped Platinum antimonide(the graphs (a),(b) and (c) are valid for three different temperatures), and (d) $Pb_{1-x}Ge_xTe$ as a function of carrier concentration in accordance with the models of Emtage and Vassilev respectively.

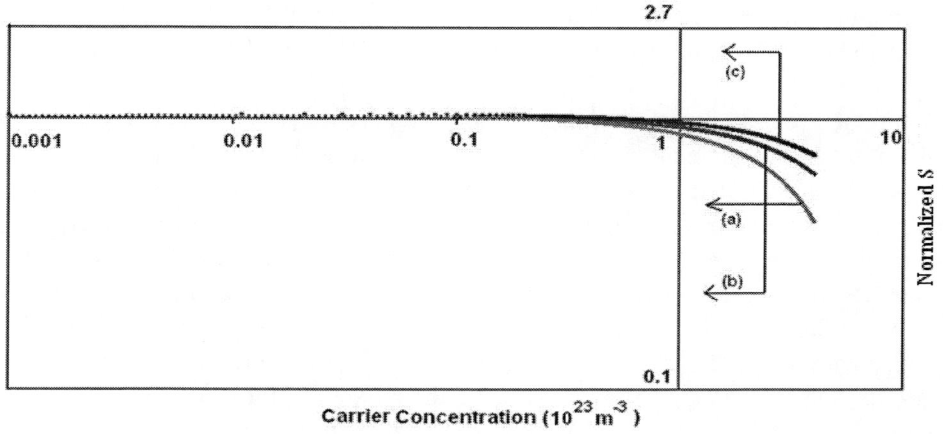

Figure 3.26. Plot of the normalized S in QDs of heavily doped GaSb as a of carrier concentration in accordance with the models of (a) Seiler et. al., (b) Mathur et. al. and (c) Zhang respectively.

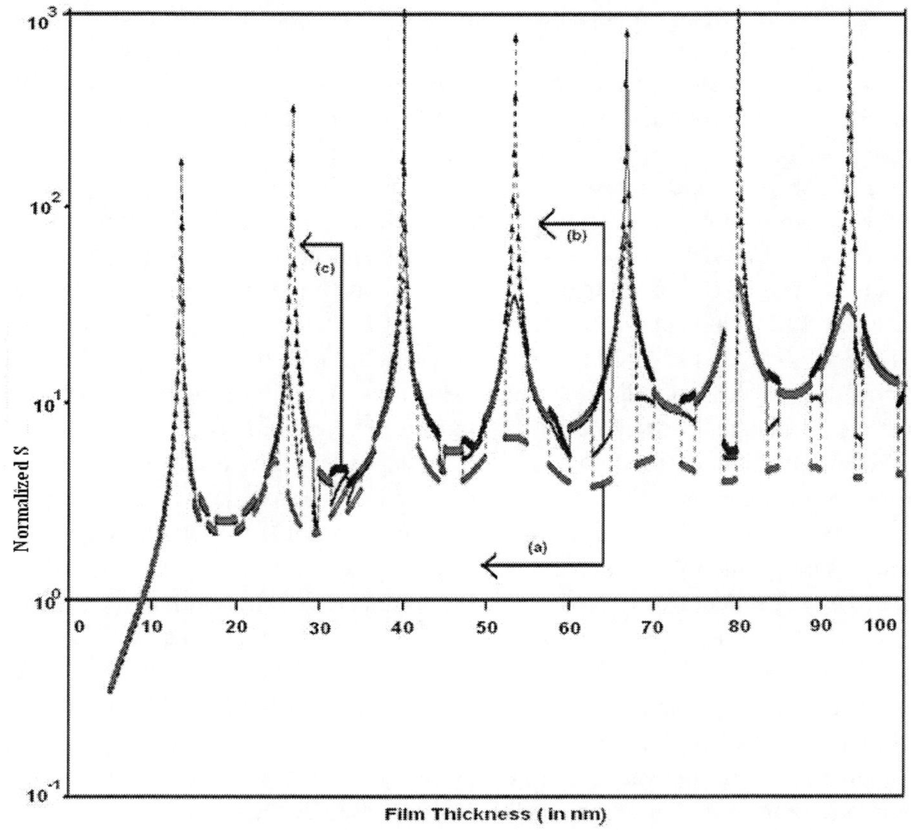

Figure 3.27. Plot of the normalized S in QDs of heavily doped Gallium antimonide as a function of film thickness for all the cases of Figure 3.26.

The Figure 3.28 illustrates the variation of the normalized S with film thickness in QDs of heavily doped PbTe in accordance with the models of Cohen, Lax, Dimmock and Bangert and Kastner as shown by curves (a), (b), (c) and (d) respectively. It appears from the figure that S for QDs of all the models of Bismuth exhibit regular oscillation with thickness, although the most prominent oscillatory lobe together with high sharp peak in the mid thickness zone is exhibited by the model of Cohen. The Figure 3.29 shows the dependence of normalized S on the carrier concentration in QDs of heavily doped PbTe for all types of band models as stated in Figure 3.28. From Figure 3.29, it appears that the S decreases with increasing electron concentration for all the models of heavily doped PbTe. The Figure 3.30 shows the variation of S on the carrier concentration for QDs of heavily doped PbTe in accordance with the model of Foley et. al (for three different values of temperature) together with heavily doped Bismuth Telluride by using the model of Stordeur et. al as shown by the curves (a), (b), (c) and (d) respectively. For small values of the electron concentration, the S in Figure 3.30 show substantial difference with each other and with increasing values of the carrier degeneracy, the S decreases and show converging tendency.

The Figure 3.31 models the variation of the S with the film thickness for all cases of Figure 3.30. From the figure, it appears that the S in PbTe and Bi_2Te_3 in accordance with the model of Stordeur et. al. exhibits dense oscillations with respect to film thickness and it also appears that the numerical values of S in QDs of heavily doped PbTe and Bi_2Te_3 in accordance with all the band models are higher than all other materials.

The Figure 3.32 shows plot of the normalized S in QDs of heavily doped (a) II-V compound (CdSb), (b) zinc diphosphide, (c) cadmium diphosphide and the plot (d) is valid for another temperature for cadmium diphosphide as a function of film thickness in accordance with the models of Yamada and Chuiko respectively From Figure 3.32, it appears that the S is largest for II-V compound. From the aforementioned discussions, we realize that the signature of 3D quantization is forthwith evident from the Figures 3.1, 3.3, 3.5, 3.7, 3.9, 3.10, 3.12, 3.14. 3.16, 3.18, 3.20. 3.22, 3.23, 3.26, 3.28, 3.30 and 3.32 for all heavily doped materials as discussed having different band structures. It can be facilely discerned from the same that the normalized S oscillates with film thickness exhibiting spikes for various values of film thickness which are totally band structure dependent. The occurrence of peaks in the said figures originates from the totally quantized energy levels of the carriers of the concerned dots. The S spectra are found bearing composite oscillations as function of the nanothickness. These are generally due to selection rules in the quantum numbers along the three confined directions. The dependence of the normalized S on the carrier concentration is manifested by the Figures 3.2, 3.4, 3.6,138, 3.11, 3.13, 3.15, 3.17, 3.19, 3.21, 3.24, 3.25, 3.27, 3.29 and 3.31 for the different materials as considered here. It can be ascertained from the same figures that the S of all the corresponding heavily doped compounds decreases with increasing carrier concentration for relatively higher values of the carrier degeneracy. Although the S varies in various manners with all the variables in all the limiting cases as evident from all the figures, the rate of variations in each case are totally band-structure dependent.

In formulating the generalized electron energy spectrum for heavily doped non-linear optical materials, we have considered the crystal-field splitting, the anisotropies in the momentum-matrix elements, the spin-orbit splitting parametersand the presence of Gaussian band tails respectively. In the absence of the crystal field splitting constant together with the assumptions of isotropic effective electron mass and isotropic spin orbit splitting, our basic relation as given by (3.25) converts into (3.29). The (3.29) is the heavily doped three-band Kane model and is valid for III-V compounds, in general.

The III-V materials are also being described in accordance with the models of Stillman et. al., Newson and Kurobe, Rossler, Palik et. al., Johnson and Dickey and Agafonov et. al. respectively. We have investigated the S in QDs of III-V comounds in accordance with the all aforementioned band models for the purpose of relative assessment and thorough investigation after formulating the respective heavily doped electron energy spectrum for each model. The S in QDs of heavily doped II-VI materials and heavily doped GaP have been studied in accordance with the undoped models of Hopfield and Rees respectively.

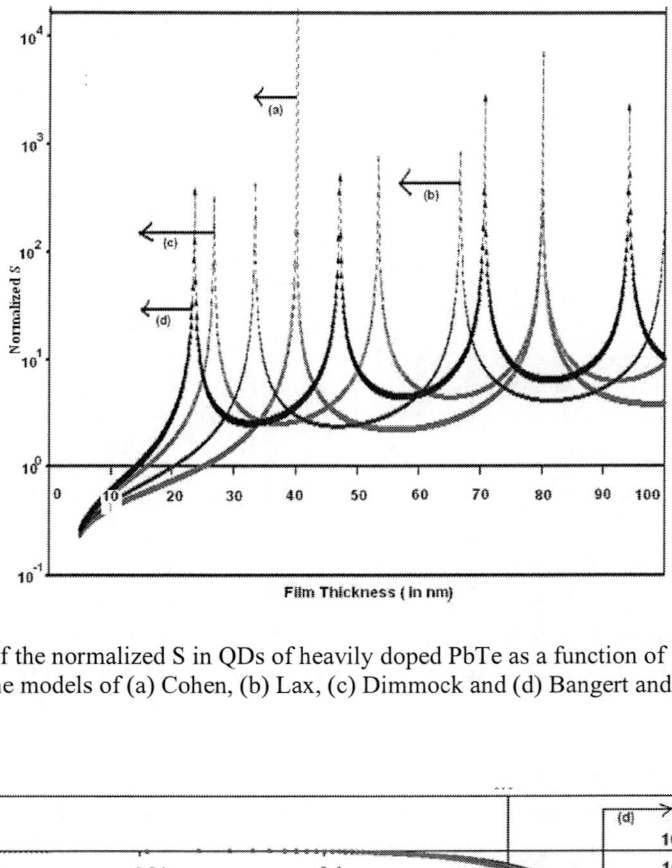

Figure 3.28. Plot of the normalized S in QDs of heavily doped PbTe as a function of film thickness in accordance with the models of (a) Cohen, (b) Lax, (c) Dimmock and (d) Bangert and Kastner. respectively.

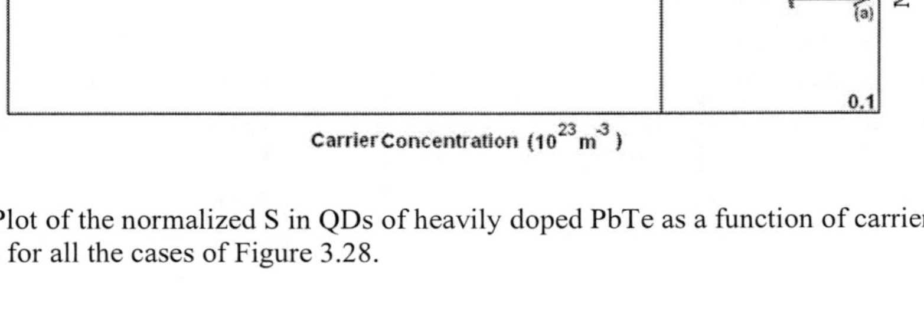

Figure 3.29. Plot of the normalized S in QDs of heavily doped PbTe as a function of carrier concentration for all the cases of Figure 3.28.

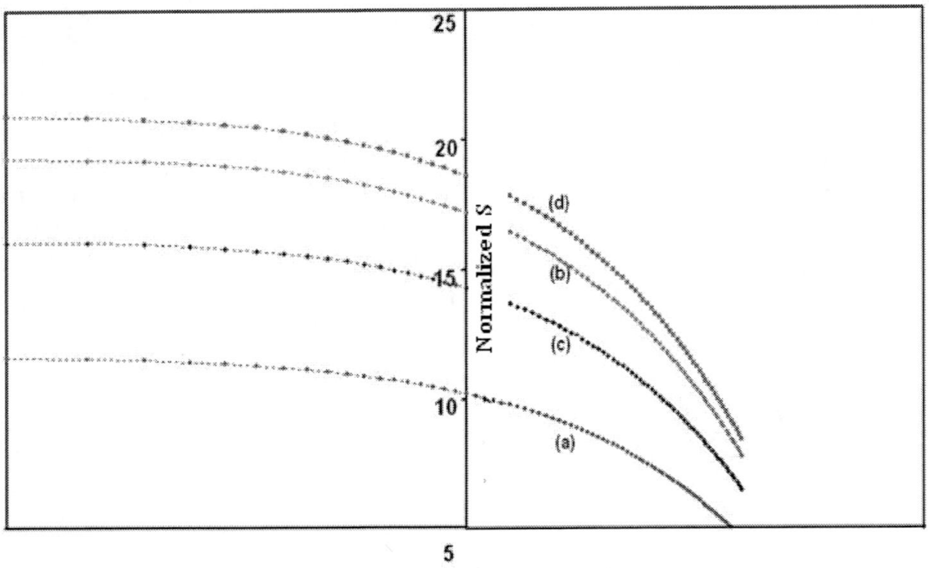

Figure 3.30. Plot of the normalized S in QDs of heavily doped PbTe as a function of carrier concentration in accordance with the model of Foley et. al.(three graphs (a), (b) and (c) are valid for three different temperature) The plot (d) refers to heavily doped Bi_2Te_3 in accordance with the model of Stordeur et. al.

Germanium is widely used since the inception of semiconductor science and S in QDs of Germanium has been presented in accordance with the newly formulated heavily doped models of Cardona et. al. and Wang and Ressler . Tellurium is an important elemental semiconductor and the S in QDs of heavily doped Tellurium has been investigated in accordance with the newly formulated heavily doped models of Bouat et. al.and Ortenberg and Button respectively In this chapter, investigations have been made of S in QDs of heavily doped $PtSb_2$ and heavily doped II-V materials in accordance with the models of Emtage and Yamada respectively. The S in QDs of GaSb has been studied by using the newly formulated heavily doped models of Seiler et. al., Mathur et. al. and Zhang respectively. The stressed compounds find extensive applications in piezoelectric systems and the S in QDs of such materials has been studied by using the model of Seiler et. al.. The S in QDs of IV-VI materials have been presented by using the models of Dimmock, Bangert and Kastner and Foley et. al. respectively. Finally, the S in QDs of heavily doped Lead Germanium Telluride, Zinc and Cadmium diphosphides and Bi_2Te_3 in accordance with the appropriate heavily doped version of the dispersion relations of the undoped models of Vassilev, Chuiko and Stordeur et. al. has been presented.

The experimental results for the verification of the theoretical analyses of this chapter are still not available in the literature. It is worth noting that our generalized formulation would be useful to analyze the experimental results when they materialize.

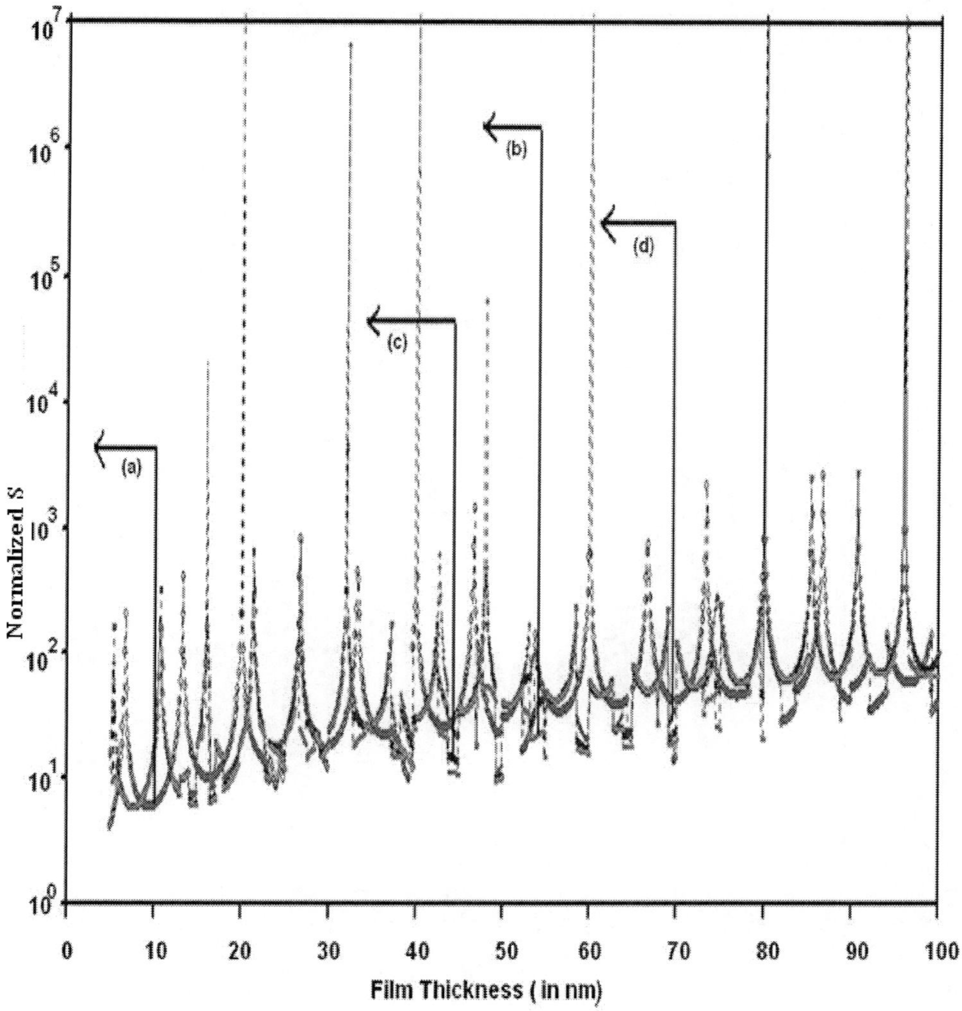

Figure 3.31. Plot of the normalized S in QDs of heavily doped PbTe and Bi_2Te_3 as a function of film thickness for all the cases of Figure 3.30.

Using the band constants from Table 1.1, the normalized S in this case in heavily doped $HgTe/Hg_{1-x}Cd_xTe$, CdS/ZnS, PbSe/PbTe and HgTe/CdTe quantum dot superlattices with graded interfaces have been plotted as a function of film thickness as shown by curves (a), (b), (c) and (d) respectively in Figure 3.33. The Figure 3.34 demonstrates the normalized S for the said quantized structures as a function of impurity concentration. The Figure 3.35 exhibits the normalized S as a function of film thickness in $HgTe/Hg_{1-x}Cd_xTe$,CdS/ZnSe, PbSe/PbTe and HgTe/CdTe heavily doped quantum dot effective mass superlattices as shown by curves (a), (b), (c) and (d) respectively. The normalized S in $HgTe/Hg_{1-x}Cd_xTe$, CdS/ZnSe, PbSe/PbTe and HgTe/CdTe heavily doped quantum dot effective mass superlattices has been plotted as a function of electron concentration in Figure 3.36.

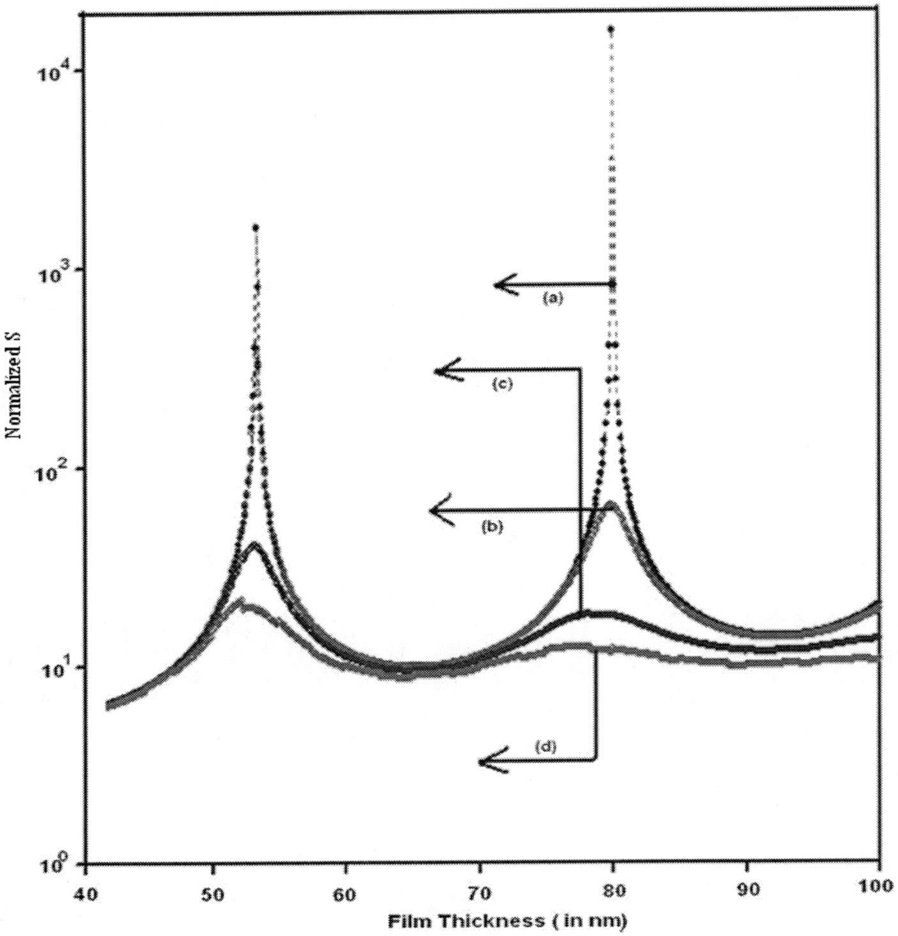

Figure 3.32. Plot of the normalized S in QDs of heavily doped (a) II-V compound (CdSb), (b) zinc diphosphide, (c) cadmium diphosphide and the plot (d) is valid for another temperature for cadmium diphosphide as a function of film thickness in accordance with the models of Yamada and Chuiko respectively.

It appears from Figure 3.33 that the S in HgTe/$Hg_{1-x}Cd_xTe$, CdS/ZnSe, PbSe/PbTe and HgTe/CdTe heavily doped quantum dot superlattices with graded interfaces increases with increasing film thickness exhibiting quantum jumps for fixed values of film thickness depending on the values of the energy band constants of the particular quantized structures. It is observed from Figure 3.4 that the S in quantum dots of aforementioned heavily doped superlattices decreases with increasing carrier degeneracy and differ widely for large values of same whereas for relatively small values of electron concentration, the S exhibits a converging behavior. From Figure 3.35, it is observed that the S in HgTe/$Hg_{1-x}Cd_xTe$, CdS/ZnSe, PbSe/PbTe and HgTe/CdTe quantum dot effective mass heavily doped superlattices oscillates with increasing film thickness. From Figure 3.36, it appears that the S of the aforementioned superlattices decreases with increasing concentration. It should be noted that all types of variations of S with respect to thickness and concentration are basically band structure dependent.

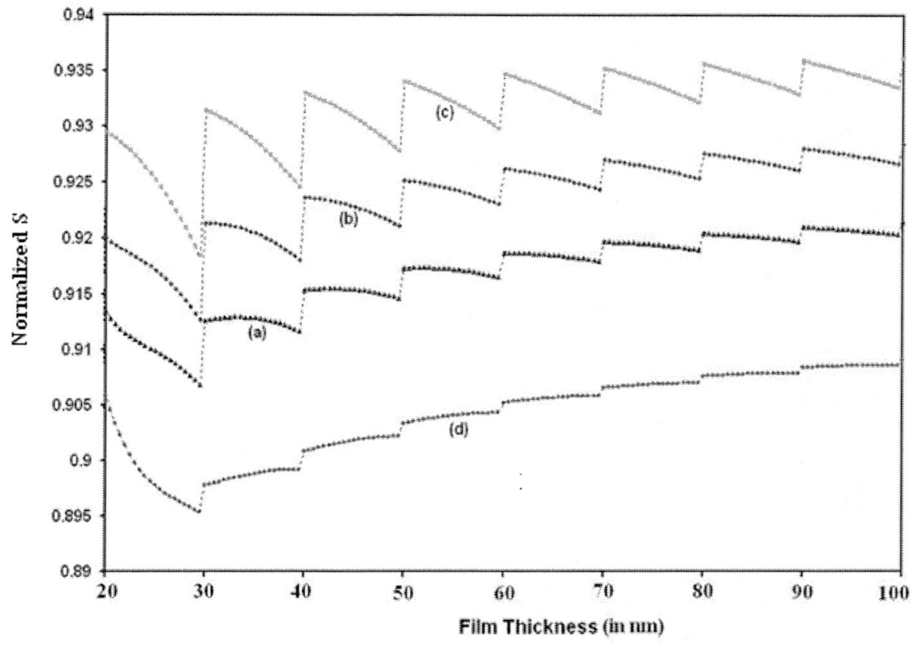

Figure 3.33. Plot of the S in heavily doped (a) $HgTe/Hg_{1-x}Cd_xTe$, (b) CdS/ZnSe, (c) PbSe/PbTe and (d) HgTe/CdTe quantum dot superlattices with graded interfaces as a function of film thickness.

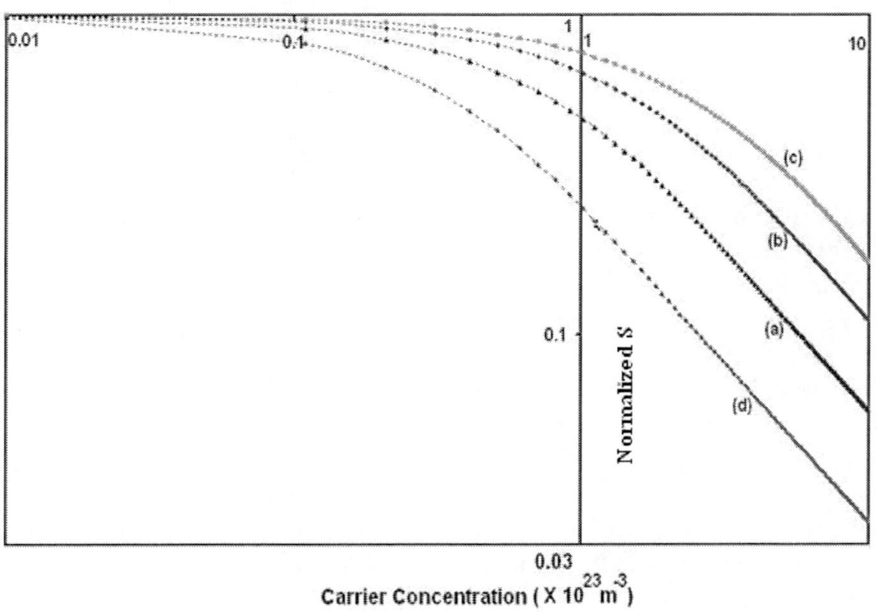

Figure 3.34. Plot of the S in heavily doped (a) $HgTe/Hg_{1-x}Cd_xTe$, (b) CdS/ZnSe, (c) PbSe/PbTe and (d) HgTe/CdTe quantum dot superlattices with graded interfaces as a function of carrier concentration.

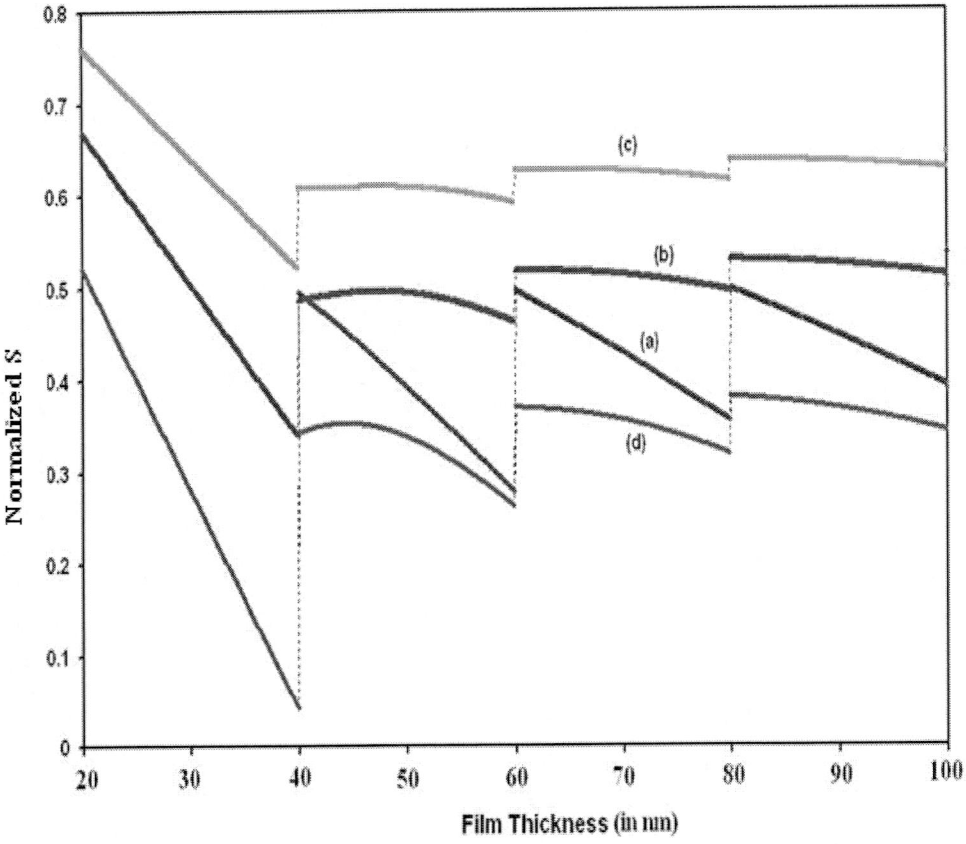

Figure 3.35. Plot of the S in heavily doped (a) HgTe/$Hg_{1-x}Cd_xTe$, (b) CdS/ZnSe, (c) PbSe/PbTe and (d) HgTe/CdTe quantum dot effective mass superlattices as a function of film thickness.

It may further be noted that the S of a two dimensional electron gas in the presence of a periodic potential has already been formulated in the literature. The SL is a three dimensional system under periodic potential. There is a radical difference in the dispersion relations of the 3D quantized structures and the corresponding carrier energy spectra of the 2D systems. From the dispersion relations of various heavily doped superlattices and heavily doped quantum dots as discussed in this chapter, the energy spectra of the various other types of low-dimensional systems can be formulated and the corresponding SLs can also be investigated. The results will be fundamentally different in all cases due to system asymmetry together with the change in the respective wave functions exhibiting new physical features in the respective cases[55]. Therefore, it appears that the dispersion law and the corresponding wave function play a cardinal role in formulating any electronic property of any electronic material, since they change in a fundamental way in the presence of dimension reduction. Consequently, the derivations and the respective physical interpretations of the different transport quantities change radically.

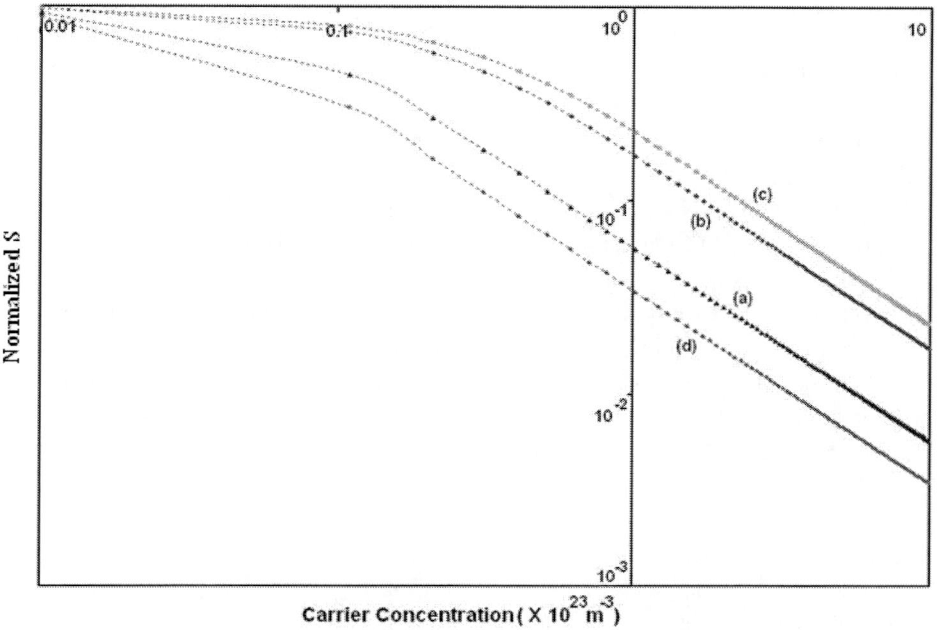

Figure 3.36. Plot of the S in heavily doped (a) HgTe/$Hg_{1-x}Cd_xTe$, (b) CdS/ZnSe, (c) PbSe/PbTe and (d) HgTe/CdTe quantum dot effective mass superlattices as a function of carrier concentration.

It is imperative to state that our investigations excludes the many-body, hot electron, spin, broadening and the allied quantum dot and SL effects in this simplified theoretical formalism due to the absence of proper analytical techniques for including them for the generalized systems as considered here. Our simplified approach will be appropriate for the purpose of comparison when the methods of tackling the formidable problems after inclusion of the said effects for the generalized systems emerge. The inclusion of the said effects would certainly increase the accuracy of the results although the qualitative features of the S would not change in the presence of the aforementioned effects. It may be noted that the basic aim of this chapter is not solely to demonstrate the influence of quantum confinement on the S for a wide class of quantized materials but also to formulate the appropriate carrier statistics in the most generalized form, since the transport and other phenomena in modern nanostructured devices having different band structures and the derivation of the expressions of many important carrier properties are based on the temperature-dependent carrier statistics in such materials.

ACKNOWLEDGMENTS

K.P. Ghatak and D. De are grateful to UGC for sanctioning the research project having the No-F 40-469/706 2011(SR) under which this paper has been completed. Besides, D. De together with K.P. Ghatak are grateful to DST, India for sanctioning the project under ''SERC Fast Track Proposal of Young Scientist'' Scheme SERC/ET-0213/2011.

REFERENCES

[1] K. V. Klitzing, G. Dorda, M. Pepper, *Phys. Rev. Lett.* 45, 494 (1980); K. v. Klitzing, *Rev. Mod. Phys.* 58, 519 (1986).

[2] J. Hajdu, G. Landwehr, In: *Strong and Ultrastrong Magnetic Fields and Their Applications,* F. Herlach (ed.), pp. 17 (Springer 1985).

[3] I. M. Tsidilkovskii, *Band Structure of Semiconductors* (Pergamon Press, Oxford, 1982).

[4] E. A. Arushanov, A. F. Knyazev, A. N. Natepov, S. I. Radautsan, *Sov. Phys. Semi.* 15, 828 (1981).

[5] S. P. Zelenim, A. S. Kondratev, A. E. Kuchma, *Sov. Phys. Semi.* 16, 355 (1982).

[6] F. M. Peeters, P. Vasilopoulos, *Phys. Rev. B* 46, 4667 (1992).

[7] W. Zawadzki, In: *Two-Dimensional Systems, Heterostructures and Superlattices,* Springer Series in Solid State Sciences, Vol. 53, edited by G. Bauer, F. Kuchar, H. Heinrich (Springer, Berlin, Heidelberg, 1984).

[8] B. M. Askerov, N. F. Gashimzcede, M. M. Panakhov, *Sov. Phys. Sol. State* 29, 465 (1987).

[9] G. P. Chuiko, *Sov. Phys. Semi.* 19, 1279 (1985).

[10] S. Pahari, S. Bhattacharya, K. P. Ghatak, *J. Comput. Theor. Nanosci.*, Invited review article, 2009 [In the press].

[11] K. P. Ghatak, S. Bhattacharya, S. Bhowmik, R. Benedictus, S. Choudhury, *J. Appl. Phys.* 103, 034303 (2008).

[12] K. P. Ghatak, S. Bhattacharya, S. Pahari, D. De, S. Ghosh, M. Mitra, *Annalen der Physik* 17, 195 (2008).

[13] K. P. Ghatak, S. N. Biswas, *J. Appl. Phys.* 70, 299 (1991).

[14] K. P. Ghatak, S. N. Biswas, *J. Low Temp. Phys.* 78, 219 (1990).

[15] K. P. Ghatak, M. Mondal, *J. Appl. Phys.* 65, 3480 (1989).

[16] K. P. Ghatak, B. Nag, *Nanostructured Materials* 5, 769 (1995).

[17] K. P. Ghatak, M. Mondal, *Phys. Stat. Sol. (b)* 185, K5 (1994).

[18] K. P. Ghatak, B. Mitra, *Il Nuovo Cimento D* 15, 97 (1993).

[19] K. P. Ghatak, S. N. Biswas, *Phys. Stat. Solidi (b)* 140, K107 (1987).

[20] K. P. Ghatak, *Il Nuovo Cimento D* 13, 1321 (1991).

[21] K. P. Ghatak, A. Ghoshal, *Phys. Stat. Sol. (b)* 170, K27 (1992).

[22] K. P. Ghatak, B. De, B. Nag, P. K. Chakraborty, *Molecular Crystals and Liquid Crystals Science and Technology Section B: Nonlinear Optics* 16, 221 (1996).

[23] K. P. Ghatak, M. Mitra, B. Goswami, B. Nag, *Molecular Crystals and Liquid Crystals Science and Technology Section B: Nonlinear Optics* 16, 167 (1996).

[24] K. P. Ghatak, D. K. Basu, D. Basu, B. Nag, Nuovo Cimento della Societa Italiana di Fisica D - Condensed Matter, Atomic, Molecular and Chemical Physics, Biophysics 18, 947 (1996).

[25] B. Mitra, K. P. Ghatak, *Phys. Letts. A* 141, 81 (1989); S. K. Biswas, A. R. Ghatak, A. Neogi, A. Sharma, S. Bhattacharya, K. P. Ghatak, *Physica E: Low-Dimensional Systems and Nanostructures* 36, 163 (2007).

[26] M. Mondal, A. Ghoshal, K. P. Ghatak, *Il Nuovo Cimento D* 14, 63 (1992); K. P. Ghatak, M. Mondal *Phys. Stat. Sol. (b)*, 135, 819 (1986); L. J. Singh, S. Choudhury, D. Baruah, S. K. Biswas, S. Pahari, K. P. Ghatak, *Phys. B* 368, 188 (2005).
[27] K. P. Ghatak, B. De, M. Mondal, S. N. Biswas, *Materials Research Society Symposium – Proceedings* 198, 327 (1990); K. P. Ghatak, *Proceedings of SPIE - The International Society for Optical Engineering* 1584, 435 (1992).
[28] K. P. Ghatak, B. De, *Materials Research Society Symposium – Proceedings* 234, 55 and 59 (1991).
[29] K. P. Ghatak, B. De, M. Mondal, S. N. Biswas, *Materials Research Society Symposium – Proceedings* 184, 261 (1990).
[30] K. P. Ghatak, *Influence of band structure on some quantum processes in tetragonal semiconductors*, D. Eng. Thesis, Jadavpur University, Kolkata, India, 1991.
[31] K. P. Ghatak, S. N. Biswas, *Materials Research Society Symposium – Proceedings* 216, 465 (1990).
[32] T. M. Tritt, M. Kanatzidis, G. Mahan, H. B. Lyon Jr. (Eds.), New Materials for Small Scale Thermoelectric Refrigeration and Power Generation Applications, Vol. 545, *Proceedings of the 1998 Materials Research Society.*
[33] C. Yu, Y. S. Kim, D. Kim, J. C. Grunlan, *Nano Lett. 8*, 4428 (2008).
[34] R. K. Willardson and A. C. Beer, *Semiconductors and Semimetals* (eds.) (New York: Academic) vol 1, p 102, 1966.
[35] E. O. Kane, *Phys. Rev.* 131, 79, 1963; Phys. Rev. B., 139, 343, 1965.
[36] V. L. Bonch Bruevich, *Sov. Phys.–Solid State,* 4, 1953, 1963.
[37] R. A. Logan and A. G. Chynoweth, *Phys. Rev.,* 131 89, 1963.
[38] C. J. Hwang, *J. Appl. Phys.* 40 3731, 1969.
[39] J. I. Pankove, *Phys. Rev.* A, 130 2059, 1965.
[40] B. I. Halperin and M. Lax, *Phys. Rev.,* 148, 722, 1966.
[41] R.A. Abram, G.J. Rees and B. L. H. Wilson, Adv. Phys. 27, 799, 1978.
[42] B.I. Shklovskii and A. L. Efros, *Electronics Properties of Doped Semiconductors* vol 45 (Berlin: Springer), 1984.
[43] E. O. Kane, *Solid State Electron.,* 28 3, 1985.
[44] P. K. Chakraborty and J. C. Biswas, *J. Appl. Phys.* 82 3328, 1997.
[45] B.R. Nag, *Electron Transport in Compound Semiconductors* (New York: Springer), 1980.
[46] P. E. Schmid, *Phys. Rev.* B, 23, 5531, 1981.
[47] Jr. G. E. Jellison, F. A. Modine, C. W. White, R. F. Wood and R. T. Young, *Phys. Rev. Lett.,* 46 1414, 1981.
[48] V.I. Fistul, *Heavily Doped Semiconductors* (New York: Plenum) ch 7, 1969.
[49] C. J. Hwang, *J. Appl. Phys.* 41 2668, 1970.
[50] W. Sritrakool, H. R. Glyde and V. *Sa Yakanit Can. J. Phys.,* 60, 373, 1982.
[51] H. Ikoma, J. *Phys. Soc. Japan,* 27, 514, 1969.
[52] P. K. Chakraborty, A. Sinha, S. Bhattacharya and K.P. Ghatak, *Physica B,* 390, 325, (2007); K.P. Ghatak, G. C. Datta and K.P. Ghatak, *Physica Scripta,* 68, 368, (2003); P.K. Chakraborty and K.P. Ghatak, J. Phys. Chem. Solids 62, 1061, (2001); Phys. Letts. A, 288, 335, (2001); *Phys. D, Appl. Phys.* 32, 2438 (1999).

[53] M. Abramowitz and I. A. Stegun, *Handbook of Mathematical Functions with Formulas, Graphs and Mathematical Tables* (New York: Wiley), 1964.

[54] I. S. Gradshteyn and I. M. Ryzhik, *Tables of Integrals, Series and Products* (New York: Academic), 1965.

[55] V. Heine, Proc. Phys. Soc. 81, 300, (1963); J. N. Schulman and Y. C. Chang, *Phys. R*ev. B, 24, 4445 (1981).

In: Quantum Dots and Quantum Cellular Automata
Editors: D. De, S. Bhattacharaya and K.P. Ghatak
ISBN: 978-1-62257-920-4

Chapter 4

TRANSISTOR LASERS WITH QUANTUM WELLS AND QUANTUM DOTS - IN -WELL IN THE BASE

***Rikmantra Basu*[1,2*]**
[1]Institute of Radio Physics and Electronics, University of Calcutta
Kolkata, India
[2] Centre for Research in Nanoscience and Nanotechnology,
University of Calcutta, Kolkata, India

ABSTRACT

Transistor Laser is a Heterojunction Bipolar Transistor (HBT) the base of which usually incorporates a Quantum Well (QW). The stimulated recombination in the QW gives rise to laser action. It can thus work both as a transistor and a laser. This paper considers the structures in which a layer of Quantum Dot (QD) is inserted in the QW (Dots-in-Well) in the base and examines theoretically if there is improvement in the performance over the usual QW structure. To put the study and comparison in proper perspectives, the usual TL structure with QW in its base (QW-TL) and the basic operation of the device are discussed first. The analytical models for terminal currents, light power output and small signal modulation characteristics for QW-TL are then presented. Similar theoretical models for TL with Dots-in-Well (QD-TL) are then developed. The model for QW-TL is tested against experimental results and good agreement obtained signifies the validity of the model. The theoretical value for threshold base current for QD-TL is found to be one order lower than the value for QW-TL. The small signal modulation bandwidth for QD-TL cannot however exceed the corresponding values for QW-TL. The analysis for QD-TL is based on excitonic models with and without excited states in the QD and a comparison is made between the results obtained from the two excitonic models. The role of separate electron-hole dynamics is also examined. The experimental values for QD-TL are not available however.

[*] E-mail : rikmantra@gmail.com

1. INTRODUCTION

Transistor Laser (TL), invented by Feng, Holonyak and coworkers in the University of Illinois, Urbana-Champaign, [1,2] is basically a Heterojunction Bipolar Transistor (HBT) in the base of which a Quantum Well (QW) is inserted. The emitter and collector terminals are used for electrical signal input and output, respectively. When the injected current into the base and subsequently into the QW is sufficient to create population inversion between the conduction and valence subbands in the QW, a coherent optical signal emerges out of the base region in a direction perpendicular to the current flow.

It appears from the above that integration of both electronic and photonic devices occurs within the same structure, i.e., the Transistor Laser. Integration of electronic devices with photonic devices was reported as early as in 1980 using a wire to integrate a transistor with a laser [3]. Subsequently reports were published about a laser transistor [4], a heterojunction field effect transistor laser [5] and a bipolar transistor that emitted light when cooled to liquid nitrogen temperature [6]. A Light Emitting Transistor (LET) based on real space transfer of hot electrons was also reported [7]. A GaAs/AlGaAs double heterostructure device functioning as a Bipolar Transistor and injection laser suitable for optoelectronic integrated circuits was reported [8]. However a device that can operate as a transistor delivering an electrical signal with gain, while at the same time giving light emission: an LET, was achieved first by Feng and Holonyak in 2004 [9] by using an HBT, and subsequently by inserting a QW in the base [10]. As mentioned already, inserting a Quantum Well (QW) in the base the group first achieved laser operation. [1,2]. As the name implies TL is a three-port light emitting device, in which the two ports are meant for electrical input and electrical output, while the third port gives optical output. As in a normal transistor, the carriers injected into the base of a TL suffer recombination as a result of which the characteristic spontaneous emission occurs. However, when the current injected into the base is sufficient to create population inversion, stimulated emission becomes dominant to give rise to amplification of light. If the medium is enclosed between a pair of Fabry Perot (FP) mirrors, coherent light comes out in the direction perpendicular to the flow of electric current. A schematic diagram of the TL is shown in Figure 1.

Since the first announcement, a number of publications have appeared in the literature describing several characteristics of TL as well as the device and systems applications [11-35]. These aspects will be covered up in a subsequent section.

Several authors have presented analytical models to predict or compare the theoretical values with the experimentally available results [36-43]. In most of the cases, these models rely on the solution of coupled rate equations for carrier and photon densities [44, 45]. A more detailed analysis by a group at the University of British Columbia (mentioned later as the UBC group) [36-40] involves solution of the continuity equation in the base and then relating the base current with laser characteristics by means of rate equations. The model developed by the UBC group has been refined by the present authors [42, 43] to be applicable to experimental conditions and the laser characteristics are obtained by using electron-photon interaction for two dimensional carriers in the QW. The model has successfully explained the available experimental data. An overview of the analytical models used by different groups and their potentials and limitations will be discussed in a later section.

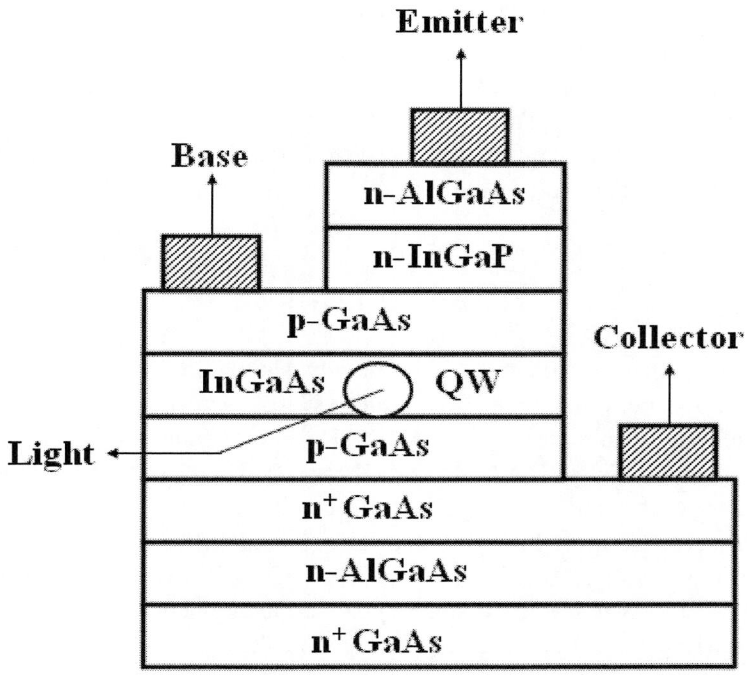

Figure 1. Basic structure of TL.

The aim of the present work is to investigate theoretically if the performance of the TL improves when the QW in the base is replaced by a layer of QDs within the QW (Dots-in the-Well). It is well known that threshold current in a QD laser is drastically reduced from the values for QW lasers [46- 49]. Also, the temperature sensitivity, and linewidth show significant improvement for QD lasers. Though small signal modulation bandwidth is expected to increase, the observed results do not corroborate it [46]. Our work is mainly concerned with the threshold current and modulation bandwidth for the QD-TL. The model used by us is an extension of our early work on QW-TL [42, 43] which successfully explained the experimental results. The model for QW-TL will be presented to prepare the groundwork for the present model. Also, since the device is quite new but holds the potential for application in several electronic: both analog and digital systems as well as in photonics including fibre optic communication, it is felt that a description of the structure and an introduction of the working principle of the device will be useful for uninitiated readers.

The present paper first describes the basic structures used and studied mainly by Feng, Holonyak and coworkers [2, 11, 12, 20, 29]. The principle of operation of the device is then explained. The analytical models to QW-TL available in the literature will then be presented in brief, followed by a detailed description of our own model. The results obtained with this model will then be presented and compared with experimental data. The work on QD-TL following this model will then be presented followed by the numerical values of the threshold base current and modulation bandwidth. A comparison with the values for QW-TL will come next. Finally the role of some intrinsic physical parameters for the QDs in limiting the bandwidth will be discussed.

The organization of the paper is as follows. In Sec 2 we describe the structure of QW-TL. A brief review of the experimental work done and important results obtained are given in Sec.

3. The theoretical models for QW-TL available in the literature are briefly discussed in Sec. 4 pointing out the limitations. Sec. 5 presents our own theoretical model for QW-TL and makes a comparison between our calculated values of threshold current and light power output with the available experimental data. This model is then extended for QD-TL in Sec. 6, in which the structure, intrinsic physical processes and calculated values of threshold base current, modulation bandwidths etc are presented. Sec. 7 gives the summary and conclusions.

2. STRUCTURE OF TL

A typical layer structure of the TL is already shown in Figure 1. For the sake of illustration we show the structure and material systems used in most of the experimental work. The higher band gap material, N type InGaP forms the emitter (E) layer of the HBT, while the lower gap material p-type GaAs is the base (B). The n-type GaAs layer beneath the p-GaAs base acts as the collector (C). Embedded within the p-GaAs base is an InGaAs Quantum Well: the active layer for light emission. The simplified band diagram of the HBT with the QW in the base is shown in Figure 2. The band gap difference between the InGaP emitter and the p-GaAs base is apportioned as conduction and valence band discontinuities in the Emitter-Base (EB) heterojunction. The thickness of the QW *(d)* in the base is very small compared with the base width (W_B).

The base region is enclosed between two plane parallel mirrors forming a Fabry Perot resonator. In most experiments the cleaved facets along the crystallographic planes act as the mirrors. The FP resonator provides the usual feedback for the lightwaves generated by stimulated recombination of electron-hole pairs in the QW. In a recent work a Distributed Feedback (DFB) structure is used in place of the FP resonator [32, 33].

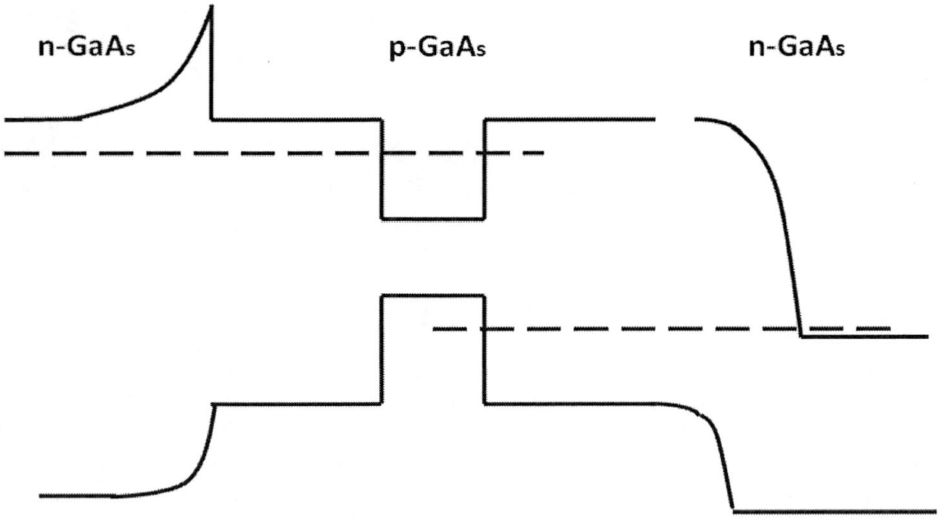

Figure 2. Band diagram of the HBT structure using a QW in base.

3. EXPERIMENTAL REPORTS

To date almost all the experimental works related to QW-TL are due to the group led by Feng and Holonyak. In this section we briefly review the work done and important results obtained.

As mentioned already, the group first reported Light Emitting Transistors by first using a HBT and then inserting a QW in the base. The laser action was reported soon afterwards [2]. In most of the samples used InGaP acts as the emitter, GaAs as the base in which an InGaAs QW is inserted and a GaAs layer is used as the collector. The room temperature continuous wave operation was first reported in 2005 [12]. The microwave operation and modulation of TL were considered [13]. The noteworthy feature of the studies is that as the threshold base current is exceeded the collector current (I_C) -collector emitter voltage (V_{CE}) curves become closer for the same increment of base current (I_B). This means that the dc current gain decreases once the threshold base current is reached. The ac current gain is also found to decrease [21]. In a number of papers the authors have also examined the small signal modulation response and provided guidelines for the enhancement of the bandwidth [16,17,22,25, 27-29]. The response shows the usual resonance observed for diode lasers [12]. It was shown that by using a tilted charge distribution in the base, the recombination lifetime can be drastically reduced and resonance free characteristics of TL and a diode laser can be achieved [29]. This proves to be useful in extending the bandwidth and bit rate of the TL. The authors have recently reported simultaneous electrical and optical output at 20 Gb/s and 40 Gb/s data rate modulation [35].

The authors also studied different material combinations [11, 18] to achieve LET or laser action at different wavelengths [13, 18, 23, 31-34] including a type II heterojunction [11]. Various application areas like electrical and optical adding, mixing and processing of square-wave signals [16], switching [19], frequency multiplication and signal mixing [25] etc. have been investigated. Franz-Keldysh photon assisted voltage controlled switching has also been achieved [23].

An equivalent circuit model has also been developed by the group [30, 31] to delineate the electrical and optical characteristics of the TL. Single mode operation using a DFB structure in place of FP resonator has also been investigated [32, 33]. Incorporation of C in the QW leads to emission at 1600 nm [34].

4. MODELS FOR QW-TL

4.1. Principle of Operation

Our studies consider an HBT consisting of *N*-InGaP/ *p*-GaAs/ *n*-GaAs layers, in which an $In_{0.2}Ga_{0.8}As$ QW is inserted in the base. The transistor operates in its normal, active mode, i.e. base-emitter junction is forward biased and the base-collector junction is reverse-biased.

In the one dimensional model the EB junction is at $x = 0$ and the BC junction is at W_B, while the quantum-well (QW) is assumed to be situated at $x = x_1$. Figure 3 shows the conduction band edge in the base and the dc excess minority carrier distribution [$\delta N(x)$] in the base region [21]. Under forward bias, the carriers injected from the emitter diffuse across the base and reach the QW. These unbounded carriers (three dimensional: 3D) may either undergo quantum capture to the bound states in the QW with a lifetime of $\tau_{cap,}$ or diffuse to the collector-base junction where they are swept out by the reverse bias. The carriers captured by the QW may escape out of it with a lifetime τ_{esc}. The 3D carriers at $x = x_1$ are located at the virtual bound states.

These states are localized at the QW, but occupy energies higher than the conduction band edge of the barrier material, and aid in the conversion of carriers from the 3-D states (n_{vs}) above the well to the 2-D states within the QW (n_{QW}) and vice versa [38,39].

The QW adds another recombination channel in the base. The recombination rate is limited by spontaneous emission below laser threshold and mainly by stimulated emission above laser threshold. The base current as a consequence consists of two components: (1) the regular base region recombination current, which is characterized by a base region recombination lifetime (τ_B) and a diffusion coefficient (D_n) and (2) the current needed to drive the laser.

Lengthwise the base of the transistor also forms a resonant cavity in which two cleaved sides of the device act as mirrors. Because of the electron recombination due to spontaneous and stimulated emission, the QW plays an important role in determining the base current. The electronic characteristics of the HBT deviate from the normal behaviour. This is evidenced by the reduction of the electrical gain, due to enhanced carrier recombination in the base region when stimulated emission occurs [13, 21].

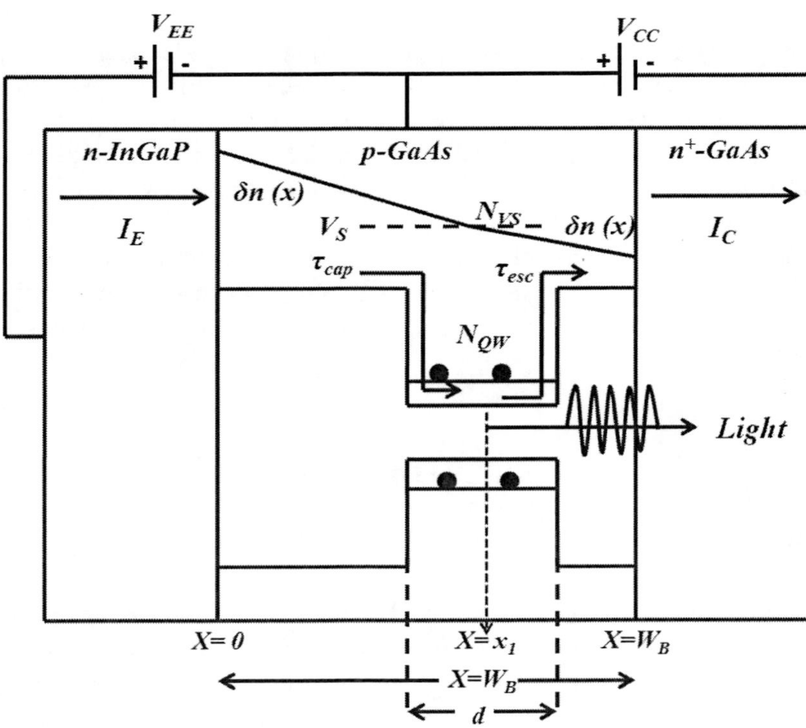

Figure 3. Schematic of carrier diffusion and quantum capture in the QW and the conduction band energy diagram of the base region.

4.2. Review of Analytical Models for QW-TL

We first give an overview of the various models presented so far to explain the characteristics of TL with a QW in the base.

The group led by Feng and Holonyak reported analytical results on the charge distribution in the base, modulation bandwidth of the TL [21] and other parameters related to it by using experimental I-V characteristics and charge control model for transistors. The optical characteristics are obtained by them using the rate equation model [44, 45]. Analytical modeling of the TL has been undertaken recently by two different groups. UBC group solved the continuity equation in the base in presence of diffusion and recombination [36-39] to obtain the charge distribution. The recombination in the QW placed in the base however occurs between quantized energy levels in the QW. They therefore used the concept of virtual states (VS) introduced for QW lasers to relate the bulk carriers with the two dimensional (2D) carriers. They also considered the capture of carriers by the QW and their escape out of it. Their expressions for the terminal currents therefore differ from the equations for normal transistors and contain the carrier and current densities related to the VS. The optical parameters are obtained from the rate equations and the value of differential gain is plugged in for numerical estimates of the threshold current, modulation bandwidth and other parameters. The structure considered by the UBC group is a Vertical Cavity Surface Emitting Laser (VCSEL). A design of MQW TL to work at 1550 nm has been presented by this group by using commercial softwares [40].

Zhang and Leburton [41], the other group proposing analytical model, relied on rate equations for both the carriers and photons. To obtain the rate equation for the carrier numbers in the QW, they introduced a capture coefficient expressed in terms of the thicknesses of the base and the QW and capture lifetime by the QW. By solving the rate equations the authors obtained the expressions for ac and transient response of the TL.

In all the theoretical works mentioned above the optical characteristics are obtained from the rate equations and the value for the differential gain available in the literature is invoked to obtain numerical values of different quantities. This method therefore cannot work for changes in several parameters, like QW thickness, QW and base materials and their combination leading to different degrees of built-in strain, altered subband energies, etc. In addition, in the work of the UBC group the QW is assumed to be placed in the middle of the base [39]. In actual experiments on the other hand, QWs are placed nearer to the BC junction [10, 12, 13, 21, 30]. Further, the charge distribution obtained in the work by the UBC group is triangular, and does not show a break at the QW as noted by the group led by Feng and Holonyak [13, 21].

In view of the above discussions, we feel motivated to formulate a more general analytical model for TL to predict its electrical and optical characteristics. The modified expressions for the terminal currents are first obtained for an arbitrary position of the QW in the base. The gain in the QW is obtained by considering subband energies and envelope functions in presence of strain, 2D density-of-states function, polarization dependent momentum matrix element, Fermi distribution function, and lineshape function. Although simplified E-**k** relationship is assumed, the theory can be modified to include results of more refined band structure calculations. The calculated values of the threshold base current and the light power output for an InGaAs QW embedded in the GaAs base agree satisfactorily with the experimental results. The charge carrier distribution profile is also consistent with the results provided by Feng, Holonyak and their coworkers.

To complete this section mention may be made of a recent study of large signal analysis of QW-TL. The work establishes theoretically wider opening of eye diagram. However, the rate equations need be solved by numerical techniques [50].

5. PRESENT MODEL

5.1. Terminal Currents

The minority carrier transport in the base may be described by drift-diffusion equation, provided the base width $W_B >> D/v_T$, where D is the diffusion coefficient and v_T is the thermal velocity [43]. Taking $m_e = 0.063\ m_0$ and $D = 26\ cm^2/s$, the length D/v_T is 5.6 nm, which is very small indeed compared to the base width ~ 100 nm in the present analysis. Further, under low level injection from the emitter into the base by thermionic emission and tunneling, the carriers occupy energy levels very close to the conduction band edge, so that they thermalize. In addition, the electric field in the base being very small, the transport is dominated by diffusion.

The time independent continuity equation for diffusion dominated transport in the base

$$\frac{d^2\delta n}{dx^2} = \frac{\delta n}{D_n \tau_B} = \frac{\delta n}{L_D{}^2} \tag{1}$$

is solved with the boundary condition and $\delta n(W_B) = 0$.

Here is the excess electron density, D_n is the electron diffusion coefficient, is the base recombination time and is the diffusion length. The dc carrier concentrations and δN_2, respectively before and after the QW, can be expressed as

$$\delta N_1 = \left[\frac{N_{V.S} - (L_D / qD_n) J_E . e^{-x_1/L_D}}{2\cosh(x_1 / L_D)}\right] e^{x/L_D} + \tag{2}$$

$$\delta N_2 = \frac{-N_{V.S} e^{-W_B/L_D}}{2\cosh(W_B - x_1/L_D)} e^{x/L_D} + \frac{N_{V.S} e^{W_B/L_D}}{2\cosh(W_B - x_1/L_D)} e^{-x/L_D} \tag{3}$$

The virtual state current density is

$$J_{vs} = qD_n \left[\frac{d}{dx}(\delta N_1)_{x=x_1} - \frac{d}{dx}(\delta N_2)_{x=x_1}\right] \tag{4}$$

From the above two equations, the terminal currents may be expressed as

$$I_E = BN_{VS}[\sinh(z_1) - \coth(z_2)\cosh(z_1)] - AJ_{VS}\cosh(z_1) \quad (5)$$

$$I_C = BN_{VS}[\tanh(z_1) - \cosh(z_1)\coth(z_2)] - AJ_{VS}\cosh(z_1) \quad (6)$$

$$I_B = BN_{VS}[\sinh(z_1) + \coth(z_1) - 2\cosh(z_1)\coth(z_2)] - 2AJ_{VS}\cosh(z_1) \quad (7)$$

where

$$B = AqD_n / L_D; z_1 = x_1 / L_D; z_2 = (W_B - x_1)/L_D$$

A is the area of the base and q is the electronic charge. The connections between the carrier densities and current densities related to the QW and the VS are made through the relations [39]:

$$\frac{J_{vs}}{qd} = \frac{J_{QW}}{qd} - \frac{N_{VS}}{\tau_S} \quad (8a)$$

and $$\frac{J_{QW}}{qd} = \frac{N_{VS}}{\tau_{cap}} - \frac{N_{QW}}{\tau_{esc}} \quad (8b)$$

In above, τ_s, τ_{cap} and τ_{esc} denote, respectively, the spontaneous recombination lifetime, capture time of electrons into the QW and the escape time of electrons from the QW to the VS. The first and second terms in Eq. (7) for the base current represent, respectively, the usual recombination current and the laser current. The recombination current is dominant below laser threshold ($J_B < J_{B,th}$) because there is no stimulated emission in the QW. Above laser threshold, the laser current is much larger than the recombination current.

5.2. Optical Characteristics

5.2.1. Strain Effect on the Band-Edge Energies

The insertion of the $In_{1-x}Ga_xAs$ QW in the undoped GaAs base leads to strain in the QW. The in-plane component of the built-in strain is given by [46, 47]

$$\varepsilon = \varepsilon_{xx} = \varepsilon_{yy} = \frac{a_{GaAs} - a(x)}{a_{GaAs}}, \quad (9)$$

where a_{GaAs} and are the lattice constants of the GaAs substrate and $In_{1-x}Ga_xAs$ QW respectively. The strain in the perpendicular direction is

$$\varepsilon_{\perp} = \varepsilon_{zz} = -2\frac{C_{12}}{C_{11}}\varepsilon, \tag{10}$$

where and are the elastic constants.

Figure 4 shows the schematic band diagram for the strained QW. The conduction band edge, the valence band edge and the band gap of the unstrained alloy are denoted respectively by and $E_g(x)$, while δE_C and δE_{HH} are, respectively, the strain induced shifts of the conduction (C) and heavy hole (HH) band edges. The energies are measured by considering $E_V = 0$. The symbols and stand for energies of the *n*-th conduction subband and *m*-th hole subband, respectively, and *U* is the band offset in the conduction band.

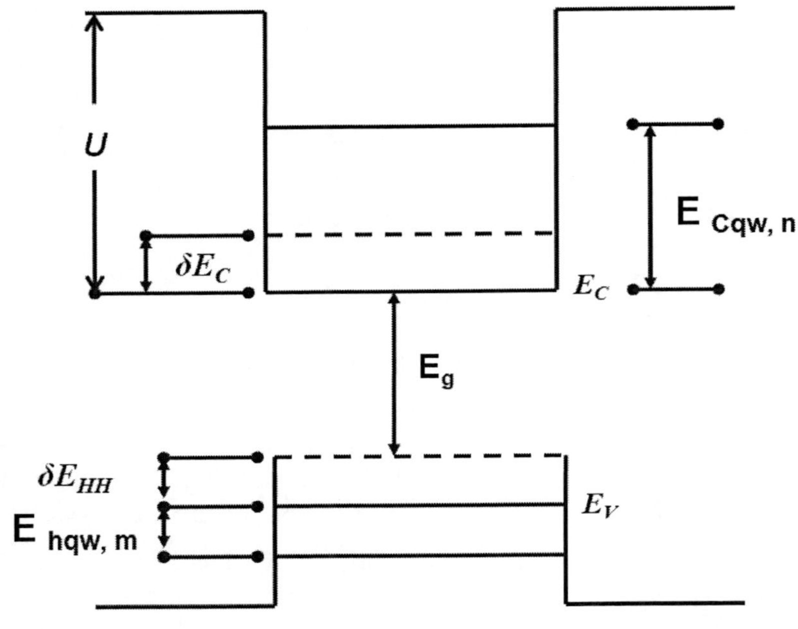

Figure 4. Band structure of the model.

For a strained $In_{1-x}Ga_xAs$ layer, the conduction band edge is shifted by

$$\delta E_C = a_c\left(\varepsilon_{xx} + \varepsilon_{yy} + \varepsilon_{zz}\right) = 2a_c\left(1 - \frac{C_{12}}{C_{11}}\right)\varepsilon \tag{11}$$

and the valence subbands are shifted by

$$\delta E_{HH} = -P - Q, \tag{12}$$

where

$$P = -a_v\left(\varepsilon_{xx} + \varepsilon_{yy} + \varepsilon_{zz}\right) = -2a_v\left(1 - \frac{C_{12}}{C_{11}}\right)\varepsilon \qquad (13)$$

$$Q = -\frac{b}{2}\left(\varepsilon_{xx} + \varepsilon_{yy} - 2\varepsilon_{zz}\right) = -b\left(1 + \frac{2C_{12}}{C_{11}}\right)\varepsilon \qquad (14)$$

In above, a_c and a_v are hydrostatic deformation potential for conduction band and valence band respectively and is the shear deformation potential. The new band edges are therefore

$$E_C = E_g(x) + \delta E_C \qquad (15)$$

$$E_{HH} = \delta E_{HH} \text{ and } E_{LH} = \delta E_{LH} \qquad (16)$$

For compressively strained layer, heavy-hole band is the top-most valence band. So the effective band gap will be

$$E_{c-HH} = E_g(x) + \delta E_C - \delta E_{HH} \qquad (17)$$

The quantum levels for the conduction and valence bands are given by

$$E_{c,n} = E_{cqw,n} + \delta E_C \qquad (18a)$$

and $E_{h,m} = E_{hqw,m} + \delta E_{bH}$, (18b)

In calculating $E_{cqw,n}$ and the finite heights of the barriers (U in Figure 2) are duly considered. The values of and δE_{bH} are calculated following equations (10, 11) considering the different types of strain.

In this case, the effective band gap should be modified as

$$E_{h,m}^{c,n} = E_g(x) + E_{c,n} - E_{h,m} \qquad (19)$$

5.2.2. Gain Spectrum

The optical gain analysis in the QW is performed by using the technique described in [46, 47, 51, 52]. It is assumed that the quantization occurs along x-direction and the wave function for electrons is expressed as where $\mathbf{k}_t$ and $\mathbf{r}_t$, are respectively the 2D wave vector and the position vector (in transverse direction) and is the envelope function for x-direction for n-th subband. A similar equation is written for 2D holes. The momentum is conserved for motion along the 2D plane (*k* conservation, which gives rise to the appearance of reduced mass m_r in the energy conservation process [46]. It is also assumed that the subbands are broadened due to different scattering processes. The expression for gain coefficient is as follows

$$g(\hbar\omega) = C_0 \sum_{n,m} \left| I_{h,m}^{c,n} \right|^2 \int_0^{\infty} dE_t \rho_r^{2D} \left| \hat{e} . p_{cv} \right|^2 \times$$

$$\times \frac{\gamma/(2\pi)}{\left(E_{h,m}^{c,n}(\mathbf{k}_t) - \hbar\omega\right)^2 + (\gamma/2)^2} \left[f_c^n(E_t) - f_v^m(E_t) \right] \tag{20}$$

where $C_0 = \dfrac{\pi q^2}{n_r c \varepsilon_0 m_0^2 \omega}$, $\rho_r^{2D} = \dfrac{m_r}{\pi \hbar^2 d}$,

ρ_r^{2D} in above is the reduced density-of-states function for 2D for quantized motion along x-direction, d being the width of the QW. *I* stands for the overlap of envelope functions for electron (in the n-th subband) and hole (in the m-th subband), is the photon energy, is the momentum matrix element for QWs and is the finite line width of the spectrum. In Eq. (20) c is the speed of light in free space, is the permittivity of free space, n_r is the background refractive index and m_0 is the free electron mass. The kinetic energy for 2D free motion, E_t, is considered to be zero for the band edge transition. The Fermi occupation probabilities are expressed in terms of quasi Fermi levels and F_v as

$$f_c^n(E_t) = \frac{1}{1 + exp\{[E_{cn} + (m_r / m_e) E_t - F_c] / k_B T\}} \tag{21a}$$

$$f_v^m(E_t) = \frac{1}{1 + \exp\{[E_{hm} - (m_r / m_h) E_t - F_v] / k_B T\}} \tag{21b}$$

We first obtain the gain or absorption spectra for various injected carrier densities. The gain shows the usual peak around the energy corresponding to the separation between the first conduction subband and the first valence (heavy hole) subband. The values of maximum gain are then plotted against injected carrier densities and from the slope of the plot the value of 'a': the differential gain constant is obtained. The gain or absorption spectra yield a value of N_{tr}, the transparency carrier density, which gives the value of current density at transparency via the relation $J_r = qdN_{tr} / \tau_s$. Then threshold current density is obtained using $J_{th} = J_{tr} + [qd / \Gamma \eta a \tau_s].[\alpha + (1/2L)\ln(1/R_1 R_2)]$, where Γ is the optical confinement factor, η is the internal quantum efficiency, α is the loss coefficient, L is the length of the device, and R_1 and R_2 are the respective reflectivities of the front and the back mirrors.

Table 1. Values of parameters used in calculation

Electron capture life time, τ_{cap}	10 ps	Escape life time, τ_{esc}	5 ns
Spontaneous emission time, τ_s	15 ns	Photon life time, τ_p	4 ps
Base recombination life time, τ_B	200 ps	Diffusion coefficient, D_n	26 x 10^{-4} m^2/s
Group velocity, v_g	8.67 x 10^7 m/sec	Loss coefficient/ unit length, α	500 m^{-1}
Confinement factor, Γ	0.03		

5.3. Results and Discussions

The values of parameters used in the calculation are entered in Table 1. The values for mass, permittivity etc of the alloy are calculated by linear interpolation of the data given in Table 2.

The choice of time constants for escape and capture of electrons in the QW is somewhat difficult; however we have chosen reasonable values after considering the simple analytical expressions given in [53,54]. The subband energies are calculated by considering the finite barrier heights between InGaAs and GaAs. The band offsets (U in Figure 2) and subband energies are evaluated by following the approximate method used in [51, 52].

Table 2. Values of parameters for alloys used in linear interpolation

Quantity	GaAs	InAs	Quantity	GaAs	InAs
Electron effective mass, m_e / m_0	0.063	0.023	Lattice constants, a_w and a_b	5.653	6.058
Split-off electron mass, m_{so} / m_0	0.15	0.049	Elastic constants, $C_{11} x 10^{15}\, dyn / m^2$	11.9	8.34
Heavy-hole mass m_h / m_0	0.38	0.26	Elastic constants, $C_{12} x 10^{15}\, dyn / m^2$	5.34	4.526
Dielectric constant, ε	12.9	15.5	Hydrostatic deformation potential for CB, a_c ,eV	-7.17	-5.08
Band gap, E_g , eV	1.424	0.354	Hydrostatic deformation potential for VB, a_v ,eV	1.16	1
Split-off energy, Δ_{so} , eV	0.34	0.41	Shear deformation potential for VB, b_v ,eV	-1.7	-1.8
Electron affinity, χ ,eV	4.07	4.88			

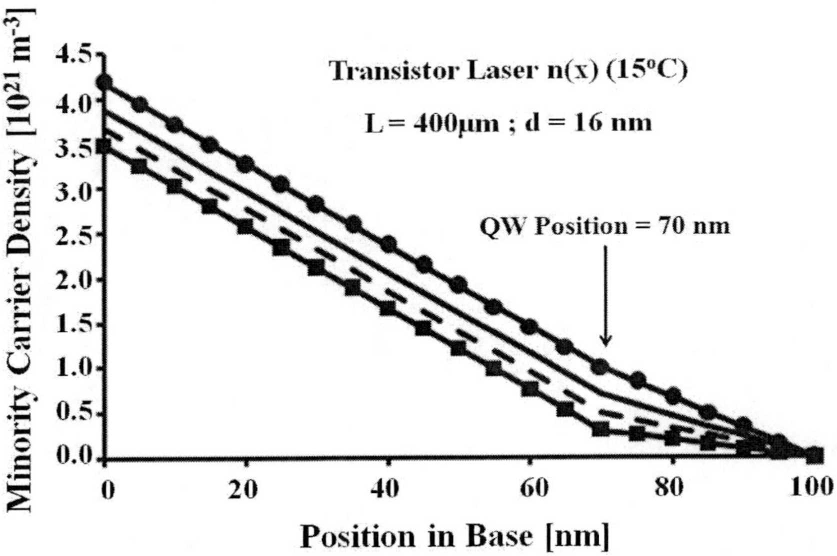

Figure 5. Variation in base with QW placed at 70 nm from the EB junction for various virtual state carrier densities (■: 3 x 10^{20} /m^3; -----: 5 x 10^{20} /m^3; —: 7 x 10^{20} /m^3 and •: 10 x 10^{20} /m^3).

We first show in Figure 5 a plot of the variation of injected carrier density in the base of width 100 nm in which the InGaAs QW is placed at 70 nm away from the EB junction. The profile is obtained by joining two straight lines of different slopes, the change of which occurs at the QW position. This plot shows the same nature as reported by Feng et al in different publications [21]. As mentioned in the work by Faraji et al [39] the plot is a simple triangle and shows no break. We also obtained similar altered slopes for = 20 and 50 nm.

Our calculated threshold carrier densities are compared with the corresponding experimental values for three different samples in Table 3. In addition, the dimensions of the QW and the base, the temperature of operation and the calculated and the experimental values of threshold base currents are entered in Table 3. A comparison is also made between the calculated and the experimental values of the emission wavelengths.

Table 3. Comparison of calculated (this work) and experimental values of threshold base current and emission wavelengths

Sample no.	Temp (K)	L (μm) and (nm)	Area (μm^2)	W_B (nm) and x_1	I_{Bth} (calculated) (mA)	I_{Bth} (experiment) with [ref] (mA)	Emission Wavelength (calculated) (nm)	Emission Wavelength (experiment) with [ref] (nm)
1	213	450, 12	4500	85/59	7.3	7.4 [13]	975	980 [13]
2	288	400, 16	4000	88/59	21.5	22 [16, 21, 30]	990	1000 [16, 21, 30]
3	298	850, 12	1870	98/59	3.6	40 [11]	1000	1006 [11]

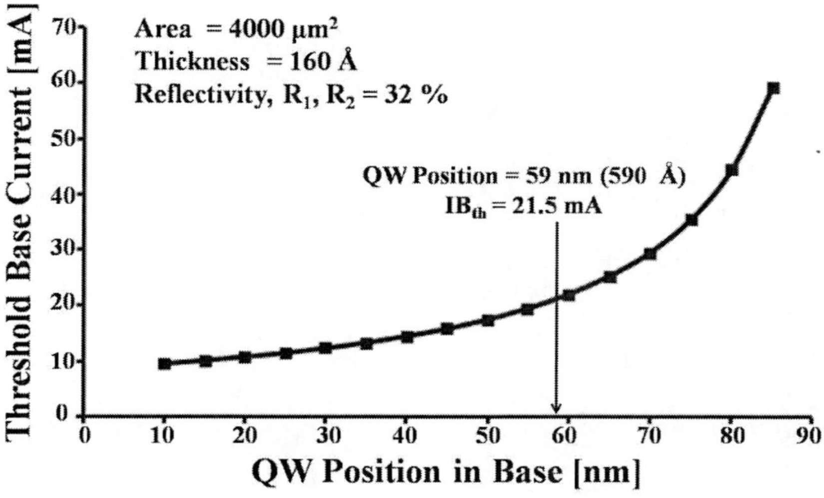

Figure 6. Variation of threshold base current for different position of QW in base.

It appears from the entries that good agreement is obtained for the calculated and experimental values of the emission wavelengths for all the three cases. This is also true for threshold base currents for the first two samples, but the discrepancy is quite large for the third sample. Note that in spite of a smaller area of the third sample, the experimental threshold base current is higher (40 mA) in comparison with the other two. This large discrepancy may be due to the fact that the quality of the sample was not high at the earlier stage of research. For all the samples the light output power shows the usual sharp rise as the base current exceeds its threshold value.

We have also estimated the values of threshold base current when the QW is placed at different positions in the base. Figure 6 shows the plot. It appears that the threshold base current is smaller the nearer the QW is to the EB junction. As the QW position approaches the BC junction, there is a sharp rise in the threshold base current. The reason for this increase may be understood as follows. The virtual state current density J_{VS} is proportional to the difference between the two slopes of electron distribution before and after the QW. As the QW moves closer to the BC junction, the difference between the two slopes gets reduced. This requires higher injection, i.e. higher base drive, to maintain the same value of J_{VS} needed to attain threshold. This trend, i.e., increased threshold base current for larger distance of QW from the EB junction, is in conformity with the results obtained by Zhang and Leburton [41]. It was predicted in [55] that the bandwidth of the TL increases as the QW is placed more close to the BC junction. The same prediction may be drawn from the expressions [Eq. (7)] in [41]. Our work provides the additional information that such an increase in bandwidth is obtained only at the cost of larger threshold base current.

The threshold base current for sample 2 in Table 3 for which the QW is placed at a distance of 59 nm from EB junction is 21.5 mA. It is interesting to note from Figure 4 that if the QW is placed at the middle of the base, as assumed in [39], the calculated value becomes 15.6 mA, which is much lower than the experimental value.

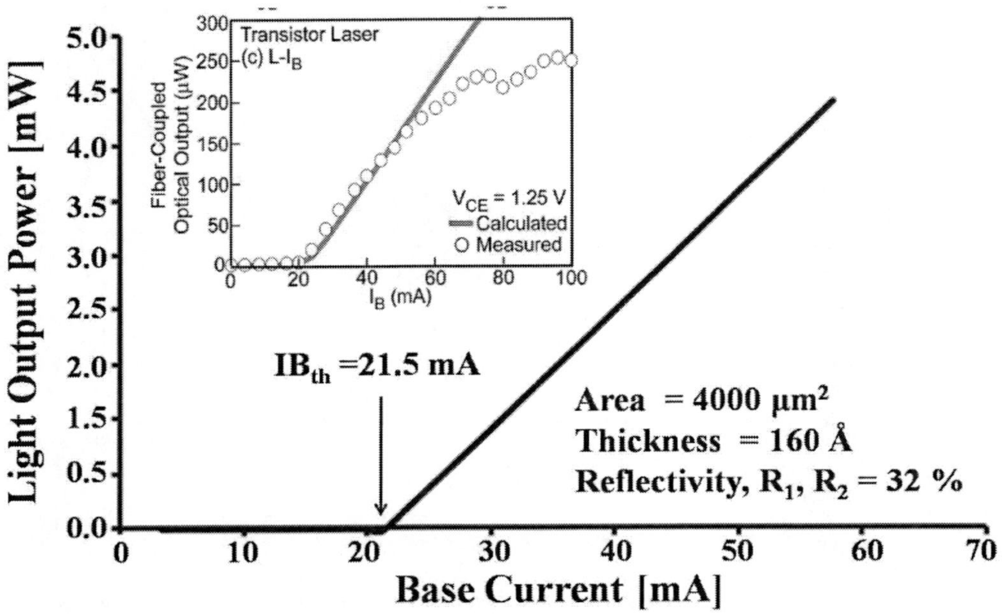

Figure 7. Variation of light output power for different base current. Inset shows the experimental results [30].

We have also calculated the light output power when the QW is placed at a distance of 59 nm from the EB junction within the base of total width 88 nm. This corresponds to the experimental situation [30].

The calculated values of light output power for different base currents are shown in Figure 7. As usual the output power increases sharply once the threshold base current (~ 21.5 mA) is reached. Our calculated value of the threshold base current agrees completely with the experimentally obtained value. The calculated power output P increases linearly with base current according to the relation [47] $P = (\hbar\omega / q)[\alpha_m /(\alpha_m + \alpha][I_B - I_{Bth}]$, where α_m denotes mirror loss. Comparing our calculated values with the experimental values shown in the inset, it is found that both the theoretical and experimental values increase linearly up to a base current of ~ 55 mA. However, our calculated values of light power are ~ 12 times higher than the experimental values in the linear region. It is to be noted that the experimental values reported in [30] are obtained by coupling the light output from the TL to a fibre, the output end of which is connected to the power meter. The coupling losses between the TL and the fiber and that between the fibre and the power meter in the experiment are responsible for the discrepancy between our calculated values and the experimental results and a total coupling loss of about 10 dB may account for this discrepancy.

The experimental light power shows a saturation behaviour when the base current is above 55 mA. Our linear theory cannot explain this saturation effect and gain compression and other phenomena should be invoked in the theory even for qualitative agreement beyond this current.

6. Structure and Model for QD-TL

6.1. Structure and Band Diagram

In most of the HBT samples used in experiments a higher band gap N-type InGaP layer acts as the emitter (E), the base (B): a p-type GaAs layer incorporates an InGaAs QW and the collector (C) is an n- type GaAs layer [11, 13, 16, 21, 30]. In our work, [56] we consider a similar structure, but in addition, assume that a layer of InAs Self-assembled QD (SAQD) layer of pyramidal shaped QDs is inserted in the InGaAs QW, which acts as the active region. The height of each QD is H and d is the thickness of active region. Base width is denoted by W_B. The structure is shown in Figure 8.

Figure 9 shows the energy band diagram of the InAs-InGaAs self-assembled QDs-in-well and the dc excess minority carrier distribution [$\delta N(x)$] in the base region. In the active region the QDs are interconnected by a continuum level associated with the wetting layer (WL) or virtual state (VS) in the GaAs base region having carrier density N_{VS} where density of states are much larger than the QD areal density [63]. In the QD ensemble, two energy levels are assumed to be present: a 2-fold degenerate ground state (GS) having carrier density N_{GS} and a 4-fold degenerate excited state (ES) having carrier density N_{ES} in the QDs.

In the one dimensional model, the EB junction is at $x = 0$ and the CB junction is at $x = W_B$ and the self-assembled QDs - in - Well (DWELL) are assumed to be situated at any arbitrary position i.e., $x = x_1$, in the base. Under forward bias, the carries injected from the emitter diffuse across the base and reach the QW. These unbounded carriers (three dimensional: 3D) may either undergo quantum capture to the bound states in the QD, or diffuse to the collector-base junction where they are swept out by the reverse bias. The carriers captured by the QD may escape out of it. The 3D carriers at $x = x_1$ are located at the virtual bound states (VS) [39] which in this case lie in the WL. These states are localized at the QW in DWELL, but occupy energies higher than the conduction band edge of the active material and aid in the conversion of carriers from the 3-D states in the base above the well to the 2-D states within the QW (N_{VS}) and vice versa.

Here a simple single-mode excitonic model is considered for the ensemble of identical, non-interacting QDs, whose populations are described in terms of e-h pairs i.e. electrons and holes are captured and emitted in pairs. With proper injection, carriers first diffuse through the base, then relax into the wetting layer (WL) or VS, and then go to the QDs. WL (VS) is commonly observed in the SAQD grown via Stranski-Krastanov mode and acts as a 2D layer with a QW like density of states. From WL or VS carriers are captured in the ES with a capture time τ_{cap}. The excitons either relax from ES to GS with an intradot relaxation time or recombine radiatively with a radiative carrier lifetime which is taken to be same for both ES and GS. Finally carriers can thermally escape in a time τ_e^{GS} form the GS to ES and in a time from the ES to the WL [48, 57, 58]. Above lasing threshold, carriers in the ground state emit photons into the lasing mode primarily due to stimulated emission.

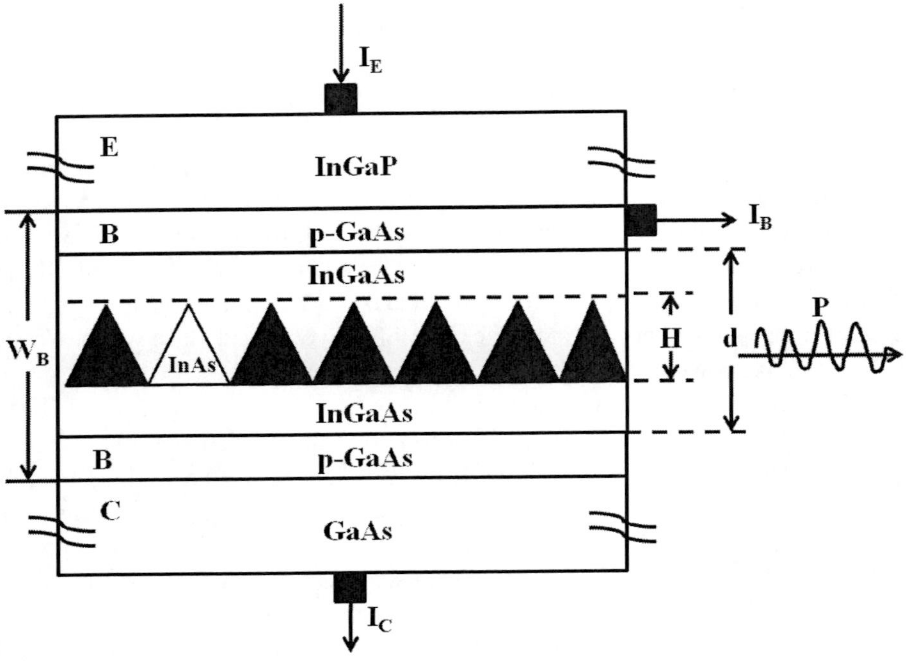

Figure 8. Simplified structure of InAs-InGaAs self-assembled QD structure in the GaAs Base.

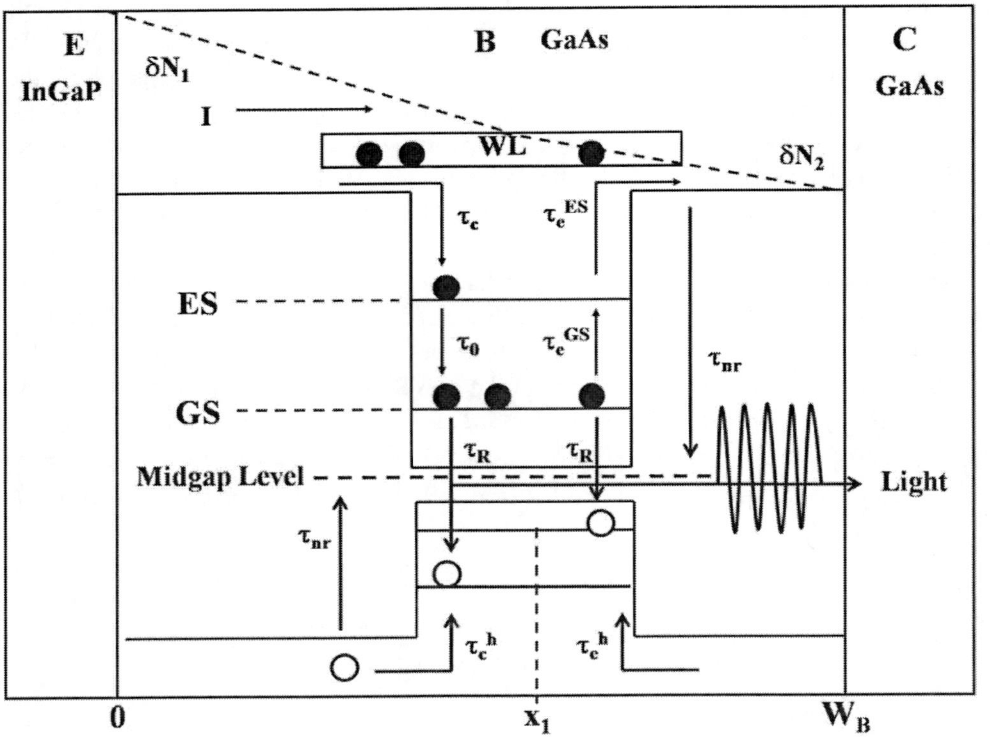

Figure 9. Simplified energy band diagram and excess minority carrier density distribution in the base.

6.2. Threshold Current and Modulation Bandwidth

The threshold current and modulation bandwidth of QD-TL are evaluated by using two different models, both of which consider electron-hole as a single entity, i.e. an excitonic model is assumed in both cases. We shall consider the model by Sugawara and coworkers [48, 57, 58] first, followed by the model by Fiore and Markus [59, 60].

6.2.1. Sugawara Model

The different transitions occurring in this model are shown in Figure 9. The following rate equations for the carrier density N_W in the wetting layer (QW in our case), carrier density N_G in the ground state and photon density N_{ph} are used.

$$dN_W / dt = \eta_{inj} I / (eV_{ac}) - N_W / \tau_S - N_W / \tau_0^* + N_G / \tau_E \tag{22}$$

$$dN_G / dt = N_W / \tau_0^* - N_G / \tau_E - N_G / \tau_S - v_G g_m N_{ph} \tag{23}$$

$$dN_{ph} / dt = v_g g_m N_{ph} - N_{ph} / \tau_P \tag{24}$$

In above equations η_{inj} is the injection efficiency, I is the injection current, V_{ac} is the volume of the active region, τ_S is the carrier lifetime, τ_0^* is the effective capture time, τ_E is an escape time from the QD ground state to the WL (QW), g_m is the modal gain of the ground state of the QD and τ_P is the photon lifetime. The expression for g_m is

$$g_m = G_{max}(2P - 1) \tag{25}$$

where G_{max} is the maximum modal gain of the ground state of the QD and P is the occupancy ratio of the carrier in the ground state given in terms of the QD density N_{QD} as

$$P = N_G / 2N_{QD} \tag{26}$$

In order to consider state filling (Pauli blocking) in the ground state of the QD the effective capture time τ_0^* is defined as

$$\tau_0^* = \tau_0 / (1 - P) \tag{27}$$

where is the intrinsic capture time.

The usual method of putting $d/dt = 0$ in Eqs. (22)-(24) is applied to obtain the steady state solutions. Under the lasing condition $N_{ph} \neq 0$, the threshold current density takes the following form

$$J_{th} = 2ePN_{QDarea}(\eta_{inj}\tau_S)^{-1}\{1 + A_2(1 - P)^{-1}\} \tag{28}$$

N_{QDarea} is the total QD area density defined as the product of the layer number and QD area density. The external quantum efficiency η_e is given in terms of internal quantum efficiency η_i by

$$\eta_e = \alpha_m (\alpha_m + \alpha_i)^{-1} \eta_i \eta_{inj} \quad (29)$$

and the current-gain relationship under non-lasing condition of $N_{ph} = 0$ is written as

$$g_m(J) = G_{max} \left\{ J' + A_2 - \sqrt{(J' + 1 + A_2)^2 - 4J'} \right\} \quad (30)$$

Here the occupation ratio of the ground state P at lasing condition is given as

$$P = (1 + 1 / v_g \tau_P G_{max}) / 2 \quad (31)$$

The other quantities are given as

$$\eta_i = 1/[1 + A_1/(1 - P)] \quad ; \quad A_1 = \tau_0 / \tau_S ; \quad A_2 = (\tau_0 / \tau_S) + (\tau_0 / \tau_E);$$
$$J' = \tau_S \eta_{inj} J / (2eN_{QDarea}),$$

and J is the injection current density.

6.2.2. Fiore and Markus Model

The light output-current characteristics and small signal modulation response of the TL are obtained by solving the coupled rate equations for carriers and photons. Here we have used the single mode excitonic model developed by Fiore and Markus [59, 60], the essence of which has been given in Figure 9 and described in Sec.6.1. In the model both the ground state and the excited state are included, but the emission is due to recombination pertaining to the ground state.

The temporal evolution of occupation probabilities can be written in a rate equation model as [59],

$$\frac{dF_{vs}}{dt} = \frac{G}{\rho_{fvs}} - \frac{F_{VS}(1 - F_{ES})}{\tau_{cap}} + \frac{4N_D F_{ES}(1 - Fvs)}{\rho_{fvs} . \tau_e^{ES}} - \frac{F_{VS}}{\tau_R} \quad (32a)$$

$$\frac{dF_{ES}}{dt} = \frac{\rho_{fvs} F_{VS}(1 - F_{ES})}{\rho_{fvs} \tau_{cap}} - \frac{F_{ES}(1 - F_{VS})}{\tau_e^{ES}} +$$
$$+ \frac{F_{GS}(1 - F_{ES})}{2\tau_e^{GS}} - \frac{F_{ES}(1 - F_{GS})}{\tau_0} - \frac{F_{ES}}{\tau_R} - 2BN_P N_D (F_{ES} - \frac{1}{2}) \quad (32b)$$

$$\frac{dF_{GS}}{dt}=\frac{F_{GS}(1-F_{ES})}{\tau_e{}^S}+\frac{2F_{ES}(1-F_{GS})}{\tau_0}-\frac{F_{GS}}{\tau_R}-2BN_PN_D(F_{GS}-\frac{1}{2}) \tag{32c}$$

$$\frac{dN_P}{dt}=4BN_DN_P(F_{GS}-\frac{1}{2})-\frac{N_P}{\tau_P}+\frac{2\beta F_{GS}}{\tau_R} \tag{32d}$$

It is assumed that the laser is operating far enough from GS saturation, thus ES photon population is neglected. The occupation probabilities are, $F_{GS}=N_{GS}/2N_D$, $F_{ES}=N_{ES}/4N_D$, and $F_{VS}=N_{VS}/N_D$, where, and N_{VS} are respectively the number of electron -hole pairs in the GS, ES and WL respectively. is total number of QDs and ρ_{VS} denotes the number of states in WL; G is the injection rate per QD per unit time, τ_{cap} is capture time, and B is related to absorption or stimulated emission cross-section. Here, capture and relaxation occur through phonon or carrier-carrier scattering and the high carrier concentrations both in QD and neighboring WL allow large Auger scattering rates. N_P is the photon number density,

The numerical solution of Eq. (32) may provide the laser temporal evolution. However, we follow the elegant method proposed by Fiore and Markus [59] by which two important key parameters: differential gain and gain compression factor are extracted. These parameters allow a direct comparison between QD TL characteristics with QW structures. To do so the rate equations for QD: Eq. (32) are recast in the following standard form for semiconductor lasers [61],

$$\frac{dN}{dt}=\frac{\eta.I}{q.V}-(R_{sp}+R_{nr})-v_g g N_p \tag{33a}$$

$$\frac{dN_P}{dt}=\Gamma v_g g N_P-\frac{N_P}{\tau_P}+\Gamma\beta R_{sp} \tag{33b}$$

where $G=\eta I/qV$ is the injection rate with active volume and internal quantum efficiency η. R_{sp} and R_{nr} are, respectively, the radiative and non-radiative recombination rates, v_g is the group velocity, g is the material gain, Γ is the optical confinement factor, N is the total carrier density defined as, $N=(N_D/V)(2F_{GS}+4F_{ES}+F_{VS})$, modal gain is, $\Gamma g=(4BN_D/v_g)(F_{GS}-1/2)$. For this reason gain becomes a function of both N and N_P.

The differential gain is defined as, $a=\partial g/\partial N|_{N_P}$, where the gain compression is defined as just the opposite of differential gain, $a_p=-\partial g/\partial N_P|_N$. Sometimes the gain compression factor is defined in terms of gain compression as, $a_p=\varepsilon g/(1+\varepsilon N_P)$. Once these two key parameters are estimated, the steady state solutions are obtained by solving equations (33), by applying simple semiconductor laser theory as followed in [59].

In case of small-signal analysis photon density has to be taken very high, since SAQD active medium is a high gain medium. Considering non-linear gain with gain compression the generalized gain expression can be written as [59],

$$g(N) = \frac{g_0 + a.(N - N_0)}{1 + \varepsilon.N_P} \tag{34}$$

where g_0 is gain at steady state and N_0 is the dc carrier density. For small-signal analysis, imposing small-signal variations as, $J(t) = J_0 + j(t)$; and $N_P(t) = N_{P0} + n_p(t)$, the small signal non-linear gain model becomes,

$$g(N) = \frac{g_0}{1 + \varepsilon.N_{P0}} + \frac{a.N(t)}{1 + \varepsilon.N_{P0}} - \frac{g_0.\varepsilon.N_P(t)}{(1 + \varepsilon.N_{P0})^2} \tag{35}$$

From the steady state solutions of Eq. (33) the small variation in gain is calculated. This according to [18] is equivalent to quasi-static approximation where intraband and interband dynamics are decoupled. The relaxation oscillation frequency, f_r and damping factor K can be calculated as [59]

$$f_r^2 = [(\Gamma a_p / \tau) + (a / \tau_p)] v_g N_P; \tag{36}$$

Thus the frequency response becomes,

$$|s(\omega)/s(0)|^2 = |H(m)| = \omega_r^4 / ((\omega^2 - \omega_r^2)^2 + \omega^2.\gamma^2) \tag{37}$$

where damping rate is, $\gamma = (1/\tau) + K(\omega_r^2 / 4\pi^2)$. At a specific bias current or bias current density J_b the modulation response at a given frequency $\omega = 2\pi f$ is defined as the ratio of modulated photon number $s(\omega)$ to the corresponding unmodulated value $s(0)$.

6.3. Results and Discussions

In our previous work we considered a 16 nm wide InGaAs QW located at x_1 = 59 nm away from the EB junction in the GaAs base, as in experiments [21,30] and found good agreement for the threshold base current and light power output with the experimental results. In the present work we assume that a single layer of InAs QD with H = 5 nm is inserted in the middle of 16 nm thick (d) $In_{0.15}Ga_{0.85}As$ QW and the total base width W_B is 89 nm (see Figure 8). The device length L = 400 μm with a width of 10 μm. The areal dot density is taken as

10^{10} /cm^2. The present work analyzes the room temperature static and small-signal operation of a QD TL having front and back reflectivities 32%, D_n = 26 x 10^{-4} m^2/sec, = 0.42 ns, = 0.2 ns, = 200 ps, B = 20/cm, β = 10^{-5}, = 8.8 ps, =3.3 ps, = 1 ps, =1 ns, v_g = 6.3 x 10^{-16} m^3, = 5/cm, and g_{max} = 10/cm [39].

Figure 10 shows the minority carrier distribution in the base with the DWELL placed at 59 nm from the emitter-base junction. The distribution shows the expected discontinuity as shown in earlier models for QW [43], a representative variation of which is already shown in Figure 5. The base charge density profile of the TL deviates from the usual HBT "Triangular" profile and is tilted more steeply before the DWELL i.e. in the emitter side than the collector side, since the stimulated recombination speeds up the overall recombination rate at QW at lasing threshold.

We next obtain the values of differential gain from gain variation data with carrier density by the divided difference method of interpolation. Figure 11 shows the light power output versus base current for the QD transistor laser (QD TL) structure obtained by using the Sugawara model and the single mode (SM) excitonic model due to Fiore and Markus [59].

It appears from Figure 11 that both the Sugawara and Fiore and Markus models give the same values of 1.3 mA for the threshold current. However, the values of light power output obtained from Sugawara model are slightly less than the values given by Fiore and Markus model. For the sake of comparison we have shown in the inset of Figure 11 the curve obtained by us [43] for a QW TL in which the active QW is placed at 59 nm away from the emitter-base junction showing that the threshold base current is 21.5 mA. The threshold with QD in well occurs at 1.3 mA, predicting a substantial reduction in threshold base current.

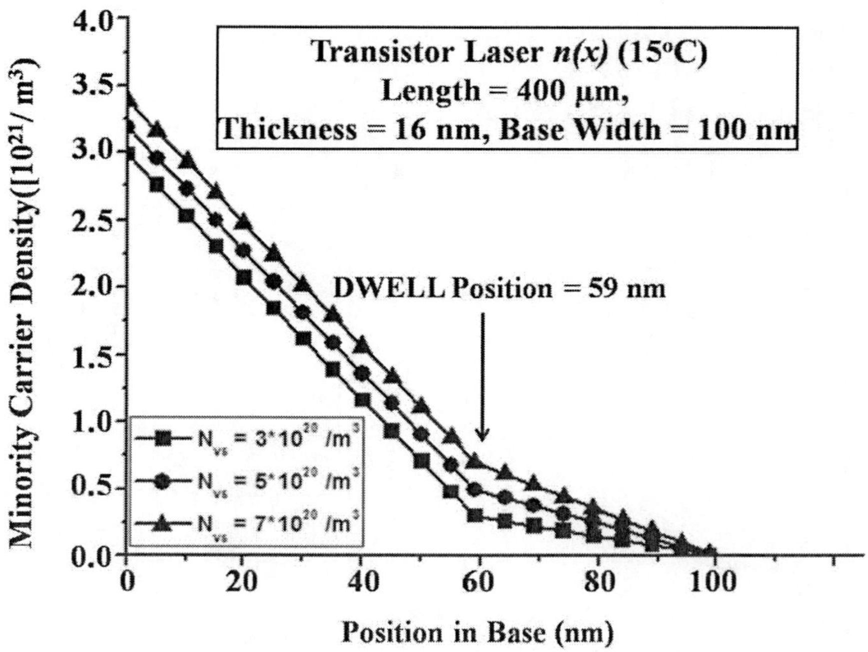

Figure 10. Minority carrier variation with base distance placing QW at 59 nm in the base having base width = 100 nm for various injected virtual carrier densities (■-: 3 x 10^{20} m^{-3};● —: 5 x 10^{20} m^{-3}; ▶-: 7 x 10^{20} m^{-3}). The profile shows a discontinuity at the DWELL position 59 nm away from the EB junction.

Figure 12 (a) shows a typical frequency spectrum of the modulation response obtained by using the single mode excitonic model by Fiore and Markus. The response is shown for three different bias currents, $I_b = 3I_{th}$, and respectively, where $I_{th} = 1.3$ mA. We have not included in Figure 12 similar response calculated by using the Sugawara model in order to avoid congestion of curves. However a comparison between the two models has been shown in Table 4 below. The response for QW TL is shown in Figure 12 (b) where mA as a comparison. This modulation response in QD TL is plotted using the resonant frequency and damping factor calculated from the obtained gain derivatives a and a_p. Modulation Bandwidths for QW TL are quite high than the respective QD cases due to absence of damping, intradot relaxation and less gain compression. Intradot relaxation rates may be improved through the claver engineering of QD energy states or by increasing the Auger, carrier- carrier and carrier-phonon scattering.

Within the region of the peak response, the carriers interact with the photon field to give a complete phase synchronism between them which leads to the general laser resonance characterized by resonance frequency f_r. The next declining part is due to the fact that the photon field lags behind that of the injection current. As modulation response, f_m, is increased beyond its peak value, $f_m(peak)$, the electron and photon fields tend to become more out of phase, resulting in damping of the relaxation oscillations and that leads to monotonic decrease in response [47].

We have made a comparison of 3dB modulation bandwidths of QD TL calculated by both the Sugawara model and the single mode excitonic model of Fiore and Markus at three bias points. The values are entered in Table 4. For the sake of comparison the calculated values of bandwidth for QW-TL for corresponding three bias points are also included.

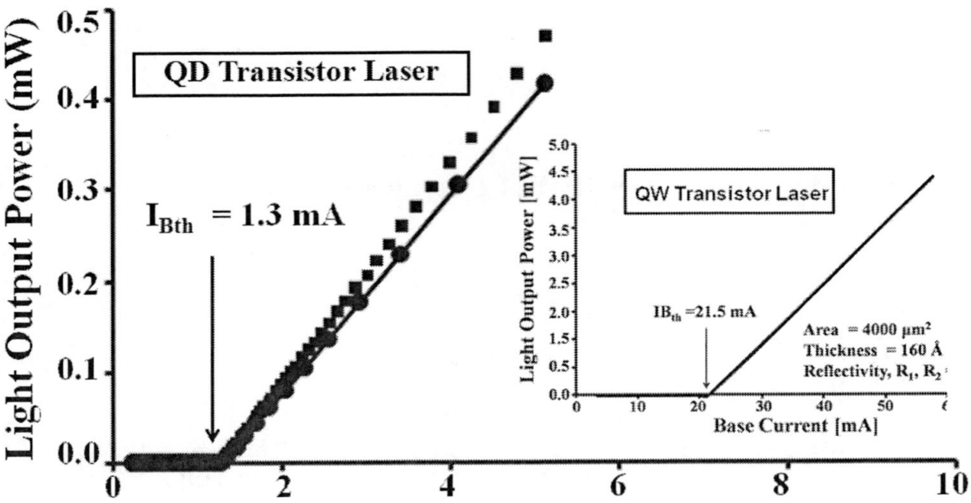

Figure 11. Light output power variation with base current for a QD TL (for a QW TL in inset of the figure) for two QD models (♦: SM Excitonic Model; ●-: Sugawara Model). Both QD model show same base threshold at 1.3 mA. QW TL has base threshold at 21.5 mA.

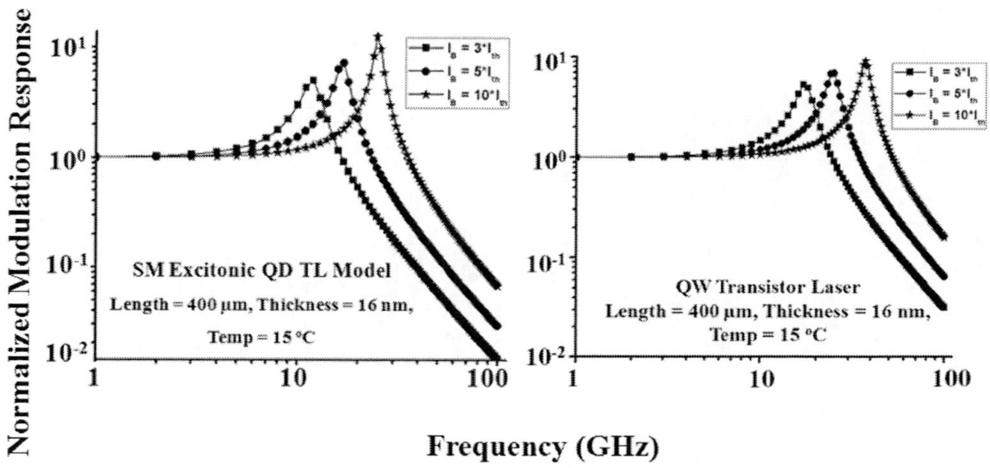

(a) (b)

Figure 12. (a) Modulation response of QD- in- Well in base of SM Excitonic model for different bias currents (--■: $3I_{th}$; --●: $5I_{th}$; --*: $10I_{th}$) with modulation bandwidths 20, 30 and 45 GHz respectively. (b) Modulation response of QW TL for different bias currents (-■: $3I_{th}$; -●:$5I_{th}$; -*: $10I_{th}$) with modulation bandwidths 30, 42 and 60 GHz respectively.

The curves in Figure 12 as well as the entries in Table 4 indicate that the modulation bandwidth for QD-TL does not achieve the high values shown by QW-TL. This trend is also consistent with prediction [22] and experimental finding for QD lasers [62, 63].

Figure 13 shows the variation of both optical bandwidth and base threshold current as a function of variable position of DWELL within the base. As the active region position moves from emitter to base, modulation bandwidth increases but at the cost of larger base threshold current. The same conclusion has been drawn earlier for the case of QW-TL as illustrated in Figure 6. The effect of increased confinement factor on the modulation bandwidth for a QD-TL is shown in Figure 14. It is demonstrated that increased bandwidth may be achieved with higher confinement factor. The theoretical expression, Eq. (36), predicts that further improvement in bandwidth may occur for larger values of differential gain, which may be achieved by using more number of layers of QDs. However, the present theory, valid for a single layer of QDs, needs substantial modifications. It may not be valid to increase the value of by simply multiplying by the number of layers to make some prediction of the bandwidth.

Table 4. Calculated values of modulation bandwidth for QD-TL using two models and the values for QW-TL

Serial No.	Bias point	Modulation Bandwidth		
		SM Excitonic Model	Sugawara Model	QW TL Model
1.	$I_b = 3I_{th}$	20	10.2	30
2.	$I_b = 5I_{th}$	30	14.5	42
3.	$I_b = 10I_{th}$	45	19.2	60

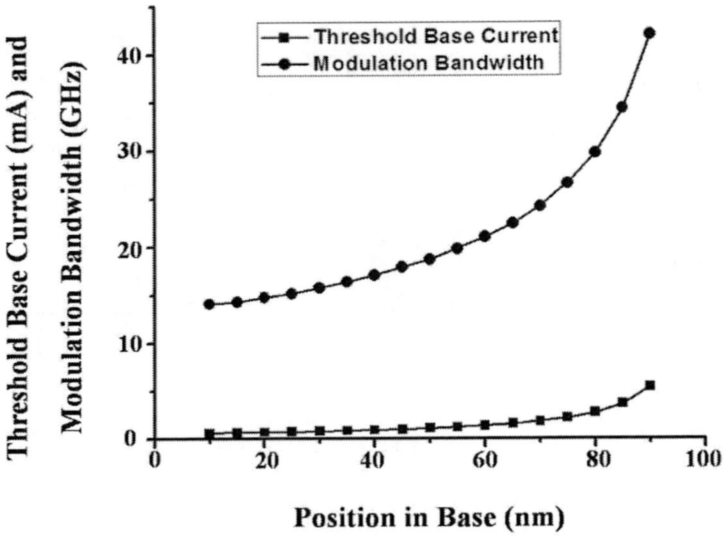

Figure 13. Variation of threshold base current (-■) and modulation bandwidth (-●) with variation of QW position in base for SM excitonic QD TL.

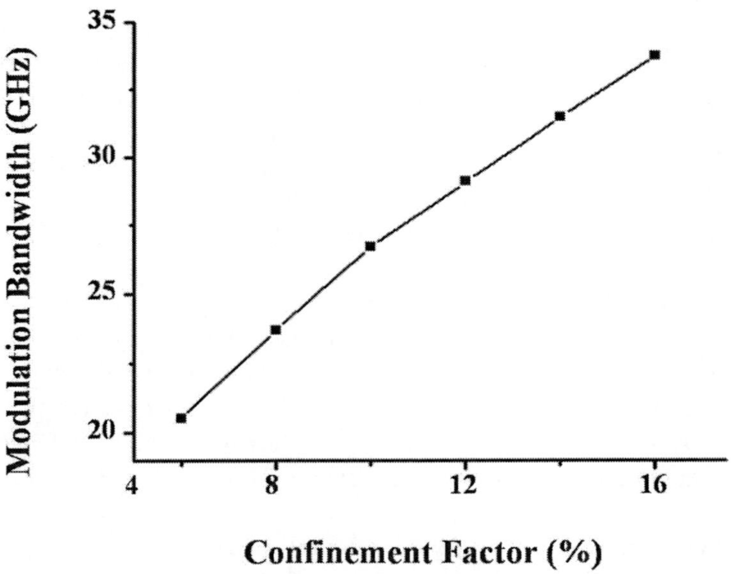

Figure 14. Expected variation of modulation bandwidth with QD confinement factor for the same structure.

6.4. Separate Electron-Hole Dynamics

Fiore and Markus [59] have also considered a separate electron-hole dynamics to obtain the modulation response of QD lasers. The analysis involves solution of seven rate equations, which involves numerical work. Their calculation demonstrates that the light output power characteristics almost coincide with the curve obtained from the excitonic model. Though the differential gain is strongly affected at low bias due to hole spreading among the closely spaced valence band states, the *K*-factor and other parameters remain almost unaltered due to limited impact of separate electron-hole dynamics. Moreover the modulation bandwidths are almost identical with the values obtained from the excitonic model. We have not made any attempt to make an analysis based on separate electron-hole dynamics that involves numerical work, but believe that the results will not differ much from the excitonic model. Such detailed analysis may be justified only when experimental data are available.

Summary and Conclusions

The present paper is primarily concerned with the theoretical models for obtaining threshold base current, light power output versus base current and small signal optical modulation bandwidths of a transistor laser. The base of the TL contains a either a QW or a layer of QDs embedded in the QW. The expressions for QW-TL are developed first and the calculated values of different quantities are found to agree with the available experimental results for InGaP-GaAs-GaAs heterojunction bipolar transistor laser with an InGaAs QW embedded in the GaAs base. The theory is then extended for a QD-TL. It is found that an order-of-magnitude decrease in threshold base current may be achieved by considering a single layer of InAs QDs. The theory for small signal modulation response for QD-TL is then developed by using two different excitonic models reported in the literature. The predicted values of bandwidth for QD-TL are not however as high as in the QW-TL structure. Experimental values for the predicted quantities are not available however for comparison. A few methods to increase the bandwidth are suggested.

Acknowledgments

The work has been supported by the Centre for Research in Nanoscience and Nanotechnology of the University of Calcutta under project no. conv/048/Nano Pr (2009). The author gratefully acknowledges helpful guidance, constant encouragement and suggestions to improve the manuscript received from Dr. (Mrs) Bratati Mukhopadhyay and Prof. P. K. Basu of the Institute of Radio Physics and Electronics, University of Calcutta.

References

[1] Walter, G.; Holonyak, N. Jr.; Feng, M.; Chan, R. *Appl. Phys. Lett.*, 2004, vol. 85, 4768–4770.

[2] Holonyak, N., Jr.; Feng, M. *IEEE Spectrum*, 2006, vol. 43 (2), 50–55.

[3] Katz, J.; Margalit, S.; Welt, D.; Chen, P. C.; Yariv, A.; Miles, H. O.; Dandridge, A. B. *Appl. Phys. Lett.,* 1980, vol. 37, 211-213.

[4] Mori Y.; Shibata, J.; Sasai,Y.; Serizawa, H.; Kajiwara, T. *Appl. Phys. Lett.,* 1985, vol. 47, 649-651.

[5] Suzuki, Y.; Yajima, H.; Shimoyama, K.; Inoue, Y.; Katoh, M.; Gotoh, H. '*Electron. Lett.,* 1990, vol. 26, 1632-1633.

[6] As reported in ref [2].

[7] Mastrapasqua, M.; Luryi, S.; Capasso, F.; Hutchinson, A. L.; Sivco, D. L.; Cho, A. Y. *IEEE Trans. Electron Dev.*, 1993, vol. 40(2), 250-258.

[8] Hasumi, Y.; Kozen, A.; Temmyo, J.; Ashahi, H. *IEEE Electron Dev. Lett.*, 1987, vol. 8(1), 10-12.

[9] Feng, M.; Holonyak, N. Jr.; Hafez, W. *Appl. Phys. Lett.*, 2004, vol. 84, 151–153.

[10] Feng, M.; Holonyak, N. Jr.; Chan, R. *Appl. Phys. Lett.,* 2004, vol. 84, 1952-1954.

[11] Feng, M.; Holonyak, N. Jr.; Chu-Kung, B.; Walter, G.; Chan, R. *Appl. Phys. Lett.,* 2004, vol. 84, 4792-4795.

[12] Feng, M.; Holonyak, N. Jr.; Walter, G.; Chan, R. *Appl. Phys. Lett.*, 2005, vol. 87, 131103 (1-3).

[13] Chan, R.; Feng, M.; Holonyak, N. Jr.; Walter, G. *Appl. Phys. Lett.*, 2005, vol. 86, 131114–131114.

[14] Dixon, F.; Chan, R.; Walter, G.; Holonyak, N. Jr.; Feng, M.; Zhang, X. B.; Ryou, J. H.; Dupuis R. D., *Appl. Phys. Lett.,* 2006, vol. 88, 012108 (1-3).

[15] Chan, R.; Feng, M.; Holonyak, N. Jr.; James, A.; Walter, G. *Appl. Phys. Lett.*, 2006, vol.88, 143508 (1-3).

[16] Feng, M.; Holonyak, N. Jr.; James, A.; Cimino, K.; Walter, G.; Chan, R. *Appl. Phys. Lett.*, 2006, vol.89, 113504 (1-3).

[17] Feng, M.; Holonyak, N. Jr.; Chan, R.; James, A.; Walter, G. *IEEE Photon. Technol. Lett.*, 2006, vol. 18(11), 1240-1242.

[18] Walter, G.; James, A.; Holonyak, N. Jr.; Feng, M.; Chan, R. *Appl. Phys. Lett.*, 2006, vol. 88, 232105 (1-3).

[19] Chu-Kung, B. F.; Feng, M.; Walter, G.; Holonyak, N. Jr.; Chung, T.; Ryou, J. H.; Limb, J.; Yoo, D.; Shen, S-C.; Dupuis, R. D.; Keogh, D.; Asbeck, P. M. *Appl. Phys. Lett*. 2006, Vol. 89, 082108 (1-3).

[20] James, A.; Walter, G.; Feng M.; Holonyak, N. Jr. *Appl. Phys. Lett.,* 2007, vol. 90, 152109 (1-3).

[21] Feng, M.; Holonyak, N. Jr.; Then, H. W.; Walter, G. *Appl. Phys. Lett.,* 2007, vol. 91, 053501(1-3).

[22] Then, H. W.; Feng, M.; Holonyak, N. Jr. "*Appl. Phys. Lett.,* 2007, vol. 91, 183505(1-3).

[23] James, A.; Holonyak, N. Jr.; Feng, M.; Walter, G. *IEEE Photon. Tech. Lett.,* 2007, vol. 19, 680-682.

[24] Dixon, F.; Feng, M.; Holonyak, N. Jr.; Huang, Y.; Zhang, B.; Ryou, J. H.; Dupuis, R. D. "*Appl. Phys. Lett.,* 2008, vol. 93. 021111 (1-3).

[25] Then, H. W.; Feng, M.; Holonyak, N. Jr. "*Appl. Phys. Lett.,* 2009, vol. 94, 013509(1-3).

[26] Then, H. W.; Wu, C. H.; Walter, G.; Feng, M.; Holonyak, N. Jr. *Appl. Phys. Lett.,* 2009, vol. 94, 101114 (1-3).

[27] Wu, C. H.; Walter, G.; Then, H. W.; Feng, M.; Holonyak, N. Jr. *Appl. Phys. Lett*, 2009, vol. 94, 171101 (1-3).

[28] Walter, G.; et al. *Appl. Phys. Lett.,* 2009, vol. 94, 241101 (1-3).

[29] Feng, M.; Then, H. W.; Holonyak, N. Jr.; Walter, G.; James, A., *Appl. Phys. Lett.,* 2009, vol. 95, 033509(1-3).

[30] Then, H. W.; Feng, M.; Holonyak, N. Jr. *J. Appl. Phys.,* 2010, vol. 107, 094509(1-7).

[31] Then, H. W.; Feng, M.; Holonyak, N. Jr., *Appl. Phys. Lett.,* 2010, vol. 96, 113509.

[32] Dixon, F.; Feng, M.; Holonyak, N. Jr. "*J. Appl. Phys,* 2010, vol. 108, 093109 (1-6).

[33] Dixon, F.; Feng, M.; Holonyak, N. *Applied Physics Letters.,* 2010, vol. 96, 241103.

[34] Huang, Y.; Ryou, J-H.; Dupuis, R. D.; Dixon, F.; Feng, M.; Holonyak, N. Jr., *J. Appl. Phys.,* 2011, vol. 109, 063106.

[35] Tan, F. et al, *Applied Physics Letters.,* 2011, vol. 99, 061105.

[36] Shi, W.; Chrostowski, L.; Faraji, B. *IEEE Photonics Technology Letters*, 2008, vol. 20, 2141-2143.

[37] Faraji, B.; Shi, W.; Pulfrey, D. L.; Chrostowski, L. *Appl.Phys. Lett.*, 2008, vol. 93, 143503 (1–3).

[38] Faraji, B.; Pulfrey, D. L.; Chrostowski, L. *Applied Physics Letters*, 2008, vol. 93, 103509 (1-3).

[39] Faraji, B.; Shi, W.; Pulfrey, D. L.; Chrostowski, L. *IEEE J. Sel. Top. Quantum Electron.,* 2009, vol. 13 (5), 594-603.

[40] Duan, Z.; Shi, W.; Chrostowski, L.; Huang, X.; Zhou, N.; Chai, G. *Optics Exp.,* 2010, vol. 18(2), 1501-1509.

[41] Zhang, L.; Leburton, J. P. *IEEE J. Quantum Electron.,* 2009, vol. 45(4), 359- 366.

[42] Basu, Rikmantra; Mukhopadhyay, Bratati; Basu, P. K. "Gain spectra and characteristics of a transistor laser with InGaAs QW in the base", 4th Intl. Conf. Computers and Devices for Communication (CODEC 09), Dec. 14-16, 2009, Inspec acc. No. 11107257 (05 Feb. 2010) pp. 1-4.

[43] Basu, Rikmantra; Mukhopadhyay, Bratati; Basu, P. K. *Semiconductor Sc. Technol,* 2011, vol. 26, 105014.

[44] Streetman, B. G.; Banerjee, S. *Solid State Electronic Devices,* 6th edn, Prentice Hall International, NJ, 2008.

[45] Statz, H.; DeMars, G. *Quantum Electronics,* Columbia University Press, New York, 1960.

[46] Basu, P. K. '*Theory of Optical Processes in Semiconductor',* Oxford Univ. Press, Oxford, UK, 2003.

[47] Chuang, S. L. *Physics of Optoelectronic Devices*, Wiley series in pure and applied optics, John Wiley, NY, 2009.

[48] Sugawara, M. 1999 *Self-Assembled InGaAs/GaAs Quantum Dots,* in *Semiconductors and Semimetals* Volume 60 (San Diego, CA, Academic Press).

[49] Bimberg, D.; Grundmann, M.; Ledenstov, N. N. *Quantum Dot Heterostructures* (John Wiley and Sons, NY, 1999)

[50] Shirao, M.; Lee, S. H.; Nishiyama, N.; Arai, S. *IEEE J. Quantum Electron.*, 2011, vol. 47(3), 359-367.

[51] Lysak, V. V.; Kawaguchi, H.; Sukhoivanov, I. A. *IEE Proc. – Optoelectronics,* 2005, vol. 152(2), 131-139.

[52] Lysak, V. V.; Kawaguchi, H.; Sukhoivanov, I. A. *IEEE J. Quantum Electron.,* 2003, vol. 41 (6), 797-806.

[53] Kan, S. C.; Vassilovski, D.; Lau, K. Y *Appl. Phys. Lett.,* 1992, vol. 61(7), 752-754.

[54] Romero, B.; Arias, J.; Esquivias, I.; Cada, M. *Appl. Phys. Lett.,* 2000, vol. 76 (12), 1504-1506.

[55] Kaatuzian, H.; Taghavi, S. I. *Chinese Opt. Lett.,* 2009, vol. 7, 435-436.

[56] Basu, Rikmantra; Mukhopadhyay, Bratati; Basu, P. K. XXXth *URSI General Assembly and Scientific Symposium*, Istanbul, Turkey, August 14-21, 2011 (paper no. DP 3.1).

[57] Sakamoto, A.; Sugawara, M. *IEEE Photon. Technol. Lett.,* 2000, vol. 12, 107-109.

[58] Ishida, M.; Hatori, N.; Akiyama, T.; Ostubo, K.; Nakata, Y.; Ebe, H.; Sugawara, M.; Arakawa, Y. *Appl. Phys. Lett*, 2004, vol. 85(18), 4145-47.

[59] Fiore, A.; Markus, A. *IEEE J Quantum Electron.,* 2007, vol. 43 (3), 287-294.

[60] Markus, A.; Chen, J. X.; Lafaye, O. G.; Provost, J. G.; Paranthoen, C.; Fiore, A. *IEEE J. Sel. Top. Quantum. Electron.,* 2003, vol. 9 1308-14.

[61] Coldren, L. A.; Corzine, S. W.; *Diode Lasers and Photonic Integrated Circuits,* K. Chang (ed) Wiley, NY, 1995.

[62] Fathpour, S.; Z. M.; Bhattacharya, P. *IEEE Photon. Technol. Lett.* 2005, vol. 17 2250-52.

[63] Yavari, M. H.; Ahmadi, V. *IEEE Photon. Technol. Lett.* 2009, vol. 15 74-77.

In: Quantum Dots and Quantum Cellular Automata
Editors: D. De, S. Bhattacharaya and K.P. Ghatak
ISBN: 978-1-62257-920-4

Chapter 5

COMBINATIONAL LOGIC CIRCUIT DESIGN USING QCA

***Ratna Chakrabarty*[1], *Kunal Das*[2,3], *Debashis De*[4,5]**
[1]Department of Electronics and Communication Engineering,
Institute of Engineering and Management, Kolkata, India
[2]Department of Information Technology,
B.P.Poddar Institute of Management and Technology, India
[3]Department of Eng. and Technological Studies, Kalyani University,
Kalyani, West Bengal, India
[4]Department of Computer Science and Engineering,
West Bengal University of Technology, Kolkata, India
[5]School of Physics, University of Western Australia,
The University of Western Australia, Perth, Western Australia

ABSTRACT

Quantum Cellular Automata(QCA) in the present days is a novel emerging Nanotechnology where the computation of the logic states are in nano level which can help to generate low power, high speed microcomputer. This technology has overcome the drawbacks of using CMOS technology for the generation of Digital Computer. In QCA information presents as either '0' or '1', with these two states information is transmitted with the proper help of clocking phenomenon. In this chapter different combinational Logic Circuits are being analyzed and the results have been compared with respect to their size and computational method using Quantum Dot Cellular Automata (QCA). The detail analysis of this chapter will help further development in combinational logic circuit with the use of QCA technology for the future generation of microcomputer.

5.1. INTRODUCTION

The QCA technology overcomes the limitation of CMOS technology like charge quantization, leakage of current, power loss etc. Shrinking transistors has been the major steps for the implementation of the new technology such as the concept of quantum dot phenomenon for low power dissipation and faster speed application in microelectronics technology. Quantum cell realized with metal islands (named as quantum dots) and tunnel junctions with the property of interdot tunneling due to capacitive action. No current flows or no power dissipate internally to the individual cells. With the advancement of the semiconductor device technology there has been an extensive research and interest in the field of Nanotechnology and Nanodevices. Considering the industrial applications and requirement, according to Moor's law in 1965, it is predicted that the number of devices integrated in a chip will be double in every 18th months. So with this exponential growth, rapid improvement with respect to power dissipation, chip size and cost of fabrication is very necessary. In the early 1970, the Field Effect Transistor (FET) has been the only choice for the high level integration technique. In 1993, the first concept was proposed by C.S.Lent et al [1] from a FET based paradigm to one of nanostructure called Quantum Dot Cellular Automata having a cell of bistable configuration represents either "1" or "0" in binary system as shown in figure 5.1.with compact size of quantum dots and with extremely low power delay device compare to conventional CMOS technology. The QCA circuit is formed by the array of quantum dot cells, which is operated without a current flow through the cell. QCA approach breaks the limitation of application of conventional CMOS technology. Quantum dots in a QCA cell are created by self assembly approach. The basic mechanism of operation of a quantum cell is columbic repulsion between dots. A quantum dot in a cell localizes charge because of its zero dimensional space structure. The barrier between the dots has to be thin enough to tunnel a charge from one dot to another in a cell by the process of quantum mechanical tunneling, with the leakage of electronic wave function it is possible for one dot electron to jump to another dot. This jump can be controlled in the time of fabrication by the physical separation of the neighboring dots or confining an electrical potential in a dot. Figure 5.2 shows the simplified schematic diagram of four dot quantum cell. Dots in the quantum cell are created by the lattice mismatch of semiconductor material such as GaAs/InAs. The basic configuration of the device usually consists of either 2 or 4 quantum dots and 1 or 2 extra electrons in the dots. If a clocking potential is added to the cell as shown in figure 5.3, the charge barrier changes the computation become possible with very low power consumption. A continuous varying clocking scheme moves the function through the line [2, 3, 4]. The computational problem in QCA can be solved by making the cell arrangement to their ground state by giving proper input which can excite the state and after that make the state to a new ground state. Making some arrangement with basic QCA logic circuits different combinational circuit can be implemented. According to Snider et al in [5], from their experimental approach QCA operation can be described based on the fabrication by electron beam lithography technique with dual shadow evaporation to form the island made up of aluminum and aluminum oxide tunnel junction [6]. From figure 5.2 QCA cell is formed by four dots D1-D4 which are coupled by tunnel junction between them, larger capacitance improve the bistability of the cell. E1 and E2 are the two electrometers. The experimental detail is given in [7, 8, 9, 10]. A small potential change in the dots being measured by the electrometer, a decrease in potential indicates decrease in conductance measured in electrometer meaning that an electron has entered in the measured dot. QCA operation started by first biasing the cell with the help of gate voltage so that an excess electron can lie between the dots either D1 or D2 and the second electron is on the switching position between the dot D3 or D4. A voltage applied between the Input gates such that V1= - V2. When this differential voltage switches from the negative to the positive value, the electron stored in position say in position D2, changes position from D2 to D1. Figure 5.3 shows the conductance of the dots between D1 and D2. When electron switches from D2 to D1 there is a drop in conductance in E2 and as a result electron moves from D3 to D4. In this way with the changes of polarization of the electrons, the operation of QCA has been performed. The charge configuration of a QCA cell having four dots can be characterized by the equation of polarization

$$P=\frac{(p1+p3)-(p2+p4)}{p1+p2+p3+p4}$$

where ρ_i is the charge on each of the quantum dot.

This polarization is not a dipole moment rather it is the measurement of charge along the diagonal of the quantum cell. The ground state is defined by the superposition of the equal weight of the two basic configurations, results the net polarization as zero. Since the cells are capacitively coupled ground state of all the cells in a binary wire has the same polarization as shown in figure 5.1. A strong stimulus may change the electron polarization. Stimulus is applied in terms of input and due to kink energy represented by

$$E_{i,j}^{kink} = \frac{1}{4\pi\varepsilon_0\varepsilon_r \, |r_i - r_j|} q_i q_j$$

where ε_0 is the permittivity of free space, ε_r is relative permittivity of material, q_i and q_j are electrical charges of cells 'i' and 'j' and $|r_i - r_j|$ is distance between two cells [11,12]. Electrons are forced to tunnel through the tunnel junction with stimulus and attained a definite polarization depending on input. Electrons are always be apart a maximum distance due to columbic interaction. Electrons are not allowed to tunnel outside the QCA cell due its confinement. Columbic impulsive force generated by QCA cell will cause a Quantum mechanical probability to impress the next QCA cell; this effect depends on 'Radius effect' [42]. There are four possible realization of the QCA cells [43] are i) The metal island QCA cell ii) The semiconductor dot QCA cell iii) Molecular QCA and iv) Magnetic QCA

5.1.1. The Metal Island QCA cell

In this chapter all the circuits are explained using metal island QCA, where aluminum island is used as the dots and they are connected with aluminum oxide tunnel junction as explained above. The area of the tunnel junction determines the operating temperature of the device operating temperature is generally 10 mkelvin which is a disadvantageous one.

5.1.2. Semiconductor Dot QCA

For the extensive availability of semiconductor material, this type of QCA device has some advantage with respect of well understood behavior and operating methods [13], but for the manufacturing of nanodevice at room temperature the operation is not much reliable.

5.1.3. The Molecular QCA

Molecular QCA has some advantage over the metal dot QCA and semiconductor QCA as the manufacturing process is little less complex than the two above methods. Here one single chemical molecule can be used as a QCA device and redox centers are used as the quantum dots which store the information. One advantage of using this QCA that it can operate at room temperature but there are various during manufacturing process such as the choosing of the material and the in the clocking mechanism.

Under two state approximation from Hilbert state functions the following equations for adiabatic states are given below.

5.1.4. Magnetic QCA

Cowburn et al proposed magnetic implementation in QCA [14], where magnetostatic interaction is possible between nanoparticles, offering bistable operation at room temperature. Nenoparticle either parallel or antiparallel movement gives the binary information either 1 or 0. One disadvantage of this device is that the dot size is much larger (100nm) than the previous types and speed of operation is very low.

$$\psi_1 = c_{a1}\psi_a + c_{b1}\psi_b \tag{5.1}$$

$$\psi_2 = c_{a2}\psi a + c_{b2}\psi_b. \tag{5.2}$$

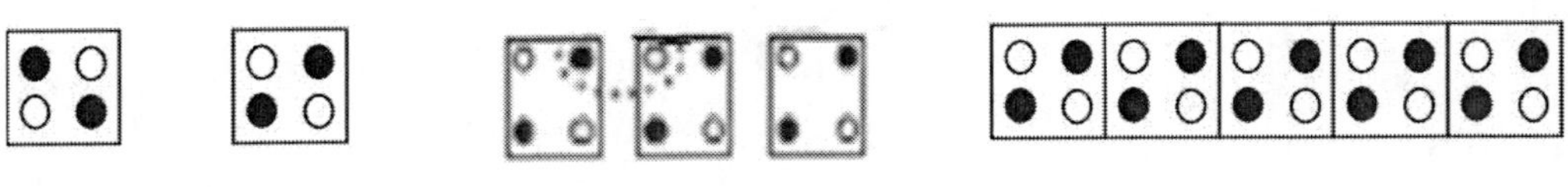

P= -1 P= +1

(a) (b) (c)

Figure 5.1. (a) Polarized cell (b) QCA cell showing Columbic interaction c) A binary wire using QCA.

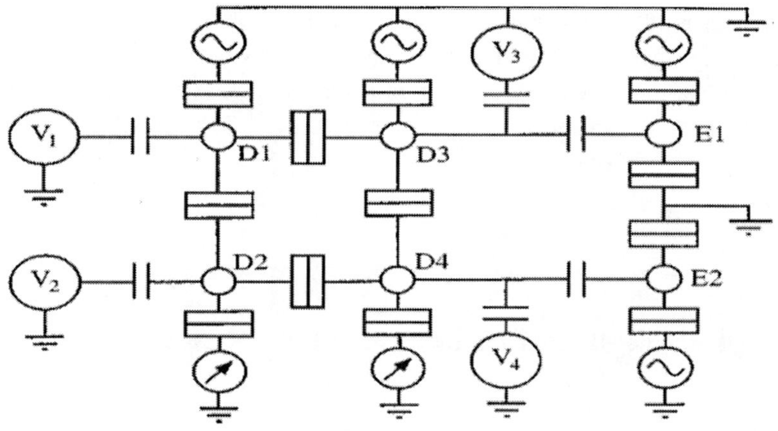

Figure 5.2. Schematic diagram of four dot quantum cell.

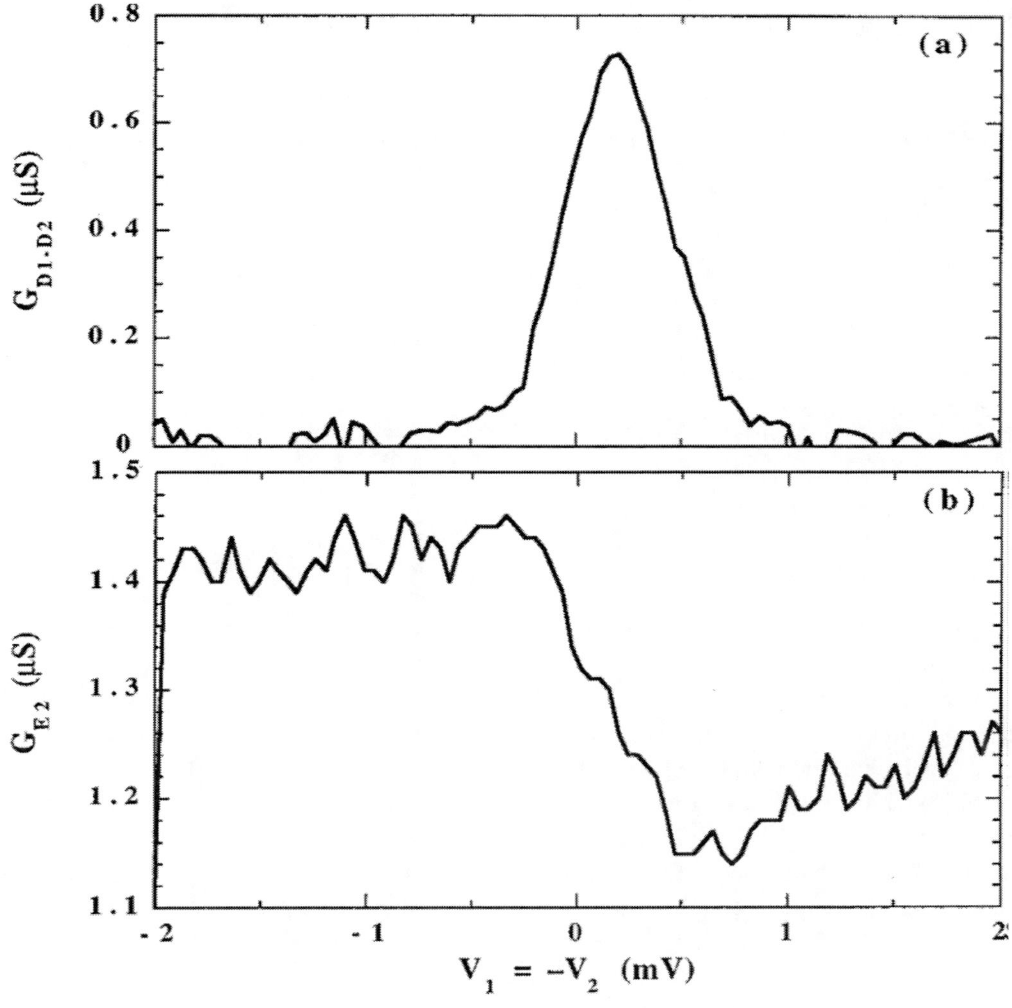

Figure 5.3. (a) Conductance through dot D2 and D1, Peak is shown when electron moves from D2 to D1. (b) Drop in Conductance when electron moves from D3 to D4.

where ψ_1 is the ground state wave function where E_1 is its Eigen Energy value, ψ_2 is the excited state wave function where E_2 is its Eigen Energy value. Ψa and ψ_b are the wave functions of state 0 and 1. The energy of the adiabatic states can be solved by two state model by using the following equation.

$$\begin{vmatrix} H_{aa} - E & \gamma \\ \gamma & H_{bb} - E \end{vmatrix} = 0 \tag{5.3}$$

where H_{aa} *and* H_{bb} are the Hamiltonians of the states 0 and 1 described as

$$H_{aa} = \langle \psi_a | H | \psi_a \rangle \qquad H_{bb} = \langle \psi_b | H | \psi_b \rangle \tag{5.4}$$

$$\gamma = \langle \psi_a | H | \psi_b \rangle . \tag{5.5}$$

where is the Electron Transfer (ET) matrix element. ET matrix can be defined as by adiabatic energy

$$\gamma = \frac{E_2 - E_1}{2} . \tag{5.6}$$

Under two state approximation equation 5.5 and equation 1.6 are equivalent. If E_1 and E_2 is solved the coefficient c $_a$ and c $_b$ for the ground and excited state can be solved by the following equation 5.7

$$\begin{bmatrix} H_{aa} & \gamma \\ \gamma & H_{bb} \end{bmatrix} \begin{bmatrix} c_a \\ c_b \end{bmatrix} = E \begin{bmatrix} c_a \\ c_b \end{bmatrix} . \tag{5.7}$$

The analogy helped locally passive and locally active molecular arrays for the integrated nano structures. This can be well understood by solving the Schrödinger equation and explanation from C.S. Lent et al [15, 16,17]. A single cell is expected to have a bistable behavior. Single cells interact to its neighbor cell and get influenced by other state. The bistable behavior and cell to cell coupling is well explained by the following figure 5.4. This bistable behavior is the basis for the application of quantum dot cells for computing the nano structure application. The cell to cell abruptness depends upon the strength of the tunneling energy to the coulomb energy for the electrons in the neighboring cells. A properly designed structure forms nonlinear coupling characteristics. No power dissipation occurs in this case.

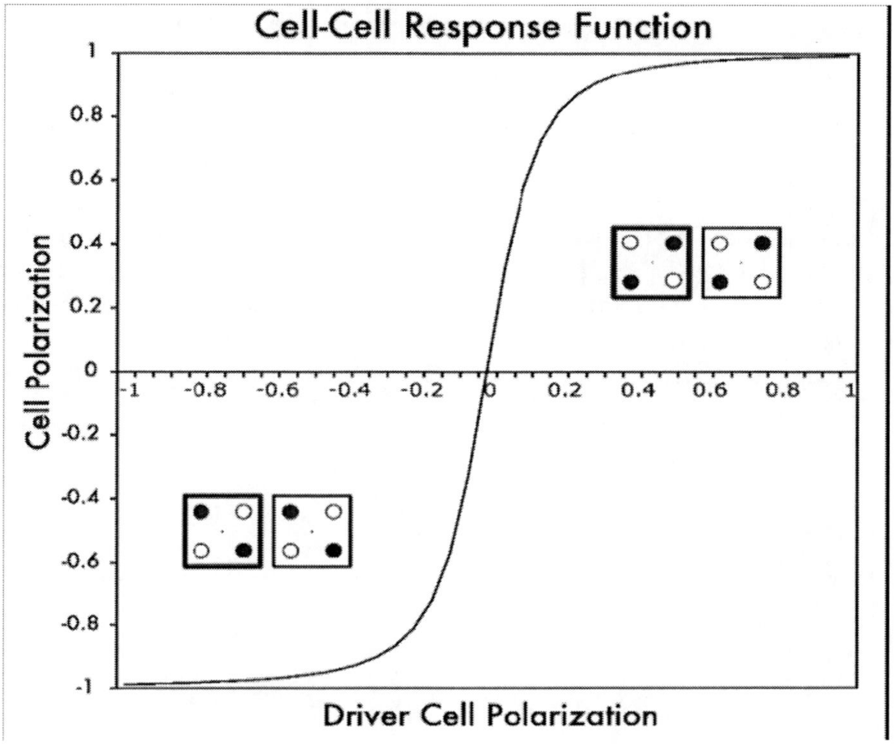

Figure 5.4. The bistable nature of a QCA cell shows the abrupt polarization shift, when the cell is subjected to an external influence coming from another QCA cell.

5.2. COMPUTING WITH QCA

In a QCA array of cells, the cells interact with their neighbors and no power dissipate like the other conventional circuit. The main concept in QCA array is that information content in the physical ground state of the system. Following steps are followed by the QCA circuit while computing Initial data is fixed by the input information given to the circuit at the edge by edge driven computation, mean setting the polarization of the input cell for delivering the energy to the system. Then the whole array is set to be relaxed or adiabatically transformed mean lowering or raising the potential barrier in the cell by applying the clock energy to the new ground state compatible with the input cells to be kept fixed. The presence of the clock signal synchronize and pipelined the architecture [18,19] Finally the results of computation are sent to the output cell by sensing the polarization of the output periphery. As no external signals are brought to the inside as no power flow from outside, internal cells cannot be influenced directly, this is the reason for ground state is used to carry the information as opposed to the state of the individual cells, thus has the advantage of noise immunity.

5.3. Influence of Thermodynamics while Computing by QCA

Ground state computing can be influenced by thermal fluctuations of the environment. Thermal noise can influence the circuit or can excite the state from ground state to the higher energy state and thus effects the computation and as a result error occur. Such errors are the relative magnitude of the excitation energy to the thermal energy kBT. External energy is inversely related to the size of the system, thus noise immunity increases for the size as become smaller and smaller. So a typical QCA cell is fabricated at cryogenic temperature while the molecular QCA (2nm) can operate at room temperature. A brief idea about the thermal behavior for molecular cell is given in [20]. Also electron transfer due to thermal excitation and optical excitation are explained both experimentally and theoretically in [21,22,23].

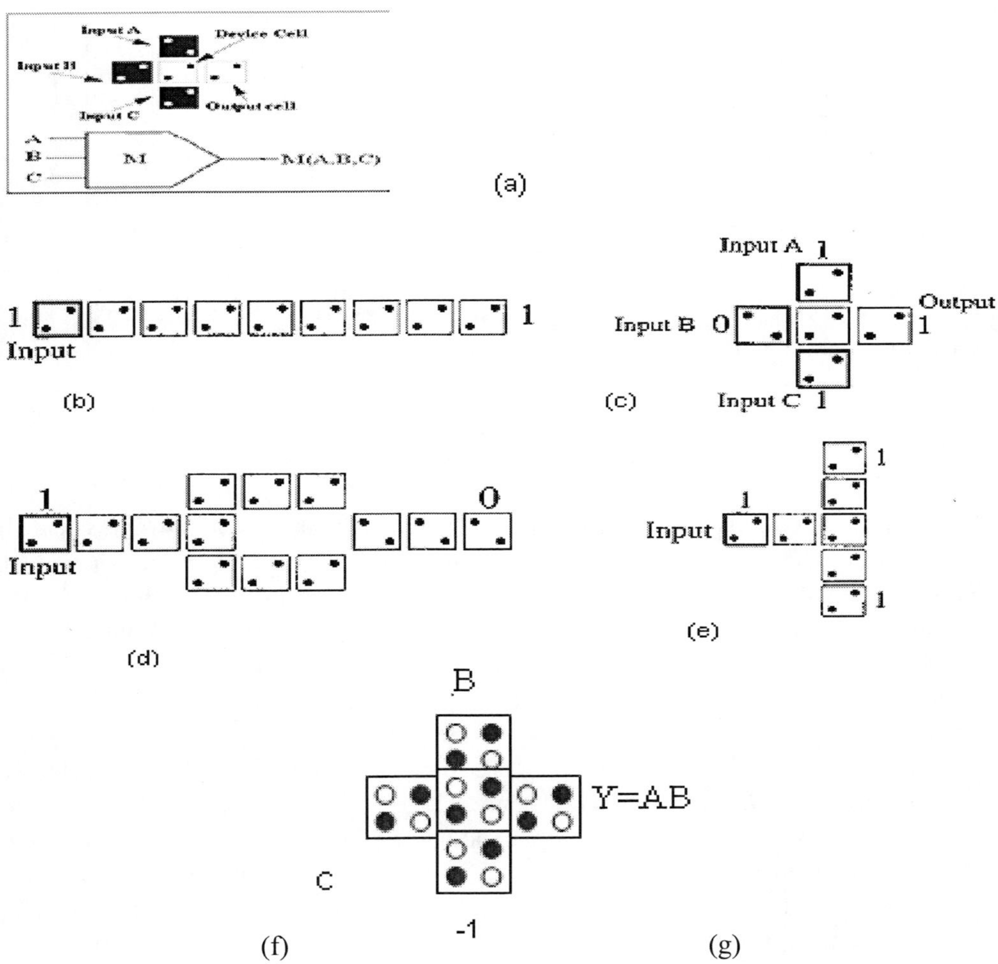

Figure 5.5. (a) and (c) Majority gate, (b) binary wire (d) Inverter (e) Fan Out (f) Majority OR gate (g) Majority AND gate using Quantum cell.

5.4. QCAD DESIGNER

For the rapid design and accurate simulation QCAD Designer tool is used for the implementation of the QCA circuits. This tool is designed and developed by the University of Calgary at the ATIPS Laboratory. The designer has three different simulation engines; among them the first one is digital logic simulator where cells are considered either null or fully polarized. The second one is a non linear approximation simulation engine which determines the nonlinear cell to cell response and the stable state of the cells in a design. The third one uses two state Hamiltonian method to form an approximation for the full quantum mechanical model in a system. Each set of simulation engine has some advantage as well as has some disadvantages also. But each simulation engine can perform a reliable and exhaustive verification as described in [24,25]. Typical Simulation parameters used in QCA are normally given as below:

Cell size are normally 18 nm, Number of samples 50.000, Convergence tolerance 0.0000100, Radius of effect 65.000000 nm, Relative permittivity 12.900000.

5.5. ELEMENTARY BUILDING BLOCKS IN QCA

The fundamental QCA logic circuits are the majority gate and the inverter. The majority gate has three inputs and one output. The output cell will polarize according to the majority polarization of the input cells. The center cell is the device cell. If A, B, C are the inputs, the majority gate computes the function M=AB+BC+CA. For any logic computing, we require universally a complete logic state. We need a set of Boolean logic gates that can perform AND, OR, INVERTER, FANIN and FANOUT operations. The combination of these is considered as universal because any general Boolean function can be implemented with the combination of these logic primitives [39]. By fixing the input polarization of one input +1 or -1 ie, logic 1 or logic 0, OR and AND logical operations can be performed respectively. Figure 5.5 shows the majority OR, majority AND and INVERTER gates realized with quantum cells In the inverter due to presence of 45 degree cell the signal alternates between the input value and its logic complement as it propagates the chain towards the output.

5.6. QCA CLOCKING SCHEME

Multi phased clocking mechanism is used in the QCA circuit to flow the information. Clock actually provides power to run the circuit. Four distinct and periodic clock signals are required for QCA circuits to operate and these signals are shifted by 90 degree of relative phase difference with each other. Clock signals are generated through an electric field which is applied to the cells to control the tunneling barrier height between the dots within a cell. The cells in each clock zone behave like a single latch. In clocking sequence the device transition occurs from a mono-stable to a bi-stable state. The clocking scheme can be accomplished by controlling the potential barriers between adjacent quantum dots. The four clocking zones can be divided as switch, hold, release and relax state as shown in figure 5.6. In switch phase the actual computation occurs according to the input. Here in this phase, inter dot potential barriers are low first and then barriers are raised and QCA cell become polarized to the state of its input driver. By the end of this clock phase, barriers are high enough and there is no possibility of electron tunneling and hence cell states are fixed. The second phase of clocking at where barriers are still high and does not allow the electron tunneling and hence output of this state can be used as the inputs to the next. In the third phase i.e. in release phase potential barrier is lowered and electron positions are configured according to the original state. The cells are in relaxed or unpolarized state which is the fourth clocking phase. So the overall polarization of the QCA cell is determined in the switch and hold phase where cells are in polarized state depending upon the neighboring cell polarizations. In release and relax phases the states are unpolarized. So QCA clocking mechanism consists of four clock signals of equal frequencies if one signal considered as reference (phase=0) others are delayed one (phase $=\pi/2$), two (phase$=\pi$), and three (phase$=3\pi/4$) quarters of periods. Signals are pushed from one clock region to another as the phase of the clock of these regions increases.

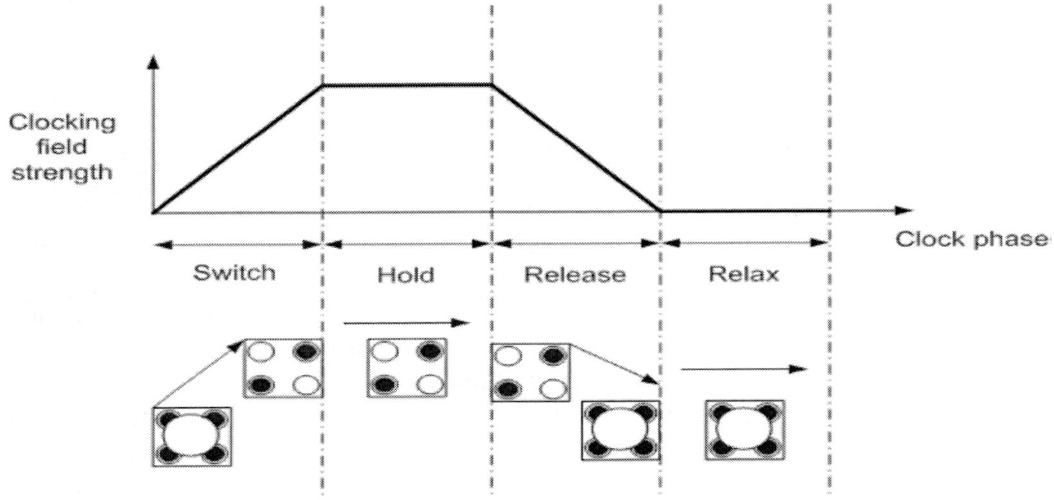

Figure 5.6. Four phases of clock signal and the associated QCA Cell polarizations

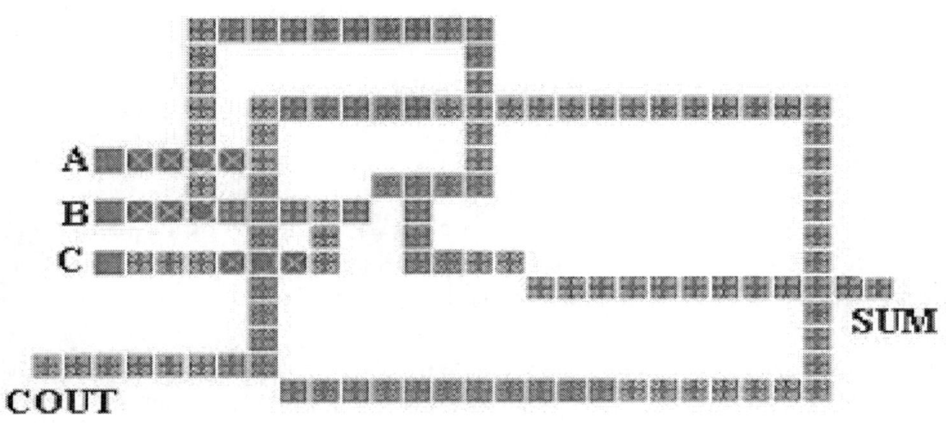

(a)

Figure 5.7. (Continued).

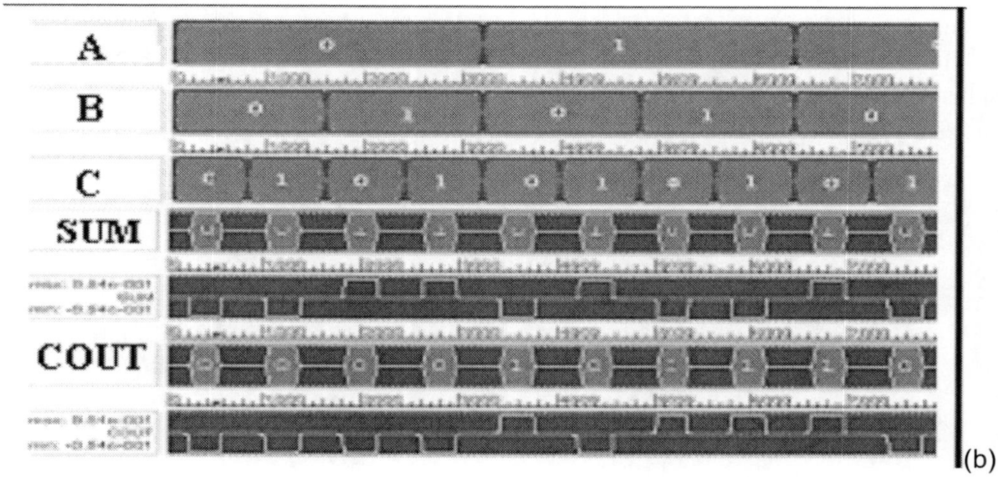

Figure 5.7. (a) Layout of one bit full adder proposed in [26] (b) simulated output.

5.7. Various Combinational Circuits Using QCA

In this chapter various combinational circuits using QCA has been described. Majority logic gates are used for all the circuit implementations using quantum cellular automata phenomenon. Various digital combinational circuits where output at any time of interval depends only on the present inputs and not on the past outputs are i) adders, ii) subtractors, iii) multiplexers and demultiplexers, iv) comparators, v) encoders and decoders vi) parity generators/checkers and vii) code converters.

5.7.1. Full Adder/Subtractor

The basic building blocks of arithmetic unit of digital computer are adders and subtractors. Vishnu C Teja et al [26] realizes one bit full adder with less numbers of hardware, they had used 124 cells to design the one bit full adder circuit with an area of 0.040176 μm $^{2.}$ Figure 5.7 shows the circuit implemented by them with the output using two clock cycles.

In [27] E.N Ganesh et al implemented another full adder circuit using majority logic method. The corresponding adder circuit and the simulated output are shown in figure 5.8. Here in the present circuit 147 cells are required which is more than the number of cells that are required in the previous design [26]. Also four clock cycles are required for the necessary output.

Figure 5.9 shows the full subtractor circuit by E.N. Ganesh et al[27] where more number of cells are required than the adder circuit but the number of clock cycles used are more than the adder circuit.

Figure 5.10 shows another subtractor designed by S.K.Lakshmi et al in [28]. Half subtractor is shown in figure 5.10 (a) where it required 77 cells with an area of 83160 nm^2 and figure 5.10 (b) shows the simulated out put of the circuit. Figure 5.10 (c) shows the full subtractor using two half adder circuits with 188 cells and an area of 208000nm ^{2} and the simulated output of the layout is shown in figure 5.10(d). According to the authors these designs can be used as the building blocks of the nano computer.

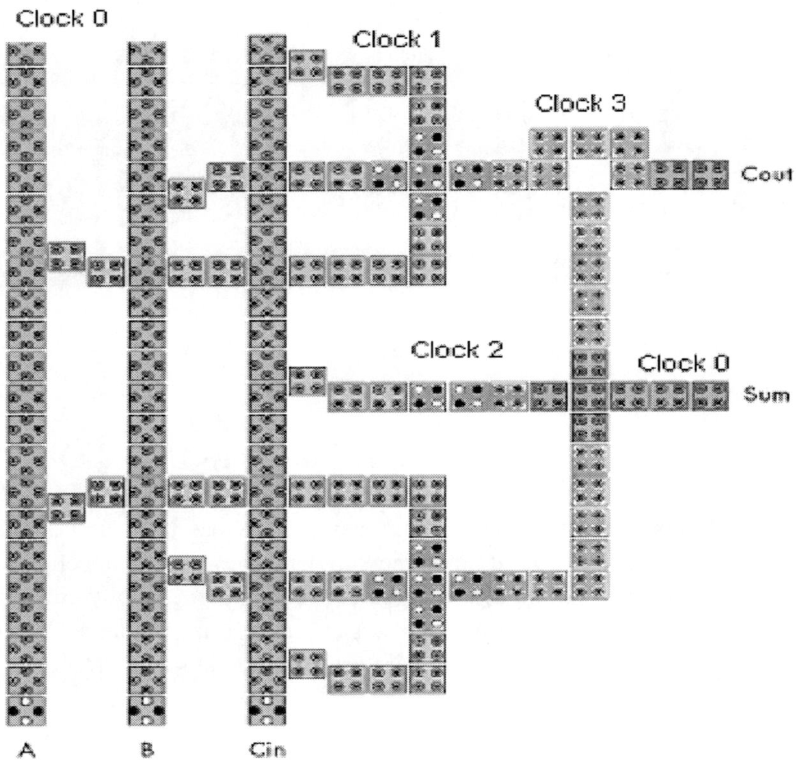

(a)

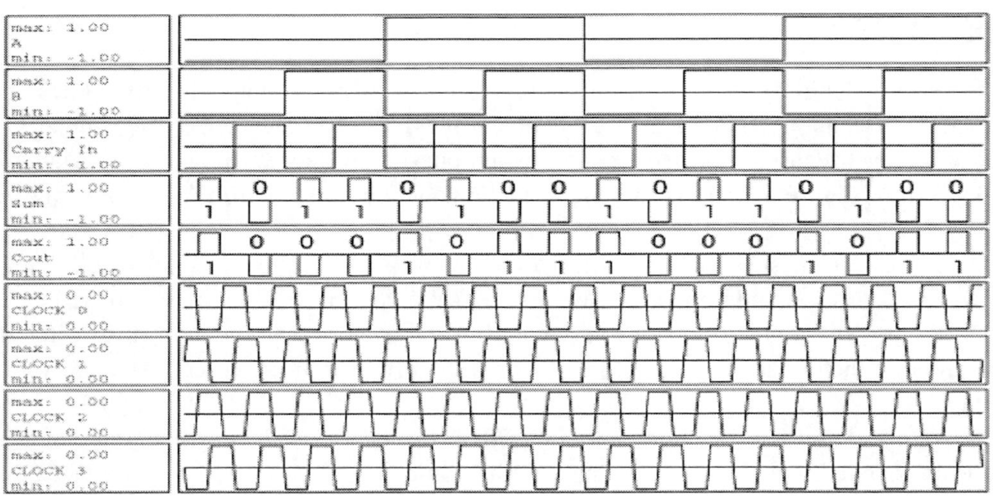

(b)

Figure 5.8. (a) QCA full adder proposed in[27] (b) Simulated output.

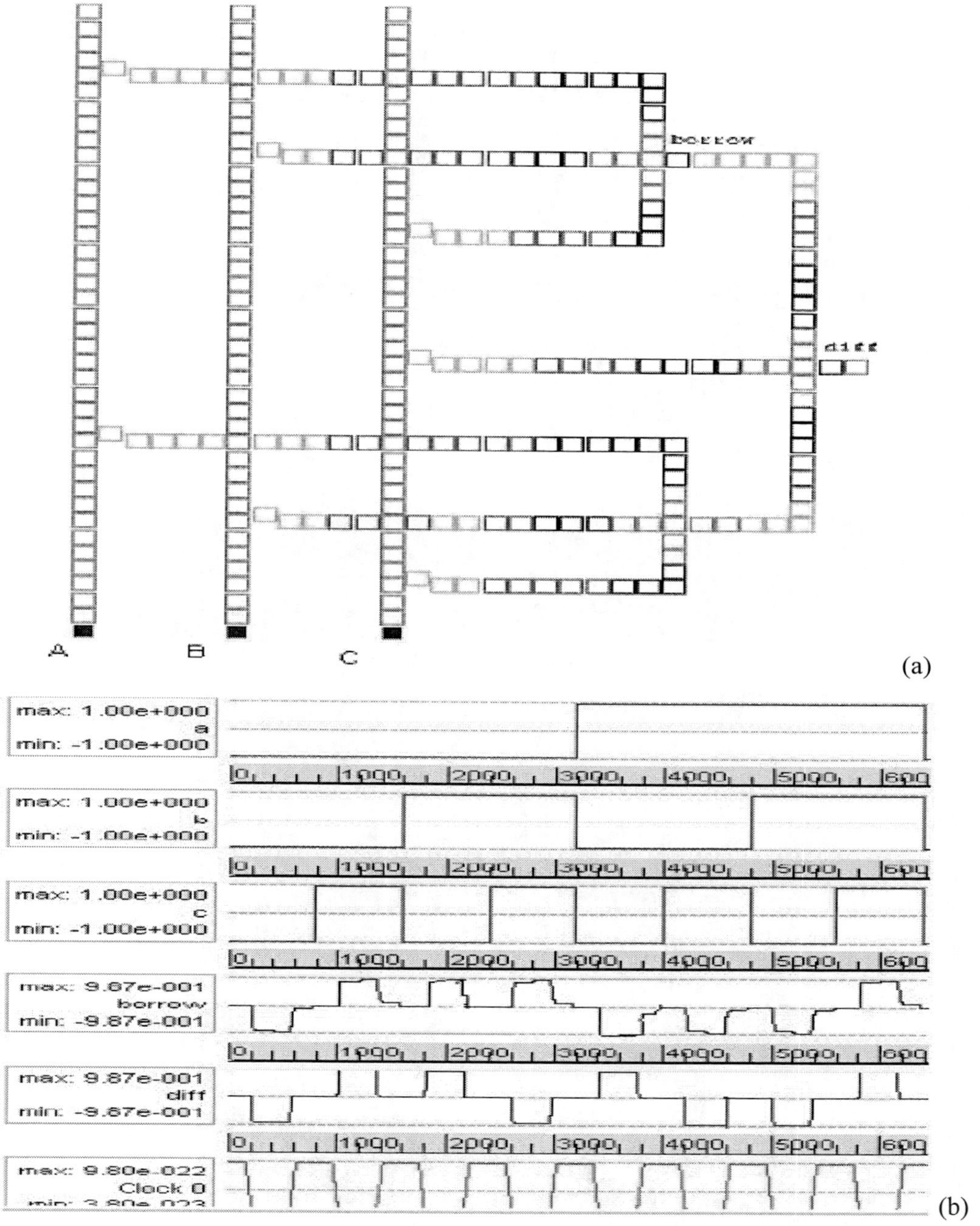

Figure 5.9. (a) QCA full subtractor proposed in [27] (b) Output simulated waveform.

P.K. Lala et al in [29] used reversible qca logic circuit to implement a ripple carry adder design which is the Simplest structure. Here full adders generates sum and carry outputs depending upon the addend bits and the carry input. Disadvantage of the circuit is carry always propagate through the circuit. Another adder circuit designed by them is carry look ahead adder which is the fastest of all other adder circuits according to them.

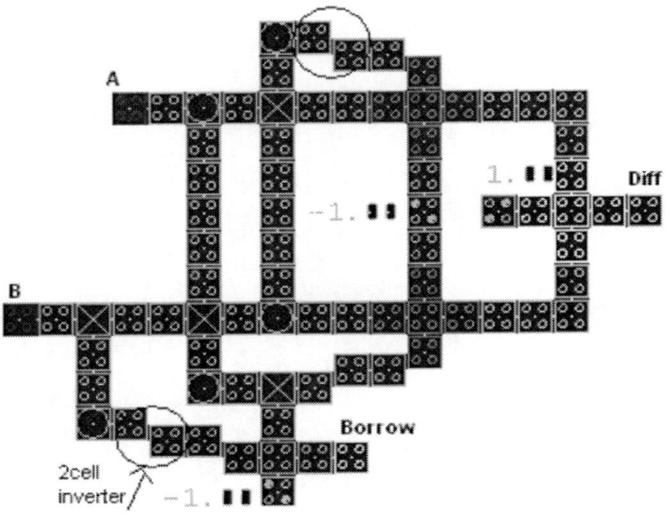

(a)

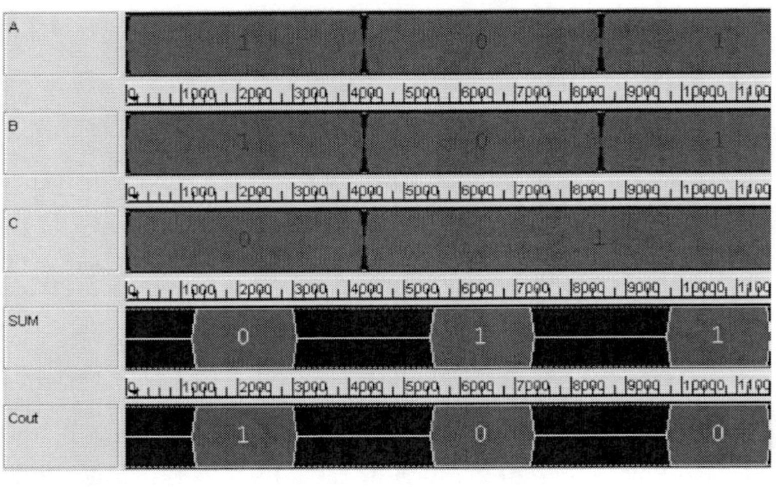

(b)

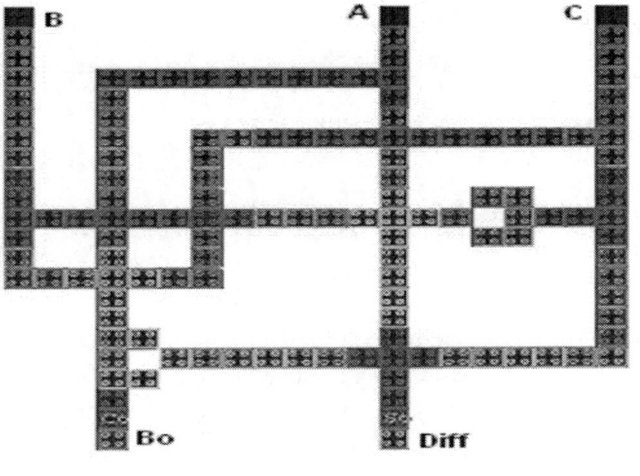

(c)

Figure 5.10. (Continued)

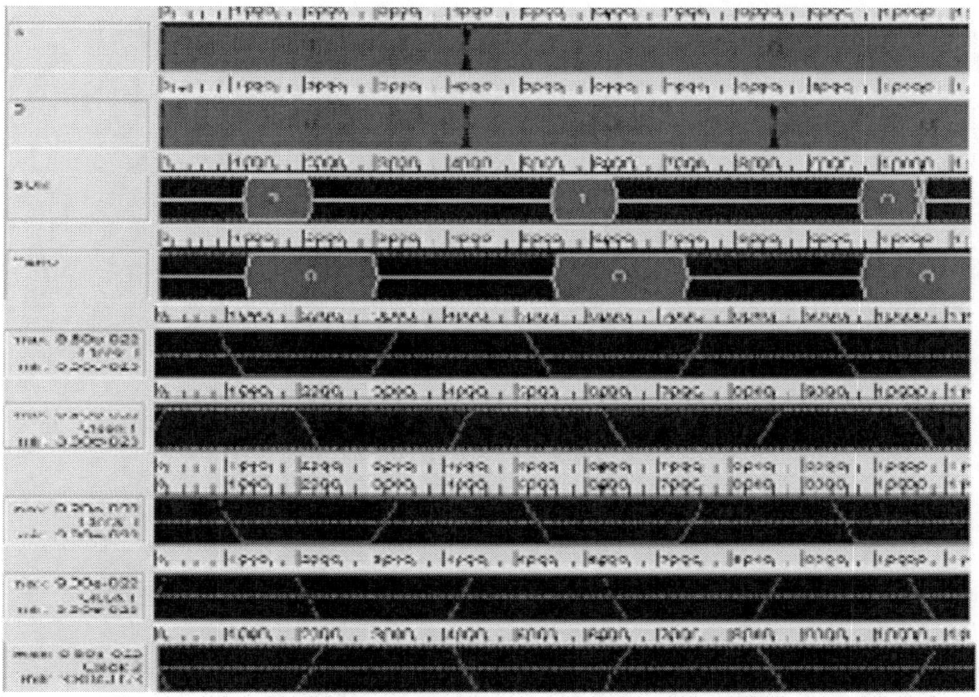

(d)

Figure 5.10. (a) QCA layout of Half subtractor circuit in [28] (b) Simulated output of the Half subtractor (c) QCA layout of Full subtractor using two half adder (d) Simulated output of the Full subtractor circuit.

They also made carry skip adder which is a compromise between the two adders. Figure 8 shows the ripple carry adder circuits with and without Fan out. Here sum bit is *Si* and carry bit is *c i+1* where inputs are a,b and carry input is c. The output expressions are

(1) Si = a xor b xor c;
(2) Ci+1 = ab + bc + ca;
(i = 0,1,2,….)

Figure 5.11 shows the circuits proposed by the authors and the improved carry skip adder circuits. Here all the circuits performances are compared with the Fredkin gate where the number of garbage output is more.

Figure 5.12 shows another complex adder circuit designed by C.S. Lent et al in [30] with 192 cells which can be used for useful computation in nanotechnology.

Figure 5.13 shows another adder circuit designed by M.Askari in [31] Here four bit adding is possible by carry out of the first full adder with the carry in of the second full adder with the adding of some delay in the circuit. Ripple carry adders are made by series connection of the full adders. The covered area of a 4 bit BCD full adder is 5.8 μm^2.

In [32] one adder circuit is designed by W.Wang et al with 145 qca cells, where crossover is highlighted which shows that cell A2 is not polarized according the input coming from cell A. Instead of this A2 is polarized to a state complementary to the cell B. This problem arise due to crossover plane, as A2 is closer to B it is polarized accordingly to B. This is called sneak noise where multiple inputs are there in QCA layout and it is known that polarization of a cell depends upon its neighbor cell and the greater the distance weaker will be the kink energy. Figure 5.14(a) shows the layout and figure 5.14(b) shows the simulated output of the design proposed by [32]. The problem of crossover for multiple input and sneak noise is overcome by K.Kim et al [33]. In crossover the vertical wire is controlled by the [clock (i+1) mod 4] and the cells in the left of the vertical wire is controlled by clock i. the input cells are controlled by the clock of [(i-1) mod4] and output cells are controlled by the [clock (i+1) mod 4.]. in the majority gate centre cell and its neighbor cells are controlled by the clock i. The adder design has three stages and each stage uses one clock cycle. The modified design proposed in [33] is shown in figure 5.14(c) and the simulated output in figure 5.14(d).

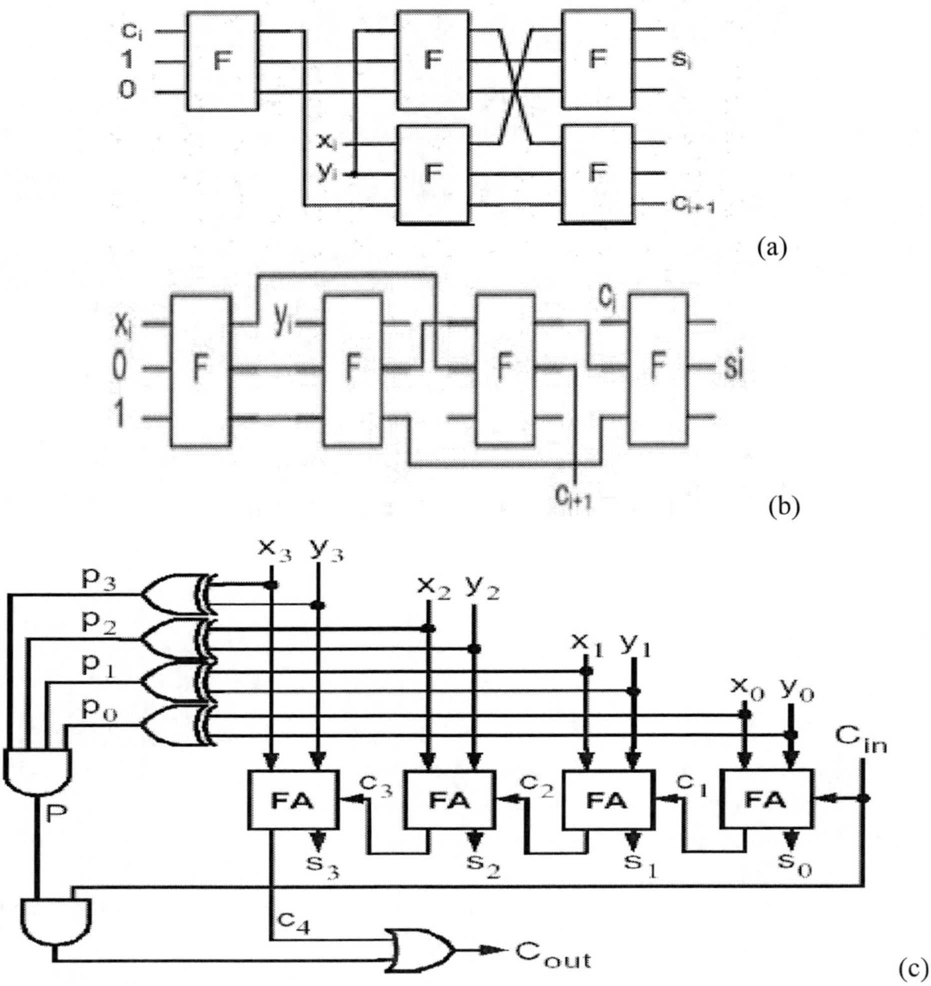

Figure 5.11. (a) Ripple carry adder with Fan Out (b) Improved Ripple carry adder without Fan Out. (c) Carry skip adder.

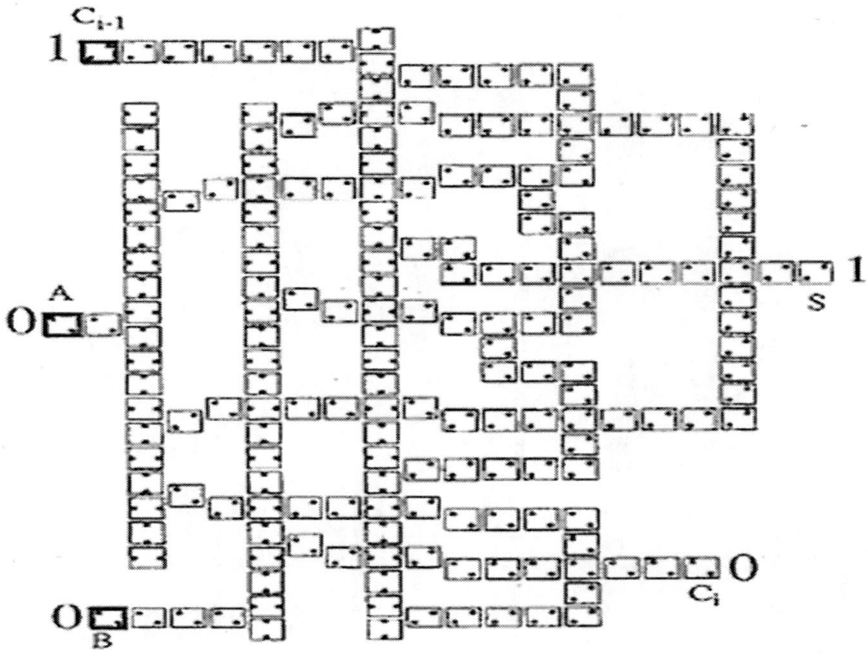

Figure 5.12. The QCA full adder with different inputs. The carry-in is one and the two addends are each zero. This makes the sum one and the carry out zero proposed in [30].

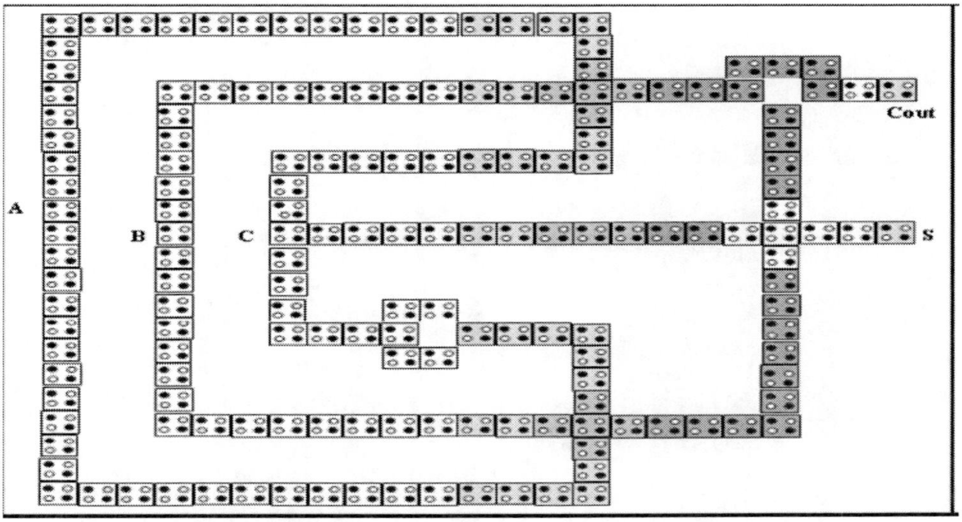

(a) (b)

Figure 5.13. (a) BCD full adder using QCA proposed in [31] (b) Simulated output.

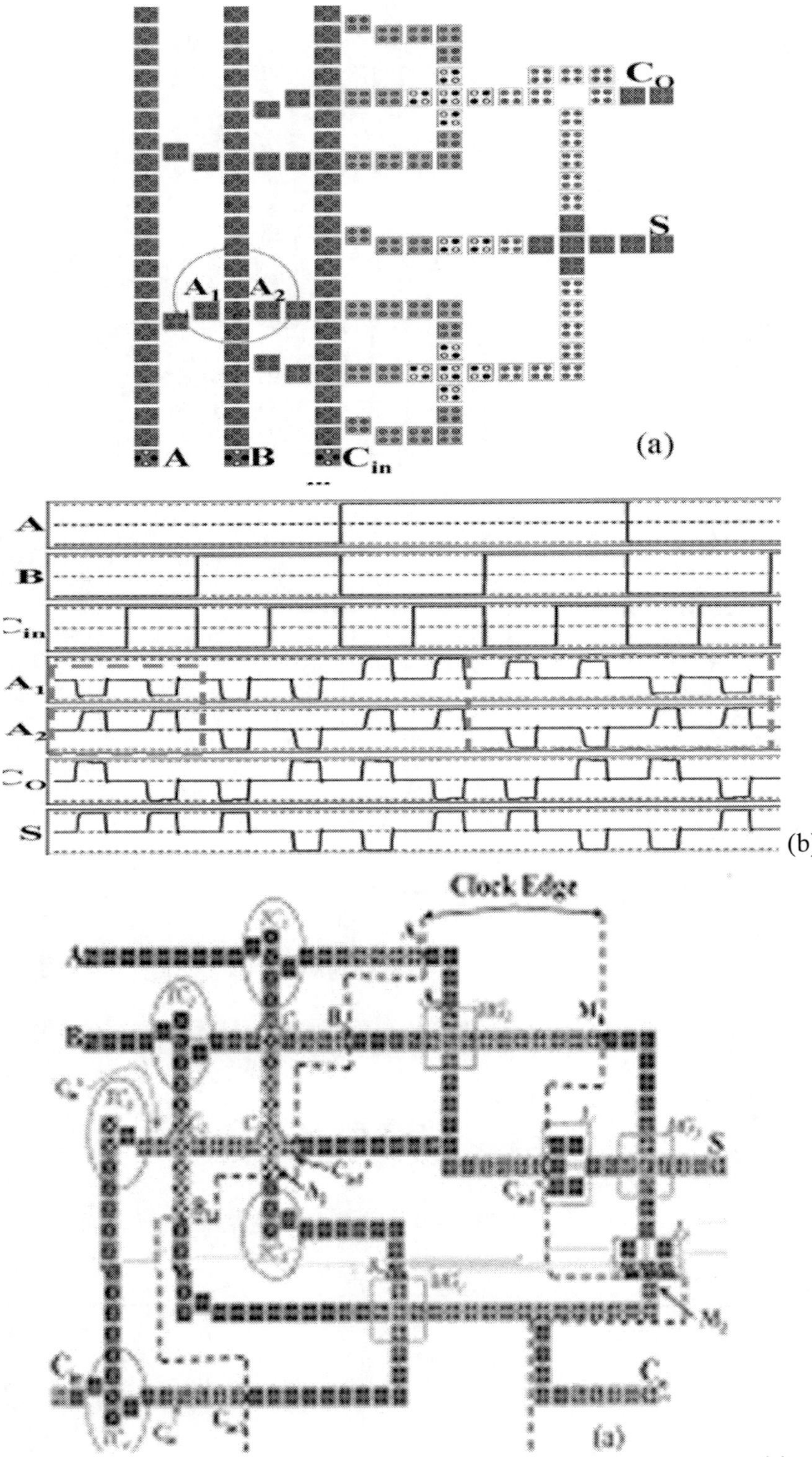

Figure 5.14. (Continued)

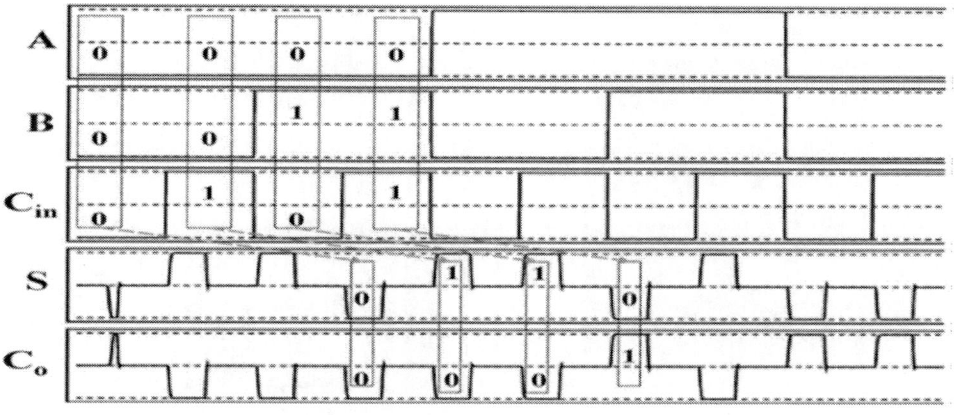

(d)

Figure 5.14. (a) QCA Layout of full adder proposed in [32]. (b) Simulated output. (c) Full Adder by [33] (d) Simulated output.

5.7.2. QCA Multiplexer/Demultiplexer

Multiplexers are one of the most important combinational circuits where a large numbers of information are transmitted via one single line. Using Quantum cellular automata multiplexers can be implemented for the low power consumption VLSI design. Here data selector selects one of the many inputs and steers the output accordingly.

In [27] E.N. Ganesh et al proposed a 4:1 multiplexer design using two 2:1 multiplexers using the majority voting method. Figure 5.16 shows the corresponding circuit and the output waveform. A, B, C,D are the inputs, Sx and Sy the control inputs and Y the output of the multiplexer.

$$Y1 = m[\ m((A,S'x,-1),m(B,Sx,-1),+1)] \tag{1}$$

$$Y2 = m[m((C,S'y,-1),m(D,Sy,-1),+1)] \tag{2}$$

$$X1 = m(Y1,S1'S2',-1) \tag{3}$$

$$X2 = m(Y2,S1S2,-1) \tag{4}$$

$$Y = m(X1, X2, +1) \tag{5}$$

From equation (1) and (2) Y1 and Y2 presents the output of 2:1 multiplexers. Equation (3) and (4) are the selection signals, selects any one of the 2:1 multiplexer and Y the final output.

A 1: 4 QCA demultiplexer also proposed by the authors where one input is transmitted over the several output lines. A, B be the select signal and D are the input signal and Y1, Y2, Y3 and Y4 are the output of the designed circuit. Figure 5.17 (a) shows the QCA demultiplexer circuit and figure 5.17(b) shows its output waveform.

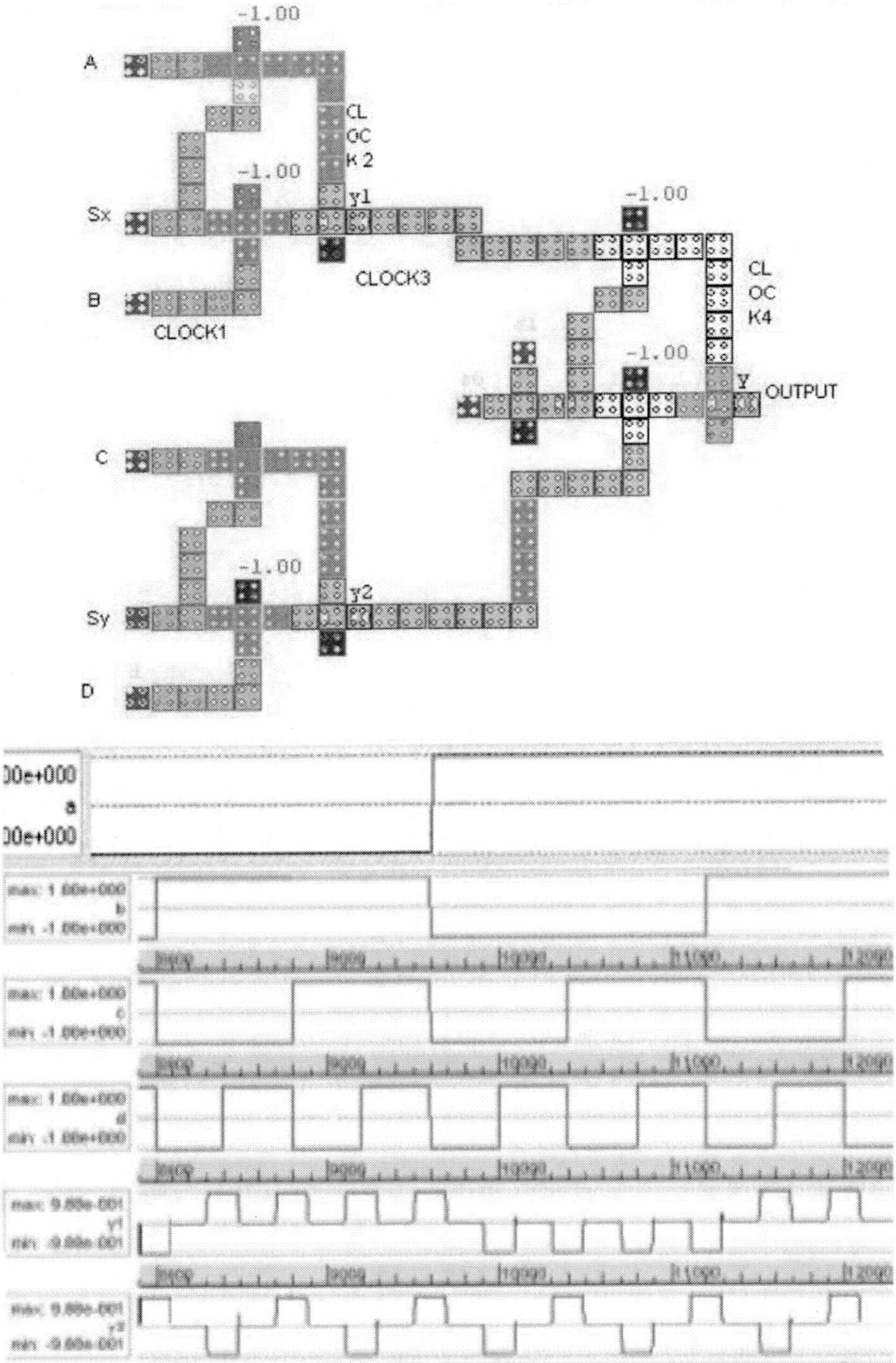

(a)

(b)

Figure 5.15. (a) A 4:1 Multiplexr proposed in[27] (b) Simulated output.

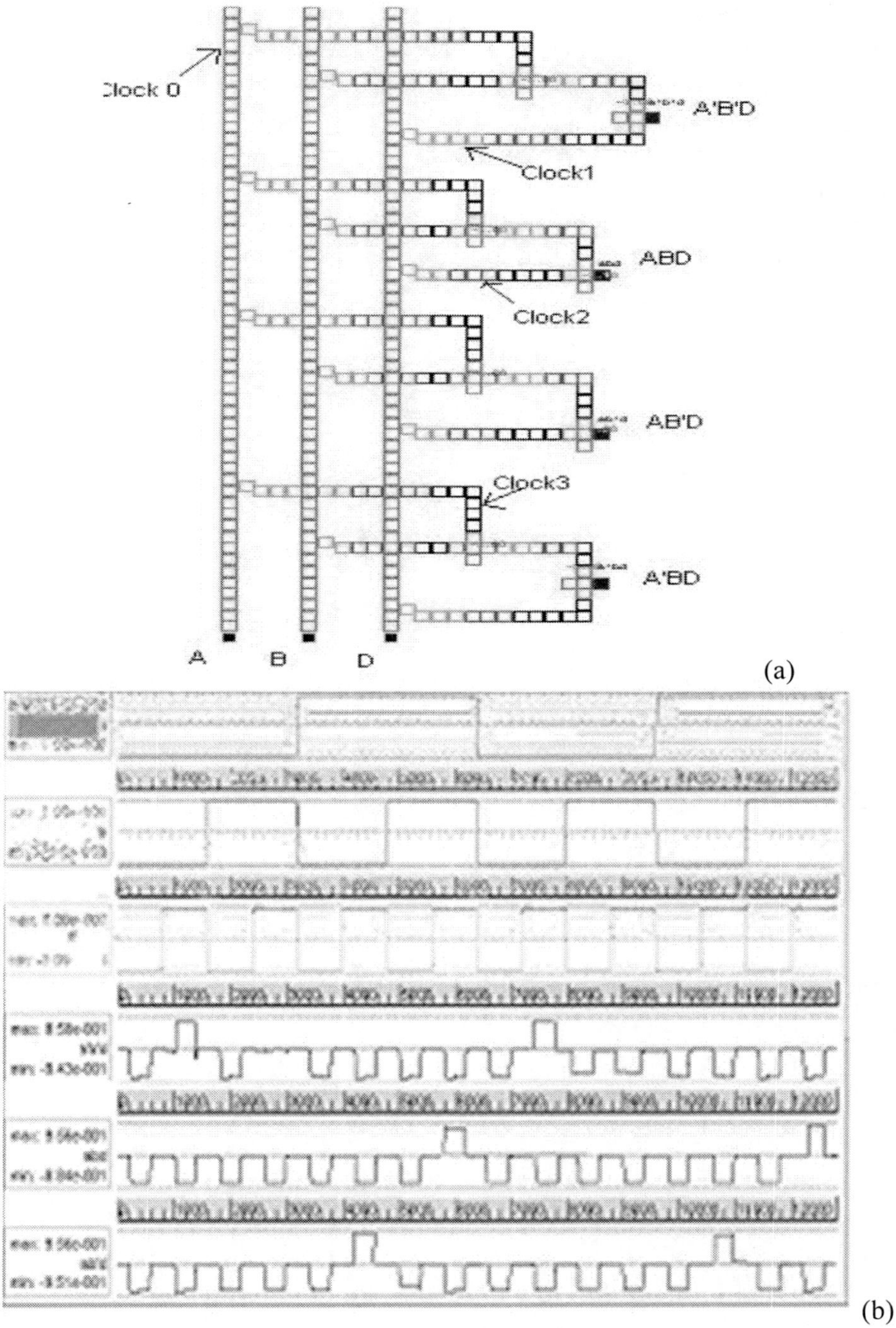

Figure 5.16. (a) QCA Demultiplexer circuit proposed in [27] (b) Simulated output waveform.

In [34] V. Mardiris et al presents a 2:1 multiplexer which can design 2^n to 1 multiplexers using the building blocks such as ($a \cdot \neg b$) block, AND block, OR block, delay block etc. the design consist of 67 cells covering with an area of 0.14 μm^2. The circuit diagram using QCAD designer tool is shown in figure 5.18.

In [35] another 2 to 1 multiplexer is proposed by Sara Hashemi et al where 36 cells are required for the design and total area used by the design is .06 μm ^{2}. Figure 5.19(a) shows the corresponding circuit diagram. Figure 5.19(b) shows the design proposed by L Sousa et al [36], where 88 cells are required for the design and the covered area is 0.14 μm ^{2} figure 5.19(c) shows the multiplexer proposed by K.Kim et al in [37] where the covered area for the design is 0.08 μm ^{2} and the area of the design is 0.08 μm ^{2} figure 5.19(d) shows the design proposed by M. Askari [38] where less number of cells are required for the design. The total number of cells required for the design is 34 and the area covered by the design is 0.04 μm ^{2}

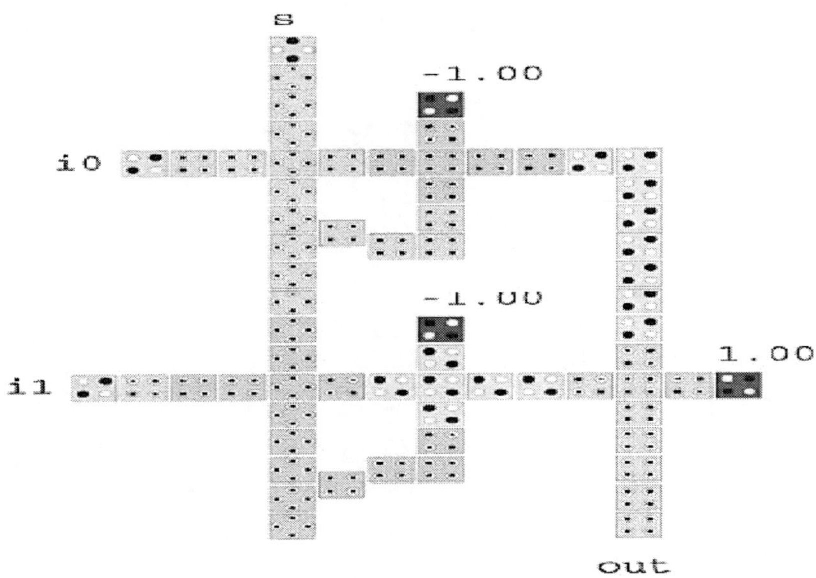

(a)

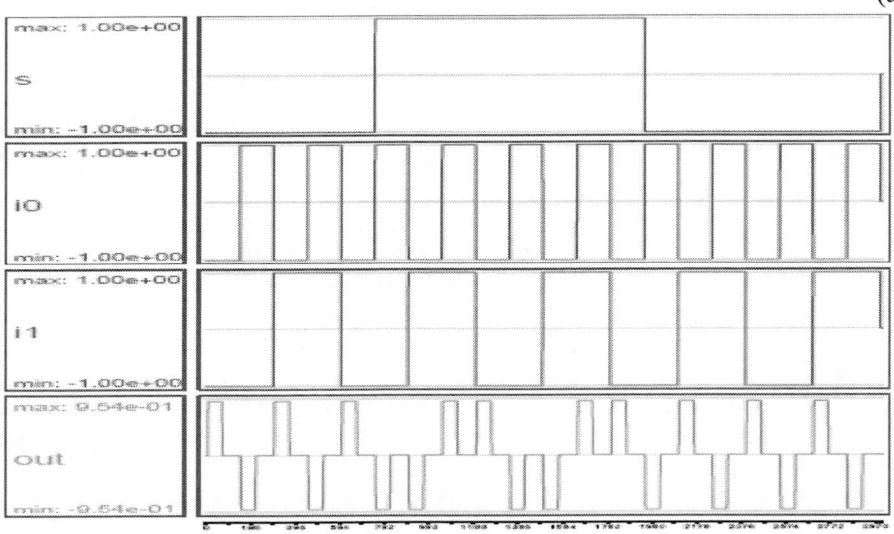

(b)

Figure 5.17. (a) Multiplexer proposed in [34] (b) Simulated output.

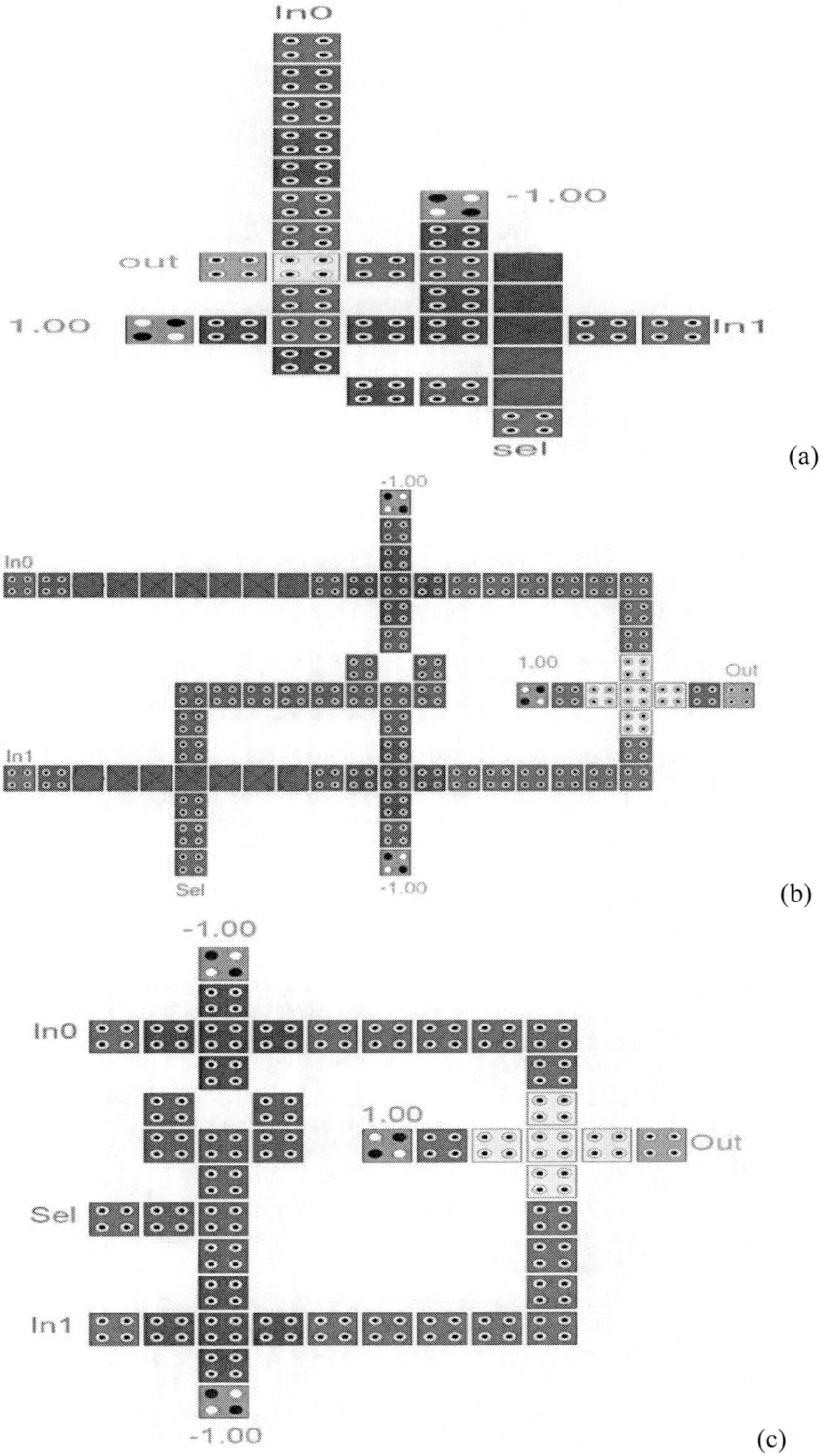

Figure 5.18. (Continued).

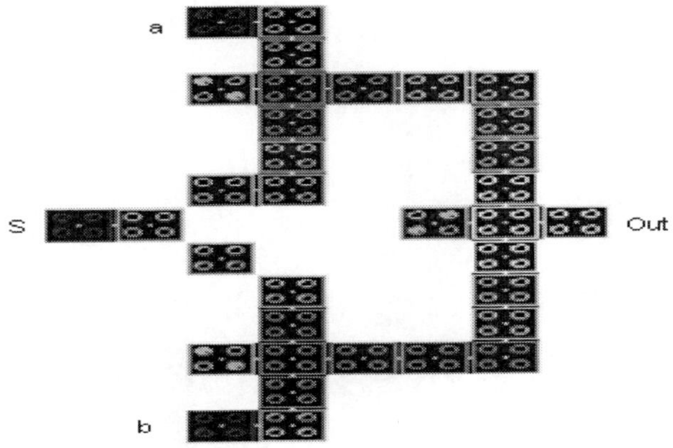

(d)

Figure 5.18. (a) 2:1 multiplexer designed in [35] (b) 2:1 multiplexer designed in [36] (c) 2:1 multiplexer designed in [37] (d) 2:1 multiplexer designed in [38].

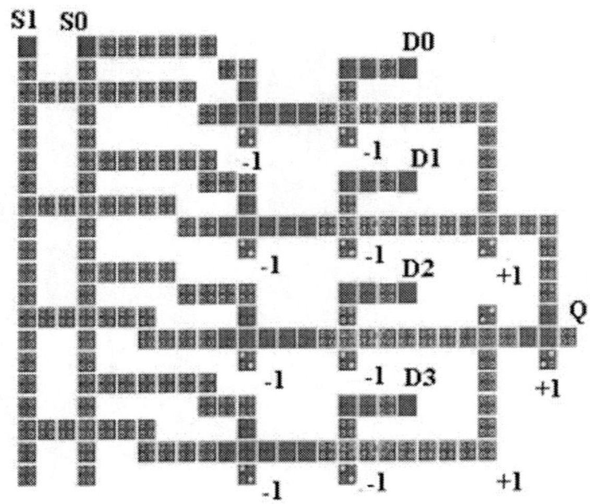

(a)

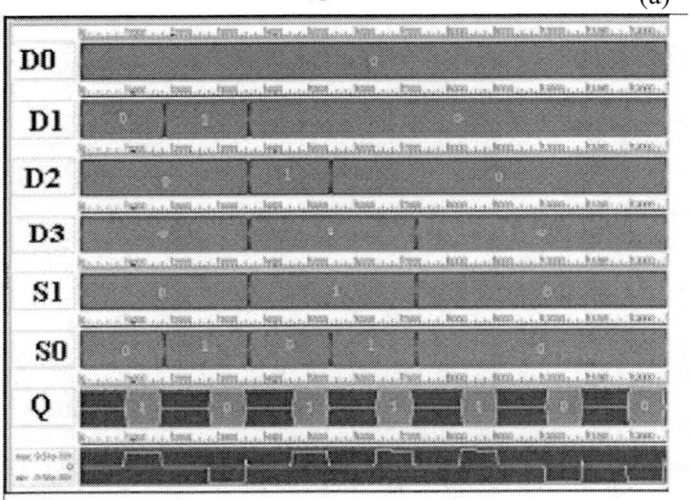

(b)

Figure 5.19. (a) 4:1 multiplexer proposed in [26] (b) Simulated output of 4:1 multiplexer.

Figure 5.19 shows another 4:1 multiplexer, proposed by Teja et al. [26], in which 246 cells are required for the design, with coverage of the area represented as 0.079704 μm^2

From the above explanation it is clear that various multiplexer circuits can be designed by quantum cellular automata using the QCAD designer tool for various digital circuit applications in nanolevel at zero temperature, with the lesser power consumption.

5.7.3. Parity Checker Using QCA

Parity checking technique is the most popular method to detect the errors in the storage bit pattern such as in memory cell, magnetic tapes etc. Using QCA parity checker circuit can be generated in nanolevel for microcomputer applications. Exclusive OR circuit is used to check the parity for a number. The layout for parity checker is shown in figure 5.21(a), proposed by V.Teja et al in [26]. According to their design 299 cell and an area of 0.096876 μm ^{2} is required. For the inputs 1010, 0001, 1110, 0011 the outputs are 0,1,1,0, as shown in figure 5.21(b). Out put is given at F, it hold 1 for odd parity and zero for even parity.

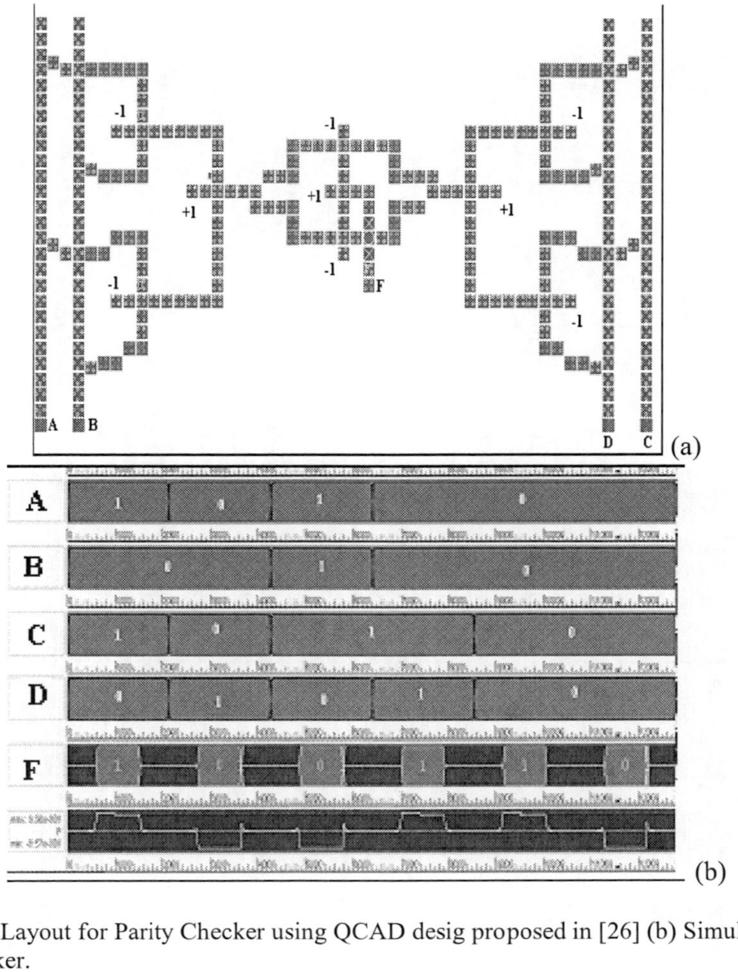

Figure 5.20. (a) Layout for Parity Checker using QCAD desig proposed in [26] (b) Simulated output of the Parity Checker.

5.7.4. QCA Encoder/Decoder

Most of the digital applications required decoding of data. Decoding is essential for digital display, demultiplexing and memory addressing etc. Using QCA combinational circuit such as Encoder/Decoder can be implemented, from different studies and analysis it can be possible to design various circuits with them. In [27] E.N.Ganesh implemented such a decoder circuit using QCA. Y1, Y2, Y3, Y4 are the output of the decoder with simple majority gate applications Figure 5.22(a) shows the design and figure 5.22(b) shows the decoded output result.

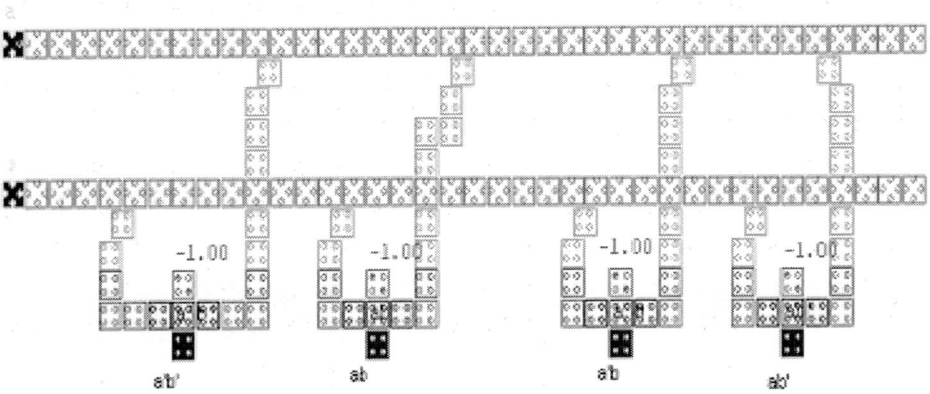

(a)

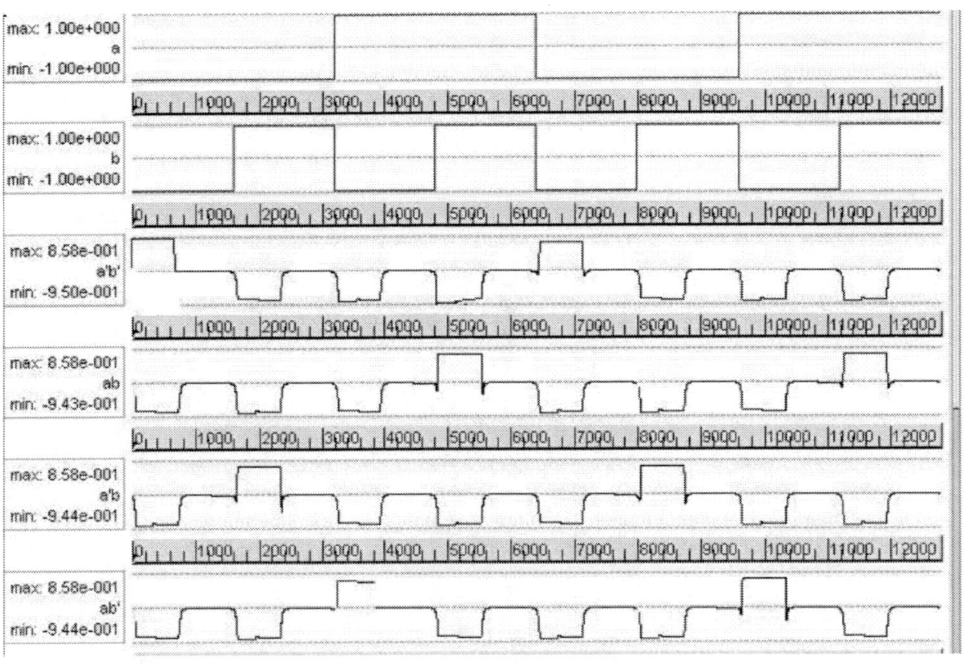

(b)

Figure 5.21. (a) QCA decoder circuit (b) Simulated output waveform.

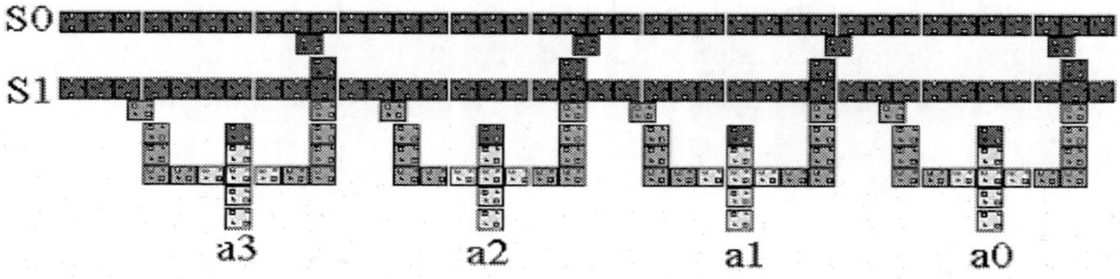

Figure 5.22. QCA decoder circuit proposed in [39].

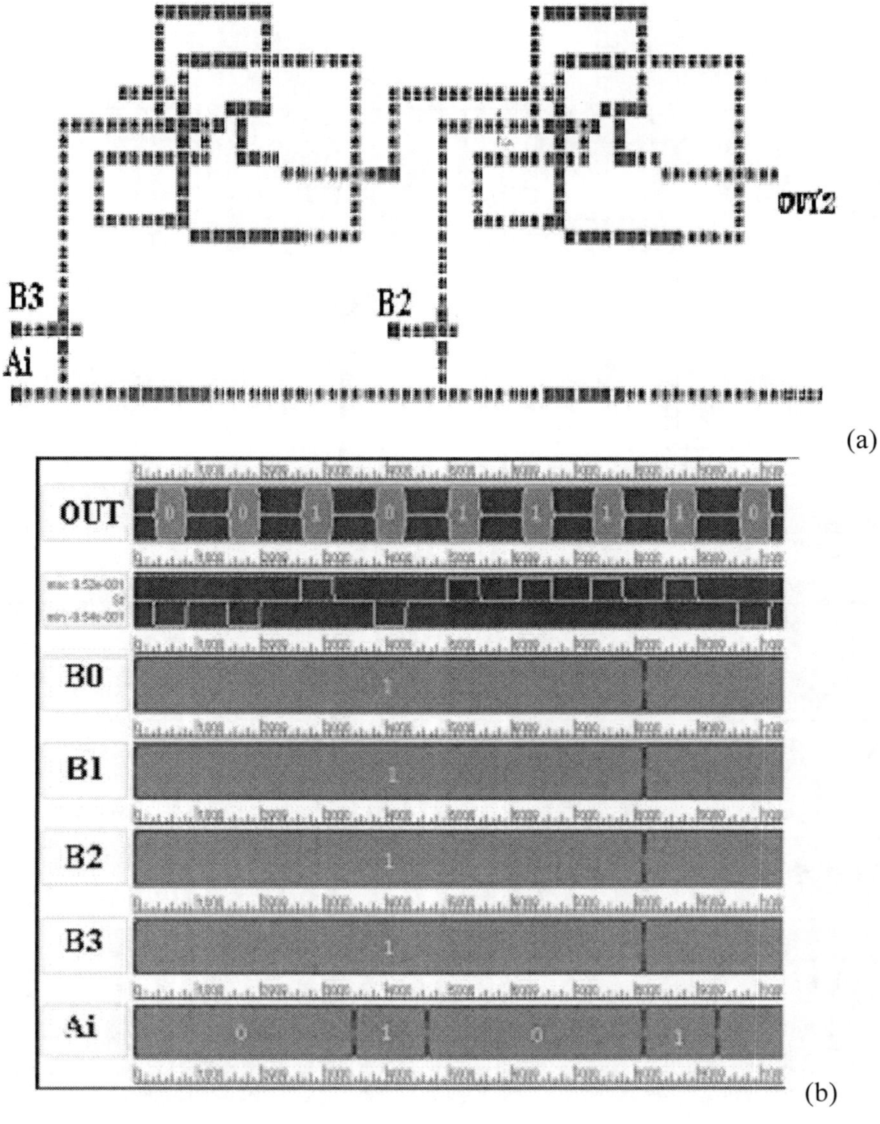

Figure. 5.23. (a) Multiplier using QCA (2 bits are shown) (b) Simulated output.

Another decoder circuit using QCA was implemented by K.Walus et al in [39] where decoder is designed for the generation of address bus of the memory cell, figure 5.23. shows the decoder circuit using QCAD designer tool According to the authors the implementation is optimal and can be used for other circuits

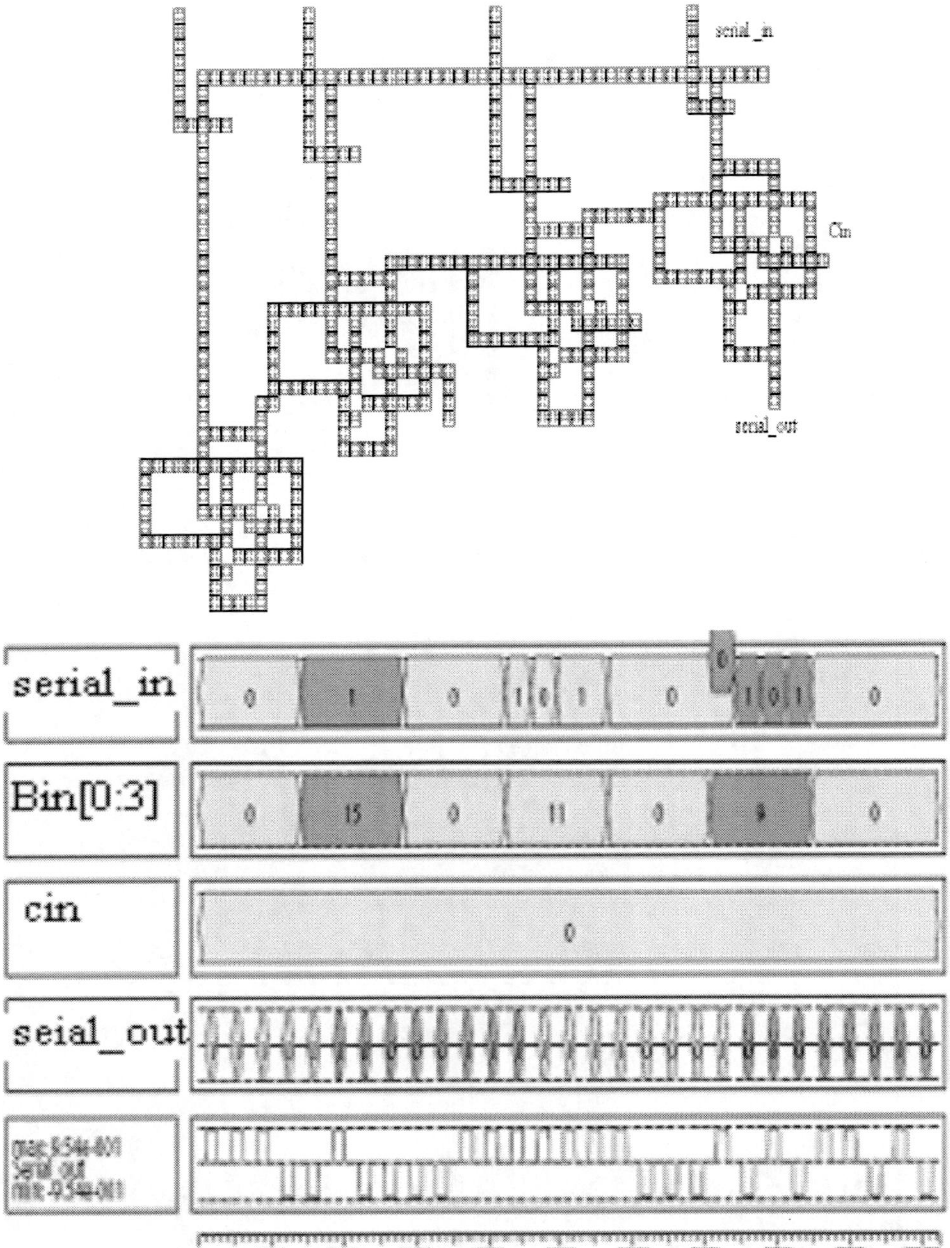

(a)

(b)

Figure 5.24. (a) Four Multiplier circuit using QCA proposed in [40] (b) Simulated Output.

5.7.5. QCA Multiplier

Multipliers are another combinational circuit which is described by V.Teja et al[26]. They implement here a 4 bit multiplication operation where one input is coming in parallel and others are coming serially, as the four bit multiplication is going here so a delay of 5 clock cycles occur here. So latches are added to control the flow of information. Figure 5.24(a) shows the multiplication operation of 1111 and 0001 where inputs in B are coming parallel and A inputs are coming serially. 853 cells are required for the design with a covered area of 0.276372 μm ^{2}. Figure 5.24(b) shows the corresponding simulated output.

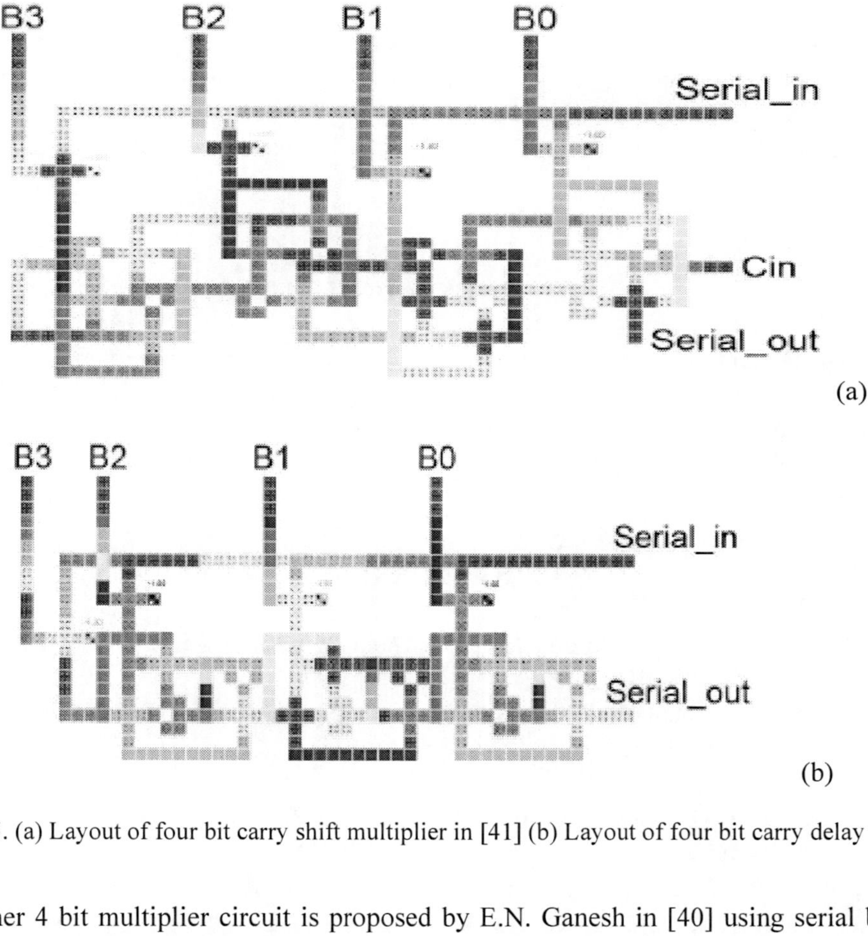

Figure 5.25. (a) Layout of four bit carry shift multiplier in [41] (b) Layout of four bit carry delay multiplier.

Another 4 bit multiplier circuit is proposed by E.N. Ganesh in [40] using serial bit full adder circuit. In serial bit full adder circuit carry in and carry out are connected internally with one clock cycle delay. Multiplier with larger word size can be implemented here by adding bits. 507 cells required for the design with area coverage of 0.61 μm ^{2}. One advantage of this multiplier is that carry propagated to the next adder and which can be used to add number of bits serially. Thus it can be used as the building block of the nanoprocessor. Figure 5.25(a) shows the multiplier circuit using QCA and figure 5.25 (b) shows the corresponding simulated output.

H.Cho et al in [41] described the multiplier circuit using carry shift multiplication with minimum delay in multiplication and carry delay multiplication with minimum latency in output. Figure 5.26(a) shows the layout design of four bit carry shift multiplier and figure 5.26(b) shows the layout of four bit carry delay multiplier circuit using QCA. Here in the carry multiplier full adder is used which is the layout of the optimized ripple carry adder and the bit serial adder is used for carry delay multiplier here carry in and carry out of the adder are connected with one clock cycle delay.

CONCLUSION

In this chapter we study and analysis various combinational logic circuits using Quantum Cellular Automata Technology which is a very effective and efficient way of manufacturing the nano devices. From the detail analysis of the above designs researchers will be benefitted in logic circuit design and in future there is a scope of reducing the area for future generation of microcomputer.

ACKNOWLEDGMENTS

K.P Ghatak and D.De are greatful to UGC for sanctioning the research project having the No-F 40-469/706 2011(SR) under which this paper has been completed besides D. De together with K.P Ghatak are greatful to DST, India for sanctioning the project under "SERC Fast Track Proposal of Young Scientist" scheme SERC/ET-0213/2011.

REFERENCES

[1] C.Lent, P.Tougaw, W. Porod and G. Bernstein, *Nanotechnology,* 4, (1993).
[2] Snider, G. et al. ICONN,(2006).
[3] Niemier, M. T., et al., In 27th Ann. *Design Automation Conference,(*2007).
[4] Kim, K., at el., IEEE Trans. *On Computer-Aided Design of Integrated Circuits and System,* 26(1),(2007).
[5] G.L.Snider, A.O. Orlov, I. Amlani, X. Zuo, G. H. Bernstein, C. S. Lent, J. L. Merz, and W. Porod, *Journal of Applied Physics,* 85(8),(1999).
[6] T.A Fulton and G.H.Dolan, *Phys.Rev*-59,Lett-59,109,(1987).
[7] A. O. Orlov, I. Amlani, G. H. Bernstein, C. S. Lent, and G. L. Snider, *Science* 277(928),(1997).
[8] I. Amlani, A. O. Orlov, G. L. Snider, C. S. Lent, and G. H. Bernstein, *Appl. Phys. Lett.* 72, 2179- 19981
[9] G. L. Snider, A. O. Orlov, I. Amlani, G. H. Bernstein, C. S. Lent, J. L. Merz, and W. Porod, *Semicond. Sci. Technol.* 13 (A130),(1998).
[10] I. Amlani, A. O. Orlov, G. L. Snider, C. S. Lent, and G. H. Bernstein, *Appl. Phys. Lett.* 71(1730), (1997).

[11] Amlani, A. Orlov, G. Snider, C. Lent, W. Porod, and G. Bernstein, *Superlattices Microstruct.* 25(1–2) (1999).

[12] C.S. Lent and P.D. Taugaw, *J. Appl. Phys.* 80(8) (1996).

[13] Walus, K,, G.A. Jullien, V.S.Dimitrov, *Proc. Asilomar Conference*,(2003).

[14] Cowburn, R. P. and M. E. Welland, *Science,* 287, (2000).

[15] P. D. Tougaw, C. S. Lent, and W. Porod, J. Appl Phys. 74, (1993).

[16] C. S. Lent, et al. " in *Proc. Eur. Conf. Circuit Theory Design*, ECCTD'97, Budapest, Hungary,(1997).

[17] W. Porod, C. S. Lent, G. H. Bernstein, A. O. Orlov, I. Amlani, G. L.Sneider, and J. L. Merz, *Int. J. Electron.* 86(5),(1999).

[18] Lent, C. S., Tougaw, P. D. and Porod, W., Proc. Workshop on Physics and Computation, Phys Comp 94, Los Alamitos, California. *IEEE Computer Society Press,* (1994).

[19] C. S. Lent and P. D. Tougaw, Proceedings of the IEEE, 85,(4),(1997).

[20] Yuliang Wang and Marya Lieberman, *IEEE Transactions on Nanotechnology,* 3(3),(2004).

[21] A.C. Ribou, J.P. Launay, M. L. Sachtleben, H. Li, and C. W. Spangler, *Inorg. Chem.* 35(3735),(1996).

[22] M. D. Newton, *Chem. Rev.* Washington, D.C, 91(767), (1991).

[23] M. D. Newton, *Coord. Chem. Rev.*,167,(2003).

[24] K. Walus, G. Schulhof, and G. A. Jullien, in Conf. Rec. 38th Asilomar Conf. Signals, *Systems and Computers,* 1(33),(2004).

[25] K. Walus, T. Dysart, G. Jullien, and R. Budiman, *IEEE Trans. Nanotechnology.,* 3(1),(2004).

[26] Vishnu. Teja, S. Polisetti and S. Kasavajalla, *Proceedings for 3rd IEEE international conference on Nono/Mocro Engineered and Molecular Systems.* Sanya, (2008).

[27] E.N. Ganesh, L. Kishore, M.J.S.Rangachar, *International Journal of Nanotechnology and Applications,* 2,(2008).

[28] S.K.Lakshmi, G.athisha, *European Journal of Scientific Research,* 53(4),(2011).

[29] P.K Lala, J.P. Parkerson, P Chakraborty, *Wseas Transactions on Circuits and Systems.* 9(9),(2010).

[30] Touggaw, P.Douglas, Lent, Craig S, *Journal of Applied Physics* 75(3),(1994).

[31] M.Askari, *European Journal of Scientific Research*.48,(2011).

[32] W. Wang, K. Walus, and G. A. Jullien, *in Proc. IEEE Nano Conf.,* San Francisco, pp.461-464 CA, (2003).

[33] K. Kim, K. Wu, and R. Karri, *IEEE Transactions on Computer-Aided Design Of Integrated Circuits and Systems,* 26(1), 2007

[34] V.Mardiris, Ch.Mizas, L. Fragidis and Chatzis, *12th WSEAS International Conference on COMPUTERS,* Heraklion, Greece, (2008).

[35] Sara Haeshmi, M.R. Azghadi, *International Symposium on Telecommunications,*(2008).

[36] T. Teodosio and L. Sousa, In 25th IEEE Norchip Conference, *IEEE,* Denmark,(2007).

[37] E.N.Ganesh, Dr.Lal Kishore and M.J.S.Rangachar, *International journal of Applied Engineering Research*,(2008).

[38] M.Askari, M. Taghizadeh, K.Fardad, *Proceedings of the International Comference on Computer and Communication Engineering Malaysia.(*2008).
[39] K. Walus, A. Vetteth, G. A. Jullien, V.S. Dimitrov, E.N. Ganesh, *International Journal on Computer Science and Engineering* 2(5),(2010).
[40] E.N. Ganesh, "Implementation and simulation of arithmetic logic unit, shifter and Multiplier in Quantum cellular automata technology." *International Journal on Computer Science and Engineering* (02)(1824-1829), (2010).
[41] Heu mpil Cho, Earl E. Swartzlander, Jr., *"Serial Parallel Multiplier Design in Quantum-dot Cellular Automata".*
[42] K. Das, D. De, *Molecular Simulation,* 37(3)(2011).
[43] Massimo Macucci, *Quantum cellular automata,* Imperial college press,(2005).

In: Quantum Dots and Quantum Cellular Automata
Editors: D. De, S. Bhattacharaya and K.P. Ghatak
ISBN: 978-1-62257-920-4

Chapter 6

SEQUENTIAL LOGIC CIRCUIT DESIGN USING QCA

***Ratna Chakrabarty*[1], *Sujata Ghatak*[2] *and Debashis De*[3,4]**
[1]Department of Electronics and Communication Engineering,
Institute of Engineering and Management, Kolkata, India
[2]Department of Computer Science and Engineering,
Sikkim Manipal Institute of Technology, Majitar, Rangpo, East Sikkim, India
[3] Department of Computer Science and Engineering,
West Bengal University of Technology, India
[4] School of Physics, University of Western Australia,
Crawley, Perth, Western Australia

ABSTRACT

In this chapter we study and analysis the Sequential logic circuits using Quantum Dot Cellular Automata (QCA). Sequential circuits can be made using metal dot tunneling QCA technology. A detail explanation about the circuits will help the researchers to know more about the sequential logic circuits using QCA technology and will help for the further development in the field of the future generation of microelectronics based devices.

1. INTRODUCTION

QCA technology and operations are discussed in detail in the chapter five. For the operation QCA proper clocking is very essential, which are of two types such as abrupt switching and adiabatic switching. Abrupt switching is not a suitable one as the inputs given to the circuit are not very fixed in all time and the circuit can reach an excited state at any time, so this is a very unsteady metastable state, also the circuit dissipate some energy when it comes to its ground state where as in adiabatic clocking system remains in its instantaneous ground state. So adiabatic clocking is preferred in the sequential circuits using QCA where information is transferred in pipelined way of transmission. Clocking scheme is discussed in [1, 2,3].

2. Sequential Circuits Using QCA

Sequential circuits are made up of various logic gates and different storage elements like registers and flip flops in which the present output depends on present input as well as on the previous output and on the state of the storage elements used in the circuit. Here in this chapter various sequential circuits have been discussed with respect to the quantum cellular automata.

2.1. Flip Flops Using QCA

Flip flops and latches are one of the basic building blocks of sequential logic circuit design. In this section we are going to discuss about the design of different types of flip-flops and latches such as D flip flop, RS flip flop JK flip flop using Quantum Dot Cellular Automata with the help of QCAD designer tool.

As we know that sequential elements can be implemented using QCA memory cells, but various problems have been raised in terms of hardware (due to its extensive control circuitry) and as a result the process become very slow in case of circuit performance. As QCA memories rely on the paradigm of memory-in-motion [4],a longer access time is required for the circuit performance using CMOS based design. From the research work of J. Huang, M. Momenzadeh and F. Lombardi [5] we come to know that latching is implicitly implemented in QCA as sequential behavior is dependent on adiabatic switching and the layout of the QCA cells. Adiabatic switching allows one to introduce timing by dividing the QCA cells in zones; this unique feature substantially affects sequential circuit performance for example, feedback paths and storage elements could be present in different locations of the layout may affect the design.

2.1.1. Design of Latches Using QCA

Latching is one of the important part of sequential QCA circuit, it depends upon on the four phase clocking phenomenon given by [5]. In D latch output follows the input in one clock cycle and output is remembered for the previous clock cycle. Usually a D-latch is implemented using cross-coupled NAND gates as shown below in figure 1 and the truth table is shown in table 1.

Here E stands for enable and D for input.

Table:1 truth table of a D latch

E	D	Q	$\overline{Q}$
0	0	latch	latch
0	1	latch	latch
1	0	0	1
1	1	1	0

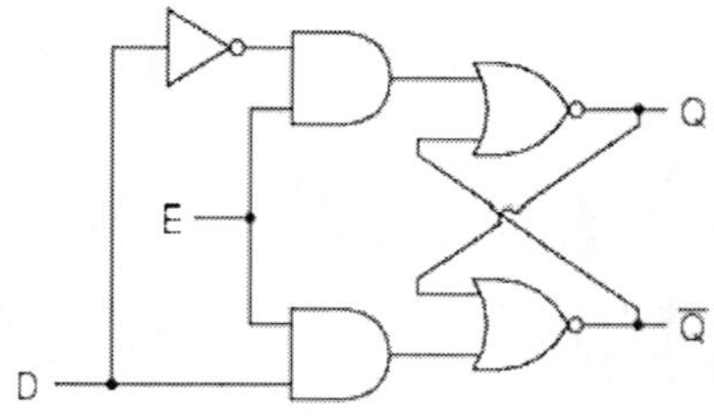

Figure 1. D latch usingNAND Gate.

The following circuit in figure 2 shows the QCA implementation of D latch in the RAM cell which acts as a memory cell according to the author in [5], the circuit has 50 cells in total showed in figure 2. The corresponding simulated output is shown in figure 3.

According to M. Askari et al [6], a simple binary wire, with four clocking zones can be used as a D latch, as shown in figure 4. With the proper timing and delay, a sequential D latch can be achieved by Cartesian layout as described in [7]. Figure 5 shows the simulation result of binary wire D latch.

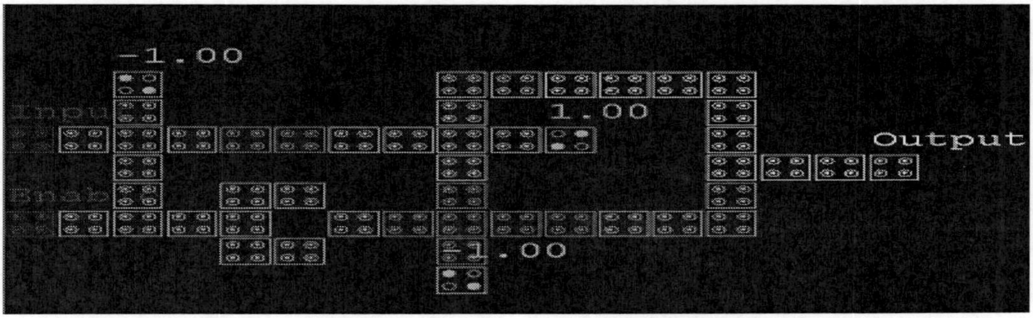

Figure 2. D latch using QCA.

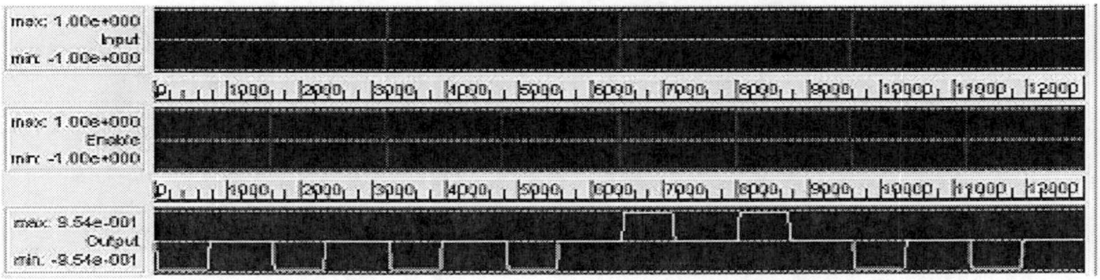

Figure 3. Simulation results of the D-Latch.

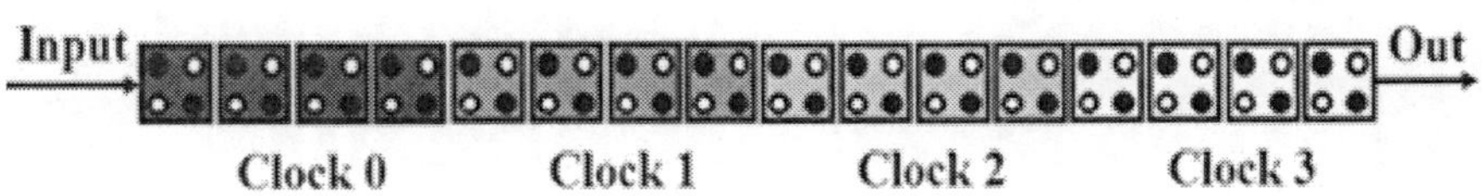

Figure 4. D latch using QCA.

Figure 5. Simulated out put of D latch.

Basically Flip-flop is the common name given to two-state devices which offer basic memory circuit for sequential logic operations. Flip-flops are heavily used for digital data storage with the bits either logic 1 or logic 0 and are commonly used in registers, which are the data storage of the device in application.

A.Vetteth et al in [8] proposed a design of D flip flop with 68 cells as shown in figure 6 where the design is used in random access memory cell with an area compatibility of 5 G bits/cm^2.Figure 7 shows the corresponding simulated output.

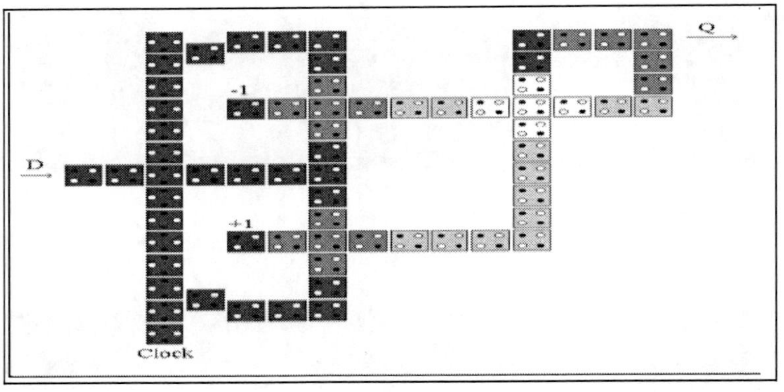

Figure 6. D Flip flop using QCA proposed in [8].

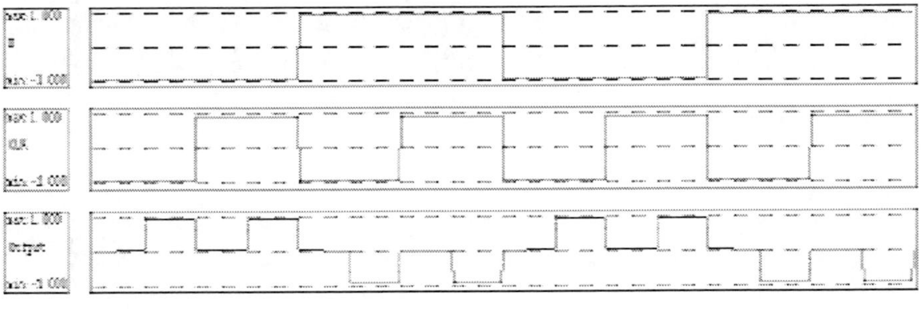

Figure 7. Simulated out put of D Flip flop.

The block diagram of master slave flip flop using digital logic is shown in figure 8, the corresponding QCA design for the circuit is shown in the figure 9, according to the author in [9], the design uses total 133 QCA cells and 8 clocking zones. One disadvantage of the circuit is that it took several iterations to come to the final output. The circuit uses negative clock edge triggering and two clock cycle delay between input and output. Simulated output is shown in figure 10.

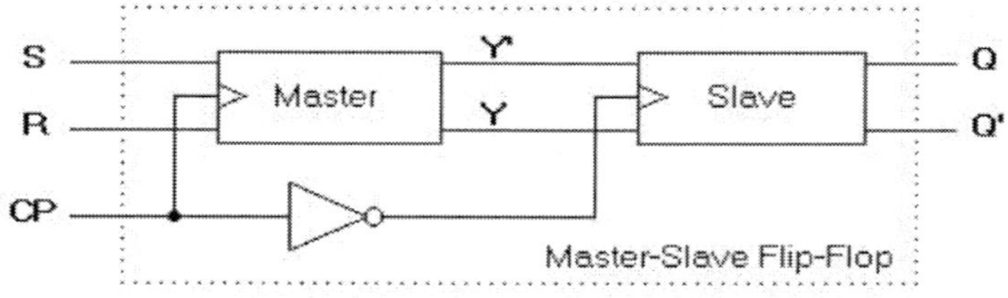

Figure 8. Master Slave flip flop.

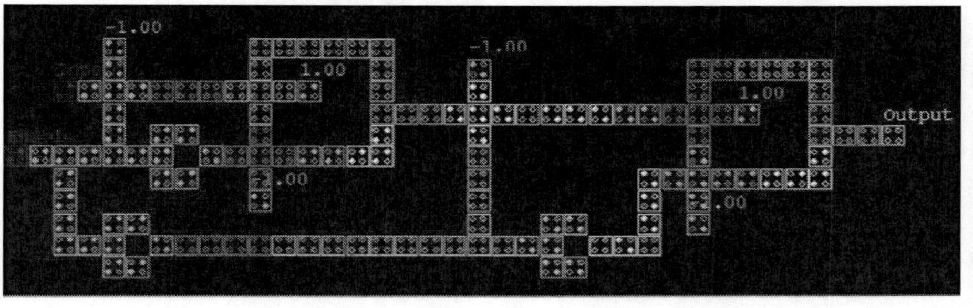

Figure 9. Negative edge triggered master/slave flip flop using QCA proposed in [9].

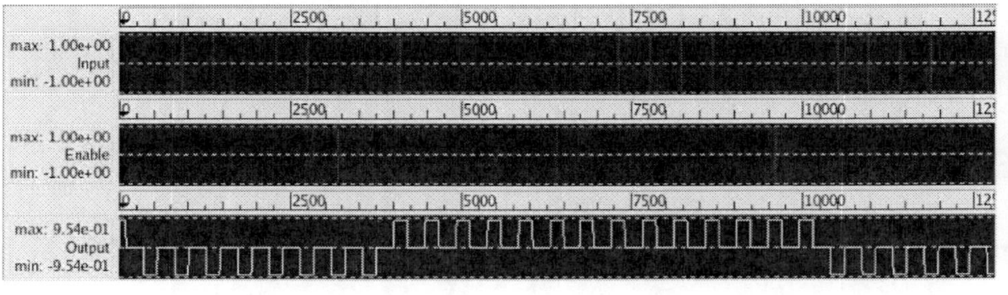

Figure 10. Simulated out put of the master/slave flip flop.

2.1.2. Design of S-R Flipflop Using QCA

According to vetteth et al[8] S-R flip flop is designed by QCA cells with 76 quantum cells altogether as shown in figure 11, the corresponding simulated output is shown in figure 12.

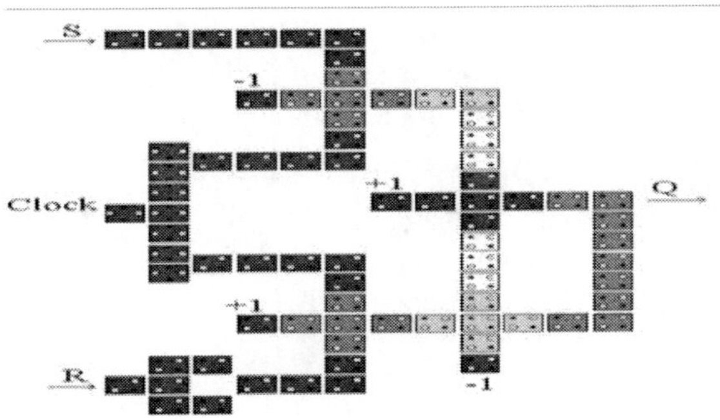

Figure 11. S-R flip flop using QCA proposed in [8].

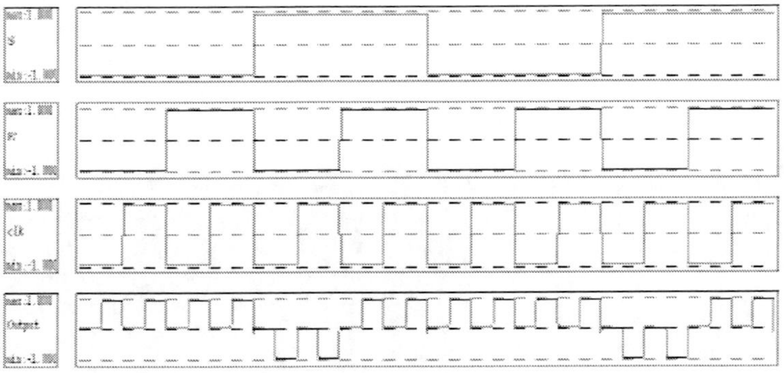

Figure 12. Simulated output of S-R flip flop proposed in [8].

2.1.3. Design of J-K Flipflop Using QCA

J-K flip flop is proposed by the same author in [8] where the QCA design is very simple with the use of 90 cells, the design layout is shown in figure 13 and the corresponding simulation is shown in figure 14, the proposed designs by Vetteth et al[8] are very simple and smaller than the similar utilization using CMOS circuits. All the circuits can operate at room temperature.

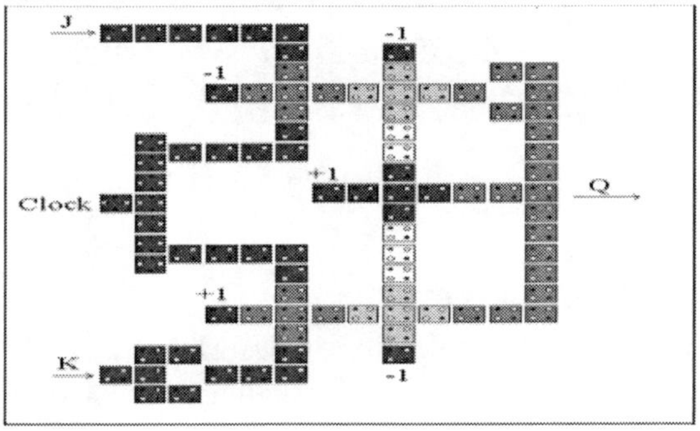

Figure 13. J-K Flip Flop using QCA proposed in [8].

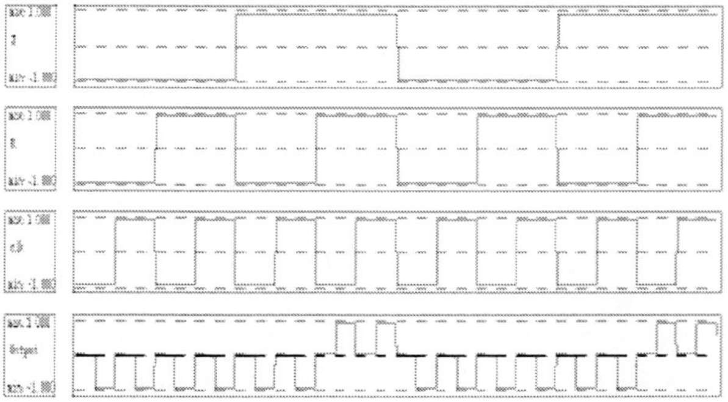

Figure 14. Simulated output of J-K flip flop proposed in [8].

2.1.4. Design of T Flipflop Using QCA

T flip flop is basically a toggle flip flop, where J-K inputs are permanently connected. It has only one input denoted as T.

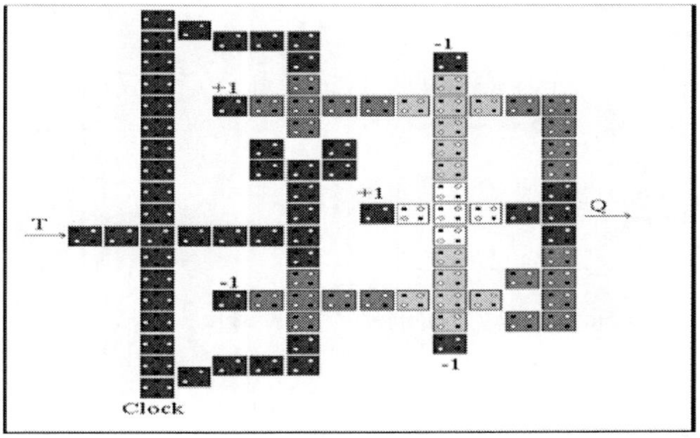

Figure 15. T Flip Flop using QCA proposed in [8].

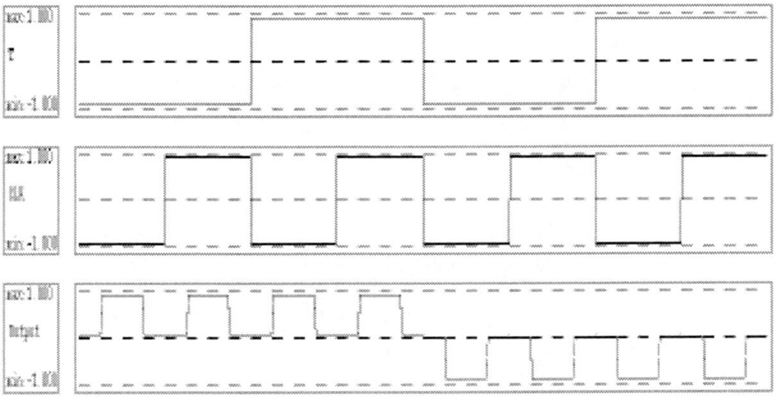

Figure 16. Simulated output of T flip flop proposed in [8].

The T flip flop is designed by Vetteth et al [8] with 92 cells, where the QCA design by the simulation tool QCAD designer is very simple and uses one clock cycle or four clocking zones. The QCA layout of the design is shown in figure 15 and the simulated waveform is shown in figure 16.

2.2. Shift Resisters Using QCA Logic Circuits

In this section we discuss about the shift registers. In register data can store or can move from one flip-flop to the other flip flop in the same register or to the other register with the application of the suitable clock, the registers which offer the data transfer operation is called shift registers.

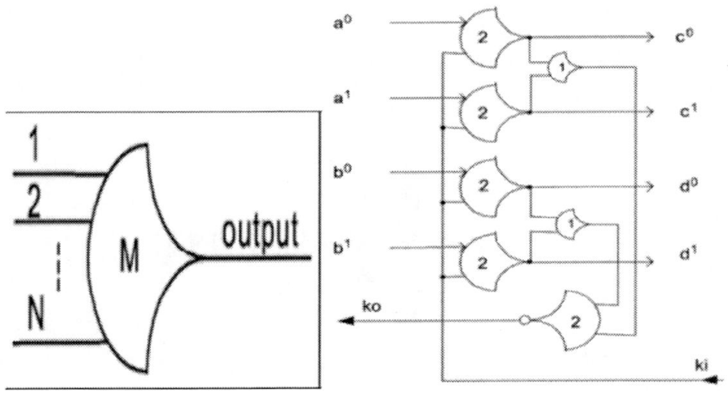

Figure 17. (a) Threshold Logic Gate (b) Two bit asynchronous register.

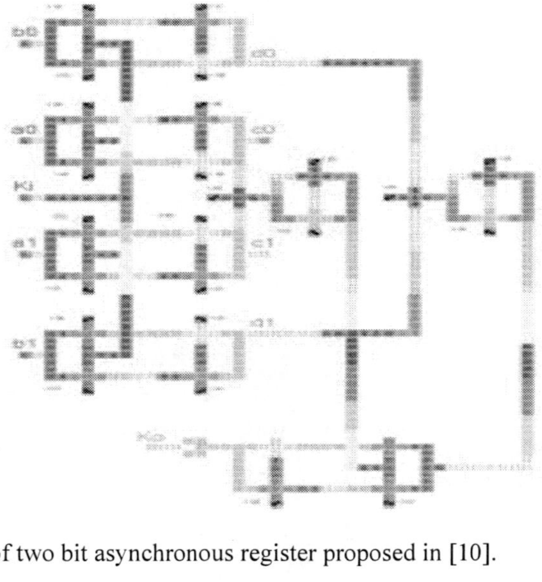

Figure 18. QCA layout of two bit asynchronous register proposed in [10].

In [10] Hamid Reza Mohaqeq et al proposed one shift register with QCA implementation using Null Convention Logic (NCL) which is a delay insensitive encoding scheme using a threshold logic gate, having two states such as null state and data state. When the output is in null state it remain in that state until the proper input is coming in the input state, similarly once the data output is there in the output state it will remain there until all the inputs return to the null state. The threshold logic gate is shown in figure 17(a) and the register layout shown in 17(b). Figure 18 shows the design layout by QCA cells. According to the author with this register delay insensitive circuits can be built.

2.2.1. Counter Circuits Using QCA Logic

A circuit used for counting pulses is known as counters, a counter consists of a group of flip flops with a clock signal. Counters are basically of two types, asynchronous counter and synchronous counter. In asynchronous counter the flip flops used in the design are not triggered at the same time, one flip flop output is used as the clock input of the next flip flop. In synchronous counter all the flip flops used in the design clocked in the same time and the propagation delay time compare to asynchronous counter is reduced in this type of design.

Z .Safikhani et al in [11] described a four bit up synchronous counter using QCA. Figure 19 shows the QCA implementation of the synchronous counter with the use of QCA majority gate.Figure 20 shows the simulated output of the synchronous counter.

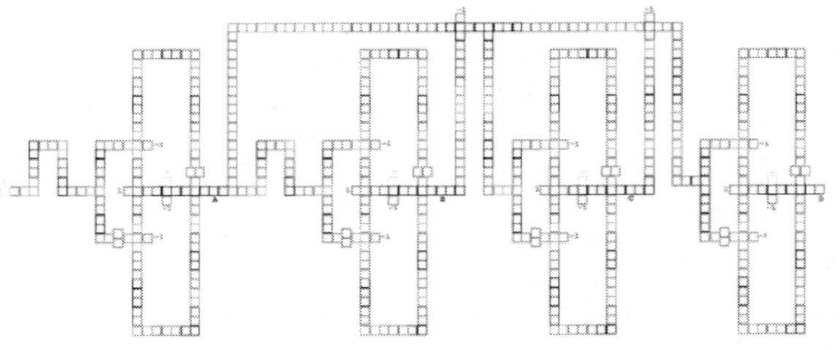

Figure 19. Layout of four bits synchronous up counter proposed in [11].

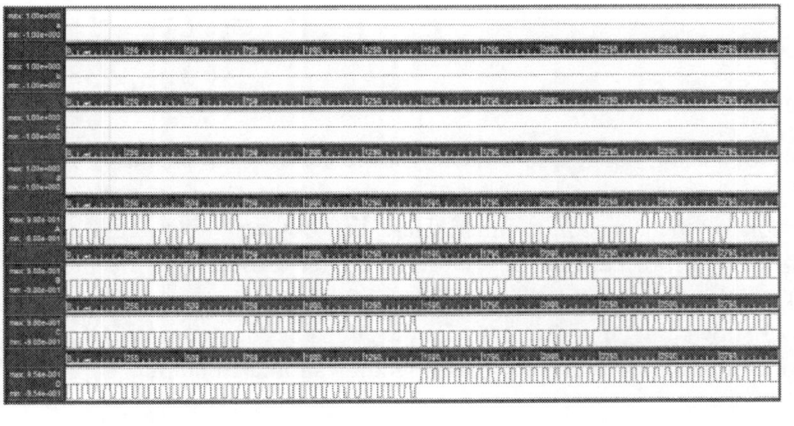

Figure 20. Simulated output of the four bit synchronous up counter from [11].

A sequential four bit Ring counter is designed by M. Askari et al in [12], in the Ring Counter input pulses can be encoded in any form of data. The data is shifted right and the output of FF3, the last flip flop is connected to the input of FF0, the first flip flop. The input is set to '1' by majority voter connection and '1' is rotated from the left to the right of the counter. Figure 21 shows the circuit diagram of the Ring counter using QCA. Figure 22 shows the corresponding simulated output of the Ring counter. The design required 86 quantum cells and the area covered by the circuit is approximately 0.11 μm^2.

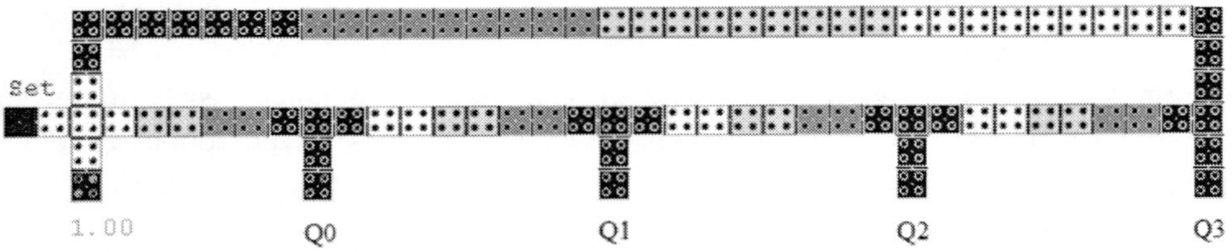

Figure 21. Four bit Ring counter using QCA proposed in [12].

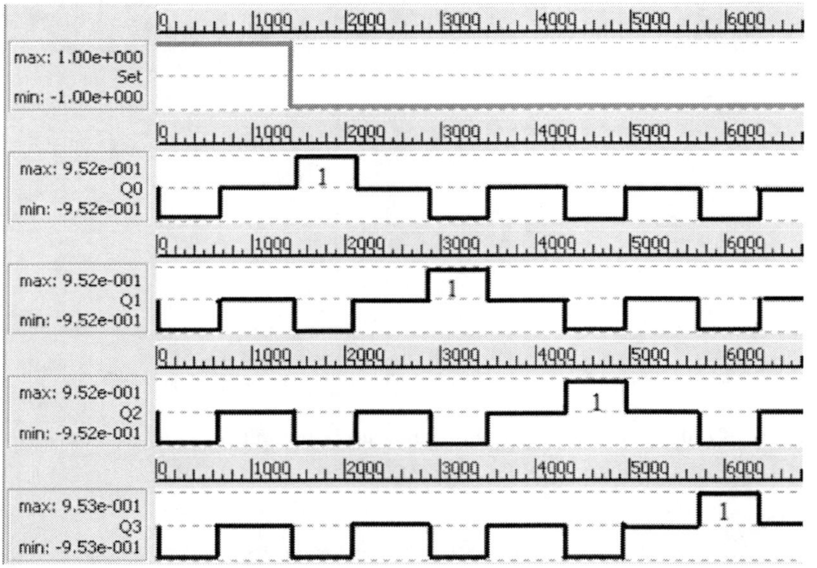

Figure 22. Ring counter circuit output in decimal form.

A four bit Johnson counter or twisted Ring counter is proposed by D.De et al in [13], where a slightly modification has done from the previous circuit designed in [12],for the implementation of the Johnson counter. Here the inverted output of the last flip flop is connected to the input of the first flip flop. The circuit uses 96 cells for the generation of the counter and the effective area is about 112884 nm2.

Figure 23 shows the circuit diagram of the Johnson counter using QCA with the QCAD designer tool and figure 24 shows the corresponding simulated output.

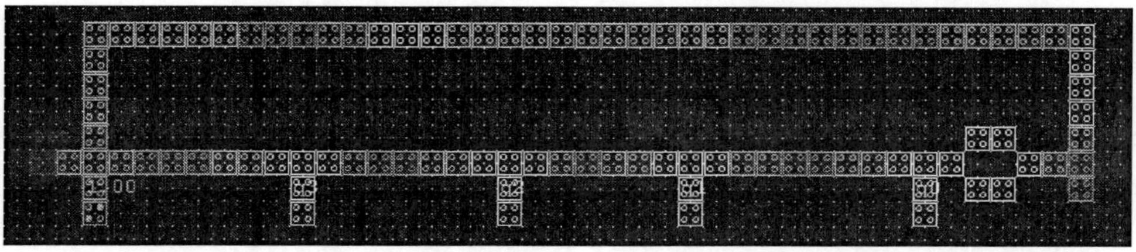

Figure 23. Four bit Johnson counter using QCA proposed in [13].

Figure 24. Simulated output of the Johnson Counter.

Conclusion

In this chapter we study and analysis various Sequential Logic circuits using Quantum Cellular Automata. This chapter will help the researchers for further design in the field of sequential logic circuit design where the area reduction can be an important study for the future generation of microcomputer in microelectronics.

Acknowledgments

K.P Ghatak and D.De are greatful to UGC for sanctioning the research project having the No-F 40-469/706 2011(SR) under which this paper has been completed besides D. De together with K.P Ghatak are greatful to DST, India for sanctioning the project under "SERC Fast Track Proposal of Young Scientist" scheme SERC/ET-0213/2011.

References

[1] Hennessy, K. and C. S. Lent, "Clocking of Molecular Quantum-Dot Cellular Automata," *Journal of Vaccum Science and Technology*,Vol. 19, No. 5,, pp. 1752-1755. 2001.

[2] Antonelli, D. A., et al., "Quantum-Dot Cellular Automata (QCA) Circuit Partitioning: Problem Modeling and Solutions," *Design Automation Conference (DAC)*, pp. 363-368. 2004.

[3] Walus, K., G. A. Jullien and V. S. Dimitrov, "Computer Arithmetic Structures for Quantum Cellular Automata," *Proc. Asilomar Conference,* 2003, available online: www.qcadesigner.ca/papers/Asilomar2003.pdf.

[4] Jing Huang, Maurizio Lombardi -Northeastern University,Department of Electrical and Computer Engineering 360 Huntington Av Boston, MA, 02115.

[5] Huang, J., M. Momenzadeh, F. Lombardi, " Analysis of missing and additional cell defects in sequential quantum-dot cellular automata," *VLSI journal (40),* pp: 503–515, DOI: 10.1016/j.vlsi.2006.08.001, 2007.

[6] Mehdi Askari, Maryam Taghizadeh,"Logic Circuit Design in Nano-Scale using Quantum- Dot Cellular Automata",*European Journal of Scientific Research* ISSN 1450-216X Vol.48 No.3, pp.516-526, 2011.

[7] Huang, J., M. Momenzadeh, F. Lombardi, 2007, "Design of sequential circuit by quantum- dot cellular automata*,"Microelectronics Journal (38),* pp: 525-537, DOI: 10.1016/j.mejo.2007.03.013.

[8] Anoop Vetteth,Konrad Walus,Vassil S.Dimitrov,Graham,A.jullien "Quantum dot cellular automata of Flip Flops",*ATIPS laboratory.*

[9] Apoorv Khurasia, " Quantum Cellular Automata", Pulkit Gambhir May 22, 2006.

[10] Hamid Reza Mohaqeq and Abbas Vafaei, "Quantum-dot Cellular Automata Serial Adder Design Exploiting Null Convention Logic", *World Applied Sciences Journal* 4 (5): 655-660, ISSN 1818-4952, 2008.

[11] Z. Safikhani, " Overview of Quantum Dot Cellular Automata, as replacement for CMOS – VLSI technology", *Trends in Advanced Science and Engineering* ISSN: 2251-6557, 2011.

[12] Mehdi Askari, Maryam Taghizadeh, "Logic Circuit Design in Nano-Scale using Quantum-Dot Cellular Automata",*European Journal of Scientific Research,*pp.516-526,vol-48,ISSN 1450- 216X,2011.

[13] R.Chakrabarty, D.De, K.K.Ghosh, "Design and Analysis of Quantum-Dot Cellular Automata Johnson Counter", *spsitm,* 2011.

In: Quantum Dots and Quantum Cellular Automata
Editors: D. De, S. Bhattacharaya and K.P. Ghatak
ISBN: 978-1-62257-920-4

Chapter 7

TILE BASED APPROACH TO DESIGN LOGIC CIRCUIT AND IT'S DEFECTS ANALYSIS FOR QUANTUM DOT CELLULAR AUTOMATA

***Kunal Das*[*,1,4], *Debashis De*[2,3] *and Mallika De*[4]**
[1]Department of Information Technology,
B.P.Poddar Institute of Management and Technology, Kol,India
[2]Department of Computer Science and Engineering,
West Bengal University of Technology, Kolkata, India
[3]School of Physics, University of Western Australia, Perth, WA, Australia
[4]Department of Eng.and Technological Studies, Kalyani University,
Kalyani, West Bengal, India

ABSTRACT

In this chapter an attempt is made to introduce the logic synthesis with tile nanostructure in QCA. Tile nanostructure supposed to be most empirical and robust in nature. Logic design with QCA tile structure is considered here. There are two possible type of QCA tile have been discussed. Depending on the population of all QCA cell, the QCA tile is classified into Fully populated and Non-Fully populated type. The studies on fully populated and non-fully populated QCA tile have been discussed in this chapter. Each type of QCA tile structures are also classified into active and passive type. The cascading effects on QCA tile structure have been represented here for functional development. The Coupled Majority–minority voter (CMmV) is most promising QCA tile structure which is actually used for designing n- variable symmetric functions. The Defect in QCA tile nanostructure is also discussed. There are several possible defects that may occurs during fabrication are considered here. The extra dot defect in QCA cell does not produces an error but only increases numerical complexity. The extra cell deposition and Missing cell deposition are discussed for different QCA tile structure.

[*] Email: kunaldasqca@gmail.com

1. INTRODUCTION

It is well known that the Digital computers are based on CMOS technology has conceived two innovate ideas, namely information are represented in terms of binary ‘1’ and binary ‘0’ together with the fact that the electronics charge state is used to represent information in terms of current switch. Because of device size minimization, the several limitations have been noticed on the CMOS technology in VLSI design such as charge quantization, leakage of current, power loss etc. The promising alternative of CMOS paradigm is the Quantum dot cellular Automata (QCA) which is used to representthe information in binary ‘1’ and binary ‘0’ in terms of Electronics charge configuration. In 1993,C.S.Lent et al. first introduce the theoretical Quantum dot Cellular Automata [1,2].In early 1999,C.S.Lent et al. describe the experimental approach to design QCA cell with GaAs[3].The dynamic behavior of QCA also been discussed with the help of the hart tree approximation[4]. Quantum mechanics also involved tofinding out the cell size and dot radius of a single QCA cell. Quantum dot simple a charge container and it is 3 dimensionally confined. The Quantum dots (QD) can classified into metal, semiconductor andmolecular dots respectively. A metal dot can have thousands of electrons and is fabricated in insulating substrate which will form small Metal Island. The small metal islands arebeing connected with tunnel junction to form a QCA cell [5]. QCA cell consist of four Quantum dots made with GaAs, placed at four corner of a square shape QCA cell. Each quantum dot can have multi number of electrons in it although in a cell, the design is such that two extra free electrons can reside within the cell. Due to coulombic interaction effect the two electrons within the cell have the choice to stay diagonally as shown in figure 1. This is known as polarization which is either P=’+1.00’ or P=’-1.00’ shown in figure 1.

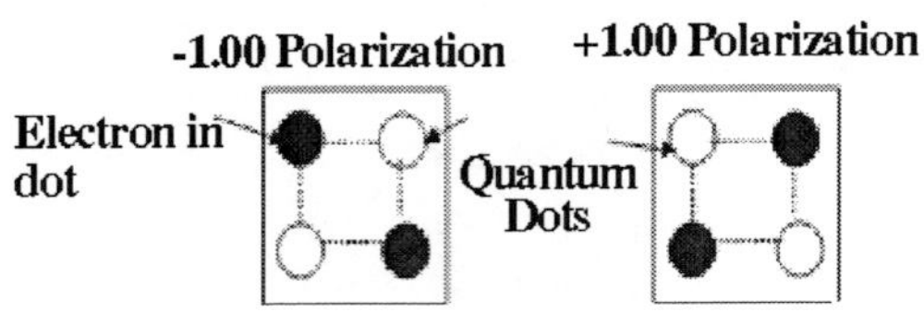

Figure 1. QCA cell consisting with four QDs connected with tunnel junction. Cell polarization ‘-1.00’ and ‘+1.00’ respectively.

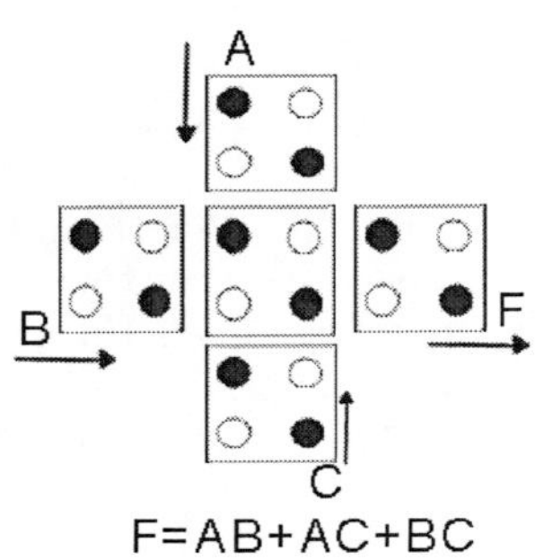

Figure 2. Majority Voter MV, A, B and C are three input QCA cell, F is output cell and middle cell is driver cell. The logic function is expressed as F=AB+BC+CA.

The basic gate is first introduced in QCA is the Majority Voter (MV). The three inputs cells are surrounding with the device cell and one output cell as shown in figure 2. The MV (A, B, C) functionality can be expressed as MV (A, B, C) =AB+BC+CA. The wire andinverter are being introduced in early 2001. There are several basic building blocks like AOI, NNI and Modified NNI, which enrich the logic implementation [6], are shown in figure 3. In 2009, a reduce size AOI gate has been introduced [8] (54% reduction in size in terms of figure3) and shown in figure 4(a).

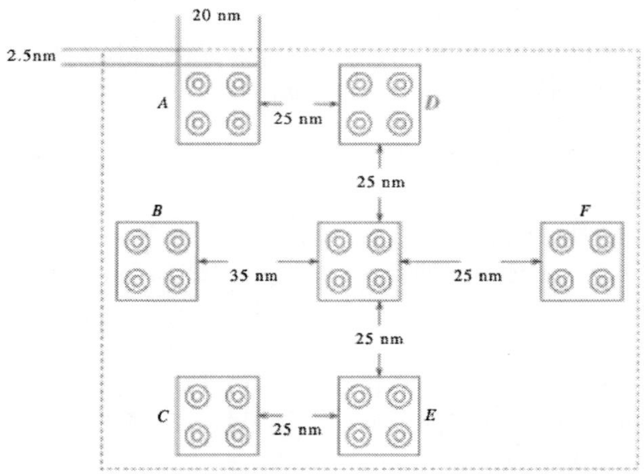

Figure 3.AOI gate with effective area125nm ×115nm [6].

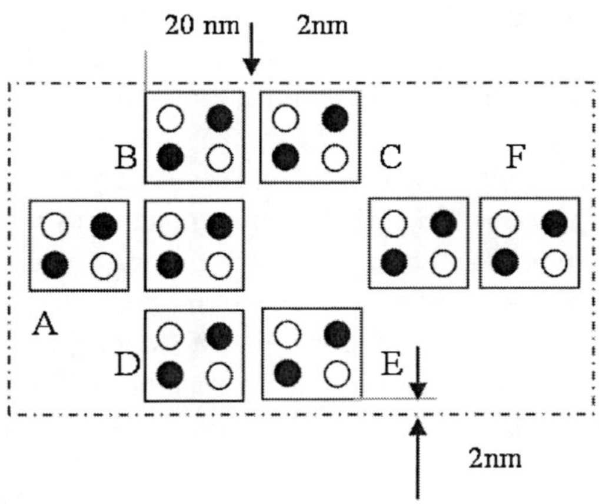

Figure 4(a). Modified AOI gate with the effective area 110nm × 68 nm [8].

The 3×3 Grid QCA nanostructure becomes interesting and another way to design logic circuit. 3×3 grid is most empirical due to its robustness and versatile logic explicabilities. 3×3 grid or tile structure can be classified in terms of active and passive categories. The active tiles are those which perform some computational work. In case of passive type tiles there is no computational part took place rather it processed by wire. Active type tile can be classified into following types (i) Orthogonal tile (ii) Butterfly tile or triple fan-out tile(iii) Five input MV tile and (iv) Coupled Majority-minority tile[13, 19]. The above tiles are shown in figure 5. The passive tiles are classified into four types (i) double fan-out tile (ii) triple fan-out tile (iii) baseline tile and(iv) symmetrical tile. In the figure 6, the passive type tiles are also shown. In terms of presence of QCA cell, it can be classify into two types namely, fully populated and Non-fully populated. The application of QCA tile is introduced [13, 17, 19, 20] to improve the design issue of the same. Orthogonal tile can be used as the majority voter functionality and Coupled Majority-minority voter tile is used to implement n variable symmetric functionality [13,19].Five input majority voter tile can be used to implement 1 bit binary adder with minimized size and less clocking zones. Multiple fan-outsare suitable to design sequential circuit, like the flip-flop and triple fan-out butterfly tile is applicable for that design issue.

Defects in fabrication are challenging part of any VLSI design. QCA design is not also out of this list [10-20]. The defectsprobable to occur are being identified and if the QCA structure can have high tolerance level, the defects can easily be eliminated. Generally, the defects may occur in chemical synthesis phase and deposition phase. In deposition phase, there may be (i) extra cell deposition (ii) missing cell deposition and (iii) misplacement of cell deposition or displacement of cell deposition can occur. In the chemical synthesis phase, the possible defect that may cause faults are (i) extra dot (iii) missing dot and (iii) extra electrons more than two are confined into the cell. In this chapter, we shall describe 3×3 tile structure in QCA and the defect and fault impact on diverse tile structure have been discussed.

1.1. Basic QCA

A basic QCA cell can be represented as coupled dot system with four dots, positioned at the four corners of a square. Two extra free electrons are confined within the cell althoughpractically, total 4M+2 numbers of electrons in QCA cell may present, where M is the integer. The two extra electrons are positioned at diagonal dots of a square shaped QCA cell by the influence of coulombic interaction. The electrons are polarized either +1.00 polarizations (binary 1) or -1.00 polarizations (binary 0) as shown in figure 1. The four dots (called sites) QCA system connected with Al tunnel junctions such that electrons can tunnel from one dot to anotherdot positioned vertically or horizontally and not in diagonally. Fabrication of metal tunnel junctions with electron beam lithography and double angle shadow evaporation technique was explored in [5]. The tunnel junctions are formed oxidizing the first layer of Al in situ prior to depositing the second layer. The experimental setup was in 1999, by C.S.Lent et al.[3] and M. Macucci reported Silicon-On- Insulator (SOI) technique for QCA fabrication in 2003[4].

The polarization of electrons can be described with the help of Quantum mechanics. The polarization P defined as

$$P = \frac{(\sigma_1 + \sigma_3) - (\sigma_2 + \sigma_4)}{(\sigma_1 + \sigma_2 + \sigma_3 + \sigma_4)} \tag{1}$$

where σ_i is the expectation value of number operator on site 'i' for ground state eigen function

The $|\psi_0\rangle$ represent the ground state define as

$$\sigma_i = \langle \psi_0 \mid \hat{n}_i | \psi_0 \rangle$$

$$|\psi_0\rangle = \sum_j \psi_j^0 |\phi_j\rangle \tag{2}$$

where $|\phi_j\rangle$ is jth basis vector and is coefficient of that basis vector. The basis vector for 4 sites system can be 16 different possible (2^4=16) vectors as shown in bellow.

$$|\phi_0\rangle = \left| \begin{matrix} 0001 \\ 0001 \end{matrix} \right\rangle, |\phi_1\rangle = \left| \begin{matrix} 0010 \\ 0001 \end{matrix} \right\rangle, \ldots, |\phi_{15}\rangle = \left| \begin{matrix} 1000 \\ 1000 \end{matrix} \right\rangle. \tag{3}$$

The basis vector notation is explained with example; consider a spin-up electron on second site and a spin-down electron on fourth site $|\phi_k\rangle = \left| \begin{matrix} 0100 \\ 0001 \end{matrix} \right\rangle$.

According to polarization 'P' the electrons are positioned as shown in figure 1. A strong stimulus such as heat, voltage or photon can change the electron polarization. Stimulus is applied in terms of input and due to knik energy represented by

$$E_{i,j}^{kink} = \frac{1}{4\pi\varepsilon_0\varepsilon_r \mid r_i - r_j \mid} q_i q_j \tag{4}$$

where ε_0 is the permittivity of free space, is relative permittivity of material, q_i and are electrical charges of cells 'i' and 'j' and is distance between two cells[3,4]. Electrons are forced to tunnel through the tunnel junction with stimulus and attained a definite polarization depending on input. Electrons are always be apart a maximum distance due to coulombic interaction. Electrons are not allowed to tunnel outside the QCA cell due its confinement.Coulombic impulsive force generated by QCA cell will cause a Quantum mechanical probability to impress the next QCA cell; this effect depends on '*Radius effect'*.

1.2. Defects in QCA

Based on several observation it is anestablished fact that QCA fabrication is suffered from high probable of defects. Defects are considered mainly on deposition phase; successfully chemical synthesized QCA cells are deposited on the substrate. In this phase, there are three common defects that had been analyzed are (a) Extra cell deposition i.e. extra QCA cell/s is/are deposited than the original requirement of cell arrangement,(b) Missing cell deposition/ Un-deposited cell deposition i.e. the QCA cell/s is/are not deposited as required in original design and (c) Displaced /Misplaced cell deposition i.e. QCA cell/s is/are misplaced from the exact position of deposition. These three types of defects may cause major errors in QCA logic circuit manufacturing. Hence, device or gate design using QCA are required a permissible defect tolerance on the above-mentioned defects such that the device no longer losses it's characteristics. Another type of defects that are probable to occur is chemical phase defects in which the individual dots are created. First step is to dot creation. Second step is to fabrication of tunnel junction and free electron injection in dot. In perfect manufacturing, we must hold four dots at four corners of the square and four tunnel junctions for tunneling the electron from one QD to another due to coulombic interaction, which is shown in figure 1. In this phase, defects are generally may arises are (a) Extra/Missing dot fabrication (b) extra/ missing electron injection.

1.2.1. Extra Dot

Let us assume that during chemical synthesis phase an extra dot is fabricated. To analyze the effect of extra dot in QCA cell, it is consideredthat the extra dot is fabricated at centre of QCA cell as shown in figure 4(b).[19]The possible polarization can be described in terms of expectation value where 'i' represent i^{th} site expectation value. The Hamiltonian for extra dot (total five dots) cell can write according to extended Hubbard model

$$H^{cell} = \sum_{i,\sigma} (E_0 + V_i) n_{i,\sigma} + \sum_{i>j,\sigma} t_{i,j} (\hat{a}^\dagger_{i,\sigma} \hat{a}_{j,\sigma} + \hat{a}^\dagger_{j,\sigma} \hat{a}_{i,\sigma})$$
$$+ \sum_{i} E_Q n_{i,\uparrow} n_{i,\downarrow} + \sum_{i>j,\sigma,\sigma'} V_Q n_{i,\sigma} n_{j,\sigma'} / |R_i - R_j| \qquad (5)$$

Steady state problem this Hamiltonian is used in Schrödinger equation.

$H^{cell} |\psi_i\rangle = E_i |\psi_i\rangle$ Where $|\psi_i\rangle$ is i^{th}eigen state of Hamiltonian and Ei is corresponding eigen value. These eigen states are found using the site-ket basis vector for five sites (for 5 dots) and two electrons of opposite spin. The basis vectors are

$$|\phi 1\rangle = \left| \begin{matrix} 00001 \\ 00001 \end{matrix} \right\rangle, |\phi 2\rangle = \left| \begin{matrix} 00001 \\ 00010 \end{matrix} \right\rangle, \ldots, |\phi 32\rangle = \left| \begin{matrix} 10000 \\ 10000 \end{matrix} \right\rangle.$$

These basis states, which are eigen states of the number operator of particular site within the cell. Now we calculate Hamiltonian matrix $H_{i,j} = \langle \phi_i | H^{cell} | \phi_j \rangle$and finding eigen vectors resulting 32×32 matrix. The ground state of cell, is represented in this basis as

$$|\psi_0\rangle = \sum_j \psi_j^0 |\phi_j\rangle.$$

The possible Polarization of cell described where 5th site denotes the extra dot fabricated in cell. Where is defined with ground state and basis vector as follows where is number operator.

As a result we found that, extra one dot fabrication defect does not produce any changes in polarization of cell, rather it improves the performances but increases the numerical complexity. Hence extra dot fabrication in chemical synthesis phase will not effect in QCA device manufacturing.

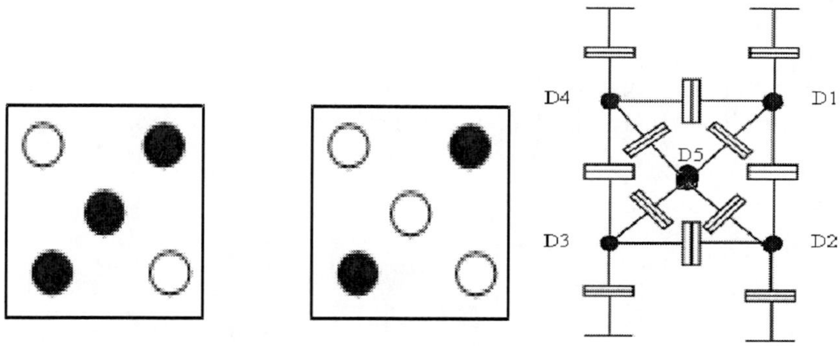

Figure 4(b). (a) Extra dot deposited QCA cell with three extra free electrons. (b) QCA cell with 5 dots with two free electrons. (c) Extra dot in QCA cell with tunnel junction due to defective fabrication.

2. Tile Based QCA

As we have already noted that the CMOS technology needs a strong alternative for Very Large Scale Integration (VLSI) design. One of the most emerging technologies in nano scale computing is QCA and it becomes focus of research interest due to low power, high operating frequencies (T Hz) nano scale device architecture.Very recently, Huang et al. [10,17],K.Das et al. [7-8] [19-20] and others introduced QCA tiles which are processed by wire of QCA to process the information. Various types of nano scale device/gate have been designed. Un-deposited cell deposition defect of different tiles are analyzed by Huang et al. [17] and in[20,25]the design of the fault tolerant devices has been discussed. The manufacturing of the QCA devicessuffer fromthe defect like extra cell / missing cell deposition and misplacement of cell at deposition phase. In Chemical synthesis phase there are several defects may occur like extra dot / missing dot deposition and extra / missing electron injection fault may cause defect.

In this section, an attempt is made to demonstrate the importance of QCA 3×3 tiles and make more promising QCA device design methodology. The 3×3 grid or tile structures can be classified in terms of active and passive type. The active type tiles are those which perform some computational work. In case of passive type tiles, no computational part took place rather it processed by wire. Active type tile can be classified into following types (i) Orthogonal tile (ii) Butterfly tile or triple fan-out tile (iii) Five input MV tile (iv) Coupled Majority-minority tile [13,19]. The above tiles are shown in figure5. The passive tile are classified into four type (i) double fan-out tile (ii) triple fan-out tile (iii) baseline tile (iv) In the figure5 the passive type tiles are also shown. In terms of presence of QCA cell namely, fully populated and non-fully populated.

3. Different Type of Tile Structure

The fundamental features of N × N fully populated (FP) grid can operate as logic was reported in [17, 19]. 3 × 3 tile is found to be most empirical due to its versatile logic operation as well as effective in terms of size. The 3 × 3 grid can be following type: (shown in figure5)

3×3 grid and orthogonal tile are shown in figure 5.a and figure 5.b respectively. Baseline tile is shown in figure 5.c. The fan-in tile, double Fan-out tile and triple Fan-out tile are shown in figure5.d, 5.e, 5.f respectively. The tile can be observed as active and passive types. If a tile implements a combinational logic function, which is to be in Sum of Product (SOP) form then it is active type tile.

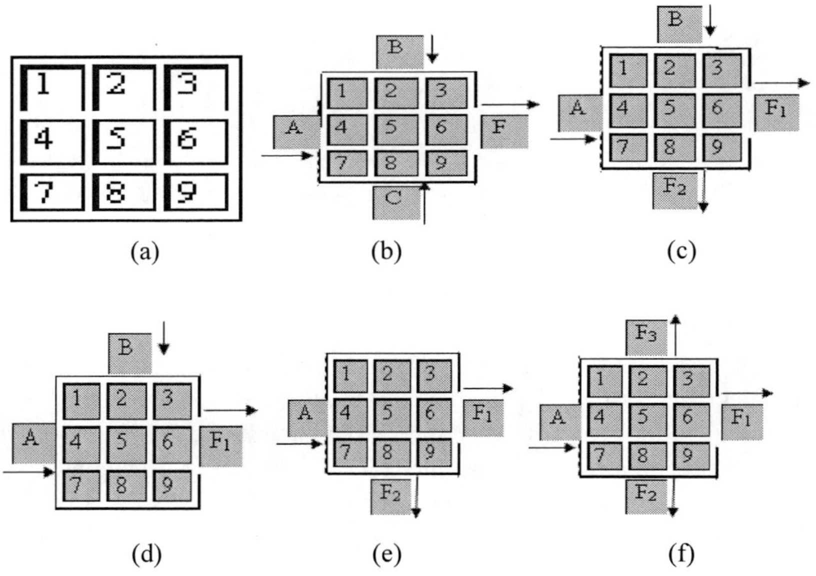

Figure 5. (a) 3×3 Grid, (b) orthogonal tile, (c) Baseline tile, (d) Fan-In tile, (e) Double Fan-Out tile and (f) triple Fan-Out tile.

In case of passive type tile, which is processed by wire and no computational part, can occur. The 3×3 tile can be discriminated into two categories 1. Fully Populated (FP): all $3 \times$ 3-grid cells are present. 2. Non Fully Populated (NFP): Some of the cells are missing which generates an active 3×3 tile –nanostructure.

Active type tile can be classified into following types (i) Orthogonal tile (ii) Butterfly tile or triple fan-out tile (iii) Five input MV tile (iv) Coupled Majority-minority tile [13, 19]. The above tiles are shown in figure 5. The passive tile are classified into four type (i) double fan-out tile (ii) triple fan-out tile (iii) baseline tile (iv) In the figure5, the passive type tiles are also shown. In terms of presence of QCA cell namely, fully populated and Non-fully populated.

3.1. Fully Populated Active Type QCA Tile

3.1.1. Triple Fan-Out Majority Voter or Butterfly Tile

The figure 6 [19] shows Triple Fan-out Majority Voter tile structure of QCA cell arrangement. It is act as three input- three outputs Majority Voter. The three inputs are A, B, C and three outputs are

$$F1=F2=F3=MV(A, B, C) =AB+BC+CA. \quad (6)$$

The simulated result explains that outputs are simultaneously drawn as Majority voter. It can be applicable to design a logic circuit that required multiple fan-outs like reversible flip-flop. It shall be useful device for implementing diverse logic in QCA that required feedback.

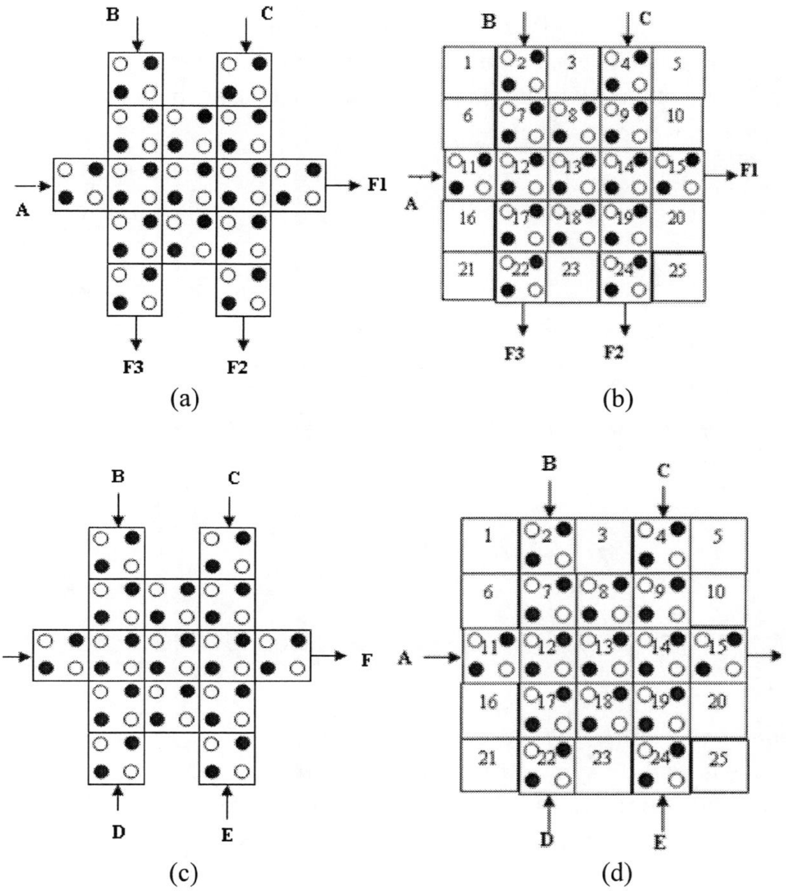

Figure 6. (a) Triple Fan-Out Majority Voter or Butterfly Tile, (b) The Corresponding 5 × 5 Grid realization. (c) Five-Input Majority Voter and (d) It's Corresponding 5 × 5 Grid realization where 1,2,3,…,24,25 represent the position where the extra / missing cell deposition may occur.

3.1.2. Five-Input Majority Voter Tile

In this subsection the aim is to describe the five-input Majority Voter tile. In figure 6.c shows the QCA tile structure, which is applicable to design five input Boolean functions. It is actually useful to design more than three variables / addressing computing or five addressing computing for QCA. It has five inputs A,B,C,D, E and one output F as shown in figure. F can be expressed as Boolean function

$$F=MV(A, B, C, D, E) = ABC+ABD+ABE+ACD+ACE+ADE+BCD+BCE+BDE+CDE. \quad (7)$$

3.2. Fully Populated Passive Type QCA Tile

3.2.1. Symmetrical Fully Populated Baseline Tile

Figure7.a, b shows the symmetrical Baseline tile, which is basicallya fully populated symmetrical baseline tile. The figure shows that the mirror rotation of the tile structure will give symmetrical result. The simulation of the above mention tile nanostructure is explained the fact that always outputs F1= B and F2= A are remain same both the cases. If the structure rotated 180 degree, it remains same.

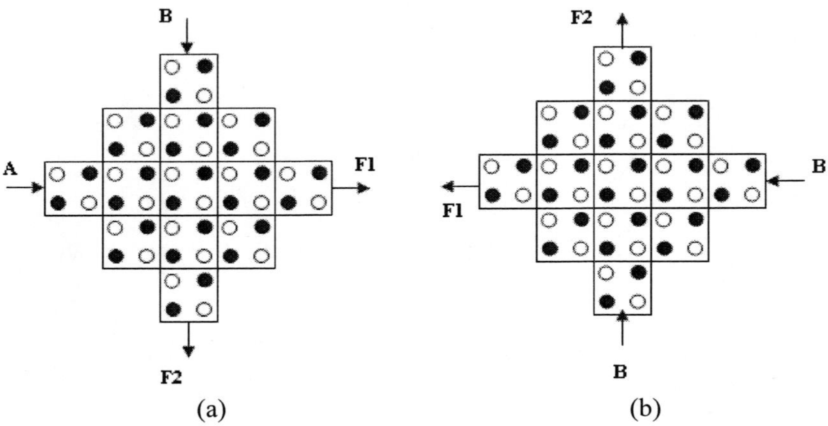

Figure 7. (a), (b) Fully-Populated Symmetrical Baseline tile where F1=B, F2=A.

3.2.2. Asymmetrical Fully Populated Fan- in Tile

In figure8 shows the Asymmetrical Fan-In tile which is represented as asymmetrical one. It shows that the mirror rotation of the tile structure will give Asymmetrical result.

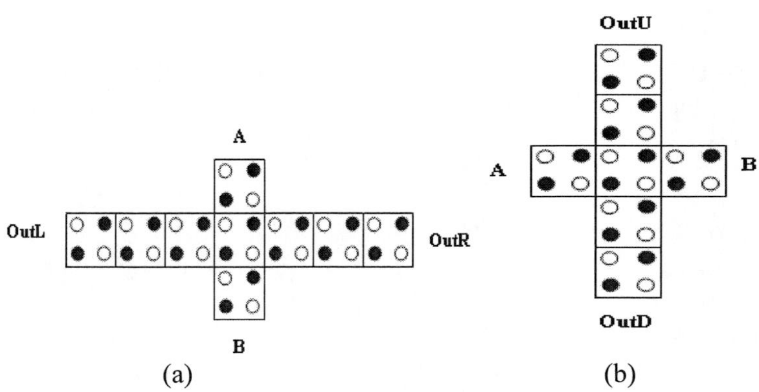

Figure 8. (a) Asymmetrical Fan-In tile in (b) A, B are two Input and Output OutL=OutR= A, in (c) A, B are two Input and Output OutU=OutD=B.

The simulation of the above mention tile nanostructure explain the fact that always outputs OutL=OutR= A and OutU=OutD=B are achieved for figure8.a and figure8.b respectively.

3.3. Non Fully Populated Active Type QCA Tile

In this section we describe Nand-Nor-Inverter (NNI) grid, 3×3 tile NNI, 3×3 tile And-Or and finally 3×3 tile And-Or- Inverter (AOI) which are belongs to non- fully populated active type QCA tile.

3.2.1. Non Fully Populated Nand-Nor-Inverter (NNI) QCA Tile

The tile base QCA Nanostructure is shown in figure.9.a[7]. It is demonstrated demonstrate as Non –fully populated active type Nand-Nor-Inverter (NNI).The figure.9.b shows that it is 3×3 NFP grid. Here, the position in the grid '2','4','6'and '8' are missing. Cell no's '1','3','10','7' and '9' acts as Input/Constant cells while in the other hand cell no's '12'is Output cell. The cell no's '5' and '11' act as device cell. The constant polarized cell D and E act as control Input. The following are shown in bellow which explains the NNI property.

Case1: if D=-1.00 and E= -1.00 polarized. Then output F=MV (D', E', MV (A', B, C')).

Now if B=-1.00 (i.e. logic '0') then output F will act as NOR gate and if B=+1.00 (i.e. logic '1') then output F act like a NAND gate. if A=+1.00 (i.e. logic '1') and B=+1.00 (i.e. logic '1') then output F act like inverter.

Case2: if D=-1.00 and E=+1.00 or vice versa and B=-1.00 then output F act as NOR gate. if D=-1.00 and E=+1.00 or vice versa and B=+1.00 then output F act as NAND gate. : if D=-1.00 and E=+1.00 and B=+1.00 as well as A=+1.00 then output F act inverter.

Case3: if D=+1.00 and E=+1.00 then no polarized output.

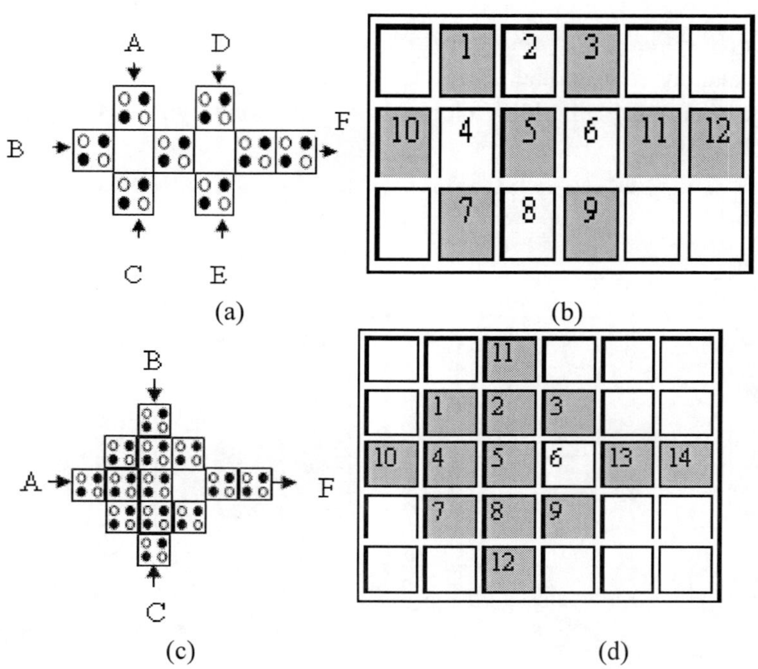

Figure 9. (Continued).

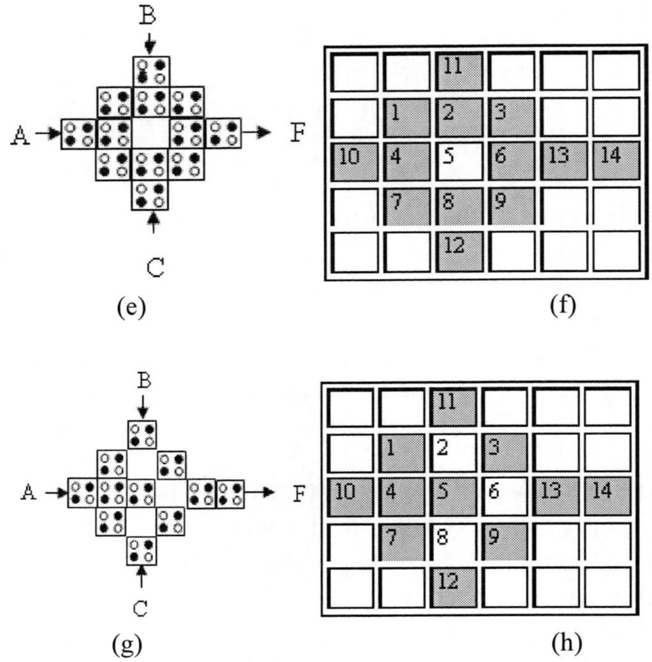

Figure 9. (a) NNI block diagram (b) NNI grid representation (c) 3×3 NNI block diagram (d) 3×3 NNI grid representation (e)3×3 AO block diagram (f) 3×3 AO grid representation (g) 3×3 AOI (h) 3×3 AOI Grid representation.

3.3.2. Non Fully Populated Nand-Nor-Inverter (NNI) 3x3 QCA Tile

Here we demonstrate another Non- fully populated 3×3 activetype tilesfor QCA [7] which must be a remarkable building block for versatile logic synthesis. The figure 9.c shows that 3x3 tile NNI gate.

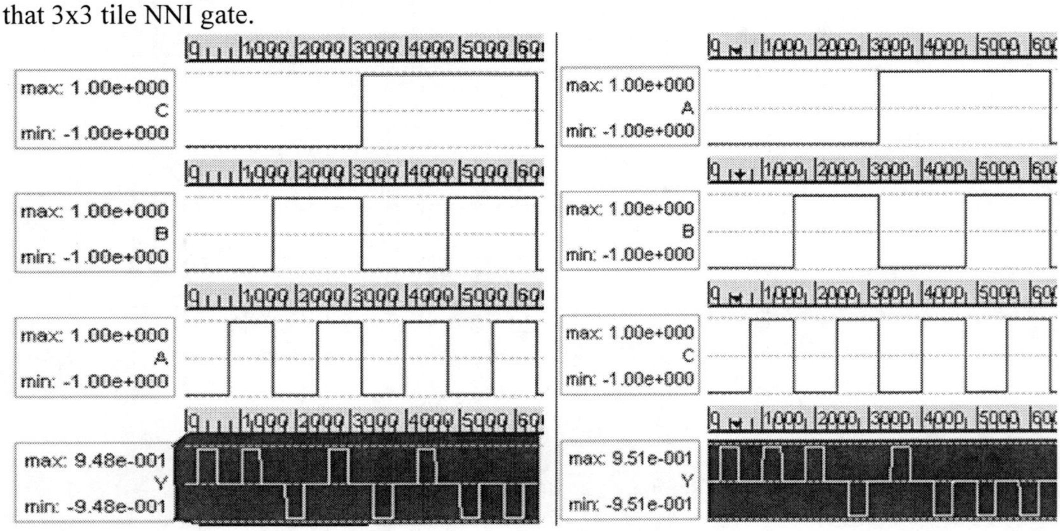

(a) (b)

Figure 10. (Continued).

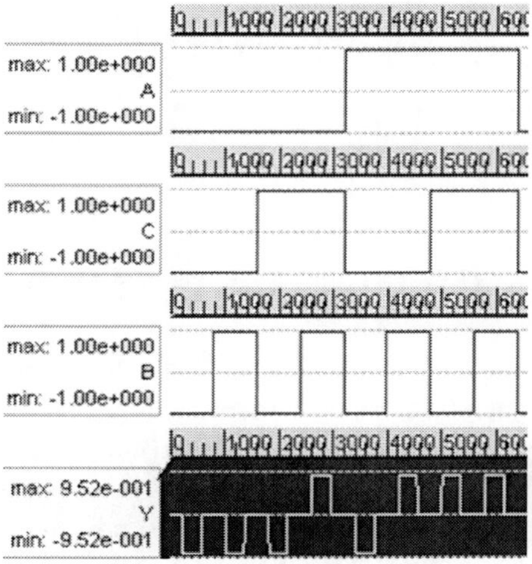

(c)

Figure 10. Simulation output (a) for NNI (b) for 3×3 tile NNI (c) for 3×3 tile AOI.

Here, cell no's '6' is only missing in 3×3 grid. Cell no's 'A','B' and 'C' are act as input/constant cell while 'F' cell act like output cell. This 3×3 grid is categories as orthogonal, active and Non-Fully-Populated (NFP) active type tile. Case 1: when C=+1.00 then output F act like NOR gate.Case2: when C=-1.00 then output F act like NAND gate. Case3: when C=-1.00 and A=+1.00 then output act like Inverter. This nanostructure for QCA was verified with the same simulator had been discussed early. The simulation results are shown in figure 10.

3.4. Non-Fully Populated Passive Type QCA Tile

Here in the figure 11.a, b shows that the symmetrical Non-Full populated Baseline tile [19].

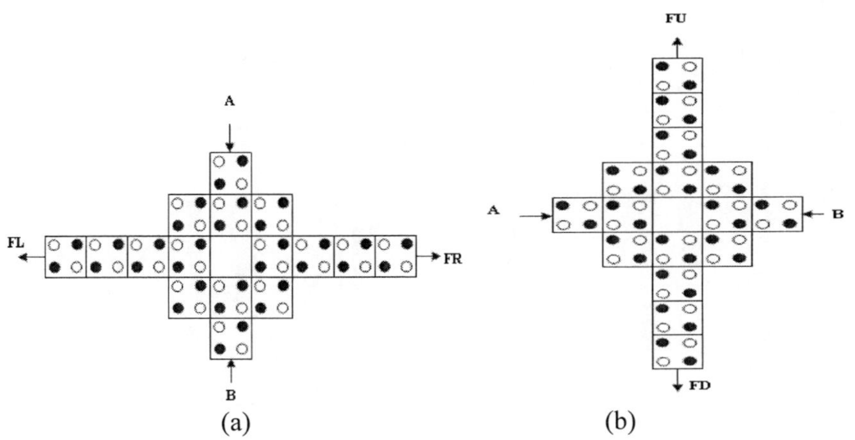

(a) (b)

Figure 11. (a), (b) Non-Fully Populated Symmetrical Baseline tiles where F L =A, FR=B and FU= B, FD=A.

The figure shows that the tiles are not fully populated. The middle cell is missing; as a result it is Non Fully populated one. It is examined that the 90 degree rotation of above mention tile structure, we get symmetrical output i.e. for the both cases we had outputs FL =A, FR=B and FU= B, FD=A.

4. Cascading Effect of QCA Tile

The Cascading Effect on QCA tile is discussed in this subsection. It was discussed [19] that two 3 × 3 QCA tile nanostructure is cascaded to design complex gate in QCA. We describe that if two 3 × 3 orthogonal tiles are cascaded then it can act as Five Input Majority voter, QCA logic designer can use 7-Input Majority voter with the mentioned effect. X-OR andX-NOR a dual gate in a single QCA structure can be design with three 3 × 3 QCA tile and three Control input are used to control X-OR / X-NOR operation, explored in subsection 2.2.2.

4.1. Cascading Two 3 × 3 QCA Tiles

In the figure12.a, we demonstrate the block diagram to cascading of two 3 × 3 QCA tiles. In figure 12.b we have shown QCA Layout to perform as Five-Input Majority voter. The four quantum dots are positioned at the four corner of each QCA cell. The simulation result of above mentioned device is achieved same as mentioned section 2.1.1. The different functional realization using Five-Input Majority voter have discussed in Table 1. When D=E=-1.00, the functional output of five input majority voter is three input AND gate and three input OR gate, when D=E=1.00.

Table 1. Different Functional realization of Five-Input Majority voter

Control Input		Functional Output
D	E	F
-1.00	-1.00	ABC
-1.00	1.00	MV(A,B,C)
1.00	-1.00	MV(A,B,C)
1.00	1.00	A+B+C

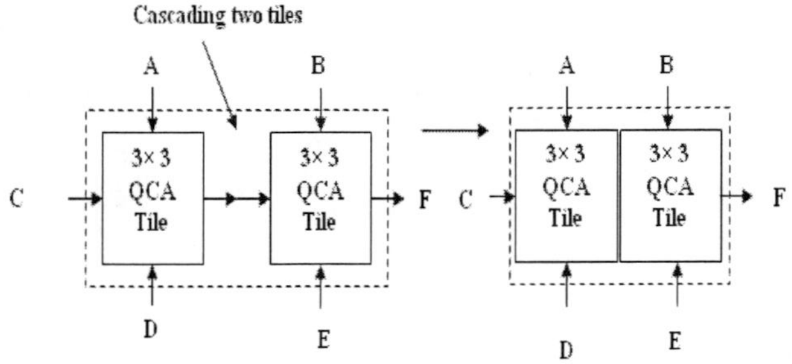

(a)

Figure 12. (Continued).

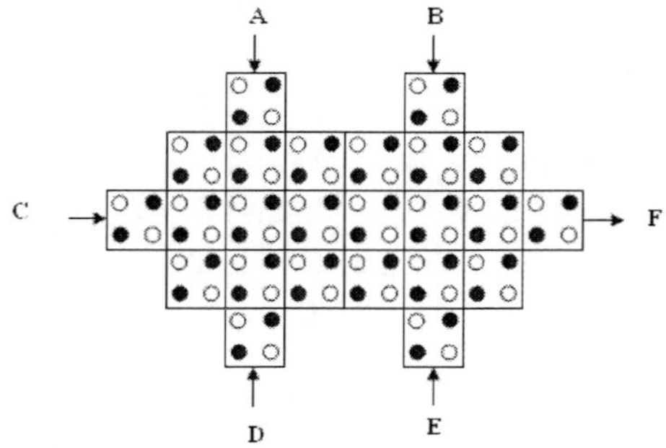

(b)

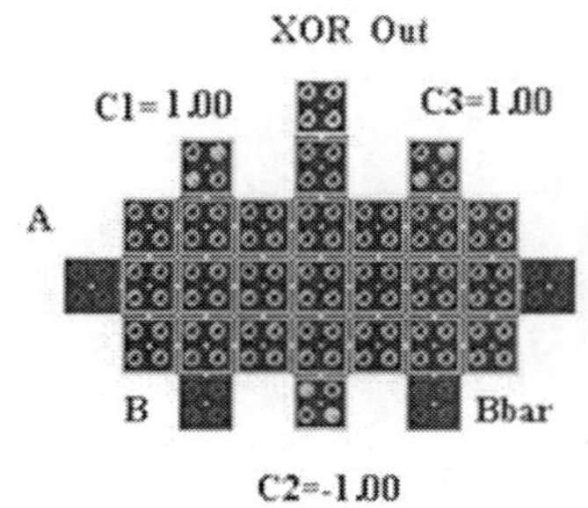

(c)

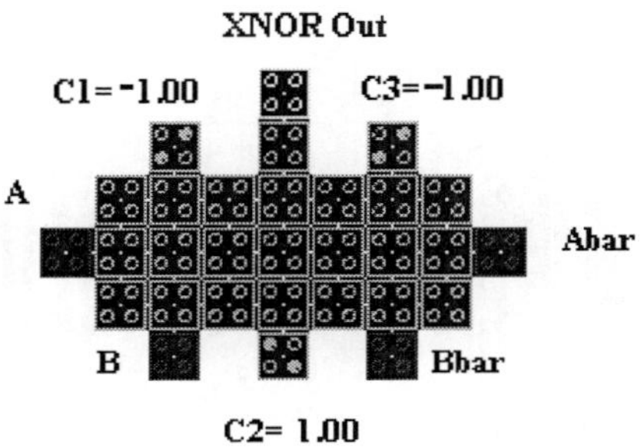

(d)

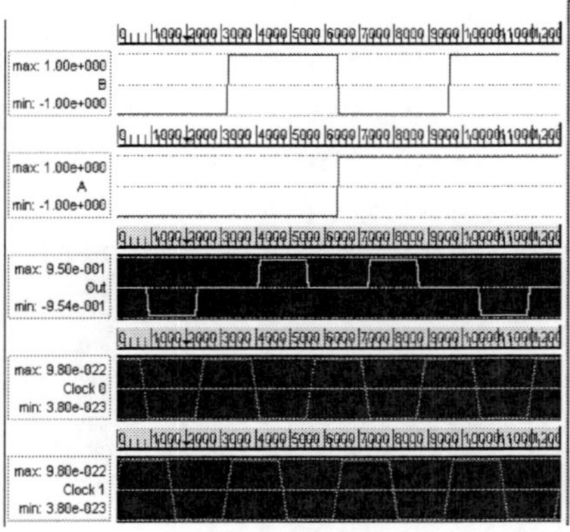

(e)

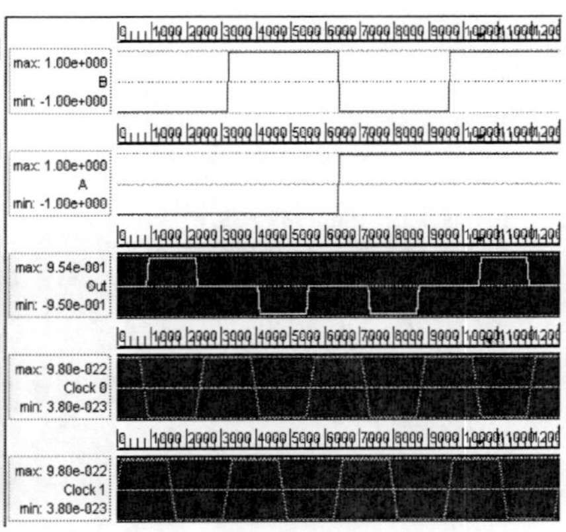

(f)

Figure 12. (a) Cascading of two 3 × 3 tiles to form Five input Majority voter, (b) QCA layout of Five input Majority voter after cascading two 3 × 3 tiles, (c),(d) QCADesigner QCA Layout for Tile based XOR and XNOR gate respectively and (e), (f) the simulated output of tile based XOR and XNOR gate respectively.

4.2. Cascading Three 3 × 3 QCA Tile to Design X-OR / X-NOR

Design and implementation of X-OR and X-NOR gate in a single QCA nanostructure is a challenging aspect. In this subsection we demonstrate that cascading of three 3 × 3 tiles can operate as X-OR and X-NOR gate [19]. The cascading of three 3 × 3 tile with a set of control signal, formed X-OR and X-NOR a dual gate which is shown figure12.c and figure 12.d respectively.

The simulated outputs of above gate are shown in figure12.e and figure 12.f. C1, C2 and C3 are the control signals, which actually control the operation of the device. The table 2 is to demonstrate the operation.For C1=-1.00, C2=1.00 and C3=-1.00 then the output Y will act as X-NOR gate andwhen C1=1.00, C2=-1.00 and C3=1.00 then the output Y will act as X-OR gate.

Table 2. Different Control input for implementing dual X-OR/X-NOR gate functionality

Control Input			Functional Output
C1	C2	C3	Y
-1.00	1.00	-1.00	X-NOR
1.00	-1.00	1.00	X-OR

5. Coupled Majority-Minority Voter

In this section, we described Fault tolerant 3 × 3 tile based coupled MmV gate. In the subsection we characterized the above gate. This Majority-minority voter is called 'coupled' because of dual outputs are driven simultaneously.

5.1. Characterization of CMmV

In the figure13.a, we have shown the 3 × 3 tile based Coupled Majority-minority-Voter. The main difference between any QCA logic circuit and coupled majority voter is that in this tile structure QD's in cell are not positioned at the corner rather they are positioned with 45◦ shift.

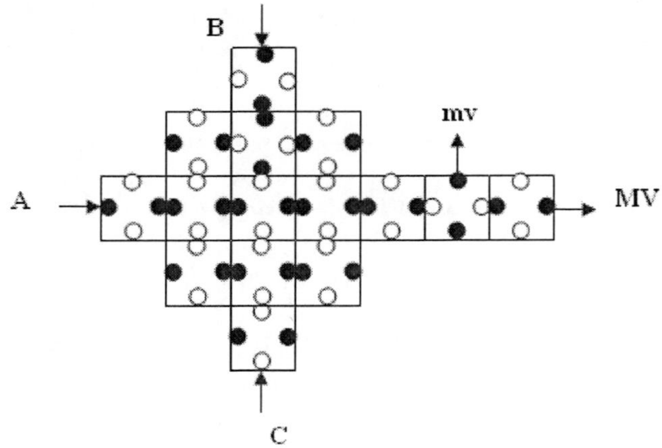

Figure 13. (Continued).

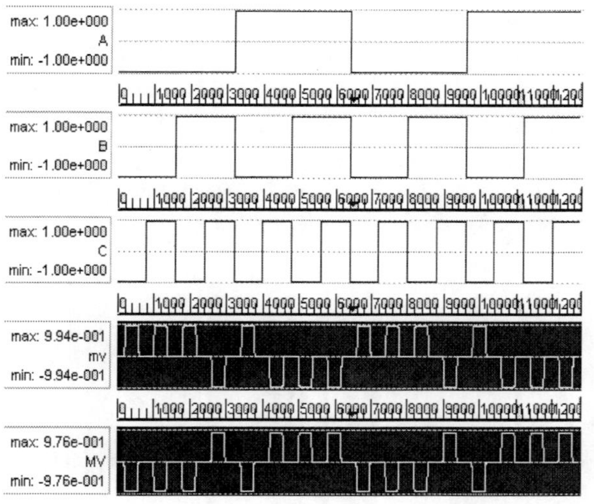

Figure 13. (a) QCA Layout of Novel 3 × 3 tile based Coupled Majority-minority-Voter (b) Simulation output for 3 × 3 tile based Coupled Majority-minority-Voter.

Table 3. Truth table for Dual o/p coupled Majority- minority gate

Dual o/p coupled Majority- minority gate Input			Output result	
A	B	C	MV	Mv
0	0	0	0	1
0	0	1	0	1
0	1	0	0	1
0	1	1	1	0
1	0	0	0	1
1	0	1	1	0
1	1	0	1	0
1	1	1	1	0

The functionality of CMmV is defined as output MV=Maj(A,B,C)=AB+BC+CA and mv=Min(A,B,C)=A′B′+ B′C′+C′A′. The truth table is shown in Table 3. The simulation result is shown in figure13.b which suggests the effectiveness for QCA logic design.

6. Logic Implementation

6.1. Flip Flop Design Using Butterfly Tile

In this subsection, we demonstrated the logic implementation with QCA tile structure [10, 19]. Triple Fan-out tile or Butterfly tile can be used for several logic implementations;

here we are given an example, designing RS Flip Flop using this tile. In the figure14.a shows the QCA nanostructure for RS-Flip Flop using two Butterfly tile. The feedback path/loop from output of one Butterfly is driven to input of another Butterfly tile. Here, we noticed the F1 output is driven as output 'Q' of RS Flip Flop. F2 output of that Butterfly tile used as feedback or loop to the input of another tile. Similarly, another output 'QBAR' is driven from another Butterfly tile. In this nanostructure we required two clocking zone. It is observed that designing logic circuit using butterfly tile is advantageous in terms of number of QCA cell, size and number of clocking zone has been noticed.

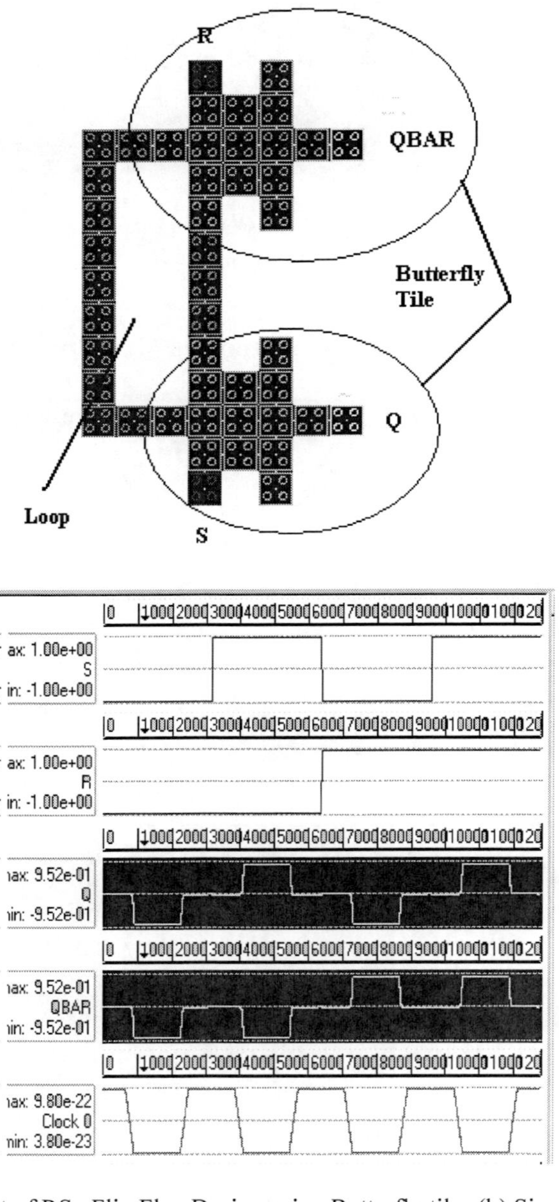

Figure 14. (a) QCA layout of RS –Flip Flop Design using Butterfly tiles (b) Simulated output for RS-Flip Flop.

One output of Butterfly tile is taken as feedback; another output is directly to output 'Q'. On the other hand using Majority voter (MV) feedback and output of RS Flip Flop are taken from same output of MV, which is complicated to design.

6.2. Three Input Full Adder Implementation

It has been discussed that Five-Input majority voter can be applicable to design circuit with less complex circuitry with respect to the circuit implementation with traditional majority voter. Here, we have shown in figure15.a the block diagram of Full Adder implementation with only single Five-Input Majority voter and a traditional Majority voter. QCA implementation has been carried out with QCADesigner. The QCA layout is shown in figure 15.b.The simulation result is also shown in figure15.c. A comparison made with the traditional adder circuit design [11] [12], shown in table 4.

Here we come into the conclusion that the 1 bit Full adder design using Five input Majority voter is more effective [19] to design with less QCA cell as well as less numbers of clocking zones. In our design we have used two clocking zones. On the other hand in case of [11] [12] four clocking zones are required.

Table 4. Comparison of 1 bit QCA Full adder

1 bit QCA Adder	Area	Clocking
Design is reported [11]	5 MV and 3 INV	4 Clocking Zones, Simple and fast
Design is reported [12]	3 MV and 3 INV	4 Clocking Zones, Simple and fast
Using Five input MV[19]	1 MV and 1 Five input MV	2 Clocking Zones, Simple and fast

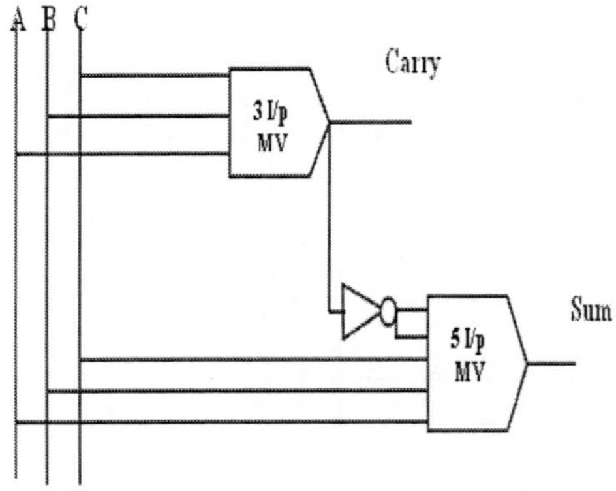

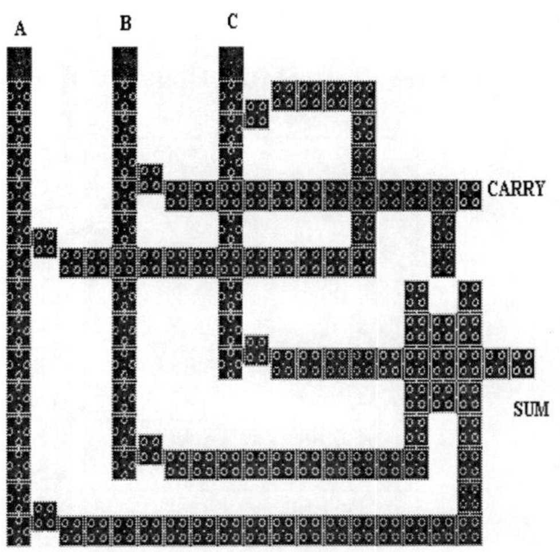

Figure 15. (Continued).

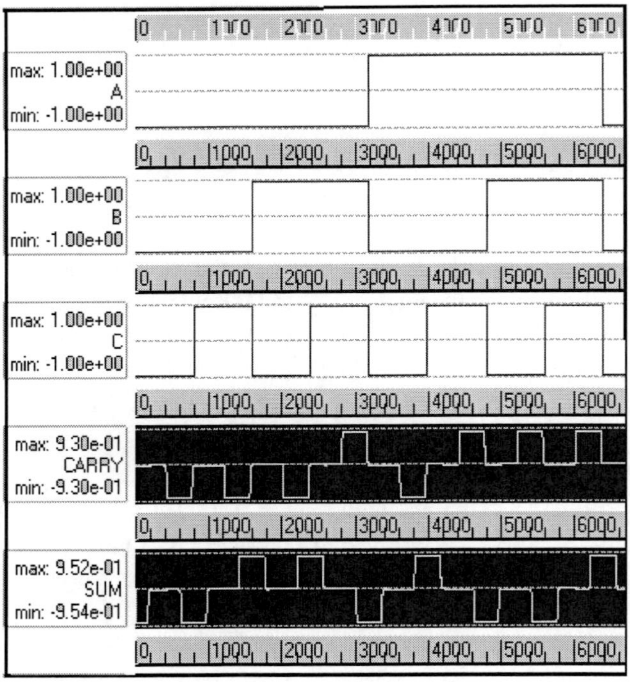

Figure 15. (a) Full adder circuit designs using Five–Input Majority voter.(b) QCA layout for 1-bit Full adder. (c) Simulated output of Full adder using QCADesigner.

6.3. Implementation of Function Using CMmV

In this subsection, we discussed functionality and logic implementation on CMmV QCA structure. We have shown in table 5 the diverse functionality that can achieve using CMmV.

Table 5.Output functionality of dual o/p coupled Majority- minority gate

Fixed Input	Output Functionality of Dual o/p coupled Majority-minority gate			
	MV output		mv output	
	Output	Function	Output	Function
C=1.00	A+B	OR	(A+B)′	NOR
C=-1.00	AB	AND	(AB) ′	NAND
A=1.00	(B+C)	OR	(B+C)′	NOR
A=-1.00	BC	AND	(BC) ′	NAND
B=1.00	(A+C)	OR	(A+C)′	NOR
B=-1.00	AC	AND	(AC) ′	NAND
A=B=1.00	C	WIRE	C′	WIRE
A=B=-1.00	C	WIRE	C′	WIRE
B=C=1.00	A	WIRE	A′	WIRE
B=C=-100	A	WIRE	A′	WIRE
A=C=1.00	B	WIRE	B′	WIRE
A=C=-100	B	WIRE	B′	WIRE

It is found that CMmV can act as Universal gate in QCA circuit. When one input, for example C is fixed then we found all logic functionality in QCA. Here, we noticed that CMmV not only truly Universal gate but also can be used for designing n- variable symmetric functions. For example, the two- variable symmetrical functions are A′ B, A+B, AB, (AB + A′B′) and (A′B + AB′) can achievable using two CMmV. Another interesting point is to note that CMmV is symmetrical 3 × 3 tile.

6.4. Logic Implementation Using Non- Fully Populated QCA Tile

In this subsection, explain logic synthesis by means of implementing thirteen standard functions for three variables [7]. These thirteen functions of three variables can represent 256 different Boolean functions. The table 6 represents the comparison of logic synthesis among MVandINV, NNI QCA tile and 3 × 3 tile AOI with respect to effective area.

Table 5A. Different 3 × 3 tile Logic primitives

Tile	Generated Logic Primitive	Category
Orthogonal	MV(A,B,C)	Active, FP
Baseline	F1=A,F2=B	Passive, FP
Fan-In	F=A	Passive, FP
Double Fan-Out	F1=F2=A	Passive, FP
Triple Fan-Out	F1=F2=F3=A	Passive, FP
Orthogonal but cell no'6'is missing	NNI	Active, NFP
Orthogonal but cell no's '5' is missing	AND, OR	Active, NFP
Orthogonal but cell no's '2','6'and '8' are missing	AOI	Active, NFP

Table 6. Logic Synthesis Using NNI and 3 × 3 tile AOI and Compare with MV

Standard Functions	Implements with MV and INV		Implements with proposed NNI		Implements with proposed 3x3 AOI	
	No's MV+ INV	Effective area	No's NNI+ INV	Effective area	No's AOI+INV	Effective area
f=ab'c	2	2	1	1.27	1	2.13
f=ab	1	1	1	1.27	1	2.13
f=a'bc+a'b'c'	3 + 2	6.334	2	2.54	2	4.26
f=a'bc+ab'c'	5 + 3	10	3	3.81	3	6.39
f=a'b+bc'	2	2	2	2.54	1	2.13
f=ab'+a'bc	4 + 2	7.334	2	2.54	2	4.26
f=a'bc+abc'+ a'b'c'	4 + 3	9.001	3	3.81	3	6.39
f=a	1	1	1	1.27	1	2.13
f=ab+bc+ca	1	1	1	1.27	1	2.13
f=a'b+b'c	3 + 1	4.667	2	2.54	2	4.26
f=a'b+bc+ab'c'	5 + 3	10	3	3.81	3	6.39
f=ab+a'b'	3 + 2	6.334	2	2.54	2	4.26
f=abc'+a'b'c'+ ab'c+a'bc	3 + 3	8.001	2	2.54	2	4.26

Here,consider 20nm ×20nm cell size. The effective area of MV was considered in [3] Ea=8100nm2. and effective area of NNI (considering 2nm avg. cell displacement) may calculated by 6×24×3×24nm2 i.e. 10300nm2.Now we consider the effective area of MV as unity Ea then effective area of NNI QCA tile is 1.27Ea.simillarly the effective area of 3×3 tile AOI is 2.13Ea. As a result It has been shown that logic implementation using NNI QCA tile and 3×3 tile AOI is an efficient way to build the logic circuit in QCA.

7. Defects on QCA Tiles

Based on several observation it is a conclusion that QCA fabrication is suffered from high probable defect. An analytical model to characterize the defects in QCA Implementation is presented in this section.

7.1. Extra Cell Deposition

$P_{f,|E|}$ is the probability of occurrence of output function with respect to '|E|' extra deposited cell. For a specific tile, the probability of occurrence is a function 'f' which is/are found due to extra/ missing cell/s deposition in the presence of |E| or |M| numbers of extra/ missing cell/s deposition can be defined as

$$P_{f,|E|} = \frac{N(f,|E|)}{|E|} \tag{9}$$

where represent number of occurrence of a function 'f' in presence of |E| extra cell deposition.

7.1.1. Triple Fan-Out Majority Voter or Butterfly Tile

The Fault Impact on this QCA tile is an attention of this subsection. It is mentionedthat defect may occur due to extra cell deposition and missing cell/undeposited cell deposition. Analysis of Fault impact is performed by exhaustive simulation for calculating all possible extra cell deposition. The table 7 shows the different functionality which may occur due to extra cell deposition at most single Cell. Result shows that 90% output of F1, F2, F3 are Majority Voter and only 10% functionality are wire. The exhaustive simulation had been performed on the Triple Fan-Out or Butterfly tile, consider '|E|' is the number of extra cell deposition where E=1,2,,10. The table 8 shows the corresponding results that are achieved. The figure16 shows the probability of occurrence of different functionality like wire, Majority voter and No polarization due to extra cell deposition on the proposed Triple Fan-Out Majority Voter. It is observed from figure 16.b that if number of extra cell deposition |E| increases up to four, probability of defect is increased but above the probability of defect gradually reduced.

Table 7. Extra Cell (at most one cell) Deposition for Triple Fan-Out Majority Voter

Different Functionality	F1	F2	F3
No Polarization	0	0	0
Wire	1	1	1
MV(A,B,C)	9	9	9
MV(A',B,C)	0	0	0
Total	10	10	10

Table 8. Extra Cell Deposition for Triple Fan-Out Majority Voter

Different Functionality	Number of extra cell deposition |E|						
f	1	2	3	4	8	9	10
No Polarization	0	0	0	4	0	0	0
Wire	1	7	23	37	32	3	0
Majority voter	9	83	89	91	40	6	1
Total	10	90	112	132	72	9	1

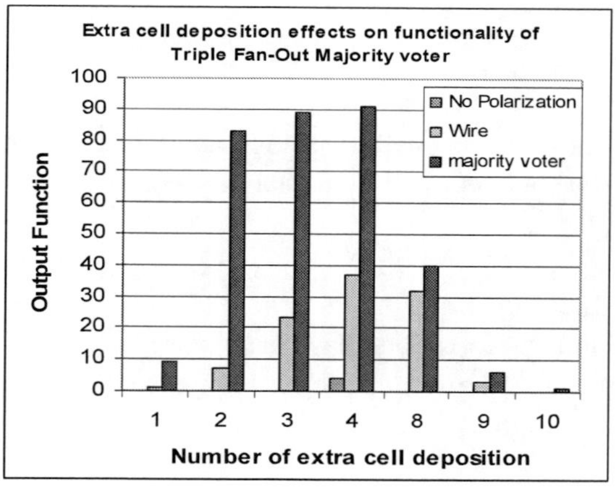

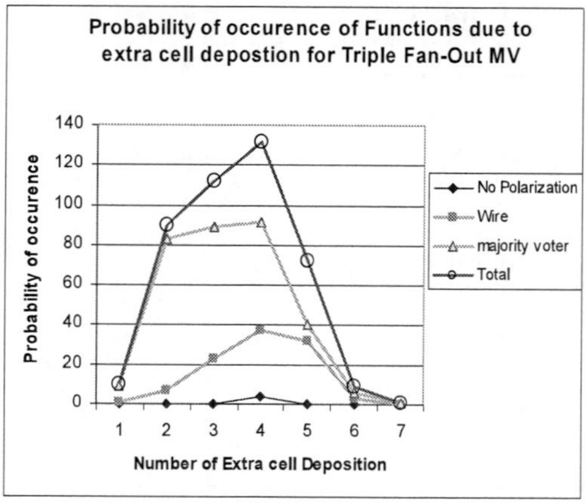

Figure 16. (a) Number of output functionality occurs due to at most one extra cell deposition on exact Triple Fan-In tile (b) Probability of occurrence output functionality of above tile for given number of extra cell deposition.

7.1.2. Five –Input Majority Voter Tile

Similarly, in case of Five-Input Majority voter tile we also analyzed the Fault Impact that is probable to occur. Table 9 shows the Extra Cell (at most one cell) Deposition for Five-Input Majority Voter tile. Approximately 70% outputs are still remaining same as with exact design but only 30% outputs are erroneous.

Table 9. Extra Cell (at most one cell) Deposition for Five-Input Majority Voter tile

Different Functionality	Output
No Polarization	0
Wire	0
MV(A,B,C,D,E)	7
MV(A,B',C,D,E)	3
Total	10

7.2. Missing Cell Deposition

Similarly, we also report missing cell deposition and Fault Impact over the triple Fan-out Majority voter or Butterfly tile and Five Input Majority voter. The results are analyzed in the following subsections.

7.2.1. Triple Fan-Out Majority Voter

Missing cell deposition at deposition phase may cause the defective fabrication of above proposed tiles. Table 10 shows the simulation results, when at most one cell is un-deposited or missing from the exact design of Butterfly tile.

Table 10. Effect on polarization due to missing Cell (at most one cell) deposition for Triple Fan-Out Majority Voter

Missing cell deposition at 5 × 5 Grid	Output Function	Polarization	
		Pmax	Pmin
‘7’	F1=F2=F3=MV(A,B,C)	+0.932	-0.932
‘8’	F1=F2=F3=A	+0.950	-0.950
‘9’	F1=F2=F3=A	+0.930	-0.930
‘12’	F1=F2=F3=C	+0.932	-0.932
‘13’	F1=F2=B,F3=MV(A,B,C)	+0.952	-0.952
‘14’	F1=-, F2=F3=A	+0.931	-0.931
‘17’	F1=F2=MV(A,B,C),F3=-	+0.920	-0.920
‘18’	F1=F2=C,F3=MV(A,B,C)	+0.956	-0.956
‘19’	F2=-, F1=F3=MV(A,B,C)	+0.934	-0.934

If the cell number '7' is missing then (as shown in fig 11) only no effect on output was noticed. All three output F1=F2=F3=MV(A,B,C) remain unaffected. Otherwise if cell number '13', '17','18' and '19' any one is missing then some output works as wire and some output works as Majority Voter, very few outputs are remain non-polarized. If cell number '8','9','12' and '14' any one cell is missing then almost all output works as wire. As a result we have 57% of Majority voter, 33% of wire and 10% remain non-polarized output, which is also explained in Table 11 and figure 17. The analysis presented here is applicable to above tile with in a single timing or clocking zone.

Table 11. Missing Cell (at most one cell) Deposition for Triple Fan-Out Majority Voter

Different Functionality	Output Functions of Triple Fan-Out MV		
	F1	F2	F3
No Polarization	1	1	1
Wire	4	5	4
MV(A,B,C)	3	2	3
MV(A',B,C)	1	1	1
Total	9	9	9

7.2.2. Five-Input Majority Voter

In case of Five-Input Majority voter missing cell deposition may cause the defects. Table 12, 13 have shown the simulation results, when at most one cell is un-deposited or missing from the exact design of above tile. If the cell number '8' and '18' is missing then only no effect on output was noticed.

Output F=MV (A, B, C, D, E) remain unaffected. Otherwise if cell number '7', '9','12','13','14','17' and '19' any one is missing then no output works as Five-Input Majority Voter, no output is remain un-polarized. The analysis presented here is applicable to above tile with in a single timing or clocking zone.

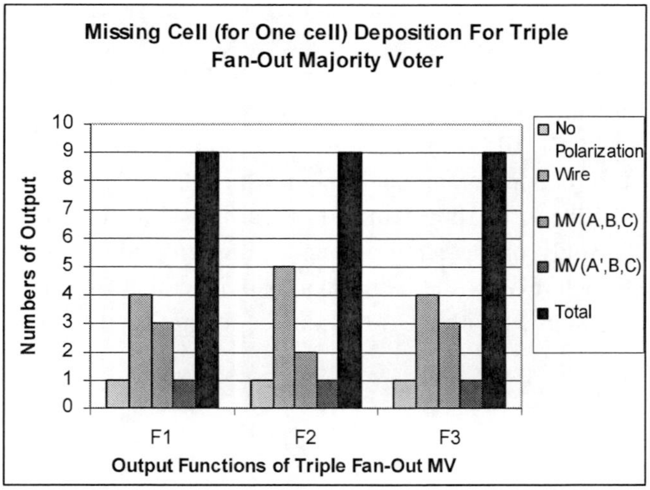

Figure 17. Number of output functionality for Triple Fan-out MV tile due to missing cell deposition.

Table 12. Effect on polarization due to Missing Cell (at most one cell) Deposition for Five-Input Majority Voter

Missing cell deposition at 5 × 5 Grid	Output Function	Polarization	
		Pmax	Pmin
'7'	F=MV (A,B',C,D,E)	+0.953	-0.953
'8'	F=MV (A,B,C,D,E)	+0.954	-0.954
'9'	F=MV (A,B,C',D,E)	+0.953	-0.953
'12'	F=MV (A',B,C,D,E)	+0.954	-0.954
'13'	F=MV (A,B',C,D,E)	+0.952	-0.952
'14'	F=MV (A,B',C,D,E)	+0.931	-0.931
'17'	F=MV (A,B,C,D',E)	+0.952	-0.952
'18'	F=MV (A,B,C,D,E)	+0.952	-0.952
'19'	F=MV (A,B',C,D,E')	+0.952	-0.952

Table 13. Missing Cell (at most one cell) Deposition For Five-Input Majority Voter

Different Functionality	Output
No Polarization	1
Wire	0
MV(A,B,C,D,E)	2
MV(A',B,C,D,E)	1
MV(A,B',C,D,E)	2
MV(A,B,C',D,E)	1
MV(A,B,C,D',E)	1
MV(A,B,C,D,E')	1
Total	9

CONCLUSION

Circuit design and logic implementation with QCA tile nanostructure is effective and efficient approach in nano-computing deign era.Many researcheshave been conceived QCA for VLSI design issue. QCA tile is most robust and empirical for circuit design. This chapter will give needful information for circuit design with QCA tile. The researcher might be interested in QCA tile due to its robust and empirical in nature. The future scope with QCA tile structure may be study on defects on coupled Majority-minority Voter. The reliability study on QCA tile structure with the help of probabilistic approach may be performed and more fault tolerant logic circuit with QCA tile may be reported.

REFERENCES

[1] C.S. Lent, P.D. Taugaw, W. Porod, and G.H. Bernstein, *Nanotechnology* 4 (1993).
[2] C.S. Lent, P.D. Tougaw, and W. Porod, *Appl. Phys. Lett.* 62 (1993).

[3] Amlani, A. Orlov, G. Snider, C. Lent, W. Porod, and G. Bernstein, *Superlattices Microstruct.* 25(1–2) (1999).

[4] C.S. Lent and P.D. Taugaw,*J. Appl. Phys*. 80(8) (1996).

[5] M. Macucci, M. Gattobigio, L. Bonci, G. Iannaccone, F.E. Prins, C.Single, G. Wetekam, and D.P. Kern, 34 (2003).

[6] M. Momenzadeh, J. Huang, M.B. Tahoori, and F. Lombardi,*IEEE Trans. Comput. Aided Des. Integr. Circuits Syst.* 24 (2005).

[7] K. Das, D. De, Int. Journal of Nanoscience, World Scientific.10 (1-2), (2011).

[8] K. Das and D. De, *IEEE Conference CODEC*-09,(2009).

[9] K. Walus, T.J. Dysart, G.A. Jullien and R.A. Budiman, 'ATIPSlaboratory QCADesigner' ATIPS laboratory, University of Calgary, Canada, Available at http:// www.atips.ca/projects/qcadesigner,(2002).

[10] M. Momenzadeh, J. Huang, and F. Lombardi, DFT 2005, 20th *IEEE International Symposium*, Washington, DC, (2005).

[11] P.D. Tougaw and C.S. Lent, *J. Appl. Phys.* 75(3) (1994).

[12] W. Wang, K. Walus, and G.A. Jullien, *IEEE Nano2003 Conference*, San Francisco, CA, (2003).

[13] S. Ditti, K. Mahata, P. Mitra, and B.K. Sikdar, *DTIS*'09,(2009).

[14] N. Jha and S. Gupta, Cambridge University Press, Cambridge,(2003).

[15] Fijany and B.N. Toomarian,*J. Nanopart. Res*. 3(2001).

[16] M. Thoori, J. Huang, M. Momenzadeh, and F. Lombardi, IEE*E Trans, Nanotechnol.* 3(4) (2004).

[17] J. Huang, M. Momenzadeh, and F. Lombardi, *Electron. Test. Theory* Appl. 23(2) (2007).

[18] M. Momenzadeh, M. Ottavi, and F. Lombardi,Proceedings of 20th *IEEE International Symposium on DFT*, 2005.

[19] K. Das, D. De, Molecular Simulation, 37(3)(2011).

[20] K. Das and D. De, *Int. J. Recent Trends Eng. Finl.* 3(1) (2010).

[21] S. Ray and B. Saha, *LNCS* 4173(2006).

[22] Sikdar. B.K et al., proc. *DTIS*'09,(2009).

[23] M.B Thoori, J.Huang, M.Momenzadeh, and F.Lombardi, *IEEE Trans.* Nanotechnol..3(4),(2004).

[24] M. Momenzadeh, Jing Huang; M.B Tahoori, Lombardi,Computer-Aided Design of Integrated Circuits and Systems, *IEEE Trans*. 24,(2005).

[25] M. B. Tahoori, M. Momenzadeh, J. Huang, and F. Lombardi, *Proc. VLSI Test Symp.*, Napa Valley, CA,(2004).

[26] W. Wang, R. Zhang, K. Walus, and G. A. Jullien, *IEEE Trans. Nanotechnol.*, 3(4),(2004).

[27] J. Huang, M. Momenzadeh, L. Schiano and F. Lombardi, in *Proceedings of IEEE conference on nanotechnology*, Nagoya, (2005).

In: Quantum Dots and Quantum Cellular Automata
Editors: D. De, S. Bhattacharaya and K.P. Ghatak
ISBN: 978-1-62257-920-4

Chapter 8

REVERSIBLE LOGIC CIRCUIT DESIGN USING QCA

***Maumita Maity**[*1], **Prasun Ghosal**[1] and **Debashis De**[2]*
[1]Department of Information Technology, Bengal Engineering and Science University, Shibpur, Howrah, India.
[2]Department of Computer Science and Engineering, West Bengal University of Technology, Kolkata, India

ABSTRACT

This chapter provides a comprehensive and self-contained introduction to reversible and quantum circuits as well as their implementation using QCA (Quantum dot Cellular Automata). Compared to the conventional methods this technology has proven itself as a viable alternative among the emerging paradigms in circuit design methodologies for its reduced power consumption with exponentially faster growth rate of computation. Thus it has found its diverse applications in low-power quantum computing and nanotechnology regime. Fundamentals like basic Reversible Logic Gates used for the design of quantum circuits are well illustrated this chapter. Following, the discussion on the design of Different Reversible logic basic Reversible logic gates has been extended. Role of QCA for implementation of such circuits and their implementations are well illustrated and lucidly represented such that the reader feels easy to follow the trend. Different fault analysis techniques like Parity Preserving, Majority Multiplexing andHardware Redundancy are described elaborately in this chapter.

1. INTRODUCTION TO REVERSIBLE LOGIC

The concept of reversible logic and zero- energy computing is first discovered in 1970. The zero-energy computing requires both reversible logic and non-dissipative hardware [1]. But the non-dissipative computing can be achieved using Reversible logic gate. Reversible

[*] Email: mou.maity@gmail.com

logic has promising applications in emerging nanotechnologies, such as quantum computing, quantum dot cellular automata, optical computing, etc. It is a promising computing paradigm in which there is a one-to-one mapping between the input and the output vectors. Reducing power becomes the main task of modern digital circuit design and reversible logic is the most attractive technology to meet this goal. Researcher Tommaso Toffoli first characterized reversible logic in his 1980's work that "Using invertible logic gates, it is ideally possible to build a sequential computer with zero internal power dissipation". The reversible circuits form the basic building block of quantum computers as all quantum operations are reversible. A Quantum Computer will be viewed as a quantum network composed of quantum logic gates; each gate performing an elementary unitary operation on one, two or more two–state quantum systems called qubits. Each qubit represents an elementary unit of information corresponding to the classical bit values 0 and 1. Reversible circuits also find application in specialized fields such as cryptography.

According to Landauer's principle [2, 3], it is possible to construct a computer using reversible circuits that can compute with arbitrarily small amounts of energy. Landauer have shown that for irreversible logic computations, each bit of information loss generates kTln2 joules of heat energy, where k is the Boltzmann's constant and T is the absolute temperature at which the computation is performed [2]. Current irreversible technologies will dissipate a lot of heat and can reduce the life of the circuit. Heat dissipation can be avoided if computation is carried out with no loss of information. Bennett showed that kTln2 energy dissipation would not occur, if a computation is carried out in a reversible way [3], since the amount of energy dissipated in a system bears a direct relationship to the number of bits erased during computation. So, according to Bennet and Landauer information loss is directly related to the power loss of a circuit. Power dissipation in Bennett clocking occurs only at the input/output ports and the internal energy dissipation can be made arbitrarily small. But the power dissipation in Landauer clocking is proportional to the number of irreversible gates in the circuit.

Therefore, Bennett clocking is more energy efficient than Landauer clocking. For any thermo-dynamical process involving a system moving from state A into state B, the change of entropy is defined by the second law of thermodynamics as

$$S(B) - S(A) \geq \int_A^B \frac{dQ}{T}$$

here S(A) and S(B) are the entropy of a system in state A (initial) and B (final) respectively, and dQ is the infinitesimal amount of heat received by the system at temperature T during the change (from state A to B). The equality sign holds for a thermodynamically reversible process [5].

Landauer has shown in [4] that a computational process that loses information cannot be thermodynamically reversible. So, a computing system must dissipate heat if its working cycle consists of an information loss. To preserve information, primitives in reversible computing must have a one-to-one onto mapping between inputs and outputs.

Reversibility can be analyzed in two respects:

1. Logical Reversibility: the bijective property (one-to-one onto function) between the input and output logic states holds. This is independent of the technology and the internal structure of the circuit.
2. Thermodynamic Reversibility: no energy is dissipated. In this case, the internal structure of the circuit must satisfy strict reversible primitives in a given technology as an implementation platform.

Thermodynamic reversibility requires logic reversibility; however, a circuit can be logically reversible, but not thermodynamically reversible.

Reversible circuits have the following advantages:

1. The reversible logic operations do not erase or lose information and dissipate very less heat.
2. Reversible logics are of high interest in low-power CMOS design, optical computing, nanotechnology and quantum computing.

Limitations of Reversible Circuit:

1. Reversible logic differs from traditional irreversible logic synthesis as fan-outs are not allowed.
2. Feedback paths are not permitted in the reversible logic.

A logic synthesis technique using reversible gate should have the following features:

i. Use minimum number of garbage outputs
ii. Use minimum input constants
iii. Use minimum number of gates

1.1. Energy Recovery Logic for Reversible Circuit

In[6] the authors propose a Reversible Energy Recovery Logic (RERL) circuit for ultra-low-energy consumption, which consumes only adiabatic energy loss and leakage current loss by completely eliminating non-adiabatic energy loss. It is a dual-rail adiabatic circuit using the concept of reversible logic with a new eight-phase clocking scheme. From the simulation result it is proved in this paper that at low-speed operation, the RERL consumes much less energy than the static CMOS circuit and other adiabatic logic circuits.

The principle of adiabatic switching is to recycle the signal energy to reduce energy dissipation. In the adiabatic circuits, two types of energy consumption exist along with leakage current loss: adiabatic loss and non-adiabatic loss. The former is unavoidable, and is inversely proportional to the transition time of the input signal [7, 8]. The later is proportional to the square of the voltage difference between the terminals of a switch when it is on, which is much larger than adiabatic during low-speed operation. The split-level charge recovery logic(SCRL)is used to solve the problem of non-adiabatic loss, but it requires too many clock rails up to twenty four even though it is a single-rail logic. Athas's logic [9] is another solution of this problem.

But this logic uses eight clock rails and does not succeed in eliminating all non-adiabatic loss. RERL is an energy recovery logic circuit that completely eliminates non-adiabatic loss. It uses eight-phase clocks, reversible pipeline connection and buffer circuit. The energy dissipation of the RERL circuit is inversely proportional to the transition time of the input signal. It is very effective than CMOS and other adiabatic logic circuits in terms of energy consumption.

2. Basic Reversible Logic Gates

Reversible computation in a system can be performed only when the system comprises of reversible gates. The major goal in reversible logic design and synthesis is to minimize the number of garbage outputs and the quantum cost. Garbage outputs refer to the unutilized outputs that are not used as primary outputs and which cannot be used as inputs for new computation, and are only needed to maintain one-to-one mapping. Each Reversible gate has a cost associated with it called Quantum Cost. The Quantum cost of a Reversible gate is the number of 2x2 Reversible gates or Quantum logic gates required in designing. Most of the reversible gates in literature are three-input three-output or four-input four-output gates. The important reversible gates used for reversible logic synthesis are Feynman Gate [10], Fredkin Gate [11], Toffli Gate [12], Peres Gate [13] and New Gate [14]. Several designs for binary Adders, Subtractors, Multipliers and Dividers are investigated based on Reversible logic. Minimization of the number of Reversible gates, Quantum cost and garbage inputs/outputs are the focus of research in Reversible logic.

2.1. Feynman Gate (CNOT Gate)

Feynman gate (FG) or Controlled-NOT gate (CNOT) is a 2x2 reversible gate having the mapping (A, B) to (P=A, Q=A ⊕ B) where A, B are the inputs and P, Q are the outputs, respectively. Figure 1 shows the block diagram of the Feynman Gate [10].

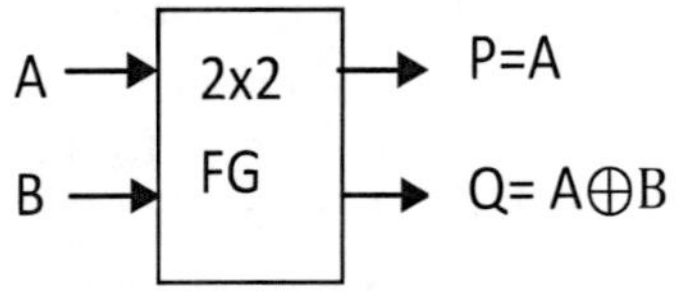

Figure 1. Feynman Gate.

2.2. Toffli Gate

Toffoli Gate (TG) is a 3x3 reversible gate. Block diagram of the Toffli gate is shown in Figure 2. Two-through means two of its outputs are the same as inputs with the mapping (A, B, C) to (P=A, Q=B, R=A ·B⊕ C), where A, B, C are inputs and P, Q, R are outputs, respectively [12].

2.3. Fredkin Gate

Fredkin gate is a (3x3) conservative reversible gate, having the mapping (A, B, C) to =A, , where A, B, C are the inputs and P, Q, R are the outputs, respectively [11]. It is a 3x3 reversible logic gate. Figure 3 shows the block diagram of the Fredkin gate.

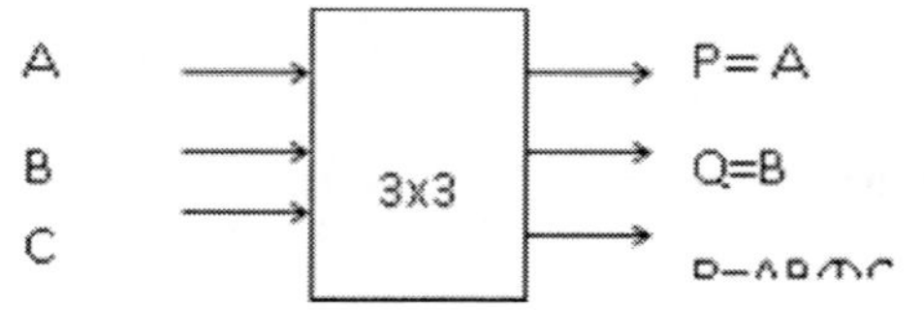

Figure 2. Toffoli Gate.

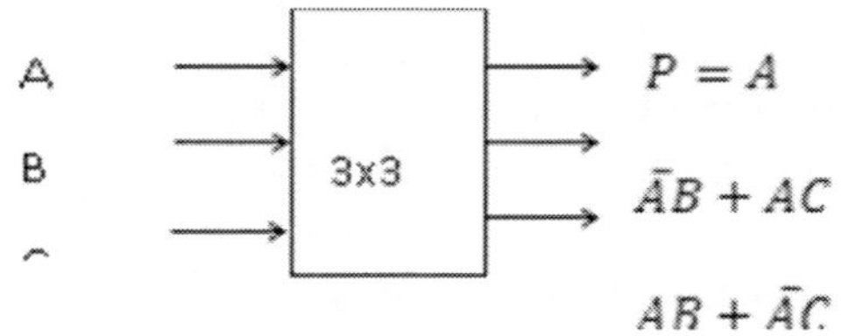

Figure 3. Fredkin Gate.

2.4. Peres Gate

Peres gate is a 3x3 reversible gate having the mapping (A, B, C) to (P=A, Q=A⊕B, R=(A·B)⊕C), where A, B, C are the inputs and P, Q, R are the outputs, respectively [13]. Figure 4 shows the block diagram of the Peres gate.

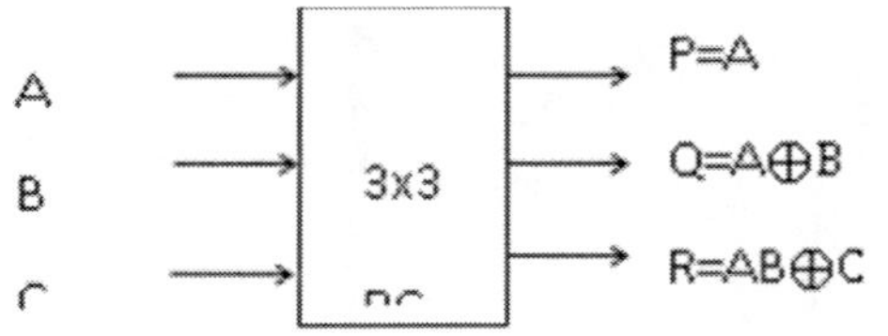

Figure 4. Peres Gate.

2.5. NFT Gate

NFT gate is a parity preserving reversible logic gate. It can be defined as I_v=(A, B, C) and O_v=(,Where I_v and O_v are the input and output vectors respectively. The NFT gate satisfies the following relation A⊕B⊕C=P⊕Q⊕R [14]. Block diagram of NFT gate is shown in Figure 5.

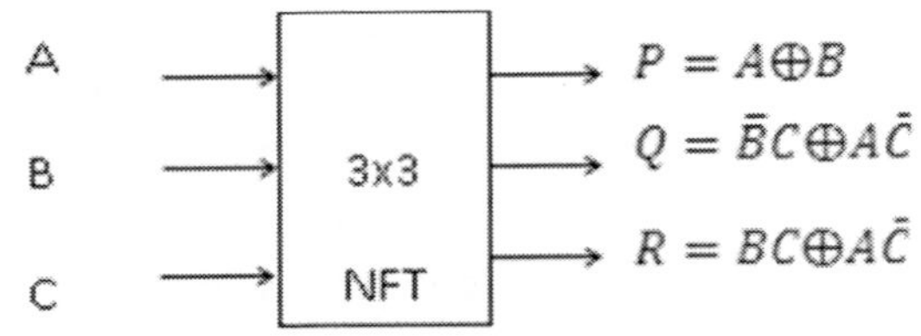

Figure 5. NFT Gate.

2.6. QCA1 and QCA2 Gate

Two new reversible gates referred to as QCA1 and QCA2 are proposed in [15].These gates are comparable with other reversible gates in terms of delay, area and logic synthesis. Reversible logic is inherently easier to test because the one-to-one onto property improves the controllability and the observability of the circuit. The gate is defined by the output functions: y1 = MV ( ); y2 = MV ( ); y3 = MV( ).

A QCA implementation of QCA1 is shown in Figure 6(a) and the second gate QCA2 is shown in Figure 6(b). Its output functions are: y1 = MV(x1, x2, x3); y2 = MV ( ); y3 = MV ( ).

QCA1 and QCA2 allow multiple MV functions to be embedded in one gate at the same time.

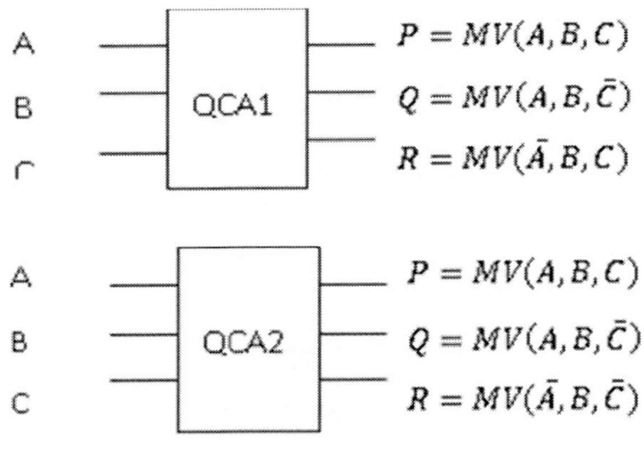

Figure 6. (a) QCA1 Gate (b) QCA2 Gate.

2.7. Different New Reversible Logic Gate Design

2.7.1. Design of RKDLG

Reversible Logic Gate (RLG) and Conservative Logic Gate (CLG) are discussed elaborately in [20].

This paper explored that reversible logic family and conservative logic family are two sets intersecting each other which are known as parity preserving reversible gates or Conservative Reversible Logic Gate.

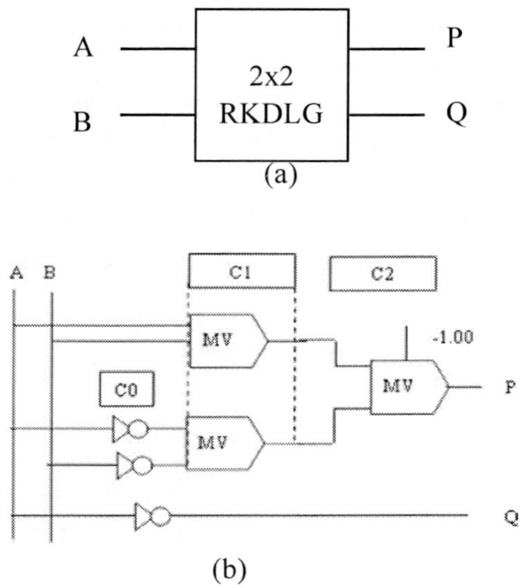

Figure 7. Symbol Diagram of 2x2 RKDLG Architectural block diagram of RKDLG [20].

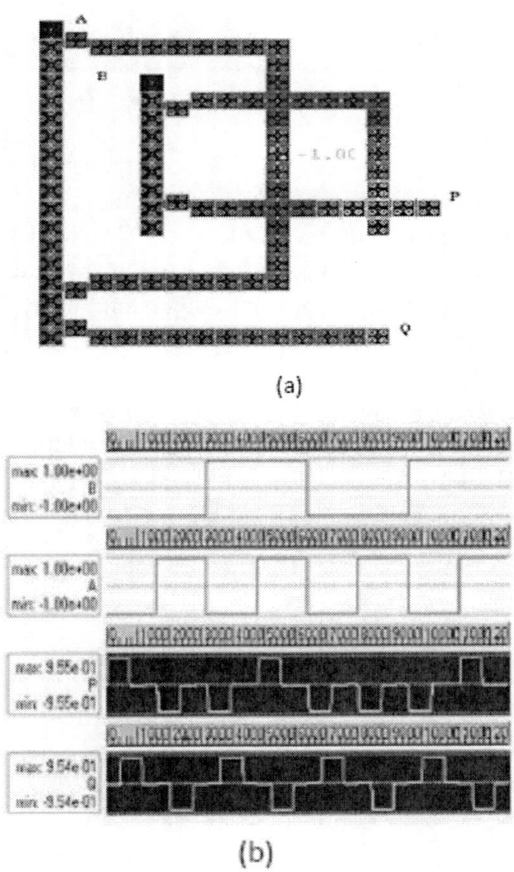

Figure 8. (a) 2 × 2 RKDLG QCA designer layout. (b) QCA designer output [20].

A novel design approach of 2 × 2 RLG is proposed as RKDLG, where "R" means "reversible," "K" means "Kunal," "D" means "Debashis," and "LG" means "Logic gate" [20]. RKDLG is a building block of k × k RLG. Figure 7(a) shows the block diagram of 2x2 RKDLG and Figure 7(b) shows the architectural block diagram of RKDLG.

Input vector of 2x2 RLG is $I_v = (I_0, I_1)$and $O_v = (O_0 = I_0 \otimes I_1, O_1 = I_0')$, where "⊗" denotes an 'XNOR' operation. This 2x2 RLG can be used to implement any kxk RLG.

QCA designer layout of the RKDLG is shown in Figure.8(a) and 8(b)shows the simulation output of the RKDLG by QCA designer. Table-I shows the input-output vector relationship of RKDLG.

Table I. Possible i/p-o/p vector for 2x2 Reversible Logic Gate [20]

Input vector	Output vector O_v									
I_v	I	II	III	IV	V	VI	VII	VIII	IX	X
I_v (0,0)	X_0	X_0	X_0	X_1	X_1	X_1	X_1	X_1	X_1	X_3
I_v (0,1)	X_1	X_1	X_2	X_0	X_2	X_3	X_3	X_2	X_0	X_1
I_v (1,0)	X_3	X_2	X_1	X_2	X_0	X_2	X_0	X_3	X_3	X_0
I_v (1,1)	X_2	X_3	X_3	X_3	X_3	X_0	X_2	X_0	X_2	X_2

2.7.2. Design of MPDRG

A new reversible logic gate named "MPDRG" is a universal reversible logic gate [30]. All the logic operations like AND, OR, NAND, NOR, XOR, XNOR can be performed by this gate. It is a three input three output reversible gate. The MPDRG gate is also used to implement different reversible latches like D-latch, T-latch, JK-latch and SR-latch which are optimized in terms of number of gates, garbage outputs and delay [31]. Figure 9 shows the block diagram of the MPDRG.

Figure 10 shows the QCA designer layout of the RKDLG and Figure 11 shows the simulation output of the RKDLG by QCA designer In this design, we have considered the cell size is 18nm x 18nm and the dot size is 5nm. For the simulation of the proposed gate, we have used the Bistable model of QCA Designer; version 2.0.3. Simulation tool [21]. QCA Designer is the product of collaboration between the ATIPS laboratory and the University of Notre Dame.

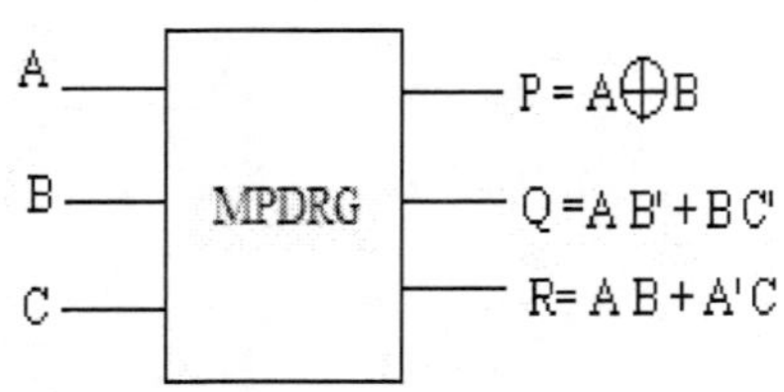

Figure 9. Block diagram of MPDRG.

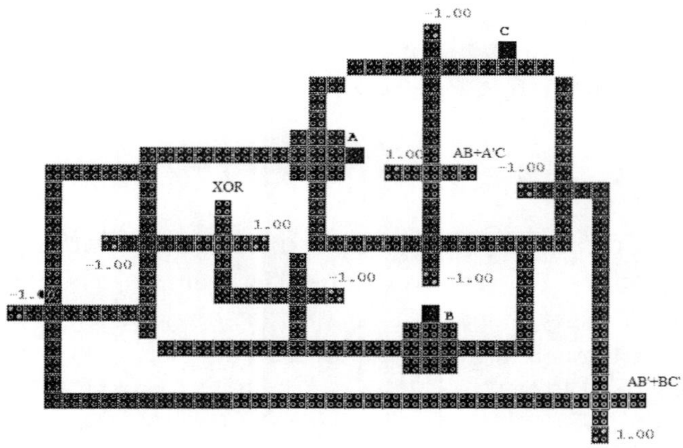

Figure 10. QCA layout of the MPDRG.

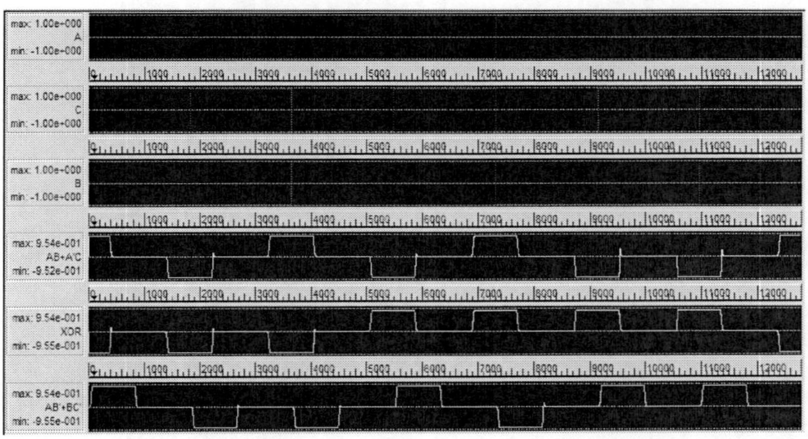

Figure 11. Simulated output of MPDRG.

3. Designing of Reversible Combinational and Sequential Circuits

3.1. Adder Circuit Design Using Fredkin Gate

A novel Full Adder circuit has been designed in [23] which have less complexity. All the basic logic gates like AND gate, OR gate can be designed using Fredkin Gate. The two input XOR gate can be designed using two FRG. Ripple Carry Adder is the simplest adder circuit. But it has some drawbacks i.e. the delay of the circuit increases linearly with the width of the circuit. So, many other adder circuits have been proposed to reduce the delay. Among them Carry Look Ahead Adder provides the best performance to reduce the delay. It has achieved the speed through parallel carry computation and for this a large number of gates have been required. Carry skip adder is another adder which is a balanced structure between the Ripple Carry adder and Carry Look Ahead Adder.

The logic equation of the Sum and Carry of the adder is:

$$S_i = X_iY_iC_i + X_i\bar{Y}_i\bar{C}_i + \bar{X}_i\bar{Y}_iC_i + \bar{X}_iY_i\bar{C}_i = X_i \oplus Y_i \oplus C_i$$

and $C_{i+1} = X_iY_i + X_iC_i + Y_iC_i$.

Special feature of the Ripple Carry Adder is if the carry generates in the first stage it ripple through all the stages. But the Carry Skip Adder reduces the computational delay. The worst case delay of the Carry Skip Adder is when the carry generates in the very first full adder circuit it ripples through all the blocks of the Full adder. Block diagram is shown in Figure 12(a) and Figure 12(b). When the most significant carry C4=1the block propagate P=1and the block carry input signal C_{in} =0 and P=1. Four bit Carry Skip Adder Block is shown in Figure 13 and Four bit Fredkin Carry Skip adder block is shown in Figure 14.

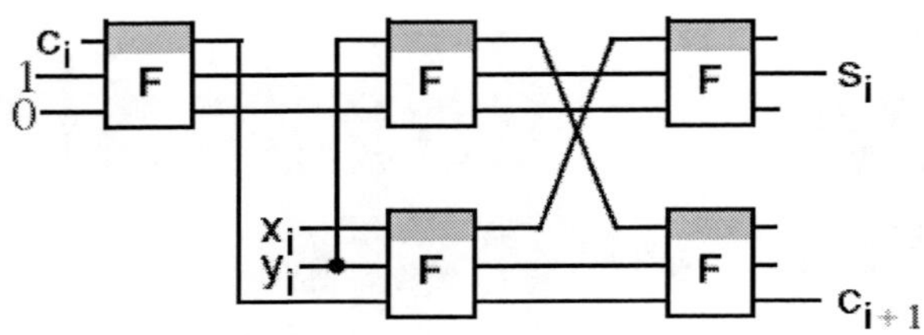

(a)

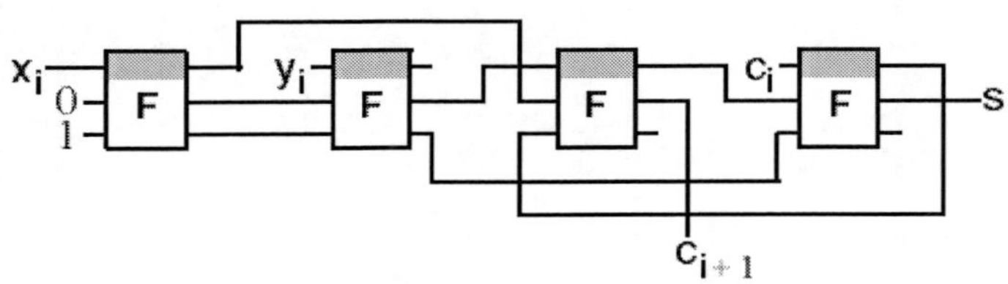

(b)

Figure 12. (a) Full Adder Circuit using Fredkin Gate proposed in [23] (b) Fredkin Full Adder Circuit avoiding Fan out.

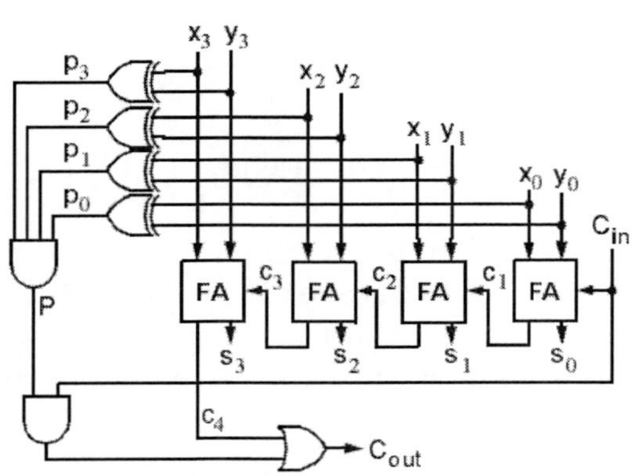

Figure 13. Four bit Carry Skip Adder Block.

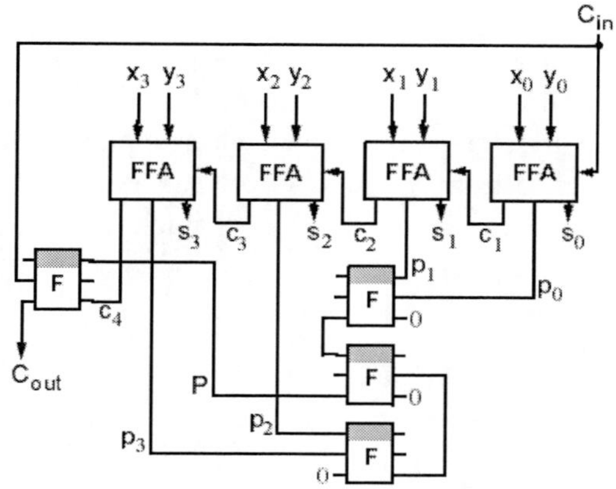

Figure 14. Four bit Fredkin Carry Skip adder block.

3.2. Design of Reversible Sequential Circuits

To design the reversible sequential circuits, conventional logic gates are appropriately designed from the reversible gates using Feynman Gate, New gate and Fredkin Gate etc. to minimize the number of reversible gates used and garbage outputs produced. The designed FFs are highly optimized in terms of number of reversible gates and garbage outputs. Fan out problem is avoided by using Feynman gate for copying the output. [16]

3.3. RS Flip-Flop

The clocked RS flip flop consists of two NOR flip flop and two AND gates. Figure 15 (a) shows the RS flip-flop designed from conventional irreversible gates. Figure 15(b) shows the RS flip flop designed from the reversible equivalent gates.

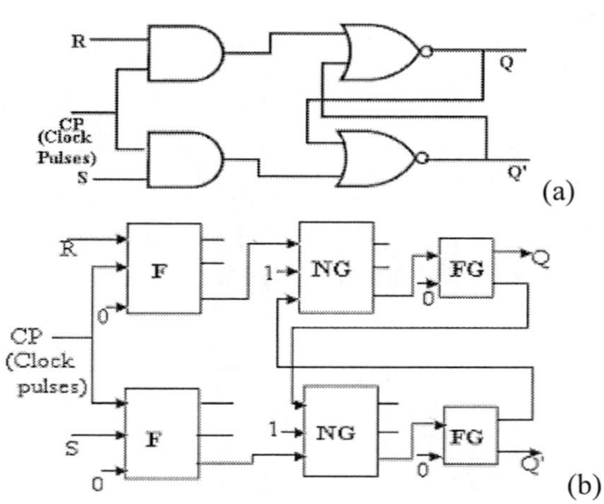

Figure 15. (a) RS flip-flop designed from conventional irreversible gates. (b) RS flip flop designed from the reversible equivalent gates.

3.4. D-Flop-Flop

The D flip flop is a modification of the clocked RS flip flop. In the D flip flop, the D input goes directly to the S input and its complement is applied as an R input. Figure 16 (a) shows the D flip flop designed from irreversible gates. Figure 16 (b) shows the D flip flop designed from the reversible equivalent gates. The characteristic equation of the D latch can be written as .

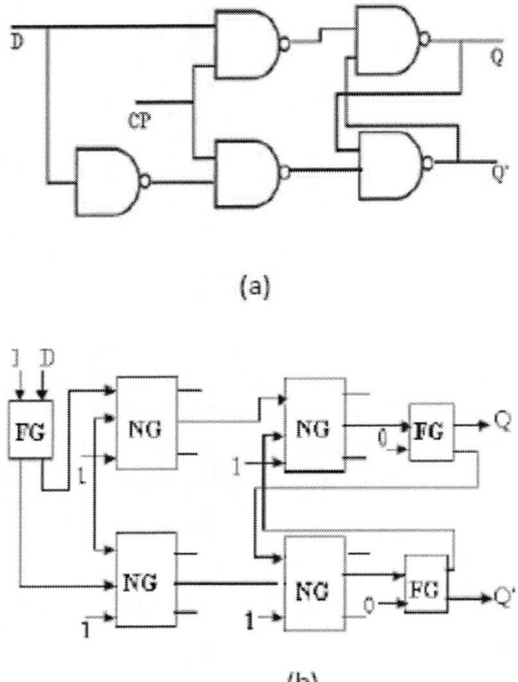

Figure 16. (a) D flip flop designed from the reversible equivalent gates (b) D flip flop designed from the reversible equivalent gates.

3.5. JK Flip Flop

The JK flip flop switches to its complement state, when inputs are applied to both J and K simultaneously. Figure 17(a) shows the JK flip flop designed from conventional irreversible gates. Figure 17 (b) shows the JK flip flop designed from the reversible equivalent gates.

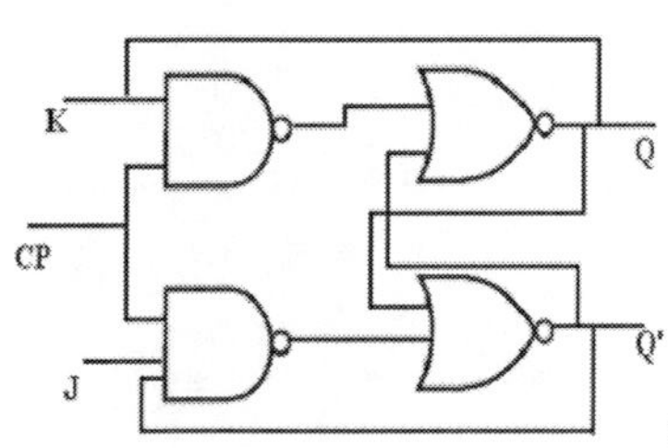

(a)

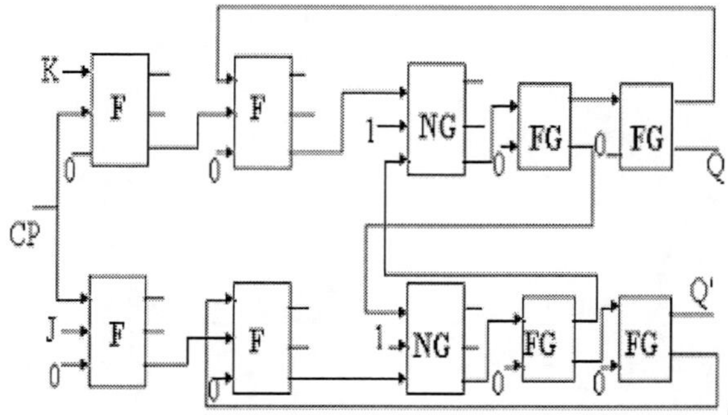

(b)

Figure 17. (a) JK flip flop designed from conventional irreversible gates. (b) JK flip flop designed from the reversible equivalent gates.

3.6. T Flip- Flop

The T(Toggle) flip-flop is obtained from the JK type if both inputs are tied together and are passed the same input signal. Figure: 18(a) shows the T flip flop designed from conventional irreversible gates. Figure: 18(b) shows the T flip flop designed from the reversible equivalent gates.

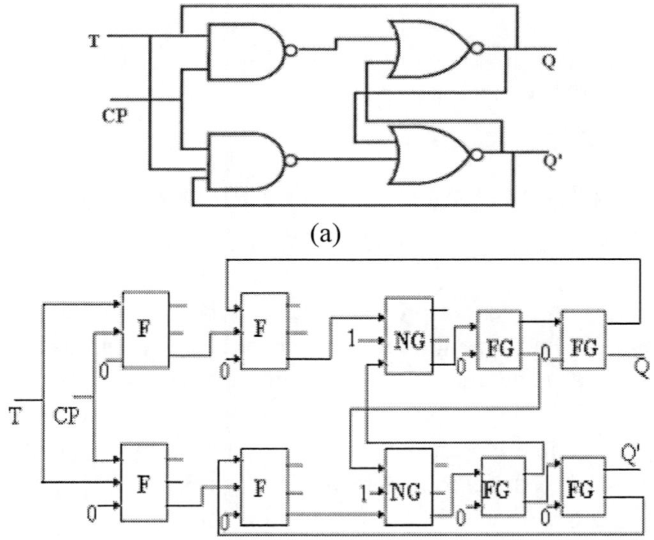

(b)

Figure 18. (a) T flip flop designed from conventional irreversible gates. (b) T flip flop designed from the reversible equivalent gates.

Master Slave Flip-Flop

A master–slave flip flop is constructed from two separate flip-flops in which one circuit serves as a master and the other as a slave. Figure: 19(a) shows the conventional master slave

JK flip-flop constructed with NAND gates. Figure: 19(b) shows the T flip flop designed from the reversible equivalent gates.

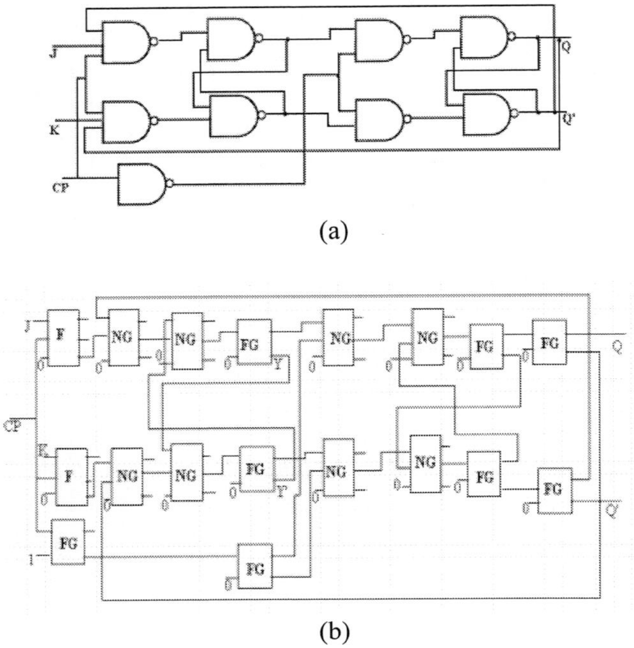

Figure 19. (a) Conventional master slave JK flip-flop constructed with NAND gates. (b) T flip flop designed from the reversible equivalent gates.

4. Fault Detection and Tolerance of Reversible Logic Gates

Fault tolerance is an essential property that enables to continuously operating a system properly, instead of present some fault in the system. If the system itself made of fault tolerant components, then the detection and correction of faults become easier and simple. In communication and many other systems, fault tolerance is achieved by parity. Therefore, parity preserving reversible circuits will be the future design trends to the development of fault tolerant reversible systems in nanotechnology. Parity checking is one of the oldest and the most widely used methods for error detection in digital systems. Parity preservation proves to be useful for ensuring the robustness of reversible logic circuits. A few parity preserving logic gates have been proposed in the literature. Among them 3x3 Feynman Double gate (F2G) [17], Fredkin gate (FRG), Haghparast and Navi have proposed another 3x3 parity preserving reversible gate, namely New Fault Tolerant gate (NFT). So, FRG, F2G and NFT gates are parity preserving since they satisfy $A \oplus B \oplus C = P \oplus Q \oplus R$. And any kxk reversible logic gate where the EX-OR of the inputs matches the EX-OR of the outputs will be parity preserving.

Table II. Truth Tale of Parity Preserving IG Gate

A	B	C	D	P	Q	R	S
0	0	0	0	0	0	0	0
0	0	0	1	0	0	0	1
0	0	1	0	0	0	1	0
0	0	1	1	0	0	1	1
0	1	0	0	0	1	0	0
0	1	0	1	0	1	0	1
0	1	1	0	0	1	1	0
0	1	1	1	0	1	1	1
1	0	0	0	1	1	0	1
1	0	0	1	1	1	0	0
1	0	1	0	1	1	1	1
1	0	1	1	1	1	1	0
1	1	0	0	1	0	1	0
1	1	0	1	1	0	1	1
1	1	1	0	1	0	0	0
1	1	1	1	1	0	0	1

A new 4x4 parity preserving reversible gate is proposed in [19], IG, depicted in Figure 20. The proposed gate is one-through, which means one of the input variable is also the output. The corresponding truth table of the gate is shown in Table II.

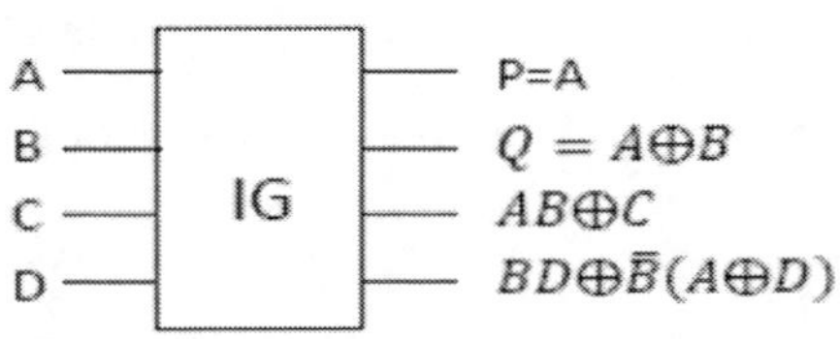

Figure 20. A new 4*4 parity preserving reversible gate IG.

4.1. Fault Detection Using Conservative Reversible Logic Gate

A gate is conservative if the Hamming weight (number of logical ones) of its input equals to the Hamming weight of its output. In case of conservative logic gate, the number of 1s (binary "1") are equal to input vector Iv and Ov, such that Ov = f(Iv)and f : M → M, where M is mapping between Iv and Ov. So, the mapping is one to one. If a gate is conservative and reversible then it is parity preserving [21]. Fredkin gate can concurrently detect permanent, as well as transient fault based on parity preserving in molecular QCA. Fault patterns in a Fredkin gate is shown in Table III.

The 3x3 conservative reversible logic gate is "CKDLG". It is better than Fredkin Gate in context of clocking zones, number of MVs and constant input for fix polarity cells, total number of cells (rotated and non-rotated cells) and finally area of two gates. Figure 21 (a) shows the block diagram of 3x3 CKDLG. 21(b) Architectural block diagram of CKDLG where P = C, Q = AB +BC +CA, R = , and C_0 and C_1 are two clocking zones.

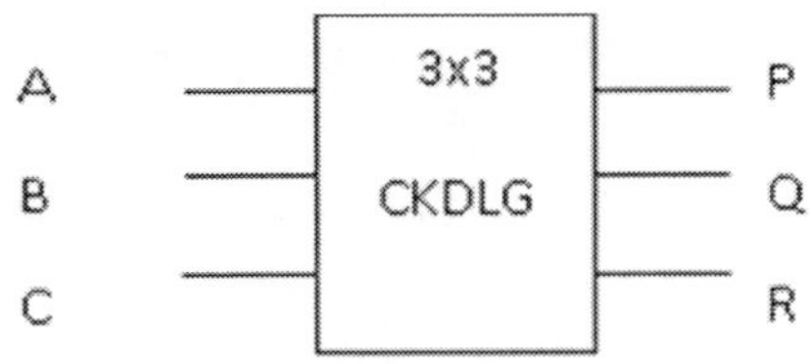

(a)

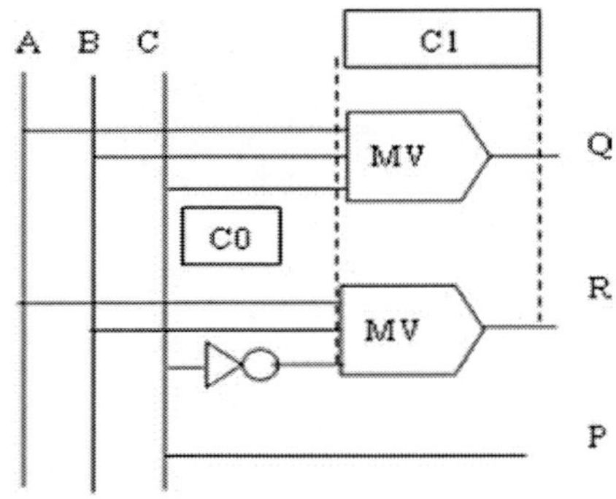

(b)

Figure 21. (a) Block Diagram of 3x3 CKDLG. (b) Architectural block diagram of CKDLG where P = C, Q = AB +BC +CA, R = AB + BC'+ C'A, and C_0 and C_1 are two clocking zones.

Table III. Fault Patterns in Fredkin Gate

Input Vector	Fault Free	Fault Patterns																			
		1	2	3	4	5	6	7	8	9	10	11	12	13	14	15	16	17	18	19	20
a0	a0	a0	a0	a1	a1	a0	a0	a1	a2	a1	a1	a0	a2	a0	a2	a2	a0	a0	a0	a2	a4
a1	a1	a1	a1	a1	a1	a0	a1	a0	a1	a1	a1	a1	a3	a1	a3	a3	a3	a1	a3	a3	a5
a2	a2	a3	a2	a2	a3	a2	a0	a3	a2	a3	a3	a2	a0	a0	a2	a2	a2	a2	a2	a0	a6
a3	a3	a3	a3	a3	a3	a2	a1	a2	a1	a3	a3	a3	a1	a1	a3	a3	a3	a3	a3	a1	a7
a4	a4	a4	a5	a5	a5	a4	a4	a4	a4	a5	a4	a4	a4	a4	a4	a6	a4	a6	a4	a6	a0
a5	a6	a6	a7	a7	a7	a7	a6	a6	a6	a6	a6	a6	a6	a6	a6	a6	a4	a4	a6	a6	a2
a6	a5	a4	a4	a5	a4	a5	a5	a5	a5	a5	a5	a4	a5	a7	a5	a5	a5	a7	a5	a7	a1
a7	a7	a6	a6	a7	a6	a7	a7	a7	a7	a7	a7	a6	a7	a7	a7	a7	a5	a5	a7	a7	a3

Conservative reversible circuits are a specific type of reversible circuits, in which there would be an equal number of 1's in the outputs as there would be on the inputs, in addition to one-to-one mapping.

Table IV. Truth Table of Fredkin Gate [22]

A	B	C	P	Q	R
0	0	0	0	0	0
0	0	1	0	0	1
0	1	0	0	1	0
0	1	1	0	1	1
1	0	0	1	0	0
1	0	1	1	1	0
1	1	0	1	0	1
1	1	1	1	1	1

Thus, conservative logic is parity-preserving, i.e., the parity of the input vectors is equal to that of the output vectors. If there is a fault in the QCA implementation of Fredkin gate, there is a parity mismatch between the inputs and the outputs, otherwise the inputs parity is the same as outputs parity.

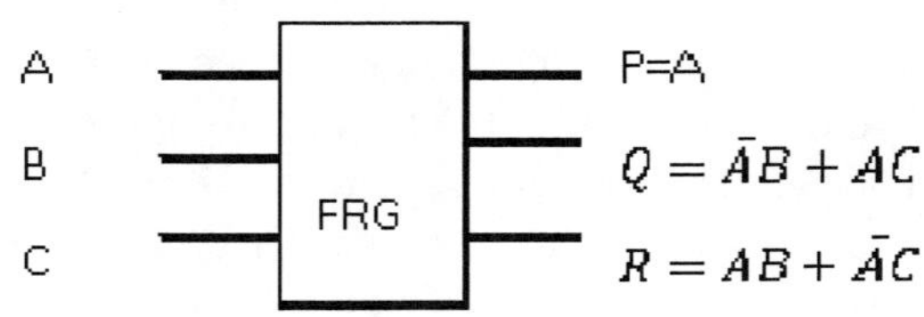

Figure 22. Block Diagram of Fredkin gate.

Any permanent or transient fault in molecular QCA can be concurrently detected if implemented with the conservative Fredkin gate. The block diagram of Fredkin gate is depicted in Figure 22 and Figure 25 shows the QCA layout of Fredkin Gate. Figure 23 depicts the Fredkin Gate implementation of OR gate and AND gate and Figure 24 depicts the Fredkin Gate implementation of XOR and XNOR gate. Truth table of the Fredkin Gate is shown in Table IV. Due to the parity-preserving property, any permanent or transient fault in molecular QCA that leads to parity mismatch can be concurrently detected. Concurrently testable latches like D latch, T latch, JK latch, and SR latch can be designed using conservative Fredkin gate.

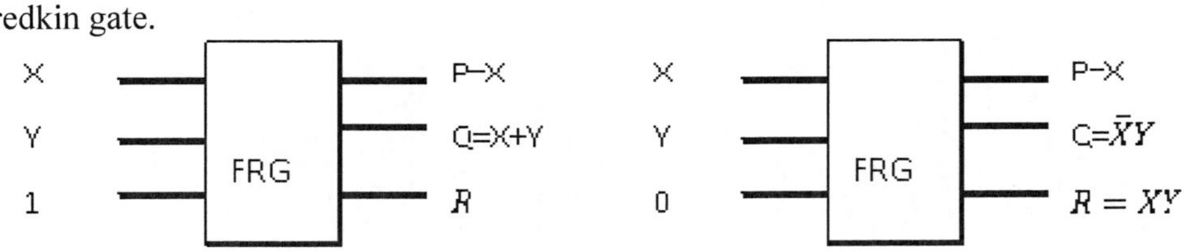

Figure 23. Fredkin Gate implementation of OR gate and AND gate.

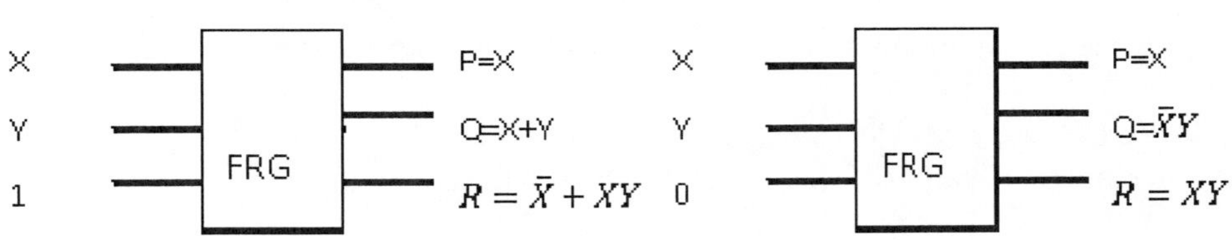

Figure 24. Fredkin Gate implementation of XOR and XNOR gate.

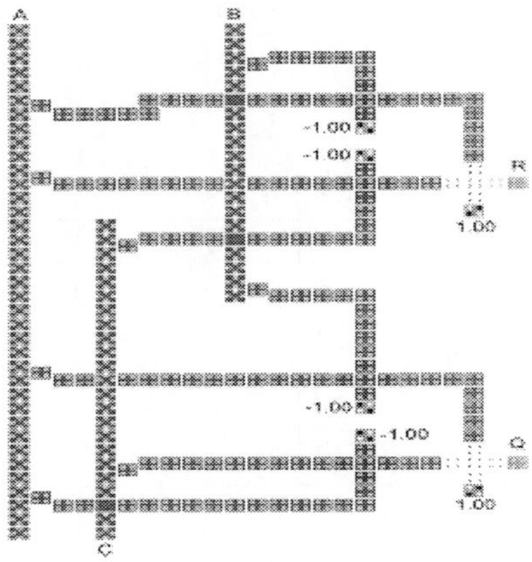

Figure 25. QCA layout of Fredkin Gate [22].

D Latch Design Using Fredkin Gate

The characteristic equation of the D latch can be mapped onto the Fredkin gate. Figure 26 (a) and (b) shows the realization of the reversible D latch using Fredkin gate.

The Testable D-latch design has two control signals, C1 and C2. In test mode, C1C2=00 will make it testable with all 0s input vectors as output T1 will become 0 and make it testable with all 0s input vectors. When C1C2=11 the output T1 will become 1 and the design will become testable with all 1s input vectors. Thus the reversible D latch can be tested with only two test vectors, all 0s and all 1s, the inherent property of conservative reversible logic.

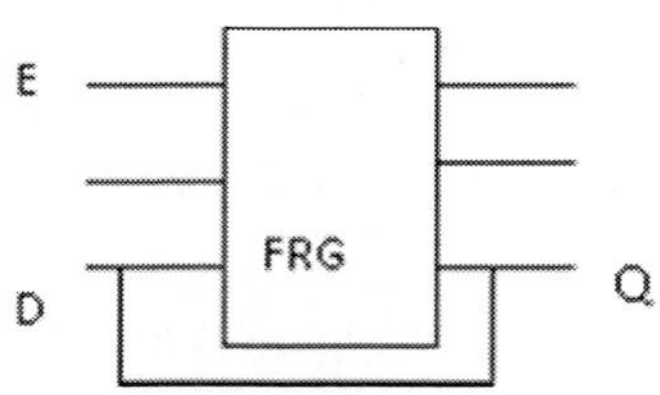

(a)

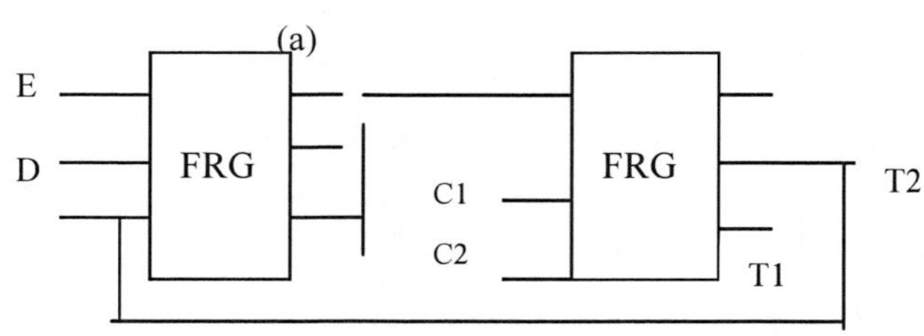

Figure 26. (a) Block diagram of D-Latch (b) Block diagram of Testable D-Latch.

T-Latch Design Using Fredkin Gate

The characteristic equation of the T latch can be written as

But the same result can also be obtained from the equation

$$\bar{Q} = (T.E)Q$$

Figure 27(a) shows the proposed design of reversible testable T latch with C1C2 as control signals as described in the section of D latch. The working of T latch in normal mode is shown in Figure 27(b) having C1C2=01.

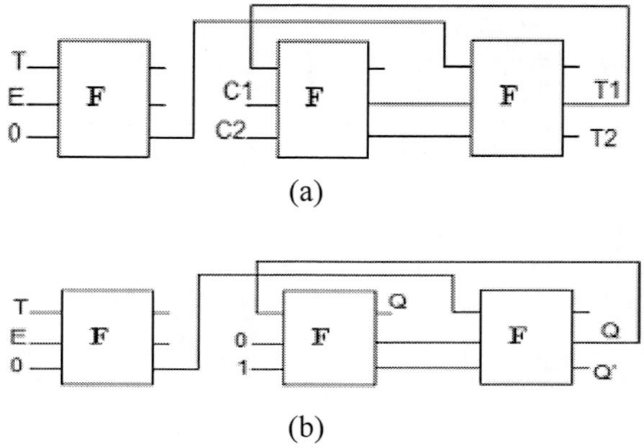

Figure 27. (a) Testable T Latch (b) Normal Mode.

JK Latch

The characteristic equation of the JK latch can be written as

$$\bar{Q} = (J\bar{Q} + \bar{K}Q).E + \bar{E}Q$$

After computing the equation $J\bar{Q} + \bar{K}Q$ it can mapped on the D Latch to design the JK Latch. Figure 28(a) shows the proposed design of testable JK latch with again C1C2 as the control signals. Its function in normal mode with C1C2=01 is shown in Figure 28(b).

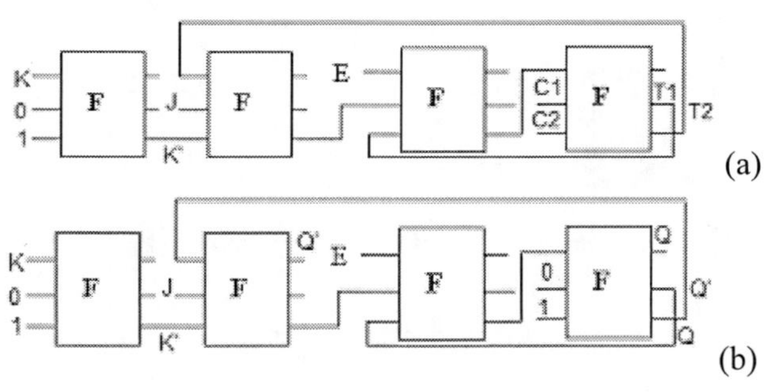

Figure 28. (a) Block diagram of testable JK latch with C1C2 as the control signals. (b) JK latch in normal mode with C1C2=01.

RS Latch

The characteristic equation of the RS latch can be written as

$$\bar{Q} = (S.E + \overline{(R.E)}Q)$$

Figure: 29 shows the proposed design of the testable reversible RS latch with C1C2 as control signal, when C1C2=11, it will work in normal mode as well as test mode.

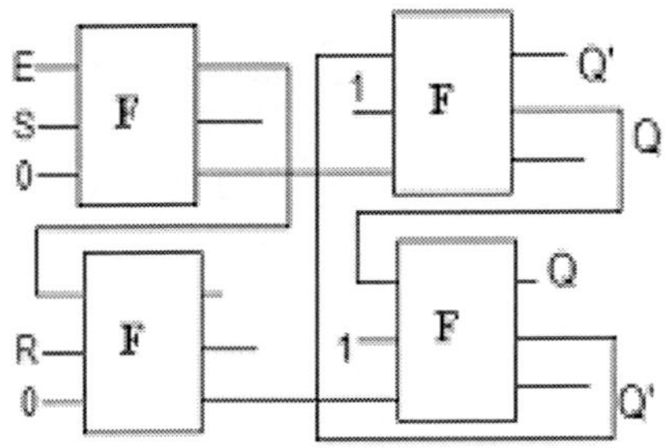

Figure 29. Testable RS Latch in Normal Mode as well as Test mode when C1C2=11.

Fault Model for Reversible Logic Circuits

A typical fault model for conventional irreversible logic circuits is the stuck-at fault model, which is based on the model that one or more wires in a circuit are stuck with some fixed logic value.

The stuck-at fault model is not adequate to represent all error effects in reversible logic circuits since it does not cover the situations where some of the reversible gates do not work correctly.

This is easily illustrated in [24]. This model can't detect all the missing Control Points using the Cross point fault model. Cross point faults in reversible logic circuits can be classified into two groups: disappearance faults and appearance faults. For a given Toffli network depicted in Figure 30 the appearance and/or disappearance of control points in a reversible gate is discussed in [24]

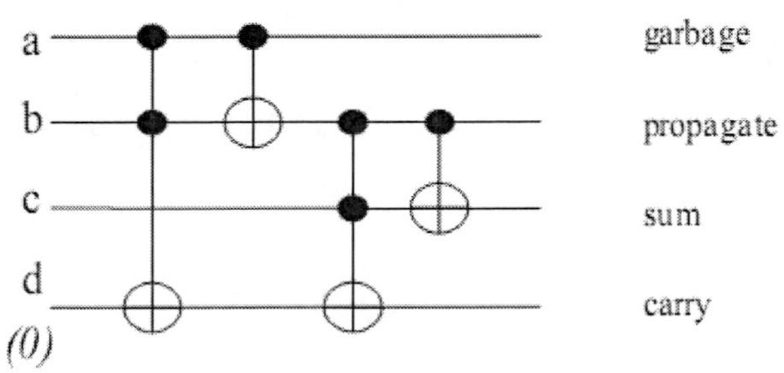

Figure 30. A Toffoli network: full adder [24].

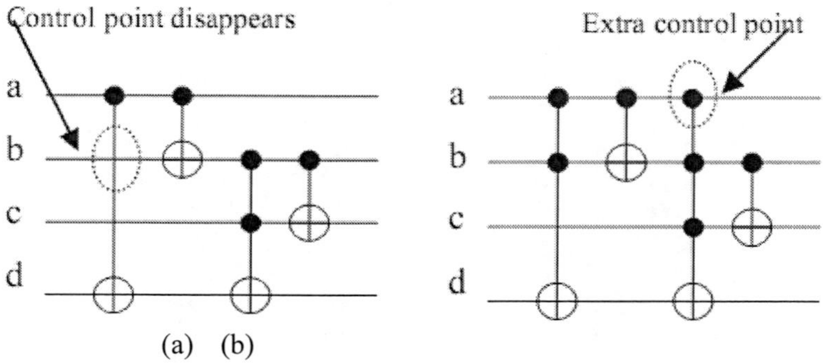

Figure 31. Three Cross-point faults (a) Disappearance fault, (b) Appearance fault.

Disappearance fault: One or more of the control points in a Toffoli gate are missing. The disappearance fault model happens to be same as the partial missing-gate fault model introduced in [25]. For example, in Figure 1, in the first Toffoli gate, a disappearance fault happens if the control point from line b is missing, causing the Toffoli gate to be changed into a CNOT gate as shown in Figure 31(a).

Appearance fault: When one or more extra control points are added to a Toffoli gate. For instance, in Figure 1, when one additional control point from line 'a' is added to the third gate, an appearance fault happens. It changes the 3-input Toffoli gate into a 4-input Toffoli gate as shown in Figure 31(b).

In [25] Testable reversible logic gates are proposed to identify various faults in online using these gates. Applications of these gates in testable ripple carry, carry–skip adders and MCNC benchmark circuits have been illustrated in this paper. Properties of reversible logic circuits under a new fault model: the cross-point fault model and fault tolerant reversible full adder design is described in [26].

Defect Characterization and Fault Tolerance of Reversible QCA

There are different fault detection techniques in QCA. They are Triple Modular Redundancy (TMR), NAND-multiplexing, Maj-MUX and Parity Checking etc. Triple Modular Redundancy (TMR) is a widely used fault tolerant technique. A TMR system as shown in Figure 32 consists of three modules, each of the modules computes the required function. Three copies of the input signals are generated and sent to a module. The outputs of the three modules are then sent to the majority voters (MVs). TMR generates a correct result at the output when no more than one module is faulty. TMRs can be cascaded to further improve the system's reliability. Signal reliability is defined to be the probability of the signal being fault free or correct [5]. If the MVs are assumed to be fault free, then every stage in the cascaded TMR system can improve the signal reliability as Rout = (Rin)3 + 3(Rin)2(1 − Rin) where Rin is the reliability of the input signal and Rout is the reliability of the output signal. The reliability of outputs of the TMR stage (Rout) is higher than the reliability of the inputs (Rin) when Rin > 50%. If this is extended to the N-Module-Redundant (NMR) scheme, then the reliability is improved as Rout = P(N−1)/2 i=0 (N i)(Rin)N−I (1 − Rin)I, when Rin > 50%. The TMR scheme is advantageous for QCA because the basic QCA device is also the 3-input MV.

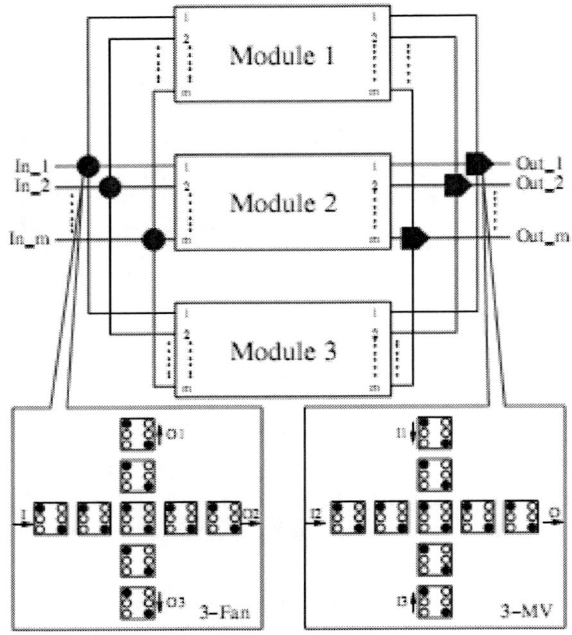

Figure 32. A TMR System in QCA [27].

Maj-MUX is the technique where the fault is detected by the Majority Voter, which is the basic device of QCA. Maj-MUX for a QCA implementation has been investigated in detail in this chapter. Fault probability improvement and signal restoration speed have been reported for Maj-MUX. It has been shown that the Maj-MUX can tolerate a higher fault rate and restore signals at lower overhead compared to NAND-multiplexing. Energy dissipation due to fault correction in majority multiplexing affects the dissipation of the overall QCA reversible circuit.

In paper [27] the testing properties of an important subclass of reversible circuits constructed from a library of reversible gates called controlled-NOT (CNOT) gates has been studied. A k-CNOT gate has k + 1 input wires and k + 1 output wires. It transmits the first k input signals unchanged, and inverts the last input signal iff the first k inputs are all 1; clearly this input-output mapping is reversible. The paper [28] investigates how to synthesis of arbitrary ternary non-reversible logic circuits by adding inputs with constant value and garbage outputs. Group theory has been also used to solve the synthesis of reversible logic circuits. Our algorithm uses the SNT (ternary Swap gate, ternary NOT gate, ternary Toffoli gate) library, by reducing the ternary non-reversible logic circuit synthesis problem to group theory representation. Here Group theory has been used to solve the synthesis process. The algorithm uses the SNT (ternary Swap gate, ternary NOT gate, ternary Toffoli gate) library, by reducing the ternary non-reversible logic circuit synthesis problem to group theory representation. The main result shows the relationship of ternary non-reversible logic circuits and the reversible circuits.

Conclusion

This Chapter has presented a brief but extensive survey over the reversible logic circuit design techniques using QCA. The designs are simulated and analyzed by QCA designer. Defect tolerance of reversible QCA circuits has been analyzed in details beside different fault detection techniques.

Different fault analysis techniques like Parity Preserving, Majority Multiplexing, and, Hardware Redundancy are described elaborately in this chapter. The energy dissipation in QCA reversible circuits is analyzed. Quantum Circuit representation and garbage output minimization of the existing reversible logic gate and some new reversible logic gate like "RKDLG" and "MPDRG" proposed recentlyare also discussed. Energy dissipation caused by fault detection and correction has been derived. Present research directions in this area focuse on the design of low power, energy efficient reversible circuits using QCA. In a nutshell, this chapter presents an extensive survey on low power, high speed logic circuit design using QCA in the era of nanotechnology.

Acknowledgments

Authors are grateful to Department of Science and Technology (DST) for sanctioning a research Project under Fast Track Young Scientist scheme, reference no.: SERC/ET-0213/2011and UGC Project named "Study of Quantum dot Cellular Automata for Circuits designing and Implementing them for high speed and low power fault tolerant Computing" under which this paper has been completed.

References

[1] Yibin Ye and Kaushik Roy, "*Energy Recovery Circuits Using Reversible and Partially Reversible Logic*", 1996.

[2] R. Landauer, "Irreversibility and Heat Generation in the Computational Process", *IBM Journal of Research and Development*, 5, pp. 183-191, 1961.

[3] C.H. Bennett,"Logical Reversibility of Computation", *IBM J. Research and Development,* pp. 525-532, November 1973.

[4] Rekha K. James, Shahana T. K., K. Poulose Jacob, and Sreela Sasi, "Performance Analysis of Reversible Fast Decimal Adders," *Proceedings of the World Congress on Engineering and Computer Science*, 2007.

[5] BOOk by F.Lombardi.

[6] "*Reversible energy recovery Logic Circuit without non-adiabatic energy loss*", Joonho Lim, Kipaek Kwon and Soo-Ik Chae, Electronics Letters, Vol. 34 No. 4, pp. 344-346, 1998.

[7] Dickinson, A.G. and Denker, J.s. "Adiabatic Dynamic Logic", *J. Solid-state Circuits*,,Vol.30, (3), pp. 311-315, 1995.

[8] Kramer, A., Denker, J.S., Avery, S.C., Dickinson, A.G., and Wik, T R.: '*Adiabatic computing with the 2N-2N2D logic family*'. Symp. on VLSI Circuits, pp. 25-26, 1994.

[9] Athas, w.c., Svensson, L.J., Koller, J.G., Tzartzanis, N., and Chou, E.: 'Low power digital systems based on adiabatic-switching principles', *ZEEE Trans. VLSZ Syst.*,, 2, (4), pp. 398-407, 1994.

[10] Feynman R., Quantum mechanical computers. *Optics News*, 11, 11-20,1985.

[11] Fredkin E. and T. Toffoli, Conservative logic, *Intl J. Theoretical Physics,* 21, 219-253, 1982.

[12] Toffoli T., Reversible computing, Tech Memo MIT/LCS/TM-151, MIT Lab for Computer Science, 1980.

[13] W. N. Hung, X. Song, G.Yang, J.Yang, and M. Perkowski. Optimal synthesis of multiple output boolean functions using a set of quantum gates by symbolic reachability analysis. *IEEE Trans. Computer-Aided Design*, 25(9):16521663, Sept. 2006.

[14] M.Haghparast and K.Navi, A Novel Fault Tolerance Reversible Gate for Nanotechnology Based System, *Am.J. Applied Sci*., 5(5),519-523, 2008.

[15] X. Ma · J. Huang · C. Metra · F. Lombardi, "Reversible Gates and Testability of One Dimensional Arrays of Molecular QCA," *J. Electron Test*, Vol.24, pp.297-311, 2008.

[16] Himanshu Thapliyal, M.B Srinivas and Mark Zwolinski, "*A Beginning in the Reversible Logic Synthesis of Sequential Circuits,*" MAPLD, 2005.

[17] B. Parhami, "*Fault tolerant reversible circuits*", in Proceedings of 40th Asimolar Conf. Signals, Systems, and Computers, Pacific Grove, CA, pp. 1726-1729, October 2006.

[18] B. Parhami, "*Fault tolerant reversible circuits*", in Proceedings of 40th Asimolar Conf. Signals, Systems, and Computers, Pacific Grove, CA, pp. 1726-1729, October 2006.

[19] Md. Saiful Islam, M. M.Rahman, Zerina Begum, Mohd. Zulfiquar Hafiz, Abdullah Al Mahmud "*Synthesis of Fault Tolerant Reversible Logic Circuit,*" IEEE, 2009.

[20] "Characterization, Test and Logic Synthesis of Novel Conservative and Reversible Logic Gates for QCA," Kunal Das, Debashis De, *International Journal of Nanoscience*, Vol. 9, pp. 201–214, 2010.

[21] C.S.Lent, P.D.TougawandW.Porod, *Appl. Phys. Lett*. 62, 714 (1993).

[22] Himanshu Thapliyal and Nagarajan Ranganathan, "*Testable Reversible Latches for Molecular QCA*," 2008 IEEE.

[23] "Efficient Adder Circuits Based on a Conservative Reversible Fredkin Gate", J.W. Bruce, M.A. Thornton, L. Shivakumaraiah, P.S.Kokate, X.Li.

[24] "Analyzing Fault Models for Reversible Logic Circuits", Jing Zhong and Jon C. Muzio, *IEEE Congress on Evolutionary Computation*, 2006.

[25] . Polian, J. P. Hayes, T. Fiehn, B. Becker, A Family of Logical Fault Models for Reversible Circuits, *IEEE Proceedings on the 14th Asian Test Symposium*, pp. 422-427, 2005.

[26] Md. Saiful Islam, Muhammad Mahbubur Rahman, Zerina begum, and Mohd. Zulfiquar Hafiz, "Efficient Approaches for Designing Fault Tolerant Reversible Carry Look-Ahead and Carry-Skip Adders".

[27] "Synthesis of Reversible Circuits for Testing with Universal Test Set and C-Testability of Reversible Iterative Logic Arrays," Avik Chakraborty.

[28] Synthesis of Ternary Non-Reversible Logic Circuits, Xiaoyu Li, Guowu Yang, Desheng Zheng.

[29] "Automatic Synthesis of Reversible Logic Circuit Based on Genetic Algorithm," Mingming Zhang, Shuguang Zhao, Xu Wang.

[30] Maumita Maity, Prasun Ghosal, Debashis De, Bishwarup Das, Universal reversible logic gate design using QCA, 2nd *International Conference on Advanced Nanomaterials and Nanotechnology*, 2011.

[31] Maumita Maity, Prasun Ghosal, Debashis De, "Low Power Reversible Latch Design Using a Novel Reversible Logic Gate," *International Workshop on Embedded Computing and Communication Systems* (IWECC), 2011.

In: Quantum Dots and Quantum Cellular Automata
Editors: D. De, S. Bhattacharaya and K.P. Ghatak
ISBN: 978-1-62257-920-4

Chapter 9

LOGIC DESIGN USING MAGNETIC QUANTUM CELLULAR AUTOMATA

Nilanjana Pradhan*[*], *Bishwarup Das and Debashis De
Department of Computer Science and Engineering,
West Bengal University of Technology, Kolkata, India

ABSTRACT

In this chapter an attempt is made to study the Magnetic Quantum Cellular Automata (MQCA) which uses the concept of magnetism. Nanometer-scale magnets are designed for digital computation in MQCA systems. We have also studied the fabrication of MQCA on silicon wafer by electron beam lithographic technique. The clocking of MQCA using Spin Transfer Torque (STT) and corresponding storage property has been discussed. The operations of MQCA devices like binary wires and majority voter operations are also briefly demonstrated .The timing analysis of Majority Voter is demonstrated by suitable simulations results using Object Oriented MicroMagnetic Framework (OOMMF) simulator. Low power consumption by MQCA using spin torque transition current fed input circuit and multilayer structures are explained. Low power MQCA voter operation using NiFe polycrystalline nanomagnets are demonstrated. The basic principle of spintronic, its application and drawbacks are also discussed in this chapter. Digital Logic using atomic scaling of MQCA is also explained.

1. INTRODUCTION

MQCA concept is a version of the field-coupled QCA architecture [1]. The original idea was introduced for a quantum dot system, in which electrons tunnel between the quantum dots under the influence of repelling Coulomb forces. When a feasible clocking method is applied this architecture can propagate and process binary information [2]. From a more general point of view, the field-coupled QCA architecture is a signal processing system built

[*] Email: nilanjana.pradhan@gmail.com

from simple, identical, bistable units that are locally connected to each other solely by electromagnetic forces. But in MQCA the magnetically-coupled, nanometer-scale magnets are assembled into the networks that perform binary computation. The nanomagnets are designed by electron beam lithography. Magnetic ferrite elements of complex shape, interconnected by windings of copper wire comprise the MQCA devices. Magnetic logic devices are very dense and continue to operate well when scaled to small sizes. In particular, unlike CMOS, they do not suffer from effects leading to intolerable power dissipation upon ultimate scaling. The downscaling of the magnets is ultimately restricted by the fact that the energy barriers separating the magnetization states have to be larger than thermal energy, k_BT, and the switching speed is limited by the precession frequency.

Lithographically designed nanomagnets are used by Magnetic RAM and MQCA. A magnetic logic involves manipulating spin-polarized electrons in magnetic semiconductors, where information is represented as either “spin up” or “spin down” electrons [3]. Nanomagnets that exhibit only two stable states of magnetization can represent digital bits. Magnetic random access memories (MRAM) store binary information in such nanomagnets. Fabrication of dense arrays of nanomagnets is also under development for application in hard disk drives. But the challenge of avoiding magnetic dipole interactions between the individual elements in the arrays limits data storage density. These interactions are utilized in the MQCA system, which is a network of closely-spaced, dipole-coupled, single-domain nanomagnets designed for digital computation. When nanomagnets are placed close to each other, two types of coupling occur (Figure 1(a)), namely, ferromagnetic (F) coupling, which occurs for magnets with their short edges adjacent and antiferromagnetic (AF) coupling, which occurs for magnets with their long edges adjacent. Due to their dipole interactions, two closely spaced, F-coupled nanomagnets prefer parallel alignment of their magnetic moments, and closely spaced AF-coupled nanomagnets prefer antiparallel alignment of their magnetic moments [3].

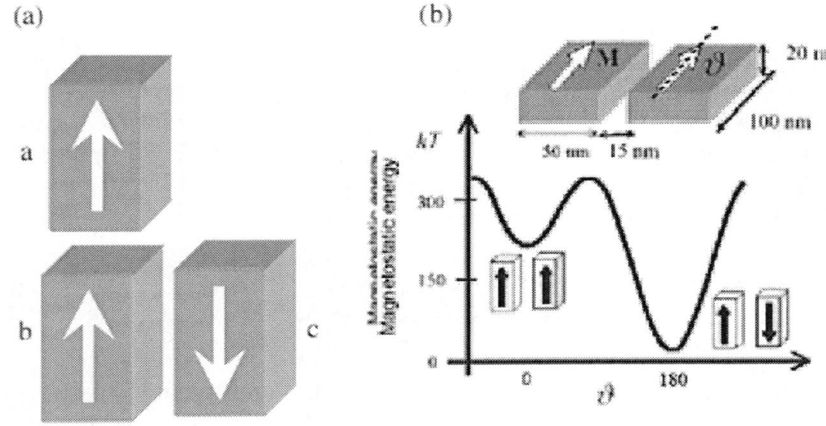

Figure 1. (a) Two types of coupling between nanomagnets. Ferromagnetic coupling (magnets a and b) and antiferromagnetic coupling (magnets b and c). All the magnets are in the ground state in the absence of an external magnetic field .(b) Energy diagram of two AF-coupled perm alloy magnets in the ground and metastable states [3].

In MQCA very low power dissipation occurs .The high integration density of functional devices, as QCA implementations do in general. MQCA can operate over a wide temperature range from sub-Kelvin to the Curie temperature.

Fabrication of the samples were done on silicon wafers by using electron-beam lithography. Thermally evaporated ferromagnetic metals are patterned. Nanomagnets are networked and imaged by magnetic force microscopy (MFM), with which individual magnetization states were distinguished and mapped. Magnetic dipole-ordering in the networks was investigated in different samples. The lengths are averagely ordered and were calculated by statistical analysis of the MFM images. These images are taken after several independent demagnetization processes. The average ordering length was found to be dependent on the shape and size of the nanomagnets and limited by defects introduced during fabrication. Two stable states of magnetization which represented digital bits are exhibited by nanomagnets. Binary information is stored in MRAM in such nanomagnets.Fabrication of dense arrays of nanomagnets is done avoiding magnetic dipole interactions between the individual elements in the arrays. This limits the data storage density and is under further development for application in hard disk drives. These interactions are utilized in MQCA which consists of a network of closely-spaced, dipole-coupled, single-domain nanomagnets designed for digital computation. MQCA cells can be clocked using spin transfer torque (STT) current. Clocking the cells require the assistance of external magnetic fields generated through current-carrying conductors. But the devices still suffer from power dissipation in the external circuits that are used for clocking. The current requirement for each of these operations is order of magnitude lesser than the corresponding current required in field-induced operations. This along with the fact that the STT current scales down with device dimensions motivated us to use these multi-layer cells as elemental cells for nanomagnetic logic computation. The cells offer the ability to individually switch and clock through STT current-induced current. The resulting magnetic configurations of the devices were imaged by both atomic force microscopy (AFM) and magnetic force microscopy (MFM). Dynamics of the magnetization processes in MQCA devices were simulated by means of the freeware micromagnetic solver, OOMMF. One important difference between MQCA and Electron Quantum Cellular Automata (EQCA) is in the way that the clocking field manipulates the barriers. In EQCA, the clock field is applied to raise energy barriers between the dots in the cell, to bring the cell out of null state, and to "freeze" the charge polarization of the cell in accordance with the input signal. When the clocking field is set to zero, no information is stored in an EQCA cell. For MQCA, the situation is the exact opposite; the application of the clocking field brings an MQCA cell to the null state. This difference shows one important advantage of MQCA over EQCA, namely the natural capacity of MQCA cells is to work as non-volatile memory elements as no power supply is required to store information [3].

2. Fabrication of MQCA Devices

Electron-beam lithography (EBL) fabricates the MQCA devices, NiFe evaporation and lift-off. Poly-methyl-methacrylate (PMMA) was first spun on the surface of an oxidized silicon wafer. After evaporation, the lift-off was done in a mixture of methylene chloride and acetone at a ratio of 8:1 respectively, at room temperature. The thickness of the patterned film was in the range of 30 to 40 nm. The magnetizing process, i.e., the application of the clocking field, was performed in the homogenous field of an electromagnet capable of a maximum 7 kOe. A very important issue for making MQCA viable is to eliminate fabrication errors that are presently limiting the size of logic gates as well as the number of nano-magnets acting together. The appropriate choice of fabrication technique must provide extremely dense arrays of precisely placed single-domain nanomagnets of uniform size and shape, and at the same time should allow the custom design of various MQCA network layouts [3].

3. MQCA Devices : Binary Wires

Information propagates along the wire entirely via magnetic interactions through the linear arrangement of nanomagents coupled to its nearest neighbors forming a "binary wire". The horizontal clocking field serves two purposes: (1) it sets the state of the horizontal input magnets and (2) for sufficiently strong clocking field forces the magnetization vectors of all nano-magnets to align horizontally (null state). However, this state will persist only so long as the clocking field is maintained. Once the clocking field is reduced to zero, the nanomagnets will return to their ground state with vertical magnetization along the easy axis. The crucial point here is that the switching behavior from the null state to one of the vertical magnetizations is strongly influenced by any additional fields, emanated by either the AF-coupled neighbors or the input magnet. While without the input magnet the probability of reaching either of vertical magnetizations is equal, this symmetry is broken in its presence [3].

4. Clocking on MQCA

If a series of nanomagnets are placed side by side, with small gaps between them, then it can be used to carry digital data. The long axis is the "easy" axis of magnetization (i.e. low coercivity), and in a ground state, the magnets polarize in either up or down directions, which can be used to represent digital logic values. Figure 2 shows such an arrangement where an external magnetic clocking field is applied along the "hard" (shorter, and high coercivity) axes.

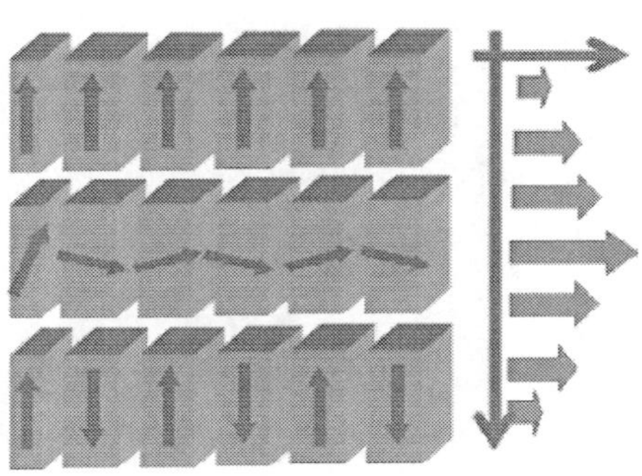

Figure 2. NMQCA wire showing polarization of nanomagnets as the external magnetic field is increased and then decreased. The resulting antiferromagnetic coupling follows the state of the input magnet on the left end [4].

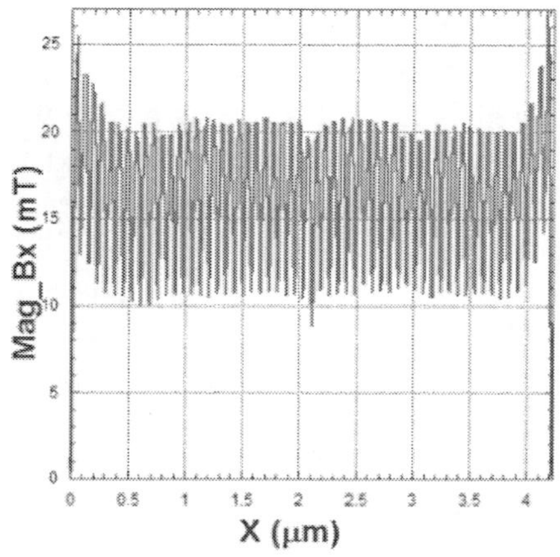

Figure 3. Magnitude of B, vs. x along y = 0.35 ptm line, when both the wires are excited with current density of 106A/cm2. y = 0.35 ptm line goes through the center of the nanomagnets [4].

When the field is increased, polarization is aligned along the hard axes of the nanomagnets, forcing the line of magnets to a "null" state, and when it is released, the line relaxes adiabatically into the ground state set by the end magnet whose coercivity is made higher than the wire magnets via a narrower geometry. Although an external field was used successfully for magnetic bubble technology, it is not a viable scheme for multiple clock phases at high frequencies. To overcome these drawbacks an on-chip clocking scheme have been designed and simulated and able to create a confined, high-speed, and sufficiently strong nulling field.

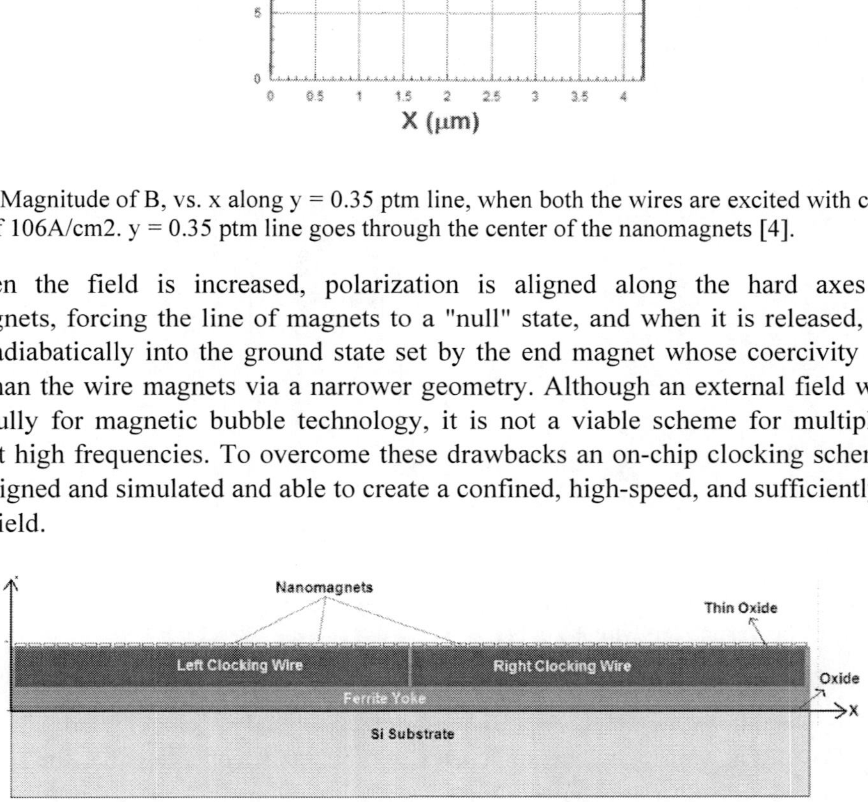

Figure 4. The cross sectional view of two blocks in the data line. The clock wires are made of copper. The oxide layers between the Si substrate and the ferrite yoke and between the copper wires and nanomagnets provide electrical insulation [4].

One clocking structure has been simulated using the MaxwellR 2D electromagnetic simulator. A cross-sectional view of a line, or "wire," of nanomagnets along with the clocking structure is shown in Figure 4. The magnetic fields generated due to currents in the clocking wires are strengthened in the nanomagnets by a ferrite yoke. The dimensions of the nanomagnets are 60 nm X 90 nm X 20 nm, with 15 nm gaps between them. As we have used a 2D simulator, each nanomagnet in Figure 4 is 60 nm in the x direction and 20 nm in the y direction. Thus, both x and y directions are hard axes for the nanomagnets.

Digital data is propagated through the QCA wire using multiphase clocking. First let us assume that there is a fixed input (up or down state of a hard nanomagnet) at the left side of the structure. During the first phase of the clock, both current-carrying wires are excited. As a result, all the nanomagnets experience a high magnetic field in the x-direction (Figure 3) that polarizes the magnets in the x-direction, nulling the state of the magnets. The field in the y direction (B_y) is small compared to B, and, hence, is not shown. In the second clock phase, the right wire and then wire to its right (not shown) are excited, while the left one is de-excited. Now, only the nanomagnets on the right side experience a high magnetic field in the x direction (Figure 5), and are nulled. The left nanomagnets relax to the ground state, and the state (up or down) of the individual nanomagnets are determined by the input on the left. Thus, the data is propagated from left to right up to the first half of the structure. In the next clocking phases, wires to the right are excited sequentially two at a time, and the data propagates from left to right. A four-phase clock is sufficient to facilitate data propagation.

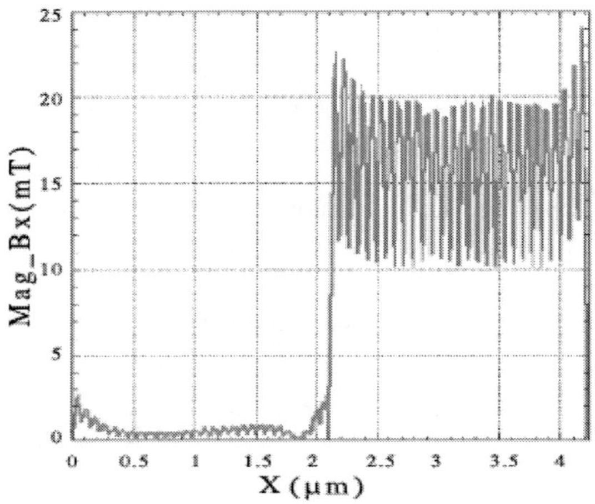

Figure 5. Magnitude of B, vs. x along the y =0.35 pim line, when the right wire is excited (current density of 106 A/cm2) and left wire is not [4].

The combined power consumption of the clock circuit and the nanomagnet switching power is very small, in the range of 1.5 μW per clock wire. This clocking circuit should be applicable to NMQCA majority gate logic as well. Thus, this technology has the potential to allow for non-volatile, ultra-low power digital logic with minimal CMOS support [4].

5. MQCA Majority Voter Feasibility Analysis

Magnetic QCA (also called NML - nanomagnetic logic) are particularly interesting for their expected low power absorption and the intrinsic memory ability, because they maintain information stored without the need of power supply. Rectangular shaped nanomagnets, small enough to be approximated as single domain magnetic devices, are used as base cells. Due to shape anisotropy nanomagnets have only two stable magnetizations, aligned to the long side of the magnet. Despite of the fact that the limited speed of these circuits can reach (about 100MHz), they can be interesting for all those applications where speed is not a key point, but where power consumption is crucial (e.g. smart sensors, biological sensors,...) [6].

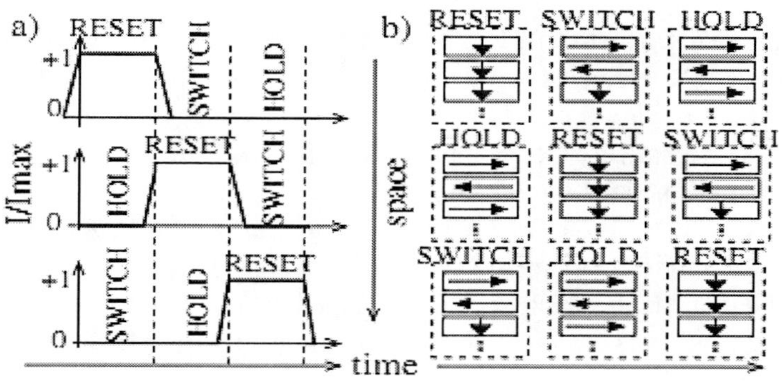

Figure 6. A) Clock signals waveform. A 120 degree phase difference occurs between every couple of phases. B) Signal propagation: At every time step magnets in one clock zone are in the SWITCH phase, those in the previous clock zone are in the HOLD phase and act as an input, those in the next clock zone are in the RESET phase and have no influence on the others [6].

For QCA circuits an adiabatic switching is required. An external field drives the cells in an intermediate unstable state, with the aim of facilitating the switching between stable states. This signal is called "clock" and, in this case, it is a magnetic field, parallel to the short side of the magnets. It is generated by the current flowing through a wire buried under the nanomagnets plane. A multiphase clock system is necessary to assure errorless information propagation, as shown in figure 6.

At every time step each clock zone can be in one of three different states: HOLD, SWITCH and RESET. When the magnetic field is applied, nanomagnets are in the RESET phase, their magnetization is directed along their short axis and they have no influence on the neighbor magnets. When the field is removed (it passes from the maximum value to zero), nanomagnets are in the SWITCH phase. They start to realign following the neighbor nanomagnets that are still in the HOLD state. When there is no field applied, at the end of the switch phase, nanomagnets are in the HOLD phase, they have a stable magnetization and influence neighbor magnets. In this way information is correctly propagated through the circuit. At the following time step this situation is replicated, but the switching zone is the next one, as shown in figure 6. Here we focus on the most relevant logic block, the Majority Voter (MV), on it physical feasibility in terms of distance, size, and aspect ratio of the magnets, using an accurate finite elements nanomagnetic simulator [7].

5.1. Majority Voter (MV) Characterization

From the experimental point of view the major limitation to nanomagnetic fabrication is the small distance between neighbor magnets. The gap between two magnets requires highly precise machines (Electron Beam Lithography, Focused Ion Beam Lithography). Moreover, the need for a so small gap has a strong impact on the quality of the patterned nano-magnets geometry. This is a strict limitation and, compared to the necessity of implementing nanometer scale elements, it has a strong impact on the experimental feasibility of this kind of circuits.

The basic structure of the MV is shown in figure 7. It is composed by 5 elements (bounded with a box in figure 7.A). Here 4 elements enclose a central element which executes the logic operation. The elements on the left and the upper and lower ones act like inputs, while the element on the right in the box is the output block. Three more blocks are present in the figure 7.A (external to the bounding box): these are magnets holding a fixed magnetization used to force the inputs of the MV in the desired state.

Nanomagnets must be forced in the RESET state applying a strong magnetic field, and, when this field is removed, elements must align themselves correctly. Here simulations replicate this situation, therefore a strong horizontal magnetic field was applied, and then verify the alignment process when the field is removed. To keep simulations as close as possible to experimental conditions, chosen as magnetic material Permalloy with a thickness of 20nm, which is a common thickness for this type of magnetic films. Magnetic dots are 50nm width and 100nm height. Figures 7.A, and 7.B show the final configuration of the MV simulation with two of the eight possible input configurations (010 and 100 for A, B and C respectively).

The logic '1' corresponds to the arrows pointed up while the logic '0' corresponds to the arrows pointed down. The magnets alignment corresponds to the MV truth table (Figure 7.C) where inputs and MV central block expected logic values are reported.

Here the design space has been explored through several simulations, changing the horizontal and vertical distances (respectively dh and dv in figure 7.B) between nanomagnetic and their aspect ratio (h/w in figure 7.B). We obtained from the simulations an operative range for the MV.

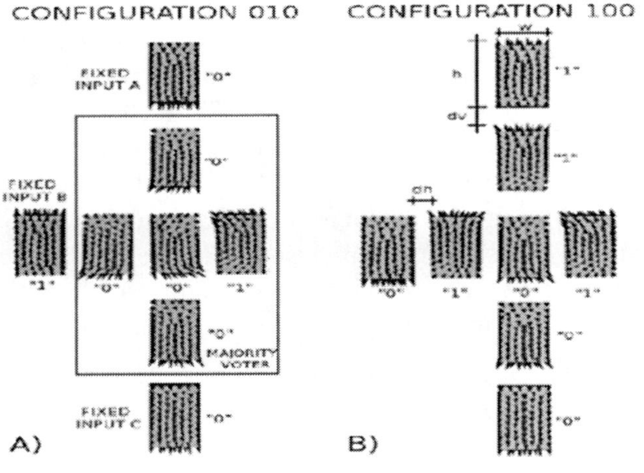

Inputs ABC	Central block
000	0
001	0
010	0
011	1
100	0
101	1
110	1
111	1

Figure 7. Majority voter simulations. A) and B) represent the MV structure in two particular conditions at the end of the simulation based on NMAG, i.e. 010 and 100 respectively. C) gives the MV truth table [5].

The shape anisotropy explains that increasing the aspect ratio the energy barrier increases, and nanomagnetic are harder to switch. As we decrease the aspect ratio the stable states become "less stable". The simulations confirm that with an aspect ratio lower than 2:1 the gate do not work in all the configurations. The technological tolerance does not prejudice the operations of magnetic QCA circuits. This technology works well with scaling down. On the other hand, even with distances of 60nm (horizontal) and 100nm (vertical), the MV is still working, and this is a very good result because these gaps are more feasible from the technological point of view, at least with the available technology. A gap of 100nm can be obtained using low-end electron beam lithography, but, it can be obtained also with high-end optical lithography. Finally, the better aspect ratio is 2:1, which grants the biggest working zone, due to shape anisotropy and energy barrier height, which is the best compromise between thermal stability and correct magnetic switch [5].

5.2. Timing Analysis

Timing analysis is most important parameters that should be analyzed. Figure 8 shows how the magnetization of the central element of the majority voter, which holds the logic function, changes in time. The different lines are related to simulations where vertical distances vary from 30nm to 80nm, while horizontal distance is fixed at 20nm. The two numbers represent the horizontal distance dh and the vertical one dv expressed in nanometers, respectively (for example M 20 40 means dh =20 nm and dv =40 nm). Figure 8.A shows a working case, i.e. the magnetization moves from 0 to a negative value, with inputs configuration 100 and an aspect ratio of 2:1. Figure 8.B shows the magnetization for the same choice of distances in case of 010 input configuration (aspect ratio 2:1). When the vertical distance is 80nm the magnetization, though initially decreasing, moves toward a wrong positive value. In the 70nm case the final result is correct, but the switching time is notably increased with respect to the other cases. In figure 8.C a table reports the switching time calculated as the delay from the instant in which the magnetization begins to move from the reset state, to the instant in which the magnetization reaches the 50% of the swing. It is reported here for all the input combinations in case of aspect ratio 2:1, horizontal distance 20nm and vertical distance 30nm, 50nm and 70nm.

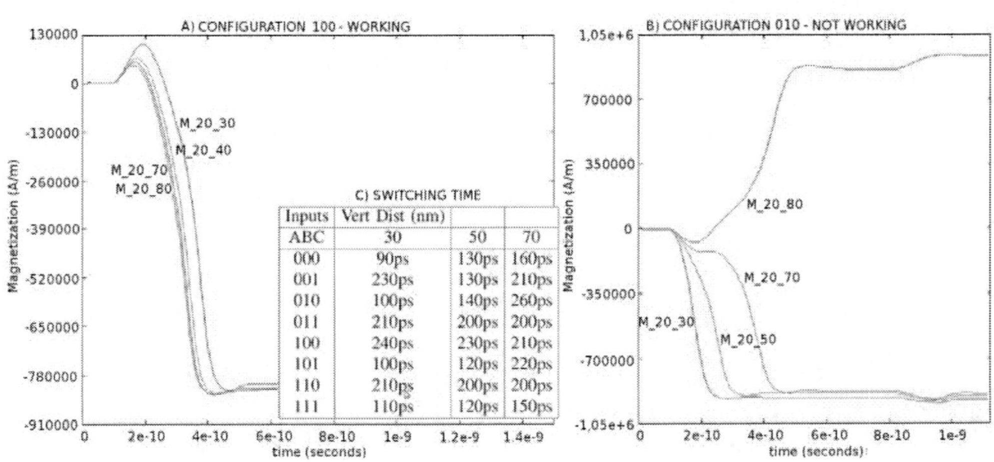

Inputs	Vert Dist (nm)		
ABC	30	50	70
000	90ps	130ps	160ps
001	230ps	130ps	210ps
010	100ps	140ps	260ps
011	210ps	200ps	200ps
100	240ps	230ps	210ps
101	100ps	120ps	220ps
110	210ps	200ps	200ps
111	110ps	120ps	150ps

Figure 8. Central MV's nanomagnet's transient magnetization and switching times for two input configurations: 100 (A) and 010 (B). Curves are related to different vertical distances (from 30nm to 80nm), while the horizontal one is fixed (20nm). Combination 010 shows a wrong magnetization in the 80nm case. Table (C) shows 50% switching time for all the combinations [5].

The switching time depends on the input configuration and on the vertical distance of magnets. This influences differently the switching time, depending on the input configuration. For example, in the 011 configuration, with the increase of the vertical distance, the switching time remains constant; but in the 010 configuration the vertical distance increment causes a proportional increase of switching time. The absolute switching time value is between 90ps and 260ps, which is the expected time evolution for this type of magnetic structures.

It is important to specify that these timing values are obtained removing the reset field abruptly, so they represent exactly.

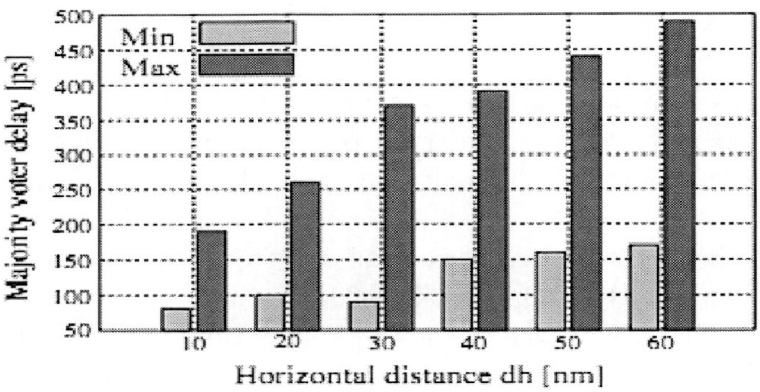

Figure 9. MV timing characterization. For each horizontal distance the minimum and the maximum delay is reported [5].

To increase the reliability and to reduce the propagation errors, the field must be removed slowly. Moreover, the switching time is only 1/3 of the clock period. This considerations lead to the expected frequency of few hundred of MHz for this kind of circuits. Therefore the timing values that we have obtained here represent the lower bound of this technology. We evaluated the switching time for all the MV input combinations. Results have been summarized in figure 9. Each value of horizontal distance the minimum and maximum delay times obtained among all the eight combinations and the possible vertical distances are shown. From figure 9 the influence of the distance on the gate delay can be clearly figured out. Increasing the distance causes a switching time rising.

If distance between the nanomagnets are increased the gate switching time is also raised, therefore distances must be changed carefully to balance technological feasibility with circuit performance.

6. Low Power MQCA

CMOS circuit designs in VLSI have reached to its limit. The obvious choice of researcher is QCA. It has been proved that information can be stored in electron spin rather than charge state. It considerably reduces the power dissipation. The principle of MQCA can be effectively implemented in spintronic devices to establish low power logic design. MQCA is a nanoscale computational paradigm that utilizes magnetic dipolar interaction and coupling to propagate and process binary information. The spin polarization is a bistable quantity and used to encode binary bits by placing the electron in the magnetic field. Spintronic devices composed of nanomagnets enable them to work at room temperature. Another method to reduce the power consumption using spin torque transition current (STT) to feed the input circuits.

In magnetic devices information is encoded in the magnetization state of ferromagnetic materials. Although commonly used for data-storage applications, there are relatively few attempts to exploit magnetic phenomena for logic functionality. One possible architecture suitable for logic using nanomagnets is the quantum-dot–cellular automata (QCA) signal-processing approach. QCA are built from simple, identical, bistable units that constitute the computational architecture. Logic gates featuring single domain magnets on the size scale of 10 to 100 nm (above the super paramagnetic limit, but within the single-domain limit) are expected to operate at room temperature because of their relatively large magnetic energies. The principle semiconductor of MQCA is used in spintronic device to build logic circuits.

Spintronics means spin electronics. The central theme is the active manipulation of spin degrees of freedom in solid state systems. Spin based electronics where the electron spin carries information where spin of the electron is being manipulated by external field. Spin unlike charge is not a scalar quantity. It has a magnitude and a polarization. The spin polarization is a bistable quantity and it is used to encode binary bits by placing the electron in a magnetic field. Spins can be controlled using electric fields. New horizons of electronics can be developed using electron spins. Effective magnetic field felt by the spins in the semiconductor can be freely controlled by the gate voltage spin has been used in various types of memory and sensors developed logic using spin. We generally think using electric field instead of a magnetic field will enable devices to work in a very small space at very high speed and with low power consumption. Ultimate application will be in quantum computers where a lot of information is stored in individual spins. Ferromagnetic material is used to build spintronic devices. Although a current hardly flows in the insulating material a quantum effect enables the current to tunnel through Tunnel magneto-resistance effect.

In addition to the electric current there is a flow of spin. Electric power is only consumed when the electron flow so if we create a situation where spin alone flows even if we don't make the electron flow, making spin alone flow consumption of power gets reduced to a large extent. Nanomagnets exhibit two states of magnetization can represent digital bits. MQCA is a network of closely spaced dipole coupled single domain nanomagnets designed for digital computation. MQCA offers very lower power dissipations together with high integration density of functional devices.

MQCA logic gate is a 3 input majority gate can operate at room temperature. Majority gate was demonstrated by employing NiFe polycrystalline nanomagnets with 70nm-120nm lateral. MQCA using ferromagnetic dots and magnetic field coupling is more advantageous compared to QCA using electrostatic devices. Logic gates featuring single domain magnets in the size scale of 100nm are expected to operate at room temperature because of the relatively high energy coupling energy. In most cases they need no extra power, other than power to switch their state. The key feature of such ferromagnetic components is magnetic hysteresis which describes the internal magnetization of the components as a function of external magnetizing force. Single domain nanomagents are the basic component programmable logic element. Propagating magnetic excitations in the magnetic nanostructures are suitable for digital signal processing. MQCA digital information is represented by the magnetization states of an individual nanomagnet. The number of stable magnetization states of a nanomagnet is determined by its magnetic anisotropy for example Crystalline or shape anisotropy. The nanomagnets are needed to be magnetically bistable so that bit values 0 and 1 can be assigned to the two states. This nanomagnet is strongly bistable as its remanent magnetization always points along the long axis due to shape anisotropy. Even though a magnetizing force can rotate the magnetization away from the long axis when the force is removed the nanomagnet switches to either of the two remanent states. Placing many of these magnets side by side along the long axis may result in an array of dipole coupled nanomagnet that favor antiparallel alignment of the magnetization directions. Low Power Magnetic Quantum Cellular Automata Realization using Magnetic Multi-Layer Structures explains Magnetic Quantum Cellular Automata (MQCA) realization using multi-layer cells with tilted polarizer reference layer with a particular focus on the critical need to shift towards the multi-layer cells as elemental entities from the conventional single-domain nanomagnets. A novel spin-transfer torque current induced clocking scheme, theoretically derive the clocking current, and have shown the reduction in power consumption achieved against the traditional mechanism of clocking using magnetic fields typically generated from overhead or underneath wires. The multilayer cell behavior in Verilog-A along with the underlying algorithm used in implementing the neighbor interaction between the cells . The paper reports the switching and clocking current magnitudes, their direction and the power consumption associated with switching and clocking operation. The simulation results from Verilog-A model of switching, clocking and neighbor interaction. Low power consumption due to spin transfer torque current induced switching and clocking along with the reasonable Magneto-Resistance (MR) distinguishing the two energy minimum states of the device, make these devices a promising candidate in MQCA realization.

Cellular Automata involving field interaction for logic computation has been the alternative computing paradigm that offers interconnect-free design architecture, and hence, provides the scope for realizing low power circuits in the nanometer scale regime. The Quantum-dot-cellular automata (QCA), a category of Cellular Automata, use Coulombic interaction among electrons to realize logic functionality. However, QCAs have been fabricated and functionally verified only in the cryogenic temperatures. Energy dissipation increased when the tunneling energy through the clock was enhanced .Also, energy dissipation was not reduced with scaling. In fact when the cell size was 10 nm, average energy dissipation was 12μeV whereas energy dissipation for a 40 nm cell was calculated to be 1.8μ eV . Molecular Cellular Automata and Atomic Cellular Automata are the other two offshoots of Cellular Automata involving field coupled computing. However, permission to use this material for any other purposes must have displayed good promise for room temperature operations; research is still under progress to develop a stable structure of the molecules. The neighbor interaction among similar cells is also a subject of study. Magnetic Quantum Cellular Automata (MQCA) has the ability to operate at room temperature. They have been successfully employed to demonstrate various logic elements including basic logic gates and horizontal and vertical wires. The two energy minimum states by virtue of their shape anisotropy enable them to represent stable binary logic. Switching between the states takes place through an external phenomenon named Clocking. Interconnect-free low power logic has been successfully demonstrated using MQCA. Traditionally the building blocks of MQCA logic were single-domain magnetic nanostructures. Clocking and the peripheral circuitries for reading from and writing into the logic were implemented using external fields that required large current densities in the wires placed below or above the cells. Multi-layer spintronic devices realizes MQCA that are capable of being written, read and clocked using spin transfer torque (STT) current. For each of these operations the current requirement is order of magnitude lesser than the corresponding current required in field-induced operations. STT current scales down with device dimensions which motivated us to use these multi-layer cells as elemental cells for nanomagnetic logic computation. The interaction between the free layers of the multi-layer devices Magnetic Tunnel Junctions (MTJs) is utilized to compute and propagate logic information.STT current for writing, reading and clocking the cells provides both controllability over the individual cells in the logic and low power logic realization.

The cell sizes are accordingly chosen to enable integration with existing CMOS technology. The integration with CMOS facilitates built-in read circuitry for reading the logic output and providing currents for clocking and writing into the cells. After a review of the existing work in MQCA we have studied the suitabilities and the drawbacks of various possible cell elements for implementing MQCA. Extensive research has been conducted in the field of MQCA. MQCA logic implementation and information propagation through ferromagnetic and anti-ferromagnetic coupling between single-domain nanomagnetic logic cells were developed at room temperature. The nanocells have demonstrated successful logic operation at room temperatures having their dimensions within the super-paramagnetic logic cells which were developed at room temperature.

Clocking the cells require the assistance of external magnetic fields generated through current-carrying conductors. However the devices still suffer from power dissipation in the external circuits that are used for clocking. Writing to the input cells have been proposed and conducted through fields either generated by input wires external to the logic or by external MTJs in close association (1 nm) with the nanomagnetic cells .Reading the output of the logic has been affected with the help of output sensors that transport the signal to off-chip peripherals for data determination. The peripheral circuitries used in writing, reading and clocking are still a subject that has not been much explored. Feasibility of simultaneous clocking and reading through the above-mentioned schemes is still an open field of study that needs further attention. A group of researchers have proposed a new all spin logic device utilizing spin transfer torque current for writing, reading and clocking the logic. Logic execution involves spin injection from a ferromagnetic conductor to a semiconductor, in this case graphene. However, the injection efficiency into semiconductors is still very low and the device has not yet been fabricated. Effective magnetic field on the free layer arising from crystalline and shape anisotropy, demagnetization field, exchange field and external field which can be in the form of coupling from the reference layer .Direction cosines of tilted polarized reference layer utilizing multi-layer cells MTJs to perform logic computation that offers a solution to the challenges faced by the traditional single-domain nanocells. The cells offer the ability to individually switch and clock through STT current-induced current. The cell sizes are limited by the super paramagnetic and single-domain limits and the underlying CMOS technology. By choosing appropriate device architecture and dimensions, within the permissible thermal stability and single-domain limits. Arrange and their durations in the range of few picoseconds, thus improving on the power and energy dissipation faced by the earlier versions of MQCAs and simultaneously targeting towards high speed logic realization.

Single-domain nanomagnetic cells compute through magneto-static interaction among cells. The size of the cells gave them their single-domain property while their shape anisotropy rendered a distinct easy and hard axis to their magnetization. The magnetization of the cells always tends to align along any of the two easy-axes directions to achieve energy minimum state. The cells were taken to their energy maximum state (aligned along any of their hard axes) using an external field in the direction of their hard axes. This phenomenon was used to suitably clock the cells using an external field pulse. When the field pulse was released, the cells settled along one of their two easy axes directions under the influence of the neighboring cells. Each of the two easy axes directions were employed to represent the two different binary logic states '1' and '0'. Logic computation was carried out using suitable placement of cells with the help of external field pulse to clock them. But this approach of logic computation using single-domain nanocells faced serious challenges in their ability to demonstrate individual control over cells during clocking. Therefore, a group of cells always required to be clocked together, hence, demanding greater area for logic implementation. The field pulses would require overhead or underneath wires and the current requirement for clocking the cells was large. This posed a hindrance to the development of low power circuits using this approach, although the cells themselves underwent adiabatic switching. In addition, the logic implemented using such elemental devices had to rely on external circuits to write to its inputs and to read its outputs. The basic multi-layer stacks comprised of two ferromagnetic materials separated by a barrier and is referred as MTJs in literature. While commonly employed in memory, they offer a possible solution to the challenges faced by the single-domain nanomagnetic cells. Like the single-domain cells, the ferromagnetic layers in the stack are also single domain and possess distinct easy and hard axes by virtue of their size and shape anisotropy. Their dimensions below 100nm, assists in the predominance of the spin-transfer torque effect on the devices over the magnetic field generated by the current through the device. The ferromagnetic layers are so fabricated that one of them has a larger coercive field than the other that renders it a behavior similar to a hard magnet in the field of interest, while the other layer behaves as a soft magnet in the same field regime. The layer with stronger magnetization is commonly referred to as the fixed layer or pinned layer or reference layer in accordance to its inclination to retain its magnetization along a specific direction while the other layer is referred as the free layer owing to its ability to easily switch its state of magnetization. However, these devices have the unique property to switch the magnetization of their free layer under the influence of spin-polarized current.

This unique property gives individual controllability over cells during switching as opposed to field pulses wherein a cluster of cells is unavoidably switched together. The spin-torque generated by the current through the device, which either aids or opposes the damping torque depending on its direction of flow.

The cells are described to be in their parallel state or logic '0' when the magnetization of their free and reference layers are aligned in the same direction while they are in their antiparallel state or logic '1' when the magnetization of their two layers are in opposite direction. A positive current, in the direction from the reference to the free layer, switches the cell to its logic '1' state, while a negative current in the opposite direction orients the cell to its logic '0' state. The existing MQCA implementation ideologies are extended to target low power applications. The half-precession switch of the multi-layer cells paves the way for ultra-fast logic development. This is the first work to report MQCA logic with individual controllability over logic elements due to STT current-induced switching and clocking.

The multilayer devices used for logic implementation eliminate the overhead in cost arising from fabrication of current-carrying wires that initially provided the requisite magnetic fields for switching [7]. The multi-layer cells facilitate integration with CMOS technology which enables the development of on chip readout circuitry. The tilted-polarizer reference layer cells provide sufficient TMR for an effective readout operation while at the same time providing low current clocking strategy and post clocking neighbor interaction. We have developed a Verilog-A model emulating the necessary behaviors, primary cell characteristics and neighbor interactions that is the key to simulating the proposed MQCA cell with CMOS circuitry. A new dimension to the existing MQCA architecture enabled future developments of ultra-fast low power logic blocks.

Quantum-dot Cellular Automata (QCA) is a computational paradigm that uses local physical coupling between nominally identical bistable building blocks (cells) assembled into arrays to perform binary logic functions. QCA offers low power dissipation and high integration density of functional elements.

Depending upon the choice of local fields causing interactions between the cells, different types of QCA are possible, such as magnetic, electronic, or optical.

Here we discuss recent developments in the field of magnetic QCA (MQCA) all-magnetic logic where planar, magnetically-coupled; nanometer-scale magnets are assembled into the networks that perform binary computation. The nanomagnets are defined by electron beam lithography. We demonstrate the operation of basic elements of MQCA architecture such as binary wire, three input majority logic gate, and their combination, and discuss interfacing such systems with conventional CMOS-based logic. Magnetic logic devices can be very dense and continue to operate well when scaled to small sizes. In particular, unlike CMOS, they do not suffer from effects leading to intolerable power dissipation upon ultimate scaling. The downscaling of the magnets is ultimately restricted by the fact that the energy barriers separating the magnetization states have to be larger than thermal energy, kBT, and the switching speed is limited by the precession frequency. These restrictions still yield impressive integration densities of 10^{10} cm^{-2},and switching time on the order of nanosecond. The presence of the external clocking field (electrical for EQCA, and magnetic for MQCA) is critical to the proper operation of QCA-based systems. It was pointed out early in the development of all-magnetic logic that “the fundamental problem is to obtain the step by step propagation of digital information along directed paths of a network of bit storage locations without degradation of the signals. In QCA this problem is solved by using a clocking field that acts as an additional source of energy (on top of the energy supplied by the input) resulting in predictable switching dynamics. Similarly to a power supply in conventional electronics, the energy supplied by a clocking field allows logic level restoration and power gain. Clocking also prevents the QCA system from being trapped in an undesirable metastable state, can be used to realize pipelined circuits, and ensures the unidirectionality of the computation. For MQCA this is achieved by applying the external clocking magnetic field along the hard axis, which puts the magnets into the logic null state; a further reduction of the clocking field to zero allows the magnetic system to settle to its ground state, which in turn depends on the polarization of the input devices [8].

7. Bridging the Gap between Nanomagnetic Devices and Circuits

In QCA data moves via nearest-neighbor interaction rather than with electric current flow. In MQCA, wires, gates, and inverters have all been realized, they operate at room temperature and estimates that if 1010 magnets switch 108 times/second, they would only dissipate about 0.1 W of power. Devices can scale and remain non-volatile provided their size/shape remains above the super paramagnetic limit.

However, binary state in nanomagnets with feature sizes below the super paramagnetic limit can be stable for around 1 ms - long enough to perform logical operations. Scaling can also decrease switching times. In MQCA, charge configurations are replaced with magnetic polarizations of single domain magnets.

A wire is just a line of magnets that are antiferromagnetically coupled to one another. The basic logic gate is based on the majority voting function – where the output of the gate is the logical value associated with the majority of the 3 input cells. Note that by setting one input of a majority gate to a logic '0' or '1', the gate will execute an AND or OR function respectively. In MQCA, the gate performs an inverting majority voting function [9].

The wire and majority gate were tested with a "clock" that took the form of a periodically oscillating, external magnetic field that drove a system to an initial state, and then controlled the relaxation of the said system to a ground state.

A magnetic field is applied along the hard (or shorter) axes of a group of nanomagnets. As an example, Figure 10(a) illustrates a line of nanomagnets that are initially in a logically correct, antiferromagnetically coupled ground state.

In Figure 10(b), the external field turns the magnetic moments of each magnet horizontally into a neutral logic state against the preferred magnetic anisotropy. In this system, this is an unstable state.

When the field is removed, the nanomagnets should relax into an antiferromagnetically ordered state but in accordance with the new polarization of the first magnet in the line (Figure 10(c)). Thus, the first magnet in this line provides the biasing field that eventually sets the state of the entire line via bipolar coupling.

Figure 10. How we might change the logical state of a wire: (a) initial, logically correct, antiferromagnetically ordered ground state; (b) a new input is applied to the line and an external magnetic field is applied to the system which polarizes magnets along their hard axes; (c) the second magnet in the line, influenced by the new input, switches to a new, logically correct state, which switches the third magnet, etc. and the applied field is removed [9].

According to the behavior of magnetic circuits, a magnetic field, H, wraps around the current-carrying wire, inducing a constant flux, φ, whose density, B, depends on the space through which it passes. In traversing a path around the clock wires, the flux threads both the magnetic materials (i.e. the ferrite and nanomagnets) and the intervening gaps (i.e. the oxide or air).Maximizing the fraction of the total distance that the flux travels in the high-permeability materials increases the field intensity, and therefore the flux density, in the remaining gaps. With this arrangement, local fields of approximately 4,000 A/m to 40,000 A/m could be produced in 2 μm wide, 4 μm long, and 200 nm thick ferrite-yoked wires with current densities of 106 A/cm^2 and 107 A/cm^2 respectively.

7.1. Analyzing Magnetic Circuit Elements

OOMMF is widely used and there is excellent correlation between simulation and experimental results.

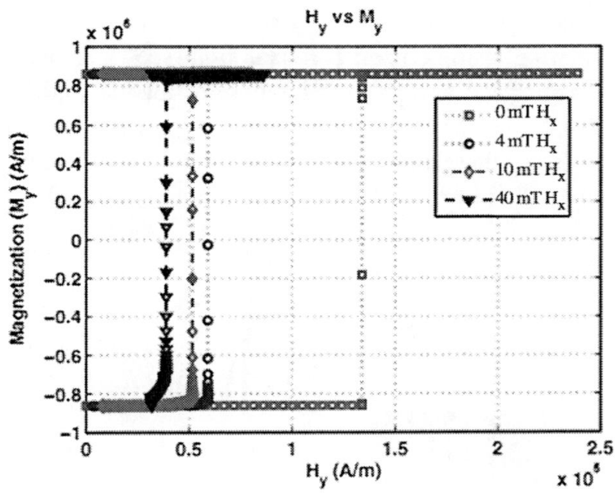

Figure 11. For a 60 × 90 × 20 nm^3 permalloy magnet in isolation, note that as the magnitude of the external field (Hx) increases, the magnitude of the biasing field (applied in the y-direction) required to flip the magnet decreases [9].

A. Magnetization State - Traditional MH Curves

For an elongated, high aspect ratio nanomagnet, in the absence of an applied magnetic field, a magnet's magnetization, M, will tend to align along its easy (i.e. long) axis (↑ or ↓) due to shape anisotropy. This bistable state can be used to encode a single bit of information where a positive M_y represents a binary 0 and a negative M_y a binary 1.

If we want to switch the magnetization state, we must apply a strong bias in the y-direction opposite to its initial magnetization. The minimum field along the y-axis required to perform this switching is $H_{bias} = (N_x - N_y)$ Ms, where N_x and N_y are the demagnetizing factors along the hard and easy axes respectively and M_s is the saturation magnetization.

As an example, an external field of approximately 130,000 A/m would need to be applied along the easy axis of a 60 × 90 × 20 nm^3 Permalloy magnet in order to change its magnetization state. That said, in an MQCA-based circuit, the biasing field required to switch the device needs to come from a neighboring input magnet. However, an equivalently sized device placed 20 nm away from the magnet that we want to switch would realistically only generate an H_{bias} of approximately 10,000 A/m when we average the y- component of magnetization in the next magnet's "footprint." The disparity between the biasing field required and the biasing field a magnet can produce can be addressed by introducing an applied field (H_x) along a magnet's hard axis (i.e. its short axis) that helps to facilitate the switching process. From Figure 11, it is clear that as H_x increases, the magnitude of the biasing field required to switch this magnet is reduced considerably.

B. Magnetization State - MQCA Representation

We present the results of two simulations (see Figure 12) performed with the OOMMF simulation suite. All simulations consider a 60 × 90 × 30 nm^3 supermalloy magnet initially oriented down (↓). H_{clock} is applied in the positive x-direction (i.e., left- to-right along the magnet's hard axis). A constant 15,000 A/m biasing field is applied along the magnet's easy axis which "tips" the magnet to the opposite polarization.

We begin by considering a single magnet in isolation (see "iso, H_{clock}" in Figure 12). The magnet is initially oriented down and its y-component of magnetization is strongly negative (at saturation M_y = -800,000 A/m). An external field (H_{clock}) is then applied along the hard axis of this magnet. As H_{clock} increases, the magnitude of the y-component of magnetization decreases – eventually reaching 0 when H_{clock} is approximately 50,000 A/m.

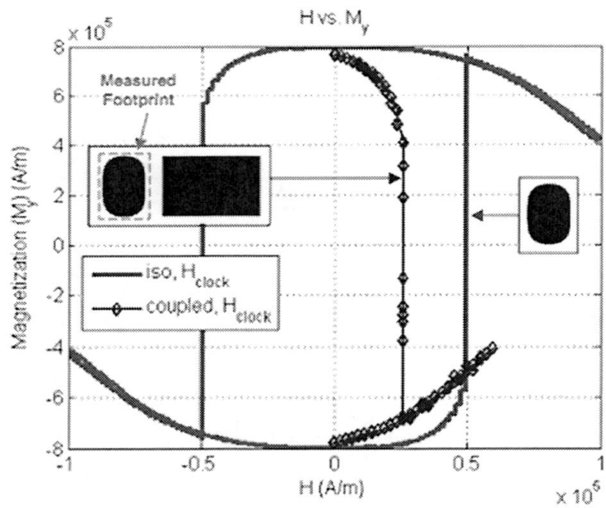

Figure 12. The simulation described by a solid line curve without markers illustrates the external field required to null a 60x90x30 nm 3 magnet in isolation. The line with markers considers the same magnet coupled with a block of magnetic material. A lower external field is required to facilitate a state change [9].

In the "iso, H_{clock}" curve in Figure 12, to increase the magnitude of H_{clock} even after M_y changes signs (indicating a ↓ to ↑ transition). The net result is that M_y is suppressed and actually begins to decrease (see the "tail" at the top-right of the "iso, H_{clock}" curve). Only when H_{clock} is eventually relaxed does M_y again saturate (time at positive 800,000 A/m). The second half of "iso, H_{clock}" in Figure 12 is essentially symmetrical to the first half.

These results suggest two very important things. First, the applied field can be too strong. In the simulation discussed above, the biasing field is externally generated, but in an MQCA-based circuit it would come from the nanomagnets in the circuit. While an external nulling field can assist with switching and the evaluation of new inputs, it can also impede dataflow by suppressing the ability of a magnet to switch its neighbor (as the biasing field generated by the magnet will be of lower magnitude). Second, the symmetry of the "iso, H_{clock}" curve suggests that the direction of H_{clock} should not affect the direction of dataflow. The state that a magnet switches to is determined solely by the biasing field of a neighboring magnet. Thus, a unidirectional clock is sufficient to facilitate bi-directional dataflow. Feedback should be possible within a single clock wire group.

The magnets and magnetic material that comprise a circuit can actually assist in the switching process which will allow us to decrease the magnitude of H_{clock}. To illustrate, we again consider a 60 × 90 × 30 nm^3 magnet, but this time add a block of magnetic material 25 nm to the right of it. The block of magnetic material has an aspect ratio that is less than 1 and its easy axis is parallel to the applied field. The first half of the simulation discussed earlier in this section (for "iso, H_{clock}") was repeated with the same 15,000 A/m bias. Looking at the "coupled, H_{clock}" curve, one can see that magnet switches when H_{clock} is approximately 26,000 A/m – a field half that required to null/switch a magnet in isolation.

When the block is placed in close vicinity to the nanomagnet, the block concentrates flux towards itself. In other words, as the block is already magnetized along the positive x-direction (due to its shape anisotropy) it increases the magnetic flux density in the x-direction in the surrounding area. This increase in flux density assists the switching process as it creates a strong local magnetic field in the x-direction. In fact, we can consider the block used in the 'coupled, H_{clock}' simulation in isolation and calculate the average field generated by the block where the magnet eventually resides. Simulation data indicates that the block provides a local field where H_x averages to be approximately 37,000 A/m. Thus, introduction of the block helps reduce the minimum applied field requirement, which ultimately reduces the external energy needed to perform the switching.

Here magnetic field produced by the block magnet and H_{clock} as H_{local} (thus $H_{local} = H_{clock} + H_{block}$). The magnitude of H_{local} clearly will depend on the block size, aspect ratio, shape, material and distance from the nanomagnet. H_{local} is an important parameter for MQCA circuit design and should be carefully tuned using all available parameters. To maximize H_{local} so that H_{clock} can be minimized, if the net field applied to a magnet is too strong, it can keep a magnet magnetized in the x-direction and impede the biasing field that will ultimately switch a neighboring device.

7.2. Shape, Size and Circuits

Size and shape will have an effect on (a) the y-component of magnetization – which Imre, et. al. leverage to represent a binary 1 or 0, (b) the magnitude and direction of the field produced by a piece of magnetic material – which affects how good of a "driver" a given magnet might be, and (c) how hard or easy it is to null a magnet and/or change its polarization – which will have an effect on the overall energy required to perform computation.

OOMMF simulations of 4 magnets in isolation be considered. Each magnet has a different aspect ratio. For each simulation, we wanted to determine the magnitude of H_{clock} required to change the magnetization state of each device. 60x90, 60x105, 60x120, and 65x95 nm^2 magnets (all 30 nm thick) were considered with the same H_{bias} applied to each. As seen in Figure 13, as a magnet's aspect ratio increases, the magnitude of H_{clock} required to switch the magnet increases as well (only two curves are included to improve graph readability).

When some material is magnetized in a given direction, the field / flux lines emerge from the North pole and return at the South pole. Some of these field lines pass through the material as well. These field lines are called the demagnetizing field as they have direction opposite to that of magnetization M, and thus they work as a demagnetizing force. The demagnetizing field in a direction is proportional to the magnetization in that direction, $H_{demag} = N_d M$, where the demagnetizing factor, N_d, depends on the size and shape of the material. In general, N_d is lowest along the longest dimension and highest along the shortest dimension. The minimum field needed to align the magnetization of a nanomagnet along its hard axis is given by $H_{clock} = (N_x - N_Y)$ Ms, where N_x and N_Y are the demagnetizing factors along hard and easy axes of the nanomagnet respectively and M_s is the saturation magnetization. As the aspect ratio of a nanomagnet increases, the difference between N_x and N_Y increases and thus the minimum H_{clock} needed to null the magnet increases.

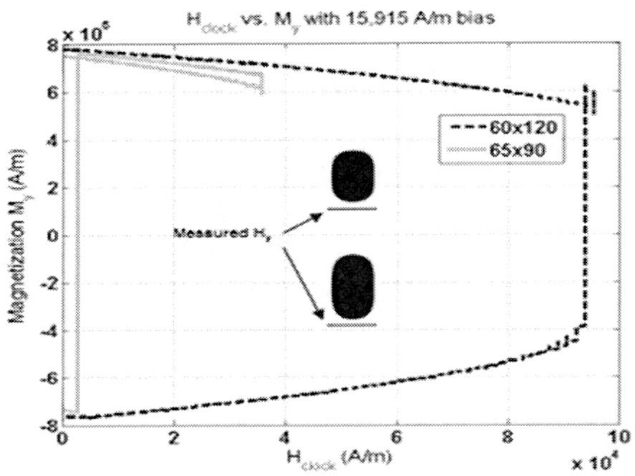

Figure 13. As a magnet's aspect ratio increases, given a similar biasing field, the magnitude of H_{clock} required to null/switch the magnet increases [9].

Simulation data use to show how aspect ratio affects how strong of a "driver" a magnet is. To quantify this, the magnitude of the magnetic field produced by each magnet by averaging the field associated with each simulation mesh point 4 nm directly below a magnet's lower edge (see insets in Figure 13 for an example) was measured. Attention to the y component of magnetization was paid (H_y − bias) because (as we will see next), it is a good measure of how strong of a "driver" a magnet is. As a magnet's aspect ratio (and M_y) increases, the y-component of the generated field increases as well. While easier to null, magnets with lower aspect ratios exhibit weaker 1s and 0s and are also weaker drivers.

For magnets with a small aspect ratio, the magnetization is not totally aligned to its easy, y-axis. As a result, it is not able to produce a strong H_{bias} along the y-direction, and consequently has a reduced polarization effect on its neighbor magnet. Also, small differences in the demagnetizing factor between the easy axis and the hard axis make the magnetization prone to variations with small field differences along the hard axis.

The above discussion is important as the magnitude of H_{bias} will have an impact on the required magnitude of H_{clock} . As H_{bias} increases, the external field required to null the magnet decreases. (One can see this trend graphically by re-examining Figure 11). From a circuit design perspective, this is important as it suggests that the overall system energy was reduced by using lower energy drive circuitry if we can generate a stronger local biasing field with magnetic material.

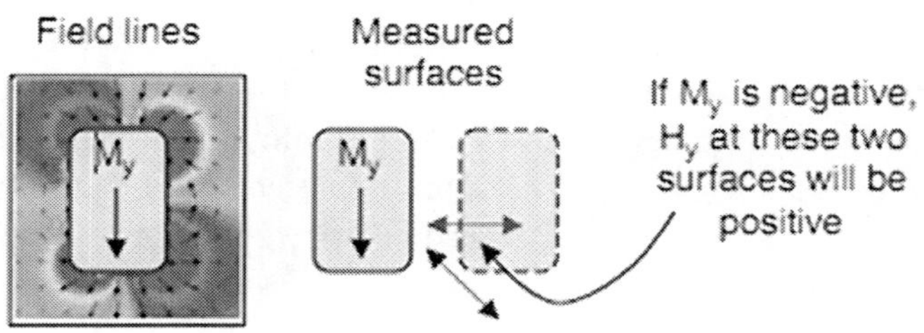

Figure 14. The magnitude of the y-component of the field produced by a 60x90 nm2 magnet is measured at two different surfaces [9].

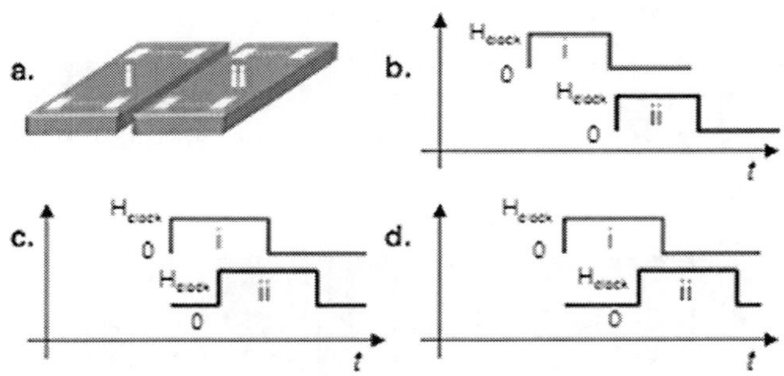

Figure 15. (a) Different groups of magnets will be controlled by different metal wires as discussed in [4]; (b)-(d) Wires can be active at different times to facilitate dataflow in the nanomagnets [9].

Obviously, a magnet's shape influences the magnitude and direction of the biasing field that it produces. In reality, the biasing field generated by a magnet will be less because it is suppressed by H_{clock} . By applying an external field to the 60x90 nm^2 magnet in isolation again without the application of biasing field we calculate the average magnitude of H_{bias} as a function of H_{clock} for the same surfaces in Figure 14. As the magnitude of H_{clock} increases, the magnitude of the biasing field generated by the magnet decreases. For example, when H_{clock} = 20,000 A/m, the average biasing field at the horizontal surface is reduced by nearly 25% to about 40,000 A/m. The average biasing field at the diagonal surface is reduced by about 60% to 25,000 A/m. Thus, one must consider the magnitude of H_{clock} when designing circuits if a specific H_{bias} is required. (Another alternative is to change the geometry of the nanomagnet to help concentrate flux – increasing the magnitude of H_{bias}).

Thus, an ideal device will essentially be easily nulled (with a low magnitude H_{clock}), but can also switch state quickly to produce a strong, local H_{bias} (even if H_{clock} is still applied). To accomplish circuit-level tasks such as facilitating dataflow in a line of magnets across clock wire boundaries and to help move data orthogonally to the direction of H_{clock} used these ideas.

8. Spintronics

Magnetic quantum cellular automata (MQCA) have been proposed as one of the types of spintronic logic. MQCA are based on bistable nanomagnet elements that can perform basic logic operations by means of magnetostatic interactions. A new field of electronics, based on the spin of electrons is known as spintronics. In spin based electronics it is not the spin charge but the spin of the electron that carries information. Spintronics involves the study of active control and manipulation of spin degrees of freedom in solid state systems. The primary focus is on spin polarization, spin dynamics. The spin of an electron is controlled by an external magnetic field and polarize the electrons. These polarized electrons are used to control the electric current. The objective of spintronics is to develop a semiconductor that can manipulate the spin of an electron.

Electronic device composed of semiconductors stores information which is transmitted by the flow of electricity in the form of negatively charged particles called electrons. The zeroes and ones of computer binary code are represented by the presence or absence of electrons within a semiconductor. In spintronics information is stored and transmitted using another property of electrons called spin. Spin is the intrinsic angular momentum of an electron and each electron is like a compass needle points either up or down. Electrons passing through nonmagnetic material have random spins and net effect is zero. External magnetic fields can be applied so that the spins are aligned (all up or all down).Here spin up represents binary ones and zeroes represents spin down respectively.

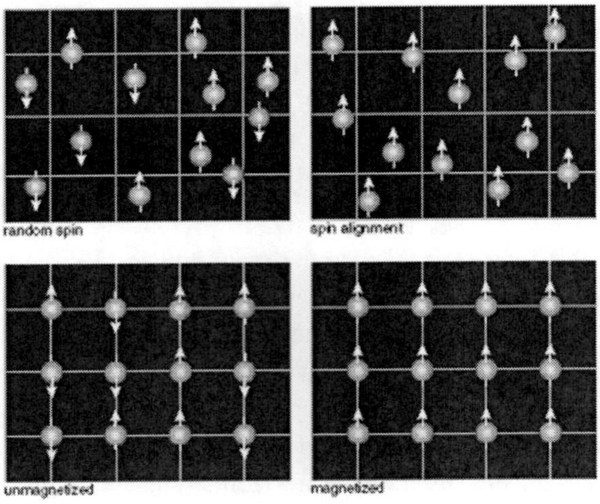

Figure 16. The Spin Alignment.

Magnetic spin properties are used in many applications such as magnetic memory, magnetic recording, etc. The ferromagnetic semiconductors developed above room temperature developed a new generation of spintronic devices. There are many advantages to this kinds, and spin-polarized transport. of information propagation. The data processing speed is enhanced and reduction of electric power consumption and increased integration density as compared to conventional semiconductor technology. The spin polarization plays an important role in information collection. The spin of electron are either aligned "up" or "down" depending upon the direction of the magnetization.

An electron have its total magnetic moment equal to the sum of its spin moment and the orbital.

$$\vec{m}_{electron} = \vec{m}_{spin} + \vec{m}_{orbit}$$

$$\vec{m}_{spin} = -\frac{q}{m_e}\vec{S}$$

$$\vec{m}_{orbit} = -\frac{q}{2m_e}\vec{L}$$

The spin moment given by $m_{spin} = -\frac{q}{m_e}S$ is twice the orbital moment.

The spin moment S can be obtained from $|S| = \sqrt{s(s+1)}\hbar$

where s=1/2 is the spin angular momentum quantum number.

8.1. Spintronic Devices

The giant magneto resistive effect (GMR) in 1988 is considered as the base of the spin-based electronics. GMR is composed of alternate layers of ferromagnetic and nonmagnetic layers. The resistance of the material is very low when the magnetic moments in the ferromagnetic layers are aligned in one direction. The resistance of the material is highest when the magnetic moments of the ferromagnetic material are randomly aligned and not unidirectional. Spin valve is a GMR based device which has a thin nonmagnetic layer is sandwiched between two ferromagnetic layers. A magnetic layer is pinned with another nonmagnetic layer and the other magnetic layer which is called free layer whose magnetization can be changed by application of a relatively small magnetic field. Magnetization in the two layers change from parallel to antiparallel alignment. The magnetizations of the two ferromagnetic layers are antiparallely coupled and they are immune to the outside magnetic fields. Magnetic tunnel junction (MTJ) is a device in which the pinned layer and the free layer are separated by a thin insulating layer (eg: Aluminium oxide).

Spintronic devices records data and store them in tiny areas of magnetized iron or chromium oxides. Tunnel Magnetoresistance (TMR) was used to the idea of a magnetic tunnel junction that has been utilized for the MRAM(Magnetic Random Access Memory).Two magnetic layers separated by an insulating layer of metal-oxide. Electrons are able to tunnel from one layer to the other when the magnetizations of the layers are aligned in the same direction. Spintronic devices combine the advantages of semiconductor and magnetic materials.

Superconducting vortices in magnetic semiconductors manipulates spin and charge of an electron. Semiconductor devices are scaled to minimum size as it increases the device speed and helps the electronics industry to explore field for processing information .The spin of an electron stores, manipulate and carry information. Spintronic applications have efficient ways to create, transport, detect, control and manipulate spin textures and currents. In a hybrid system most of these operations can be performed in a simple manner. A hybrid system consists of a superconducting film and a paramagnetic diluted magnetic semiconductor quantum well. Energy levels of a parabolically confined quantum dot in the presence of spin-orbit interaction were discovered. In a parabolically confined quantum dot the energy levels in the presence of the Rashba spin-orbit interaction (SOI) was theoretically proved. The features of some low-lying states in various strengths of the SOI are examined at zero and nonzero magnetic fields. It is shown that the spin-polarized electronic states can be more easily achieved in a weakly confined dot when SOI is greater than a critical value of the confinement strength. The presence of a magnetic field enhances the possibility of the spin polarization and the SOI leads to different energy dependence on magnetic fields applied. Furthermore, in high magnetic fields, the spectra of low-lying states show basic features of Fock–Darwin levels as well as Landau levels.

In the performance of Magnetic Quantum Cellular Automata and Limitations Due to Thermal Noise .In magnetic quantum cellular automata the operation parameters are evaluated for the purposes of reliable logic operation. The nanomagnets have their dynamics simulated via the Landau-Lifshitz-Gilbert equations. In the macrospin approximation, the switching speed does not change under scaling of both size and distances between nanomagnets. The thermal fluctuations put a limitation on the size of nanomagnets. As we consider a majority gate that features a biaxial anisotropy as a stabilizing mechanism and a uniform clocking field, the gate error rate becomes excessive for nanomagnets smaller than about 200 nm at room temperature.

The theory of quantum cellular automata uses a single spin measurement approach to correlate a spins to measure large amount of ancillary. The distinct advantage of this technique is very efficient and extent fault-tolerant. Through numerical simulation is analyzed by us we can work on the scalability. The measurement of a single (nuclear) spin is one of the most interesting challenge would bring us closer to spin based quantum computers, and have a number of applications,ranging from spintronics, to protein analysis. Several methods have been proposed, and single-spin detection has been done, and the measurement in some specific cases has also been achieved .Using entangling operations a system of N independent spins as a measurement system is coupled to the spin being measured, large correlated state is being created, one whose signal can be measured with current NMR technology. O (N) pulse sequences are required for the method presented over there in order to achieve N quanta of polarization. We expand on their ideas to create a scheme which uses O (√3 N) pulses, and is fairly robust against noise and errors, and does not rely on entanglement. The improved running time is the most important advantage is the improved running time, since the whole procedure must finish before decoherence destroys the information being measured. The method inspired by quantum cellular automata and pulse driven quantum computers is presented over here. A cubic lattice crystal with two nucleus types, A and B. We assume, for the time being, only nearest neighbor interaction. In each species nuclei A is connected only to nuclei B, and vice versa in a checkerboard fashion. The upward and downward z polarizations are referred as |+1⟩ and |-1⟩.We assume, for the time being, that the crystal is initialized to a completely polarized state with all nuclear spins in a downwards z polarization state, |-1⟩. The method then consists of bringing one corner of the crystal into close proximity to the spin we wish to measure, so as to couple the two spins. Once coupled we can use NMR rf pulses to correlate them, or swap the states. Suppose the spin we wish to measure is initially in the state $|\psi\rangle = \alpha| - 1\rangle + \beta| + 1\rangle$. After the swap the top-left vertex nuclear spin will be in the state $|\psi\rangle$. An efficient method to create a very large correlated state within the crystal is presented. Under ideal conditions, that is complete polarization, perfect pulses, and no decoherence, the method creates the state $\alpha|-1\rangle N + \beta|+1\rangle N$ using only $O(\sqrt{3}N)$ steps (each requiring a constant number of rf pulses).

In spintronic applications a single spin is measured using cellular automata techniques .Highly Idealized, atomic Scale Magnetic Quantum Cellular Automata Logic Circuits are simulated. Spintronic logic devices based on majority gates are formed by atomic-level arrangements of spins in the crystal lattice. Switching dynamics is modeled by time-dependent solution of the density-matrix equation with relaxation. The devices are shown to satisfy requirements for logic. Switching speed and dissipated energy are calculated and compared with electronic transistors. By observing the simulations it is shown that for the highly idealized case it is possible to trade off speed for switching energy and achieves lower power operation than ultimately scaled CMOS devices.

We describe a magnetic spin wave nanofabricate, which may provide a route to building reconfigurable spin-based logic circuits compatible with conventional electron-based devices. A distinctive feature of magnetic nanofabricate is that a bit of information is encoded into the phase of the spin wave signal [12]. This makes it possible to transmit information without the use of electric current and to utilize wave interference for useful logic functionality. The basic elements include voltage-to-spin-wave and wave-to-voltage converters, spin waveguides, a spin wave modulator, and a magneto electric cell. We illustrate the performance of the basic elements by experimental data and the results of numerical modeling. The combination of the basic elements leads us to construct magnetic circuits for NOT and majority logic gates. Logic gates such as AND, OR, NAND, and NOR are shown as the combination of NOT and reconfigurable majority gates. Examples of computational architectures such as a multi-bit processor and a cellular nonlinear network are described. The main advantage of the proposed magnetic nanofabricate is its ability to realize logic gates with fewer devices than in CMOS-based circuits. Potentially, the area of the elementary reconfigurable majority gate can be scaled down to $0.1\mu m^2$. We also discuss the disadvantages and limitations of the magnetic nanofabricate.

Magnetic logic gate for binary computing is demonstrated that a suitable arrangement of ferromagnetic elliptical nanodots having dimensions less than 100 nm is able to execute logical NAND and NOR operations [13]. This is named as magnetic logic gate (MLG). The NAND and NOR operations are known as universal gate which are able to execute all fundamental logical operations. For a certain range of the external magnetic field, the magnetization of the elliptical dots takes place only along the major axis because of strong uniaxial shape anisotropy. This inherent binary nature of the dots enables them to be used in the MLG [13]. The MLG possesses number of advantages including high integration density, fast operation and non-volatility.

A spintronic full adder for magnetic CPU is also developed using the spintronic technology. Spintronic devices are based on the up or down spin of the electrons rather than on electrons or holes as in the traditional semiconductor electronics devices [14]. Magnetic processors using spintronic devices in principle are much faster and with the potential features of nonvolatile, lower power consumption and higher integration density compared with transistor-based microprocessor. Full adder is one of the most important basic units of the arithmetic/logic unit for any processors. The design of the full adder determines the speed and chip-density of a processor. In this paper, a novel spintronic full adder is proposed based on novel programmable spintronic logic devices. Only seven magnetic tunnel junction elements are needed for this full adder design.

Application of mesoscopic magnetic rings for logic devices of the ground and metastable states in micron-size ferromagnetic rings as a function of the ring geometry [15]. Our goal is to evaluate the applicability of magnetic ring structures with their bi-directional ground states for applications in magnetic quantum-dot cellular automata (MQCA) structures. Magnetic tunneling junction (MTJ) based programmable logic devices have been proposed and studied for future reconfigurable and non-volatile computation devices and systems. Spin Transfer Torque (STT) based switching has advantages in device scaling compared to the field-switching mechanism. However, the previously proposed MTJ logic devices have operated independently and therefore are limited to only basic logic operations. Consequently, the MTJ device has only been used as an ancillary device, rather than the main computation device. As a result, the full benefits of MTJ based computation have not been explored. New designs are needed to accelerate the development of the MTJ based logic devices. Specifically the realization of direct communication between the MTJ devices is crucial to fully utilize the MTJ devices in the circuits to implement more advanced logic functions. In this paper, new MTJ based spintronic logic units (building blocks) for spintronic circuits using the STT switching mechanism have been proposed and investigated, which includes the designs of a basic STT-MTJ logic cell, a direct communication between the MTJ logic cells, a three-MTJ logic unit and a spintronic logic circuit acting as an Arithmetic Logic Unit. A nano-scaled spiral inductors suitable for future low-power digital spintronic circuits. The proposed nano-spiral inductors can be effectively used to generate magnetic field in a class of spintronic logic/memory circuits such as magnetic random access memory (MRAM) and magnetic quantum cellular automata (MQCA). Measured results on test-chip demonstrate the effectiveness of the proposed nano-spiral inductors and we estimate 3.36 Ã— lower write-current in MRAMs and 85% lower clocking energy in MQCAs.

A paradigm for digital logic based upon ferromagnetic nanowires is presented. NOT-gates, AND-gates, fan-outs and cross-over elements have all been demonstrated experimentally [16]. We propose an application of this architecture to nonvolatile serial memory for high density data storage in small form factor applications such as mobile phones, digital cameras and USB memory sticks. Novel hybrid CMOS-MQCA architecture is developed using multi-layer Spintronic devices as computing elements. A feasibility study is presented with 22nm CMOS where new approaches for spin transfer torque induced clocking and read-out scheme for variability-tolerance are introduced. A first-of-its-kind Spintronic device model enables circuit simulation using existing CAD infrastructure [16]. Approximately 70% reduction in energy consumption is observed when compared against conventional field-induced clocking scheme.

A generic abstraction for a binary switch is described and related to the operational properties of three physical implementations: charge, spin and optical [17]. For a system at equilibrium with thermal environment, an energy barrier is needed to separate two distinguishable states. We utilize the fundamental concept of the energy barrier to study limiting characteristics of these three classes of binary switch. Some of the technical challenges that must be overcome if spin-based binary switch are to find widespread application are discussed A spin-based logic device is proposed. It is comprised of a common free ferromagnetic layer and four discrete ferromagnetic nanopillars, each containing an independent fixed layer [17]. It has the functionality of a majority gate and is switched via motion of domain walls by spin transfer torque. Validity of its logic operation and a quantitative performance prediction are demonstrated by micromagnetic simulation. It is entirely compatible with complimentary metal-oxide-semiconductor technology. The spin relaxation from continuous iron film, to lithographically patterned submicron Permalloy disks, and to iron and nickel nanoparticles were studied by ultrafast magnetooptical technique. In the continuous film, the damping parameter is obtained through fitting the experimental data. In the lithographically confined submicron disks, the excitation due to the formation of the standing waves is found. Ultrafast spin relaxation and spin relaxation due to magnetic coupling are investigated.

Nanomagnetics and Spintronics promise the basis for Future Logic and Universal Memory and gives the review of some recent developments in the utilization of magnetism for applications to logic and memory. It provides some description of new innovations in nanomagnetics and spintronics [18]. Nanomagnetics is based on the magnetic interactions, while spintronics is primarily utilizing spin polarized currents to construct certain devices. Complementary-metal–oxide–semiconductor (CMOS) as comes towards end of its highest scalability nanomagnetics can provide a new paradigm for information process using the principles of magnetic quantum cellular automata (MQCA).A new nonlithographic method of producing reconfigurable arrays of MQCAs is introduced and/or storage bits that can be configured electrically [18]. The first mainstream spintronic nonvolatile random access memory is the Magnetoresistive random access memory (MRAM), which projects how far its successor spin transfer torque random access memory (STT-RAM) can go to provide a truly universal memory that can in principle replace most, if not all, semiconductor memories in the near future. An all-metal logic architecture based on magneto resistive structures (transpinnor) was developed using magnetic tunnel junctions(MTJs) by the application of specific logic. Both nanomagnetics and spintronics utilize spin or magnetism to provide new ways to store and process information; nanomagnetics is based mainly on magnetic interactions between nanomagnets, while spintronics is primarily associated with the utilization of spin polarized currents in memory and logic devices. A new method of fabricating MQCAs that are fully reconfigurable is proposed. These reconfigurable arrays can also be prepared as independent storage bits and by virtue of an extremely regular structure, both for logic and memory, very cost effective polymeric self-assembly manufacturing methods that can complement, and perhaps eventually replace, the expensive, top down lithographic techniques currently used to produce semiconductor logic and memory. MRAM is the first mainstream spintronic nonvolatile random access memory cell can switch between the ground states in a highly nonlinear and abrupt manner. The information transmission can be realized by binary wires [which are composed of linear arrays of cells that are coupled via the Coulomb interactions]. The QCA cell has been experimentally demonstrated in 1997 followed by the logic gate at subkelvin temperatures due to the weak strength of the Coulomb interaction. In order to operate QCA at room temperature, the scale of QD and QCA cell needs to be extremely small, which remains a challenge to the current technology. MQCAs on the other hand function based on the dipolar interactions between nanomagnets. However, it is the exchange interaction among spins inside one ferromagnetic dot that can transform the dot (~100 nm) into one single classical spin which then can be coded as B0 or B1[based on the spin direction], e.g., up versus down. The first demonstrated MQCA at room temperature using a chain of ferromagnetic circular dots was demonstrated by a group of scientists. Later, the group at the University of Notre Dame developed the MQCA using noncircular dots and showed the magnetic switching of MQCA wires with the assistance of an external magnetic field (the clocking field), the schematics of MQCA cells and wires developed by Bernstein et al. similar to the QCA, the switching between the ground states in MQCA is adiabatic. The clocking field helps to align the magnetization along the hard axis of MQCA dots and once it is removed, the spins in the dots relax into the ground state set by the initial input without falling into metastable states. A logic gate based on the majority voting function has been also demonstrated in 2006 .MQCA is scalable to rather small dimensions and the scaling limit is the thermal fluctuation limit of about 40 kT which translates to a nanomagnet size of about 5 nm for a typical ferromagnet. Below the critical size, the spins in the ferromagnetic dots can randomly flip direction due to thermal fluctuations. However, the spin state can be stable for ~1 ms even below the superparamagnetism limit, which is long enough to allow some information processes. The size scaling also reduces the number of spins in the dots which in turn increases the switching speed of the ferromagnetic domain.

One disadvantage of MQCA is its relatively low speed in comparison to a conventional CMOS device. The bottleneck is the precession speed of the domains in response to the dipolar interaction with neighboring nanomagnets. These speeds are typically of the order of nanoseconds. This limits the top speed of MQCA. In addition, the interface between MQCA and conventional integrated circuits is also complicated owning to the information conversion between the charge and the spin.

Reconfigurable Array of Magnetic Automata (RAMA) is developed using a self assembled thin-film array of magnetic nanopillars that can be configured into MQCA architecture. These nanopillars are located at the intersections of a very large and regular cross-point structure with wires below the nanopillars running one way, and wires on the top running perpendicular to the wires below, as illustrated in Figure 3.

The nanopillars are magnetized and the selection and geometry of the ferromagnet ensure that the magnetic moments of the nanopilllars point either up or down, perpendicular to the substrate. It has already been demonstrated that a random array of up and down polarized ferromagnetic pillars ($CoFe_2O_4$) embedded in a ferroelectric or multiferroicmatrix (e.g., $BiFeO_3$) can have their magnetizations rotated from being perpendicular to the pillar (and the film) surface to being in-the-plane of the film with the application of a modest electric field. These structures can be fabricated using polymeric self-assembly methods for both the nanopillars in the matrix and for the crossbar nanowires .

A MRAM Memory MRAM is based on the concept of using the direction of magnetization to store information, and magneto resistance (variation in resistance with magnetization) as information readout. It was originally invented in the late 1980s to replace bulky and heavy plated wire radiation hard memories. The basic memory cell is a dual stripe of anisotropic magneto resistive (AMR) layers separated by a nonmagnetic spacer.

AMR materials were also used in the read head of magnetic recording hard disk drives (HDD). AMR is a change in the resistance of ferromagnetic conductors depending on the angle between the magnetization and the current. The magnitude of this effect is only about 2% for the most common magnetostriction-free NiFe or NiFeCo alloys suitable for device applications.

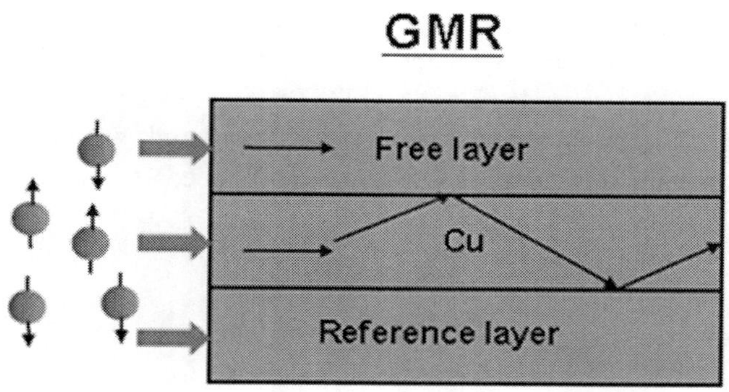

Figure 17. GMR.

The low resistance of metallic AMR thin- memory elements and low magneto resistance resulted in an AMR–MRAM memory architecture that required the connection of a large number of memory elements in series with a large pass transistor, hence the read signal was further degraded. RAM was a niche nonvolatile (i.e., retains memory when power is off) memory device for specialty radiation hard application when it was first invented. The advancement of the MRAM technology benefited greatly from the rapid improvements and revolutionary discoveries in magneto resistive materials fueled by the need to enhance magnetic recording areal density. In the late 1980s, giant magneto resistance (GMR) was discovered. It was immediately realized that this not only had the potential to enhance the performance of HDD, but also would improve MRAM read performances.

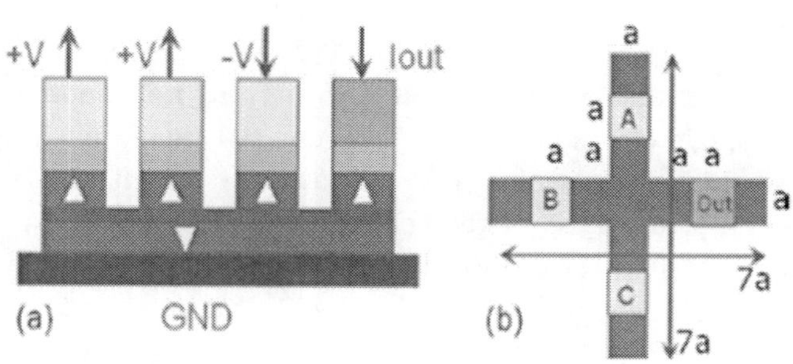

Figure 18 Cross Section Of Nanopillars.

The GMR films consist of two magnetic layers separated by a Cu spacer, and there is a magneto resistance (MR) ratio of 6% initially and later more than 10% with improvements. MRAM storage cells using spin valves and pseudo spin valves were proposed. The memory states were defined by the direction of the magnetization free layer being either parallel or antiparallel to that of the pinned or hard magnetic layer. Magnetic fields are not required during read operation for these new memory modes. The all-metallic thin films require many elements to be connected in series with a pass transistor, which degraded the read signal and speed. In the early 1990s, higher tunnel magneto resistance (TMR) was discovered in MTJ materials. Many efforts were used at adapting MTJs for HDD read head applications for higher areal density. Some pioneering groups, partially funded by the U.S. Defense Advanced Research Projects Agency (DARPA), thus pursuing MTJ materials for MRAM applications. The resistance of MTJ material is easily tunable by varying the tunnel barrier thickness to match the impedance of the semiconductor isolation devices, so it enabled, for the first time, one MTJ element and one transistor bit cell (1TMTJ) architecture. The magneto resistance of an MTJ (TMR ratio) is much higher than for GMR material, and the read signal of MTJ-based MRAM is greatly improved as a result. Read speed in the range of 10 ns became feasible as a consequence of the improved signal levels.

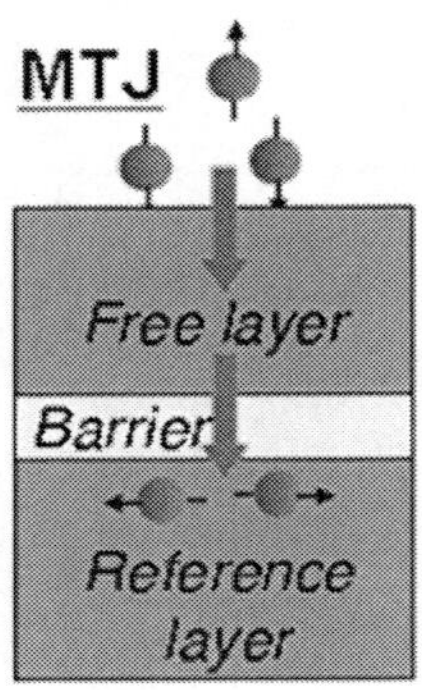

Figure 19. MTJ.

Standard MRAM is written by flow of orthogonal currents which creates the required orthogonal magnetic fields at the cell at their cross point. The writing current required per line is of the order of 5–10 mA or larger. The pulsing of large current is controlled by a large pass transistor. The only feasible option to minimize memory chip footprint is to share this large write current over many rows or columns of cells. The bits At the cross points get written the bits get written, but the bits that experience half of the writing field under just one current line are half selected. In a single free layer where a magnetic switching field is applied bits depend sensitively on the exact bit shapes since shape anisotropy is the main contributor to the free layer magnetic anisotropy, and have a large variation in a large array of memory. As a result, some bits may experience half-select disturb switching problems. This was one of the key problems blocking the development of MRAM memory products. A synthetic antiferromagnetic (SAF) free layer was invented which is used as a new storage mode and resulted in a rotational switching of the free layer with a greatly enhanced writing operational window. A spin-based logic device consists of a common free ferromagnetic layer and four discrete ferromagnetic nanopillars, each containing an independent fixed layer.

The writing of standard MRAM is accomplished by flowing orthogonal currents to produce the required orthogonal magnetic fields at the cell at their cross point. The writing current required per line is on the order of 5–10 mA or larger: a very large pass transistor is required to control the pulsing of this large current. The only feasible option to minimize memory chip footprint is to share this large write current over many rows or columns of cells. The bits at the cross points get written, but the bits that experience half of the writing field under just one current line are half selected. The magnetic switching fields of a single free layer bits depend sensitively on the exact bit shapes since shape anisotropy is the main contributor to the free layer magnetic anisotropy, and have a large variation in a large memory array. Some bits may experience half-select disturb switching problems. This was one of the key problems blocking the development of MRAM memory products. A new storage mode using a synthetic antiferromagnetic (SAF) free layer was invented and resulted in a rotational switching of the free layer with a greatly enhanced writing operational window.

A spin-based logic device is composed of a common free ferromagnetic layer and four discrete ferromagnetic nanopillars, each of which consists of an independent fixed layer. It functions like a majority gate and is switched via motion of domain walls by spin transfer torque. Its logic operation and a quantitative performance prediction are demonstrated by micromagnetic simulation. It is compatible with COMES technology. SPIN-BASED devices are one of the alternative computing technologies listed in the International Technology Roadmap for Semiconductors. Research in spintronic has resulted in fascinating fundamental physics discoveries and in proposals for several spintronic devices. Some devices belong to the class with electric inputs and outputs but have a spin degree of freedom involved in their operation: the Datta-Das spin modulator, the spin field-effect transistor, and the magnetic-tunnel-junction-based logic. Spintronic devices defined by spin inputs and outputs, e.g., the magnetic cellular automata (MQCA), the magnetic domain wall logic, the all-spin-logic (ASL) device; the domain wall majority gates (DWMGs), the spin-wave-bus devices, and the spin gain transistor. In this second class of devices, it is necessary to convert spin to electric outputs and vice versa, in order to communicate with the electronic circuits. However, within the spintronic circuit, the computational variable is stored, and the signal is passed from one device to another in the spin form. A new spin logic device belonging to the second class—the spin torque majority gate (STMG)—is proposed here. The STMG structure can be fabricated using a relatively simple process flow and is fully compatible with complimentary metal–oxide–semiconductor (CMOS) integration. STMG structure a promising candidate for implementing nonvolatile logic in the future. The STMG has three inputs and one output. The output of the majority gate assumes the same logical state ("0" or "1") as the majority of the three inputs.

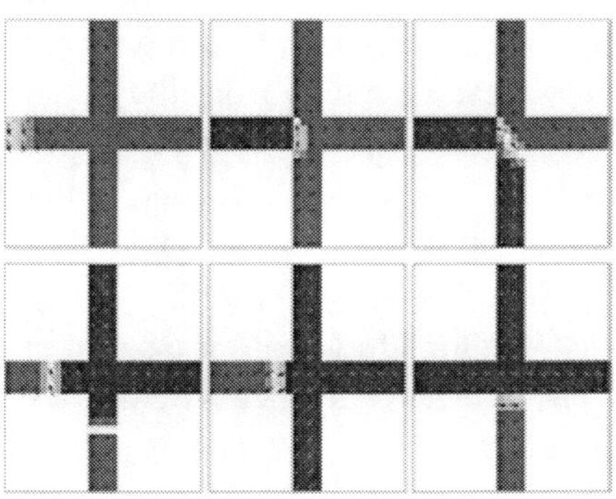

Figure 20. Perpendicular magnetization in a CROSS STMG.

Scheme of the cross section of a STMG with perpendicular magnetization. White triangles designate the magnetization direction. The bottom magnetic layer forms the common free layer. The state of the three input nanopillars (on the left) is defined by input voltages, and that of one output nanopillar (on the right) is defined by its magneto resistance. The middle layers with magnetization in nanopillars are corresponding fixed layers. Schematic of the top view of the cross-STMG. The squares designated by "A," "B," and "C" are input nanopillars. The directions of writing current are shown. The square designated by "Out" is the output nanopillar. The minimum width of arms, the size of pillars, and gaps between them is a. Free layer is switched by spin torque to a state determined by the majority (at least two out of three) of current passed into its inputs. The stack of layers and materials in a STMG are similar to that in a spin transfer torque random access memory (STTRAM) and therefore are entirely compatible with the CMOS process. However, in the STMG, the inputs are represented by separate nanopillars electrically isolated from each other. The structure bears some resemblance to a three-terminal STTRAM. Unlike the STTRAM, the STMG can implement complicated logic functions. Even a single majority gate has the functionality of reconfigurable AND or OR logical function of any two inputs with reconfiguration done by setting the third input. Each of the input nanopillars and the output nanopillar has separate fixed layers which are pinned by the antiferromagnetic layers on top of them. The polarity of the voltage applied to each of the input nanopillars specifies the logical state of the input: "plus" voltage = "1", "minus" = "0". We designate polarities as follows: A = plus, B = plus and C = minus corresponds to (ppm). "Minus" input voltage to a specified nanopillar drives current from bottom to top and forces the magnetization in the free layer below that nanopillar to align parallel with the magnetization of the fixed layer. For "plus" voltage, the current forces the magnetization below the nanopillar to align opposite to the fixed layer. These local influences interact with each other in the common free layer and set the alignment of the free layer to the alignment of the majority of nanopillars. Once the direction of magnetization settles close to its steady-state value, the input current can be turned off, and the direction of magnetization can be determined from the magneto resistance of the stack underneath the output nanopillar. Higher magneto resistance of the antiparallel magnetization corresponds to, e.g., "1," and lower magneto resistance of the parallel magnetization corresponds to "0." The magnitude of the magneto resistance is detected with a conventional sense amplifier. It is advantageous to use materials with perpendicular magnetic anisotropy for the ferromagnetic layers because they have a smaller critical current than in-plane magnetized layers. For such materials, shape anisotropy does not affect device operation. Therefore, one can separate the areas of inputs from the output and place the output on the periphery of the device, arriving at the cross-configuration. Cross-STMG can be concatenated to produce more complicated circuits, e.g., an adder, without the need to convert the magnetization signal into the electric one. In the late 1950s, several systems based on Josephson junctions were developed using the theory of majority gate networks but were subsequently overtaken by CMOS. Our simulations are based on the Landau–Lifshitz–Gilbert equation for magnetization. The effective magnetic field is calculated from the gradient of magnetic energy comprised of the demagnetization energy, material anisotropy, and exchange energy. Simulations are performed using the NIST simulator OOMMF incorporating spin transfer and field-like torques (FLTs). The simulations do not include the noise in magnetization due to thermal fluctuations and thus correspond to effectively zero temperature. We neglect the in-plane component of the current in the free layer due to thick metal layers below it, which provide the low-resistance current path. We assume rectangular current pulses starting simultaneously for all inputs, since the typical jitter delays in driving CMOS circuits are much shorter than typical spin torque switching times. The patterns of magnetization at various times illustrate the nature of switching. In this example, the initial state of magnetization under the input nanopillars is "000"; the (ppm) polarity of voltages results in the "110" state of the inputs and, thus, the output state "1." The areas with magnetization pointing up are separated from those pointing down by clearly. It may be seen that the current perpendicular to the plane is effective in moving domain walls, and it has been shown elsewhere that perpendicular currents are, in fact, more effective than current along the ferromagnetic wires . The input spin torques at first creates the magnetic domains and then moves their walls over their respective arms of the cross. The majority of the input torques win over the middle of the cross (either flipping or preserving magnetization there).In this process, the domain walls sometimes retreat, resulting in magnetization's return to original values. However, in the end, the domain wall propagates to the output arm of the cross, and the majority of inputs enforce their magnetization direction at the output.

Since the relative magnetization projection varies from 1 to −1, the switching speed is defined as the inverse time from the onset of the current pulse to the last instant that magnetization crosses −0.8. Note that switching does not occur over a range of medium and higher currents. This happens when magnetization waves, reflecting from the ends of the arms of the cross, converge to near the middle of the cross and form pinned domain walls. In order for the domain wall to move, it needs to become longer, which costs energy. Pinning can be alleviated by using materials with a smaller anisotropy. The critical current corresponds to current density Jc = 0.4 MA/cm^2 per input, related to the total area of the free layer. Since scalability of devices is necessary for logic, we examine the dependence of switching on the size of the devices. Current density is used to compare them fairly. It is found that larger (40-nm) devices suffer from domain wall pinning and do not switch at higher values of current, while the problem disappears as they are scaled to smaller sizes. Also, the critical current density increases for smaller devices, because they have a different aspect ratio (arm width to thickness) and, thus, a smaller out-of plane component of shape anisotropy. We have also studied the influence of various factors—the Oersted magnetic field (OMF) produced by the current in the nanopillars and the FLT on the switching of the STMG. It shows that accounting for both factors produces faster speed for smaller currents and alleviates problems with domain wall pinning at larger currents. STMG is different from MQCA and DWMG in that it does not require the external clocking magnetic field. This is an important advantage both from the energy consumption and the packaging design standpoint. It is also distinct from ASL which also uses spin torque, but it plays a different role there—to maintain a nanomagnet in an unstable state until a signal arrives to switch it. In ASL, the signal is passed as spin-polarized current, which is known to be fraught with challenges. In STMG, the signal is passed as a wave of magnetization. In conclusion, a spin logic device—the STMG—is proposed, and its logic operation is verified by micromagnetic simulation. The main advantages over the existing CMOS logic are smaller area, low power, nonvolatility, reconfigurability, and radiation hardness. The disadvantages are slower switching speed and, thus, higher switching energy. Furthermore, STMG also has advantages over other spin-based logic devices (including ones having the majority gate architecture) in simplicity of operation and CMOS compatibility. It promises a better feasibility of its demonstration. A spin transistor using only nonmagnetic materials that exploits the characteristics of bulk inversion asymmetric(BIA) and symmetric quantum wells. We show that extremely large spin splittings due to BIA are possible in InAs/GaSb/AlSb heterostructures, which together with the enhanced spin decay times in quantum wells demonstrates the potential for exploitation of BIA effects in semiconductor spintronic devices. Spin injection and detection is achieved using spin-dependent resonant interband tunneling and spin transistor action is realized through control of the electron spin lifetime in an InAs lateral transport channel using an applied electric field Rashba effect [20]. This device may also be used as a spin valve, or a magnetic field sensor. A number of semiconductor spintronic devices have been proposed in recent years that rely on the energy splitting between electron spin states arising from structural inversion asymmetry, also known as the Rashba effect. Among these device concepts, those involving only nonmagnetic materials are especially attractive since they avoid the complex materials issues and unwanted stray magnetic fields associated with the incorporation of magnetic contacts, and because their operation relies on applied electric fields only, which may be modulated at considerably higher rates than magnetic fields. The 6.1 Å lattice constant family of heterostructures ~AlSb/GaSb/InAs offers substantial advantages for such applications because of the high electron mobility of InAs and the large spin splitting characteristic of these heterostructures. However, in devices relying entirely on the Rashba effect, a tradeoff exists between the spin splitting and the spin relaxation time (T1) due to characteristics of the associated crystal magnetic field. In semiconductor heterostructure, the magnitude and direction of the wave vector-dependent crystal magnetic field determines the size of the spin splitting and the rate of spin relaxation through precessional decay. The Rashba effective magnetic field lies in the plane of the heterostructure and perpendicular to the electron wave vector. In this case, regardless of the choice of the nonequilibrium spin orientation to be used within a specific spintronic device, promising designs that incorporate a large Rashba spin splitting will suffer from rapid precessional spin relaxation, placing serious limitations on feasible device architectures. Here we propose a spin transistor using nonmagnetic materials that exploits the unique characteristics of bulk inversion asymmetry semiconductor heterostructures. Since the BIA crystal magnetic field in symmetric quantum wells is oriented approximately in the growth direction for all electron wave vectors, in devices based on such structures there is a natural choice of quantization axis for spin along which precessional spin relaxation is suppressed. In the present work, we demonstrate that large BIA spin splitting are possible in InAs/GaSb/AlS heterostructures, exceeding reported Rasbha spin splitting in this system and in InGaAs/InAlAs heterostructures. The device, utilizes spin-dependent resonant interband tunneling in an InAs/AlSb/GaSb heterostructure for both the generation and detection of spin polarized electrons. Spin transistor action is realized through the application of an external electric field to a symmetric InAs two-dimensional electron gas between the injector and detector, which reduces the spin relaxation time through the Rashba effect yields unpolarized carriers following lateral transport. The electronic structure and spin relaxation times are calculated using a nonperturbative The physics of spin transfer torques in magnetic devices provides the mechanism of spin transfer torque, and review the theoretical and experimental progress in this field. The electrons that carry charge current in electronic circuits also have spins [21]. In non-ferromagnetic samples, the spins are usually randomly oriented and do not play a role in the behavior of the device. However, when ferromagnetic components are incorporated into a device, the owing electrons can become partially spin polarized and these spins can play an important role in device function. Due to spin-based interactions between the ferromagnets and electrons, the orientations of the magnetization for ferromagnetic elements can determine the amount of current. By means of these same interactions, the electron spins can also influence the orientations of the magnetizations. Transition-metal like ferromagnets helps in creating spin polarized currents arise.

The existence of spin transfer torques occurred in the late 1970's and 1980's, with Berger's prediction that spin transfer torques should be able to move magnetic domain walls, followed by his group's experimental observations of domain-wall motion in thin ferromagnetic ~1 ms under the influence of large current pulses [23] . The phenomenon did not attract a great deal of attention at the time, largely because very large currents (up to 45 A!) were required, given that the samples were quite wide on the scale of mm. However, with advances in nanofabrication techniques, magnetic wires with 100 nm widths can now be made readily, and these exhibit domain wall motion at currents of a few mA and below.

Spin Transfer Torque, and Magnetic Dynamics discusses Thus far we have discussed the dynamics of magnets in the absence of the spin transfer torque. Spin transfer torques arise whenever the spin-angular momentum through a sample is not constant, but has sources or sinks. This happens, for example, whenever a spin current is filtered again by another magnetic thin film whose moment is not collinear with the rest. In the process of filtering, the second magnet necessarily absorbs a portion of the spin angular momentum that is carried by the electron spins. Changes of spin angular momentum also occur when spin polarized electrons pass through a magnetic domain wall or any other spatially non-uniform magnetization distribution.

In this process, the spins of the charge carriers rotate to follow the local magnetization, so the spin vector of the angular momentum changes as a function of position. In either of these cases, the magnetization of the ferromagnet changes the spin angular momentum by exerting a torque on the owing spins to reorient them, and therefore the owing electrons must exert an equal and opposite torque on the ferromagnet. This torque that is applied by non-equilibrium conduction electrons onto a ferromagnet is what we will call the spin transfer torque. Its strength can be calculated either by considering directly the mutual precession of the electron spin and magnetic moment during their interaction (an approach discussed in the article by Haney, Duine, and MacDonald) or by considering the net change in the spin current before and after the interaction (the approach we will emphasize). Our discussion in this Section will consist of two parts. First we will consider how it is that a spin-polarized current can apply a torque to a ferromagnet. This will be straightforward since a torque is simply a time rate of change of angular momentum, considerations of angular momentum conservation can be used to relate the spin transfer torque directly to the angular momentum lost or gained by spin currents. We will use two simple toy models to illustrate some of the physics involved in this process. The second part of our discussion will describe how to incorporate the spin transfer torque into the equation of motion for the magnetization dynamics. This step of the argument will involve some more-subtle points, related to the connection between the magnetization of a ferromagnet and its total angular momentum. To explore these points fully, we will consider how one might derive the equation of motion for the magnetization, dM=dt, within a rigorous quantum mechanical theory.The primary quantity on which we will focus our interest will be the spin current density Q. This has both a direction in spin space and a direction in real space, so it is a tensor quantity. For a single electron, the spin current is given classically by the outer product of the average electron velocity(v_s) and spin density Q = v_s,for a single-electron wave function.

8.2. Drawbacks of Spintronics

Understanding the role of structural disorder on spin polarization in $CeMnNi_4$ is a huge thrust in contemporary scientific research is on the new paradigm of electronics viz., spintronic, utilizing the spin degree of freedom of the electron [23]. This additional degree of freedom promises fundamentally new advances in device industry in terms of nonvolatility, increased data processing speed, decreased electric power consumption, and increased integration densities. The success of spintronic is conditional upon detectable spin polarization at room temperature. Though several materials have been theoretically predicted for large polarization, a major bottleneck for their practical realization is nanoscale coherence length of spin current, which makes them ultrasensitive to the existence of structural disorder or defects. Therefore, the largely detrimental role of structural disorder on polarization has been well addressed problem in spintronic science. Interestingly, a few recent experiments defied this conventional and overemphasized detrimental notion of disorder by demonstrating disorder-induced ferromagnetism. This could have interesting implications for the development of spintronic materials that had been long toiling on minimizing defects. The formation of defects being intrinsic and thermodynamically favored, this would be considerably easier than the conventional scheme by reducing defects of developing spintronic materials. Further, new functionalities employing disorder can be conceived. The interest in the present work was motivated by the recently experimental reported enhancement in polarization.

The unusual electronic structure of this material emulates near half-metallic character, marked by a pseudogap at the Fermi level for both the spin channels. Antisite disorder ~5 – 10 % is ironically favored in this material by identical atomic radii of the component metals Mn-Ni. Earlier work on half metals reports the detrimental role of such antisite disorder on polarization.

Probing the degree of antisite disorder predictably Mn-Ni in this material using x-ray absorption fine structure XAFS described later and recalculating polarization electronic structure on the basis of the XAFS derived bond parameters. Neutron diffraction, though indicated large disorder for Mn atom, could not conclusively identify the origin of this disorder.The uncertainties in the neutron-diffraction results entail from the large number of variables and could be significantly reduced by employing an element-specific technique, such as XAFS.

XAFS is the modulation in the absorption spectrum of an atom due to the presence of neighboring atoms. By analyzing XAFS, one can obtain detailed information on the species, bond length R, coordination N, and disorder Debye-Waller factor DWF of the neighboring atoms. The abundance of antisite disorder may subsequently be derived from the coordination information. Ni K edge rather than Mn K edge has been excited to extract the above information. Since Ce L1 edge ~6.549 keV closely follows the Mn K edge ~6.5399 keV, it would superimpose on the latter, thus rendering the Mn K-edge XAFS unusable. Our analysis of the data clearly resolves the existence of Mn-Ni antisite defects; their abundance to be 6% i.e., one site per supercell; the defect-induced disorder in bond-length distribution, and their thermal evolution. Revised electronic structure calculations, based on XAFS results, raised the polarization drastically to 47% closer to the experimental value.

Both transmission electron microscopy and photoluminescence investigations show highly defective material directly after the implantation process. Upon annealing to 800 °C, the implanted Mn remains in the ZnO nanobelts and the matrix recovers both in structure and luminescence. The produced high-quality ZnO: Mn nanobelts are potentially useful for spintronic. The overall morphology in size and length of the ZnO nanobelts did not change upon the implantation process, as observed by SEM. On the other hand, the density and structure of the defects introduced into the nanobelts has been investigated in detail using high resolution TEM. A continuous nanosize stacking fault throughout the nanobelt is visible in the middle region and is common for this type of ZnO nanobelts due to the synthesis process. The implanted nanobelt shows a rough edge in contrast to non-implanted nanobelts, which is due to the sputtering of individual surface atoms during implantation. Furthermore, it is obvious that the implantation process created no amorphous areas even with such a high implantation fluence. The appearance of an opposite dislocation pair is very reasonable, because no additional planes can be introduced to the overall lattice with an implantation process. In addition to the planar defects, the areas in-between indicate the presence of point defects like interstitials and vacancies.

Challenges for semiconductor spintronic, high-volume information-processing and communications devices are at present based on semiconductor devices, whereas information-storage devices rely on multilayers of magnetic metals and insulators. Switching within information-processing devices is performed by the controlled motion of small pools of charge, whereas in the magnetic storage devices information storage and retrieval is performed by reorienting magnetic domains (although charge motion is often used for the final stage of readout). Semiconductor spintronic offers a possible direction towards the development of hybrid devices that could perform all three of these operations, logic, communications and storage, within the same materials technology. By taking advantage of spin coherence it also may sidestep some limitations on information manipulation previously thought to be fundamental.

9. Suitability for Digital Logic and Scaling of Atomistic Magnetic QCA

Electronic logic circuits represented by silicon CMOS transistors scaled for the past 40 years to reach nanoscale device sizes. Currently the search is on for alternative, "beyond CMOS", logic devices, which might have certain advantages compared to electronic logic and will enable further scaling of logic circuits. Spintronics (spin-based logic) is one such possibility. However, schemes for spintronic devices are often proposed and pursued without the analysis of their suitability for digital logic. Here we analyze one type of spintronic devices, magnetic quantum cellular automata (MQCA), verify that it satisfies necessary requirements, and determine its scaling trends.

The logical state, '0' or '1', in a spintronic device is defined by the positive or negative projection of a spin or magnetization on a preferred axis (set e.g. by material or device anisotropy). In contrast to previous work considering magnetization in microscopic discs, we envision the logic state carried by atomic spins in a crystal lattice, e.g. Mn ions in the ferromagnetic oxide LSMO, $La_{1-x}Sr_xMnO_3$, which have localized spin S=3/2. The signal propagates as spin waves along patterned wires of LSMO. MQCA rely on majority gates. For each computation cycle, all the spins are initialized by the global magnetic field ("clocking") to a metastable state, orthogonal to the preferred axis. Input spins are set, e.g. by spin torque effect and are not perturbed by the clocking field. After that the clocking field is switched off and the computation is allowed to unfold. As the spin waves propagate through the whole circuit, it reaches its steady state. Then the output spins are read off, e.g. by the tunneling magnetoresistance effect.

The operation of such devices modeled by numerical solution of the density matrix equations for spins on a lattice including interaction of neighboring spins via indirect exchange and spin relaxation due to coupling to the environment. Exchange coupling (30meV) and the spin relaxation time (10fs) determine the speed of computation. The indispensable requirements, or "tenets", for digital logic are in fact satisfied for MQCA is demonstrated:

1. Non-linear characteristics (related to the signal-to-noise ratio). The angle of spins in the output wire proves to be a sharp S-shaped function of the angle in the input wire, similar to the CMOS inverter characteristic.
2. Power amplification (gain>1). It is demonstrated by the functional 1:3 fan-out, i.e. the device in which 1 input wire transfers its logical state to 3 output wires. The energy for gain comes from the clocking field.
3. Concatenability (output of one device can drive another). The output variable (angle) has the same nature and range as the input variable. Change of spin direction was shown to launch a spin wave to the next circuit stage.
4. Feedback prevention (output does not affect input). Simulations demonstrate that a change of spin in the output wire after the majority gate reached its steady state causes a very little change in the input wires.
5. Complete set of Boolean operators (NOT, AND, OR, or equivalent). Propagation in a wire creates a standing spin wave with alternating directions, which can be used as an inverter. The majority gate includes the OR or AND logical functions of its any two inputs, depending on the fixed value of the third input.

The operation of majority gates for various sizes are simulated (i.e. various widths of wires) and the switching time and dissipated energy is determined. This indicates that spintronic devices are slower than electronic transistors of the same size, but they consume less energy to process a bit. The energy-delay product for both electronic and spintronic devices prove to scale along approximately the same line.

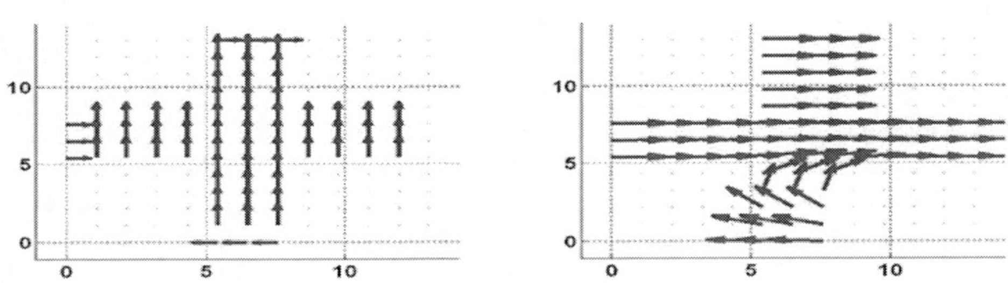

Figure 21. Spin orientations in the cells of MQCA. Majority gate with inputs initially set to ‘0’, ‘1’, ‘1’ from the bottom clockwise. After clocking at t= 0ps (left) and t= 2ps (right). The output reaches the steady-state value ‘1’ [24].

CONCLUSION

This chapter has indicated the great progress made in the field of magnetic QCA and Spintronic in the past few years. Magnetic QCA is the latest technology due to its ability to operate at room temperature than quantum cellular automata which operates at ultra low temperature. It is also suggested over here that spintronic which will attract the attentions of the research community in the following years.Further works on Magnetic QCA logic circuits will allow the development of new generation magnetic random access memories and extend the implementation of magnetism towards logic. Progress in the manipulation of single electron spin is a happening concept in the field of Spintronic.

ACKNOWLEDGMENTS

Authors are grateful to Department of Science and Technology (DST), India for sanctioning the project under "SERC Fast Track Proposal of Young Scientist" Scheme SERC/ET-0213/2011 and UGC major project entitled "Study of quantum dot cellular automata for designing circuits and implementing them for high speed and low power fault tolerant computing" under which this chapter has been completed.

REFERENCES

[1] C. S. Lent, P. D.Tougaw, W. Porod and G. H. Bernstein, "Quantum cellular automata," *Nanotechnology* 4, 49 (1993).

[2] I. Csurgay, W. Porod, and C. S. Lent, "Signal processing with near-neighbor-coupled time-varying quantum-dot arrays," *IEEE Trans. On Circuits and Systems I*, v 47, pp 1212-1223 (2000).

[3] Orlov, A. Imre, G. Csaba, L. Ji, W. Porod and G. H. Bernstein "Magnetic Quantum-Dot Cellular Automata: Recent Developments and Prospects," *Journal of Nanoelectronics and Optoelectronics*, vol.3, pp. 1–14, 2008.

[4] Mohmmad T. Alamt, Michael Niemier, Wolfgang Porod, Sharon Hu, Michael Putney, Jarett DeAngelis, and Gary H. Bernstein "*On-Chip Clocking Scheme for Nanomagnet* QCA",v 9,pp 348-351,IEEE, 2007.

[5] M.Vacca, D.Vighetti, M.Mascarino, L.G.Amaru, M.Graziano and M.Zamboni "*Magnetic QCA Majority Voter Feasibility Analysis*",pp 229-232, IEEE, Trento, 2011.

[6] G. Csaba et al. "Simulation of Power Gain and Dissipation in Field-Coupled Nano-magnets", *J. of Computational Electronics*, Springer, Vol. 4, 2005.

[7] T. Fischbacher et al. "A Systematic Approach to Multiphysics Exten-sions of Finite-Element-Based Micromagnetic Simulations: Nmag", *IEEE Transactions on Magnetics*, 43, 6, 2896-2898 (2007).

[8] Low Power Magnetic Quantum Cellular Automata Realization Using Magnetic Multi-Layer Structures, *J. Das, Syed M. et al.*, Austin.

[9] Michael Niemier, X. Sharon Hu, Aaron Dingler, M. Tanvir Alam, G. Bernstein, and W. Porod "*Bridging the Gap between Nanomagnetic Devices and Circuits*", CA, IEEE,pp 506-513, October,2008.

[10] Carlos A. P´erez-Delgado, Michele Mosca,Paola Cappellaro,and David G. "*Single Spin Measurement Using Cellular Automata Techniques*", Phys.Rev, v 97,Sept.,2006.

[11] S. Patil, A. Lyle et al "*Spintronic Logic Gates for Spintronic Data Using Magnetic Tunnel Junctions*", Amsterdam, pp 125-131,October,2010.

[12] Khitun, A. Mingqiang Bao et al." *Spin Wave Magnetic NanoFabric: A New Approach to Spin-Based Logic Circuitry*",v 44,pp 2141-2152,Sept.2008.

[13] S A Haque, M Yamamoto, R Nakatani, Y Endo " *Magnetic logic gate for binary computing*",Japan,,Sc. Technol, pp 79-82,2004

[14] H. Meng; J. Wang et al *"A spintronic full adder for magnetic CPU",IEEE,* v 26,pp 360-362,June 2005.

[15] Zhou, L., Orlov, A., Csaba, G., Bernstein, G.H., Porod, W., Metlushko, "Application of mesoscopic magnetic rings for logic devices" IEEE,pp 137-139,August,2004.

[16] Cowbum, R.P.; "*Digital logic using magnetic nanostructures*",IEEE Solid state research Connference '2005,pp 249-252,Sept. 2005.

[17] Hutchby, J.A.; Zhirnov, V.V.; Cavin, R.K.;"An Energy Barrier Framework for the Limits of Charge, Spin and Optical Switches",IEEE,Shanghai,pp 1049-1053, Oct. 2006

[18] Stuart A. Wolf, Jiwei Lu, Mircea R. Stan et al. "*The Promise of Nanomagnetics and Spintronic for Future Logic and Universal Memory*",v 98,pp 2155-2168, USA,2010

[19] Stuart A. Wolf, Jiwei Lu et al. "*Beyond CMOS: RAMA A Revolutionary Approach to Logic and Storage*",University Of Virginia,

[20] K. C. Hall, Wayne H. Lau et al. "*Non-magnetic semiconductor spin transistor*",USA,pp 1-8,Jun,2009.

[21] D.C. Ralph, M.D. Stiles "*Spin transfer torques*", pp 1-23,USA,February 2009.

[22] "*Journal of Magnetism and Magnetic Materials*" (2007)Volume: 320.

[23] Debdutta Lahiri, S. Khalid, et al. "*Understanding the role of structural disorder on spin polarization in CeMnNi4 using XAFS*",v 82,pp1-6,October 2010.

[24] Dmitri E. Nikonov, George I. Bourianoff, and Paolo A. Gargini "*Suitability for Digital Logic and Scaling of Atomistic Magnetic QCA*" IEEE, pp163-164,CA,2008.

In: Quantum Dots and Quantum Cellular Automata
Editors: D. De, S. Bhattacharaya and K.P. Ghatak
ISBN: 978-1-62257-920-4

Chapter 10

Reliability Analysis of QCA

Debarati Dey*[*], Pradipta Roy and Debashis De
Department of Computer Science and Engineering
West Bengal University of Technology
Kolkata, West Bengal, India

Abstract

In the era of advancement in Nano-technology, the device dimension has been reduced rapidly. The reliability analysis of a nano device circuit is very important aspect in the field of Nano-technology due to the device complexity, increase of power dissipation and short channel effects greatly influenced on circuit performance. These difficulties of nano-device performance have been discussed in this chapter about the probability analysis of switching activity of an Electrostatic Quantum-dot Cellular Automata (EQCA) cell. In this chapter we derive the Boolean difference of the nano-scale circuitry. Reliability analysis is essential for an error free logic circuit design. Since it is very difficult to find out the reliability of a nano-scale device due to fabrication defects, in this chapter the algebraic method is used to find the reliability using Probabilistic Transfer Matrix (PTM). The concepts of Boolean Difference and Signal Activity are being introduced in this context.

Keywords: PTM, ETM, ITM, Boolean Difference, Signal Activity, Switching Activity

1. Introduction

The emerging trend in nano-scale logic design is Quantum Dot Cellular Automata (QCA), where a unique cell is used to implement both the logic gate and interconnections. In QCA architecture a quantum cell behaves as a switch, due to the movement of electrons from one dot to another. Many logic devices can be made using this switching quantum dot cells.

[*] Email: debaratidey@yahoo.com

According to the Moore's law the number of transistors in an integrated circuit will be doubled in every eighteen months. Due to this incremental numbers of transistors leads to decrement the device dimensions in nano-scale. But as the device dimension decreases, the possibility of error occurrence is increased, and thereby introduces self contradictory system. Incidentaly, many types of defects occur in quantum dot cells and these are discussed in section 3.2.1 [7, 39, 40].

It is essential to know about the reliability of a nano-scale circuit. If the output of any circuit varies from one circuit to another then that particular circuit needs to evaluate its reliability. The output of the Circuit under Test (CUT) needs to be compared with its ideal counter part. In this chapter we use the probabilistic technique to analysis the switching activity of the logic devices [3, 7,].

Since it is difficult to find out the actual switching state of any quantum cell, hence we derive the equations of each device in terms of probability. Before that we need to derive the PTM for each logic device. Ideal Transfer Matrix (ITM) can be derived for each ideal logic device or in this case for the majority voters and thus Element Transfer Matrix (ETM) can easily be formed using these two matrices [1]. Using these matrices and kronecker[1] product one can easily derive the reliability of each logic devices [7]. All these procedures are described in section 5.3. As the devices are scaled down the different vital issues like hot electron effect, tunneling phenomenon, leakage current components and other allied effects will prevail in feature. Therefore it is very much desirable to know reliability of the circuit and systems using quantum cells. Fijany and Toomarian concentrated their efforts in dealing with how a QCA cell can be modified into a "block majority gate" cell where multiple number of QCA cells are arranged in a rectangle and operate as a single cell [7, 25]. This cell is found to be more resistant to the cell misalignment, cell omission and cell rotation like defects [7]. This is one of the examples of increase the reliability and hence reducing the error. Dysart has also shown in his dissertation that how reliability can be achieved and enhanced using probabilistic methods [7].

2. Related Work

Primarily two approaches have been derived to increase reliability in QCA devices. The first approach is, a programmable logic array (PLA) for QCA has been presented by M. Crocker, X. S. Hu and M. T. Niemier [7, 26, 27]. Their work utilize a new parameter Effective Defect Rate (EDR) to get details about the malfunction of a QCA cell and also shown that only displacement like defect affects the cell malfunctioning whereas not all the defects will cause the malfunctioning of a QCA cell [7, 26, 27]. Using probabilistic EDR the fault can be analyzed in PLAs. The yield of PLAs can therefore be computed as a function of PLA cell probabilities [7, 26, 27].

The second approach is to use hardware redundancy with time redundancy of computing with shifted operands [27] for QCA circuits. In this process shifted operands and three times add function is used to find out the redundancy hence the hardware complexity for this circuit is increased. But instead of these three additions they perform the process in time step, henceforth the complexity is reduced [27-29]. To enhance reliability of a system generally two types of architectures are proposed for fault tolerance [7]. The common technique in nano electronics is aka programmable or reconfigurable architecture and another is the application of hardware replication and redundancy [7-10].

Reconfigurable architecture was developed to enhance reliability and to reduce hardware complexity. A simple example of reconfigurable architecture is PLAs, which are generally used for computing of sum of product function and another useful architecture is Field Programming Gate Array (FPGA), which is used to implement a simple logic function circuit to a microprocessor [7, 26, 27]. An overview is given by M. Crocker, X. S. Hu and M. T. Niemier [26, 27]. Two methods are implemented in hardware redundancy, one is Von Neumann multiplexing and another is N-modular redundancy [7, 32].

N-modular redundancy was first proposed by Von Neumann [32]. Triple Modular Redundancy (TMR) and Quintuple Modular Redundancy (QMR) were developed by Timothy J. Dysart [7]. If the original system reliability is low then each circuit is divided into slices or cascaded in TMR method [7] to enhance circuit reliability [7]. In QMR single level module is proposed by Timothy J. Dysart [7]. In case of probabilistic methods for reliability analysis Probabilistic Gate Models (PGMs) are also come into account [33, 43]. Low power operations and lower operating voltage lead to the large error rate in nano scale devices [35]. This motivates to analysis reliability with probabilistic approach [7, 19, 33, 35].

To produce defect-free nano scale devices, it is important to realize Automatic Test Pattern Generation (ATPG), which distinguishes between defect-free output patterns with defective pattern by generating input sets that are used to make the output of the circuit under test to be different. Boolean difference is a new algorithm for combinational logic circuits to find out the error at each node of a circuit [35].

S. Krishnaswamy, George F. Viamontes, Igor L. Markov and John P. Hayes have indicated regarding the probabilistic approach to find out the reliability of logic gates [3, 30]. They also found the following two definitions based on probabilistic method to analyze reliability. Definition 1: Given a circuit A with i inputs and k outputs, the probabilistic transfer matrix for A is a $2i \times 2k$ matrix M whose inputs and outputs are $M(a,b) = P$ [outputs $= b$ |inputs $= a$]. [30]. Definition 2: An error-free circuit has a PTM called an ideal transfer matrix (ITM), in which the correct logic value of each output occurs with probability 1 [30].

3. Advancement of QCA

It may be noted that Konard Juse first shown the CMOS gate length can be reduced below 0.1 μm [5, 10]. Continuously shrinkage of gate length in CMOS leads to the several adverse effects like short channel effect, hot carrier injection effect, Drain Induced Barrier Lowering (DIBL), punch through etc. This entire phenomenon affect to the CMOS performance. Hence, it is essential to nullify all these effects to get high performance nanoscale device. Though all these effects cannot be nullify in CMOS fabrication but its effects can be diluted to some extent. In QCA these effects are totally abolished as because of its features [11]. Moore's law describes a long term trend in the history of semiconductor device fabrication as well as in the field of computer architecture in nano scale [10]. Continuous shrinkage in the device dimension leads to the evolution of nano science and nano technology. In QCA, generally we talk about the working principle of each cell through software simulation. It is difficult to fabricate each quantum-dot with the diameter of 5 nm and also keep the distance between two dots is 20 nm [9].

3.1. QCA Architecture

QCA architecture is based on the simple inherent technology, where four quantum-dots are placed in the four corner of a square shaped cell and coupled by tunnel barrier.

The cell is therefore, can be biased and if there are two excess electrons among the four-dots, then Coulomb repulsion will force the electron to move to the opposite corner. Therefore, two energetically polarized equivalent cells are obtained. These two polarizations can be labeled as logic '0' and logic '1' [4]. The arrangement of these cells is made in this way so that the polarization of one cell sets the polarization of nearby cell. The most important advantage of QCA devices are that the simplified interconnections. In EQCA, the cells communicate only with the nearest neighbors through the passing of stray charge [5, 12, 13, 23]. A basic functional cell is then configured like the Figure 1.

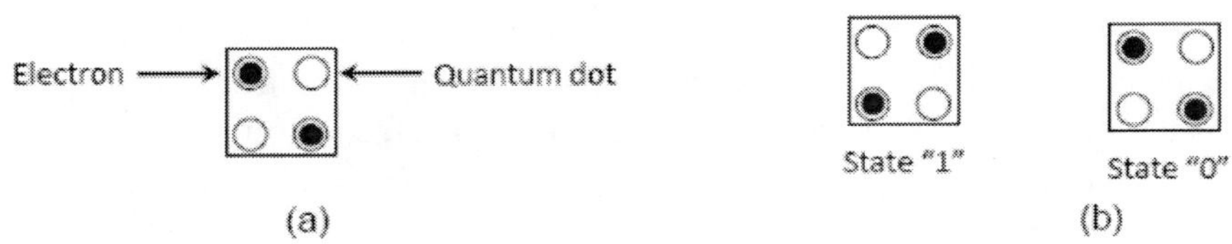

Figure 1. (a) A simplified diagram of four-dot QCA cell. (b) The two possible states of polarization of four-dot QCA cell [10].

Different types of QCA:

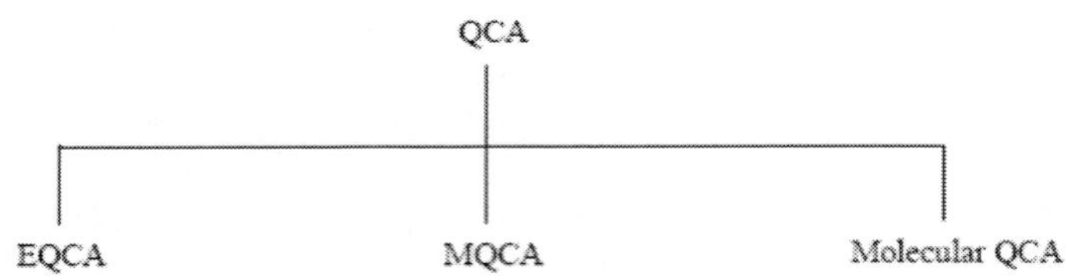

Table 1. Truth table of 3-input majority voter MQCA [6]

Logic State of Input Magnets	Logic State of Central Magnets	Logic State of Output Magnets
ABC	Y	F
000	0	1
001	0	1
010	0	1
011	1	0
100	0	1
101	1	0
110	1	0
111	1	0

MQCA can be formed using ferromagnetic materials, though in case of EQCA we have used metal dots. In case of MQCA the information are carried out through the magnetization state change of the ferro-magnetic dots shown in the Table 1 [6, 15, 18, 19]. Hence, the logic state of the central magnet is to be determined by the majority voter MQCA logic gate. Therefore, the output combination is obtained, which shows the anti-ferromagnetic property with respect to central magnet [6]

We can use III-V compound semiconductors in formation of quantum-dot, like, Gallium Arsenide (GaAs), Aluminium Gallium Arsenide (AlGaAs) etc [12, 13].

In MQCA, information is carried out by the magnetic interaction of the dots. The shapes of these dots are various types. Nano-magnets exhibit two distinct stable states which are used to store binary information. There are mainly two binary states of the nano-magnets, bit '0' and bit '1'. The bit value '0' is assigned to magnetization direction down along the vertical axis of ferromagnetically coupled input magnets and the central magnet. The value '1' is assigned to magnetization up [17, 18]. The arrangement of these nano magnet cells is made in this way so that the polarization of one cell sets the polarization of nearby cell. The most important advantage of MQCA devices is their simplified interconnections. A basic functional magnetic quantum dot cell is then configured like the Figure 2 [17, 18, 36].

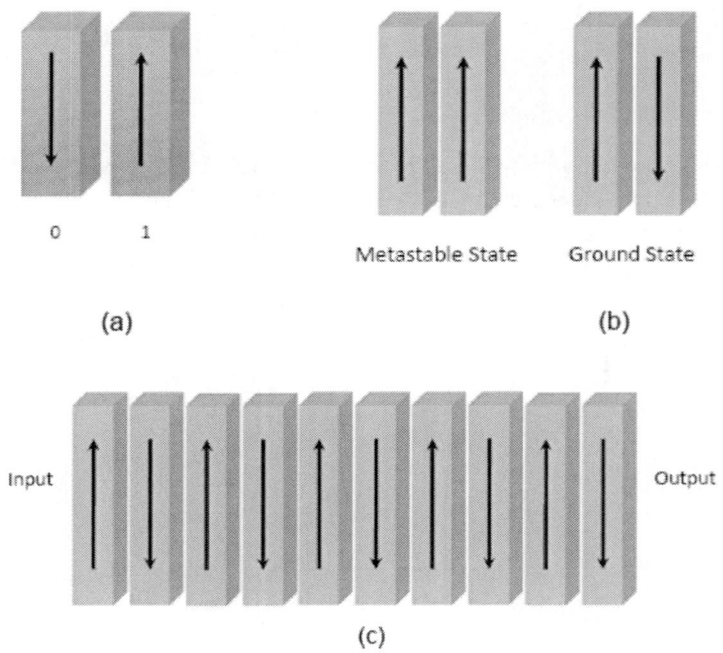

Figure 2. (a) A simplified diagram of MQCA cell. (b) The two possible states of polarization of MQCA cell. (c) Basic structure of nano-magnetic wire [6,36].

MQCA can be formed using ferromagnetic materials [17]. The hysteresis graph of nano magnet is shown in the Figure 19 [17, 18, 36]. If there is any type of fabrication defects like erroneous cell alignment, missing of cells then this erroneous information is carried out to the next cell.

It is proposed that the PTM framework for capturing non-deterministic behavior in MQCA logic gate circuits. This probability concept provides a concise description of both normal and faulty behavior of the circuit. It can also be found that the reliability metrics can be derived from this PTM framework [1, 4]. It can be considered that the PTM algebraic representation to check the disorder or dropped cells or any kind of asymmetry in the cell's co-ordination. PTM can be used to compute the reliability performance for the nano scale devices [1, 3, 20].

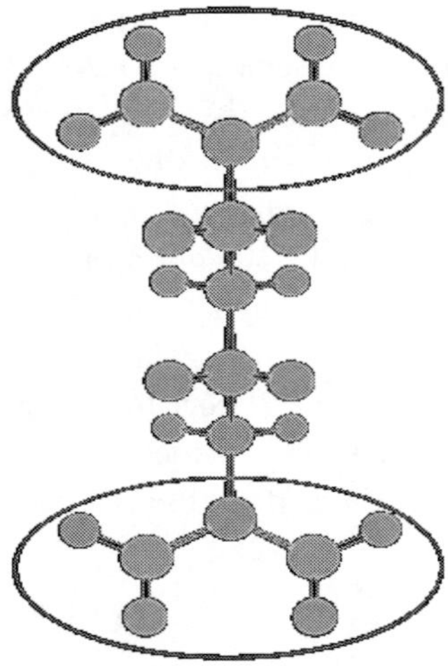

Figure 3. The Molecular QCA system [37].

A QCA cell is taken in which charge is localized on specific sites and can tunnel between those sites to construct Molecular QCA is shown in Figure 3 and to construct this feature a single molecule is required. The role of the dots is then played by redox sites [37], with tunneling paths provided by bridging legends. This phenomenon is first proposed by Aviram and later studied by Hush et al. A majority voter logic gates can be easily implemented by using Molecular QCA. The truth table for a three input Molecular QCA majority gate is depicted in table 2. The geometry of a Molecular QCA majority voter is shown in Figure 4.

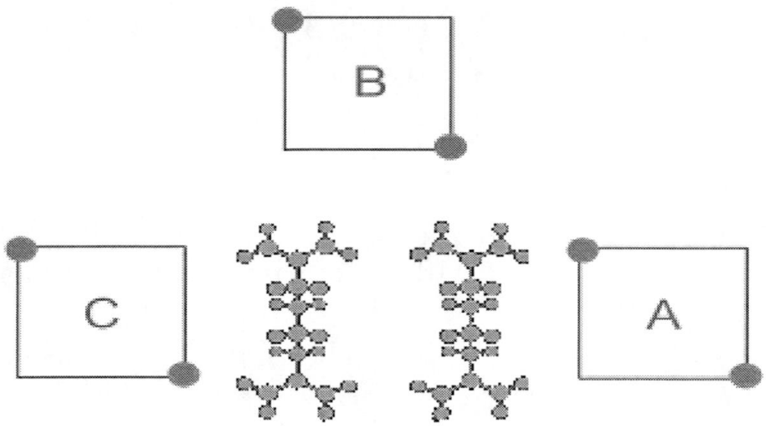

Figure 4. Geometry of a molecular QCA majority gate. Inputs A, B, and C are quadruples. This part of the molecule simulates the configuration of charge in other QCA molecules [37].

Table 2. The truth table for a three input Molecular QCA majority gate. [20, 37]

	A	B	C	Output
AND gate	0	0	0	0
	0	0	1	0
	0	1	0	0
	0	1	1	1
OR gate	1	0	0	0
	1	0	1	1
	1	1	0	1
	1	1	1	1

3.2. Defects in QCA

There are several types of errors can occur whenever the device dimensions reduced to the order of nanoscale. Generally it is considered that for square cell, four quantum dots are placed to the four corners of a square with the diameter of 5 nm and 20 nm spacing between them [9, 22, 29, 38]. There are also two free electrons dynamically change their locations from one dot to another. If the stray charge distribution within the dots is not symmetric or if the electronic stray charge do not spread evenly within the quantum dots then these phenomenon leads to error. Due to this error, the output of each cell becomes erroneous because of that the induction effect of those cells leads to the error. In some cases, due to fabrication draw back the dimension of the cells and spacing between them may not be symmetric. Due to this reason the charge tunneling between them is not uniform [25-28, 38, 40]. The charge polarization ‘P’ of a QCA cell can be described as [39]

$$P = \frac{(P_1+P_3) - (P_2+P_4)}{P_1+P_2+P_3+P_4}$$

where P_i is the charge on each quantum dots and ‘i' is labeled as the Figure 5. As the stray charge can be varied in EQCA due to the movement variation of electronic charge distribution then the value of P_i may vary from cell to cell, hence we obtained faulty result. The reliability of these devices can then be checked with the help of PTM theory which is defined in the section 5.3 [40].

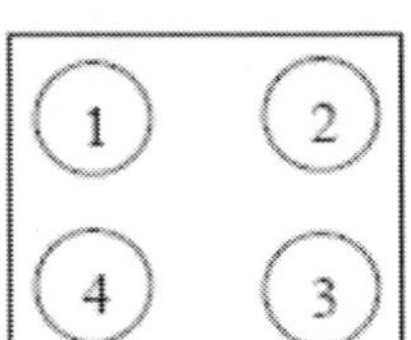

Figure 5. Quantum –dots position in a QCA cell [39].

Now, if the stray charge distribution among the dots is not uniform for a pair of cell then polarization effect of each dot may change, as a result the whole polarization of each QCA cell P is changed. This leads to the erroneous output of a system. A small polarization change in one cell induces a large polarization change in its neighbor [25-28].

3.2.1. Different Types of Defects Occur in QCA

In this section we described different types of error that are usually occurs due to non-deterministic nature of QCA devices. The defect occurs mainly due to fabrication fault. As a result whole circuitry starts malfunctioning. Some of the types of defect are discussed below [38, 41]:

1. *Cell Displacement:* It occurs when the distance of separation between the cells becomes greater than the exact distance. For example, in Figure 6(a) the exact distance of separation between the cells is *'d'*, which is perfect fabrication. On the other hand, Figure 6(b) shows the defective fabrication where the distance of separation *'d1'* is greater than *'d'*. Due to this variation of gap the cells become malfunctioning [38, 41].
2. *Cell Alignment:* If the cells are not properly aligned (suppose instead of normal alignment or vertical alignment it is horizontally aligned) at their proper location then the ultimate output will be varied with its actual non-faulty output. It actually indicates the deviation in the cell location along with its direction. The Figure 6(c) shows that the vertical misalignment of the cell 2 is *d1* whereas it should be equal to zero [38, 41].
3. *Cell Omission:* If any cell is missing from its pre-specified location due to fabrication defect in the layout then also the final output will become erroneous [38, 41]. The cell omission is shown in Figure 6(d).
4. *Stray charge defect:* If the stray charge is not uniformly spread within the quantum dots then it leads to an error. This erroneous information then carried out to the next cell. This phenomenon is referred as stray charge defect [1, 7]. This defect is shown in Figure 6(e).
5. *Rotation defect:* If any cell-aligned oppositely within a wire then this phenomenon leads to rotational defect [7, 38, 40].

QCA devices are suffering from defects during their manufacturing process. It has been observed that during the quantum dot synthesis and deposition phases of QCA devices the defects can occur more frequently.

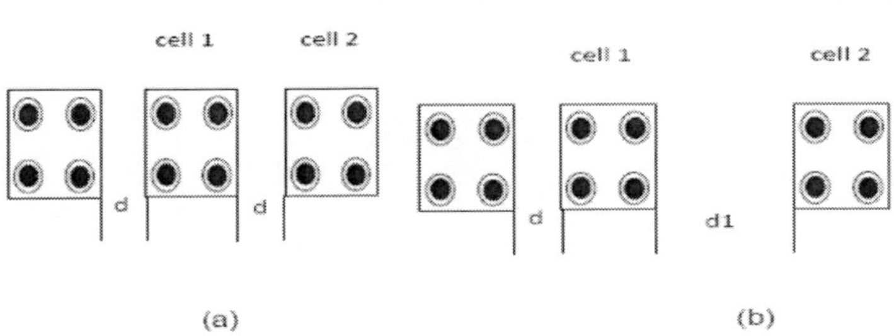

Figure 6. (Continued).

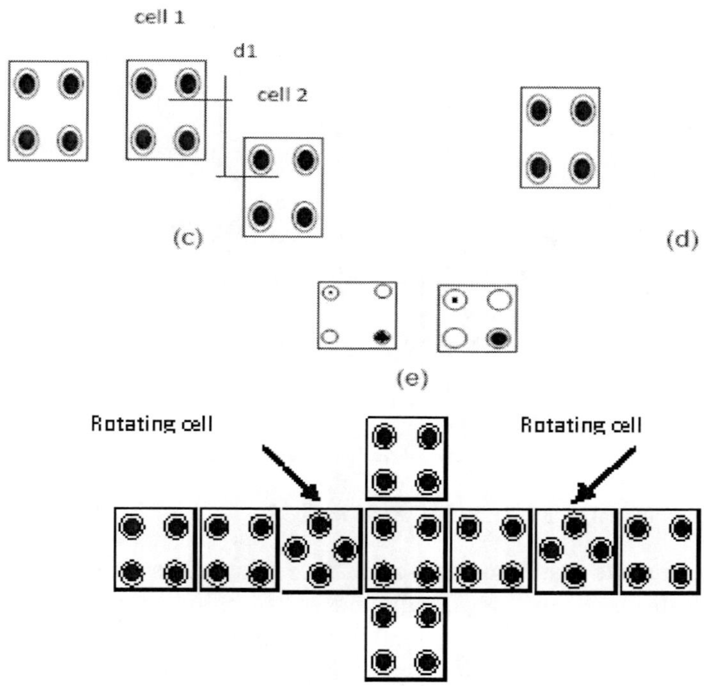

(f)

Figure 6. Three types of misplacement: (a) defect-free, (b) displacement, (c) misalignment, (d) omission, (e) stray charge defect and (f) rotation defect [7, 38, 40].

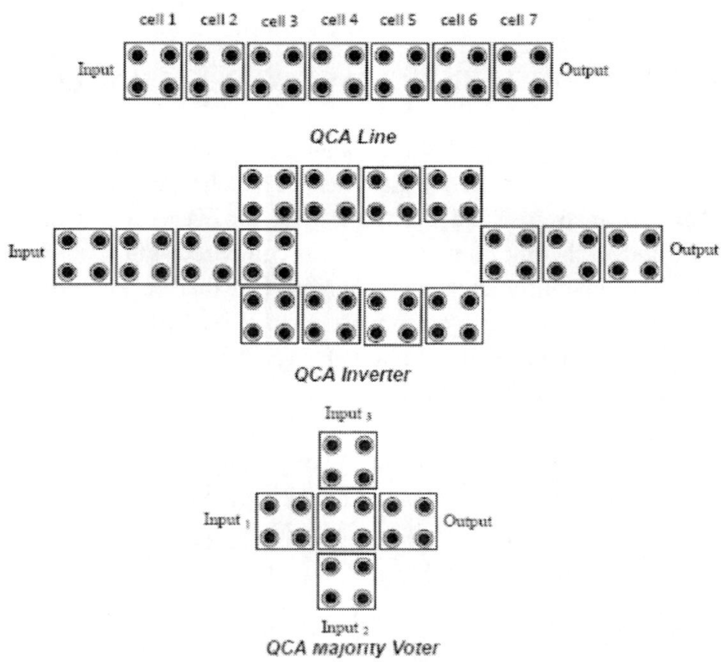

Figure 7. (Continued).

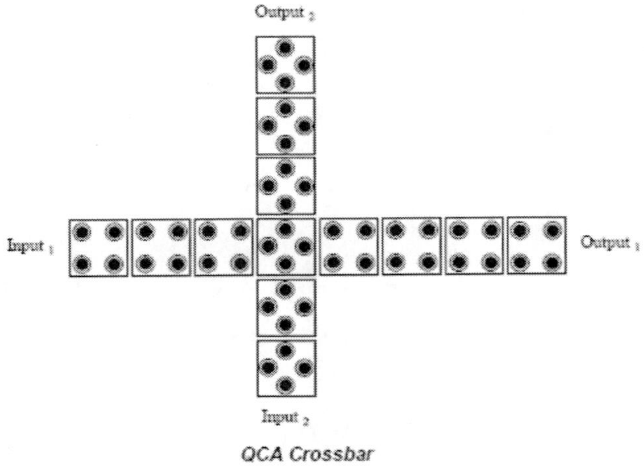

Figure 7. Statistical channel models for different QCA devices [38].

A cell may have extra dots or missing dots or missing electrons in the synthesis phase, on the other hand the cell displacement, cell misalignment, cell omission and cell rotation may occur in the deposition phase. Due to the difficulty in precise control of the cell location at nanoscale range the defects caused by cell misplacement will be dominant in QCA. Several types of cell misplacement that may occur are shown in the Figure 6.

All these defects hamper the normal functioning of a quantum cell. These phenomenon leads to the reliability analysis of the functional cell as well as the device or the system. The various statistical channel models for different types of QCA device is shown in Figure 7.

4. Reliability in Science and Technology

In the nano-technology regime, reliability is an important aspect, which is useful in every field of science and technology. Reliability has to be checked for any circuit before its application. If there are many replicated circuits then for every circuit the outcome is to be taken into account and each time their outputs are compared with another. In this way the inherent consistency of reliability analysis is increased. If there is any type of abnormality is shown then we can say the circuit is erroneous. As many as time of repetition is done, the probability of error free outcome will increase. Hence we get more reliable device circuitry and also get almost error free outcome [33].

1. To get more reliable output 'normal distribution' is followed by the researchers. More generous results can be obtained by taking the average value of all the outputs.
2. When the results are continuously varied widely then it is obvious that the circuit is unreliable hence imperfect design of the circuit is obtained.

In early days Teramac – like system [2] will be the model system for non-CMOS nano-device architecture. Fabrication of computer architecture converges to nano-scale regime; reliability is becoming a significant feature for circuit analysis. Researchers need to analyze the reliability of both the QCA cell as well as the whole circuitry. Reliability of basic circuitry in QCA is more challenging aspect rather than the individual cell. [33].

4.1. Relation between Reliability and Statistics

Same results can be obtained by the scientists every time due to the relative prediction in terms of physical entities. In nuclear physics experiment or inorganic chemistry experiment all time anyone may get the same result. But in case of chaos, ecology or social science experience some experimental results may vary due to the randomness, natural fluctuations etc.

Figure 8. Reliability Vs Validity.

The reliability is immensely associated with statistics. An experiment is said to be successful if the reliable results are obtained. If anyone performs the same experiment again and again he/she will get the same statistical value. Comparison is to be made between the reliable outcomes of the same experiment and the most reliable value is to be chosen as almost error free result [42]. The relation between reliability and validity is shown in Figure 8.

4.3. Need of Reliability Analysis

As complementary metal–oxide–semiconductor (CMOS) technologies are scaled down into the nanometer range the non idealities and errors must be come into the focus. These non-idealities are therefore can be overcome to achieve a successful analog and physical circuit design [40].

The nature of these non idealities in CMOS design in nano-scale has been discussed in the technical articles. These non-idealities are namely hot carrier injection, drain induced barrier lowering (DIBL), leakage current, stress and lithographic effects limiting matching accuracy, electro-migration effects limiting conductor lifetime and mobility effects, limiting device performance and chip power dissipation limits etc. If these non-idealities are not coming into the feature then the individual circuits are more energy efficient [40].

The problems in electronic design and automation industry are basically due to the lack of facility in analog design and simulation tools. In the design phase, sometimes post layout simulation tools are used but that is not useful whereas computer aided design simulation tools are slow and commonly not used by analog circuit designers [40]. In the modern CMOS design technology in nano-scale the short channel factors affect circuit operations; hence methods of migrating or overcoming them are equally important. In the field of research on Quantum-dot Cellular Automata (QCA), the most challenging aspect is related to its performance analysis. In nano scale design it is difficult to get fault free circuits. As because the device dimension reduces continuously, power dissipation, drops of quantum-dot cells, deformities in the stray charge distribution within each dot of QCA lead the defects in nano scale device modeling [4, 8]. Among these, reliability testing of these nano scale devices is one of the challenging tasks [43].

In this chapter we have dealt with Electrostatic Quantum-dot Cellular Automata (EQCA) and Magnetic Quantum-dot Cellular Automata (MQCA). In some cases, we observe that when we make a large circuit in nano scale, some cells may either dropped out or the stray charge among the dots may vary. Thus, we have found some probabilistic method to find out the circuit reliability. With the help of Majority Voter (MV) we have seen ITM, PTM and ETM [1, 3] for two-input and three-input circuit. From these results we can easily analysis the reliability performance of the MV logic gates for higher number of inputs. Our aim is to build a reliable zero defect system in nano scale [1, 3].

5. Metrics Used for Reliability Analysis

There are several metrics are used for reliability analysis in nano-scale design. Generally PTM framework is to be done for capturing non-deterministic behavior in MV logic gate circuits. This probability concept provides a concise description of both normal and faulty behavior of the circuit. We can also find out the reliability metrics from this PTM framework [3]. We have considered PTM algebraic representation to check the disorder or dropped cells or any kind of asymmetry in the cell's co-ordination. PTM can be used to compute the reliability performance for the nano scale devices [1, 7].

5.1. PTM Framework

The PTM framework has been proposed by Smita Krishnaswamy, George F. Viamontes, Igor L. Markov and John P. Hayes to capture non-deterministic behaviour in logic circuits. Concise descriptions of both normal and faulty behaviour are provided by PTM and well-suited in reliability and error susceptibility estimation. Recursively build larger PTMs can be easily achieved from smaller gate PTMs by a few simple composition rules that based on connectivity. To obtain the tensor product operation and matrix multiplication PTMs for gates in parallel and in series are combined respectively. To calculate accurately the joint output probabilities in the presence of reconvergent fan-out and inseparable joint input distribution PTM is used. We can use PTMs to encode as Algebraic Decision Diagrams (ADDs) to improve computational efficiency. For newly defined matrix operations such as, eliminate variables, eliminate redundant variables which aid in the numerical computation of circuit PTMs. Some equivalent ADD algorithm can be developed. PTMs are used to evaluate circuit reliability and derive polynomial approximations for circuit error probabilities in terms of gate error probabilities. PTMs are used to analyze the effect of logic and electrical masking on error migration. If logic masking is ignored then the overestimate errors may happen by an order of magnitude. Electrical masking is incorporated by computing error attenuation probabilities based on analytical models into an extended PTM framework for reliability calculation. Susceptibility measures are defined to identify gates whose errors are not well masked [1, 3, 7, 8, 20, 32].

5.2. Introduction and Background to PTM Theory

With the advancement of digital device technology the non-deterministic circuit behavior can be determined [32, 33]:

1. Short channel effects and process variation are increased in deep sub-micron level CMOS VLSI design.
2. Probabilistic reliability analysis can be done in CNT (Carbon nano tube) and sub micron level CMOS design.

For many years the circuit testing is treated with circuits' probabilistic reliability analysis. Many researcher have dealt with the problem of signal probability estimation [Parker and McCluskey 1975; Ercolani et al. 1989; Savir et al. 1983], the idea is based on random pattern testability. As there is no transition state and the probability of a signal is 0 or 1 that gives some indication of the difficulty in controlling the signal that is the main idea to achieve probabilistic analysis of reliability. On the other hand, to calculate signal probability, we generally deal with complex probabilistic failure modes, transition states analysis error propagation and their effects on circuit behavior and their reliability. Circuit reliability estimation involves computing single output distribution along with the output error probability for each input pattern. When the input distribution is fixed then simply computing the output distribution does not give the overall circuit error probability because each gate has input pattern dependent error probability. For example, as shown in the Figure 10 that the output distribution is unaffected, but the wrong output is paired with each input though only the Ex-OR gate has an output bit-flip error as in this case we never compute the transition state analysis between the switching from '1' to '0' or '0' to '1'. Then we have to calculate the error associated with each input vector separately [1, 3, 20, 29, 30, 32].

Now if we consider the circuit as in Figure 9 which shows that the error probability p = 0.1 for each gate. For input combination 000 the error probability of the circuit is 0.244. From study we can show that the error probability of the circuit is 0.205 for input combination 111. The overall error rate of the circuit can be measured as the sum of the error probabilities weighted by the input combination probabilities. For the circuit shown in Figure 9 the circuit error probability is 0.225 for the uniform input distribution. Combine probabilities of input combinations are useful to capture correlations among inputs rather than individual inputs [1, 3, 20, 32].

Using the PTM framework, we can find out the circuit reliability and other aspects of probabilistic behaviour. Algebra is formed to represent circuits with probabilistic failure modes (e.g. variation in pattern generation at gates which is an example of dependent error probabilities).

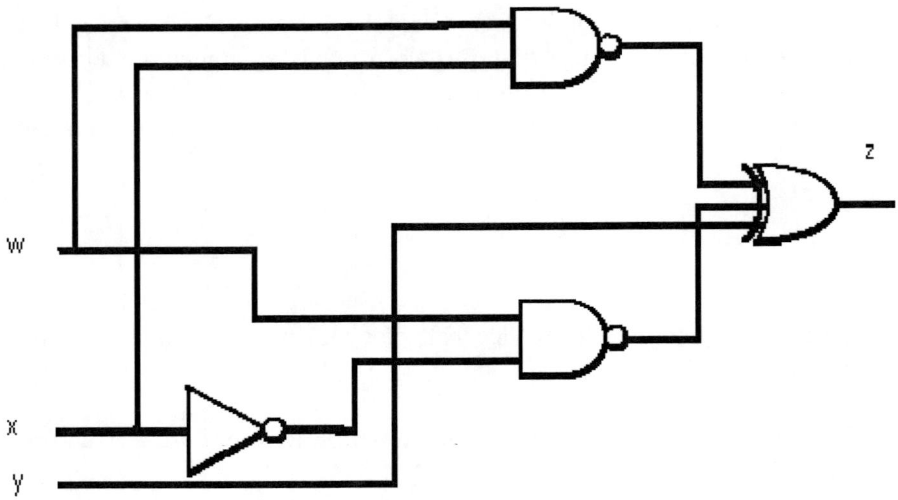

$(F_2 \otimes F_2 \otimes I)(I \otimes swap \otimes NOT_p \otimes I)(NAND2_p \otimes NAND2_p \otimes I)(XOR3_p)$

Figure 9. Sample logic circuit and its symbolic PTM formula [3, 30].

	ITM 0	ITM 1	PTM 0	PTM 1
000	1	0	0.756	0.244
001	0	1	0.244	0.756
010	1	0	0.756	0.244
011	0	1	0.244	0.756
100	0	1	0.295	0.705
101	1	0	0.705	0.295
110	0	1	0.295	0.705
111	1	0	0.705	0.295

ITM PTM

Figure 10. ITM and circuit PTM where the probability of error for each gate is 0.05% [1, 3, 7].

The PTM methods implicitly capture signal correlations caused by reconvergent fan-out. To determine the impact of path-based cumulative effects such as glitch attenuation and logic masking on error propagation, the above mentioned methods are generally used [1, 3, 20, 32].

Here the PTM methods are described first and then ADD based method is implemented. When we represent the truth table for any given circuit, it shows the full range of I/O combinations for a logic function. The truth table is basically a matrix of 0/1 whose binary row and column indices correspond to inputs and outputs respectively [1, 3, 20, 32].

As we know that there are many errors may happen during the formation of circuits using quantum-dot cells, so by using the algebraic representation, it is much easier to find out the fault and compute reliability [1, 3, 20, 32].

For general framework limited scalability is one of the main factors that capture complex circuit behaviour. Therefore, the PTM framework is introduced to use Algebraic Decision Diagrams (ADDs) to compress matrices. Various derived ADD procedures are used to compress directly and combine the PTM in order to obtain the circuit PTM [3]. Figure 10 shows the ITM and PTM for the circuit. In this example, PTM representation shows the probability of 0.05% of an output bit-flip. Several appearance of the identical value in the PTM shows that PTMs are generally compressible. To remove the multiple occurrence of the same value we can implement ADD method to get only one copy of distinct value of PTM entry. In several cases ADDs contain exponentially fewer nodes than the number of entries in the explicit matrix. In this case linear algebraic process which is faster and useful than the ADD exponential matrix technique can be used often [1, 3, 20, 32].

To develop efficient method of ADD procedure for PTM to operate directly on the compressed form is a major challenging task. PTM based calculation is used in matrix multiplication algorithm. In a paper, Bahar et al. proposed a method of ADD procedure for matrix multiplication which involves taking a series of dot products using the APPLY operation. Nowadays, a lot of effort is given to calculate the soft error rate of a digital circuit [3]. The logic masking, electrical masking and latching-window masking are the various masking methods that are mostly proposed technique for primary model masking mechanism [3]. For this work, the PTM framework provides a matrix based mathematical formula. Using PTM we can easily obtain the approximate electrical masking. The main goal is to build technology independent logical modeling. To verify symbolic check and reachability analysis several probabilistic approaches have been applied so far. These methods are mainly concerned with verifying finite state machines whose transitions are nondeterministic. The circuit specific failures modes are generally not addressed by them but they are very useful to analyze Von Neumann's NAND-multiplexing fault tolerant architecture [1, 3, 20, 32]. We can conclude the followings:

1. PTM framework generally gives the symbolic representation in the form of matrix [3].
2. Rigorous computation and compression methods obtained the encoding of PTMs into ADDs [3].
3. Heuristics effects that enable the PTM-based algorithms implementation [3].

The application of PTMs to logic-level reliability analysis and gate hardening against soft errors is shown in Figure 10.

5.3. PTM Theory

Let A be the matrix of order m×n for any arbitrary logic function y and each entry of *A (i, j)* where '*i*' is the input ($i=i_0 i_1 ... i_{m-1}$) and '*j*' is the output ($j=j_0\, j_1 ... j_{n-1}$) for the successive entries into the matrix. If i_0, i_1, i_2, ……, i_{m-1} produce output j_0, j_1, j_2, ……, j_{n-1}, then we consider '*1*' for that, otherwise we consider '*0*'. These representations are considered as ITM. Using ITM, we can easily obtain PTM. PTM can also be formed using truth table represent-tation. Now for each input combination of *'i'* we will get a certain output combination of *'j'*. If we do not get so then it will generate an error. We can also derive ETM from the Kronecker[1] product of ITM and PTM. To calculate reliability we can use this ETM. We consider another row vector *'A'*; which denotes that the error is equally distributed among the inputs. Some new matrix operations such as eliminate variables, eliminate redundant variables and computations of fidelity are introduced by S. Krishnaswamy et al [1, 3, 7, 8, 32].

5.4. Example of Reliability Calculation

To calculate reliability, we have to divide the circuit into basic gate level, where each level contained one component. For each level Kronecker[1] product is to be calculated. Thus we obtained PTM for each level. PTM at these levels then multiplied by circuit inputs to outputs using normal matrix multiplication to generate final circuit PTM.

[1]Kronecker product denoted by ⊗ and it is block matrix multiplication with a S=X×Y matrix and T=A×B matrix. Then their Kronecker product C=S⊗T [1][8].

$$\underset{\text{PTM}}{\begin{array}{c} \\ 00 \\ 01 \\ 10 \\ 11 \end{array}\begin{array}{c} \begin{array}{cc} 0 & 1 \end{array} \\ \begin{pmatrix} 1-p & p \\ 1-p & p \\ 1-p & p \\ p & 1-p \end{pmatrix} \end{array}} \otimes \underset{\text{ITM}}{\begin{array}{c} \begin{array}{cc} 0 & 1 \end{array} \\ \begin{pmatrix} 1 & 0 \\ 1 & 0 \\ 1 & 0 \\ 0 & 1 \end{pmatrix} \end{array}} = \underset{\text{ETM}}{\begin{array}{c} \begin{array}{cc} 0 & 1 \end{array} \\ \begin{pmatrix} 1-p & 0 \\ 1-p & 0 \\ 1-p & 0 \\ 0 & 1-p \end{pmatrix} \end{array}}$$

$$A \begin{pmatrix} 0.25 \\ 0.25 \\ 0.25 \\ 0.25 \end{pmatrix} * \begin{pmatrix} 1-p & 0 \\ 1-p & 0 \\ 1-p & 0 \\ 0 & 1-p \end{pmatrix} = [\,0.75(1-p) \quad 0.25(1-p)\,]$$

Figure 11. Computation of PTM, ITM and ETM for two-input majority voter logic gates. Reliability for this circuit is obtained 25% [1, 3, 7, 8].

$$\underset{\text{PTM}}{\begin{array}{c} \\ 000 \\ 001 \\ 010 \\ 011 \\ 100 \\ 101 \\ 110 \\ 111 \end{array}\begin{array}{c} \begin{array}{cc} 0 & 1 \end{array} \\ \begin{pmatrix} 1-p & p \\ 1-p & p \\ 1-p & p \\ p & 1-p \\ 1-p & p \\ p & 1-p \\ p & 1-p \\ p & 1-p \end{pmatrix} \end{array}} \otimes \underset{\text{ITM}}{\begin{array}{c} \begin{array}{cc} 0 & 1 \end{array} \\ \begin{pmatrix} 1 & 0 \\ 1 & 0 \\ 1 & 0 \\ 0 & 1 \\ 1 & 0 \\ 0 & 1 \\ 0 & 1 \\ 0 & 1 \end{pmatrix} \end{array}} = \underset{\text{ETM}}{\begin{array}{c} \begin{array}{cc} 0 & 1 \end{array} \\ \begin{pmatrix} 1-p & 0 \\ 1-p & 0 \\ 1-p & 0 \\ 0 & 1-p \\ 1-p & 0 \\ 0 & 1-p \\ 0 & 1-p \\ 0 & 1-p \end{pmatrix} \end{array}}$$

$$A \begin{pmatrix} 0.125 \\ 0.125 \\ 0.125 \\ 0.125 \\ 0.125 \\ 0.125 \\ 0.125 \\ 0.125 \end{pmatrix} * \begin{pmatrix} 1-p & 0 \\ 1-p & 0 \\ 1-p & 0 \\ 0 & 1-p \\ 1-p & 0 \\ 0 & 1-p \\ 0 & 1-p \\ 0 & 1-p \end{pmatrix} = [\,0.5(1-p) \quad 0.5(1-p)\,]$$

Figure 12. Computation of PTM, ITM and ETM for three-input majority voter logic gates. Reliability for this circuit is obtained 50% [1, 3, 7, 8].

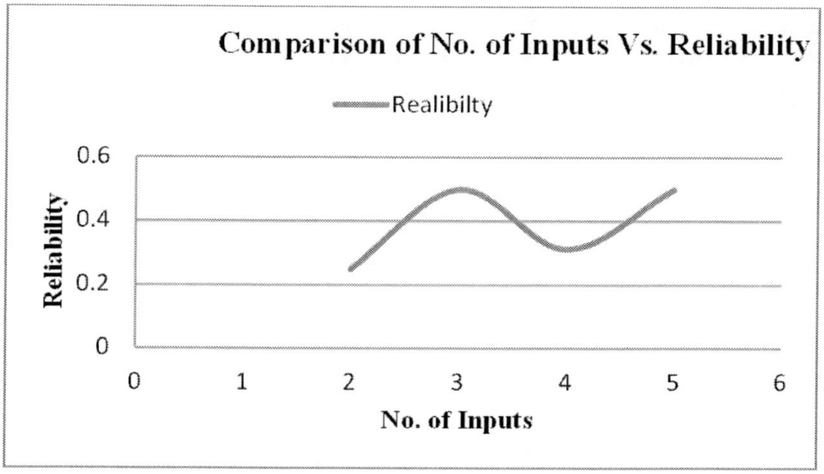

Figure 13. Comparison graph of No. of Inputs vs. Reliability for majority voters.

To generate the reliability for a specific circuit this PTM is again multiplied element-wise with its ITM to get ETM. To calculate reliability ETM is then multiplied with matrix A which provides a row vector. The elements of this row vector are the probability of error or error rate *'p'* which is equally distributed among the inputs [1, 3, 7, 8, 32].

In the following section the PTM, ITM and ETM are observed for different inputs MV and also their reliability are realized. Some results are given in Figure 11 and Figure 12.

For four and five input majority voter reliability are obtained 31.25% and 50% respectively. Hence, we can derive the comparison graph of majority voters for different inputs with respect to reliability. The graph is shown in Figure 13.

6. Circuit Analysis

We have dealt with several logic circuits taken randomly. In the Figure 14 we have find an interesting phenomenon. In this case we have made one comparison between the actual outputs of the circuit with its MV output. Surprisingly, we have found that the computable reliability of that particular circuit for its actual output with respect to its MV is lesser for the three inputs. In case of three inputs MV reliability is 50% where as for normal output or we can say for ITM of the particular circuit is 37.5%. The result is shown in the Figure 15 [19-24].

The graph shown in the Figure 13 is a comparison approach of reliability with the number of inputs of logic device. This graph shows that the reliability is increased for odd numbers of inputs comparing with even numbers of inputs. This variation in reliability analysis can be obtained for two, three and five input majority voter logic circuit device.

6.1. Flow Chart Representation of Principle of Reliability Analysis

In case of four inputs reliability of the actual circuit output ITM will decrease compare to MV output ITM. We can assume that in case of odd number of inputs reliability is increased.

But the number of '0' and '1' are equal in the output for odd number of inputs. The reliability remains 50%, which are shown in Figure 15 and Figure 12 for that circuit's output ITM as well as their majority voter ITM.

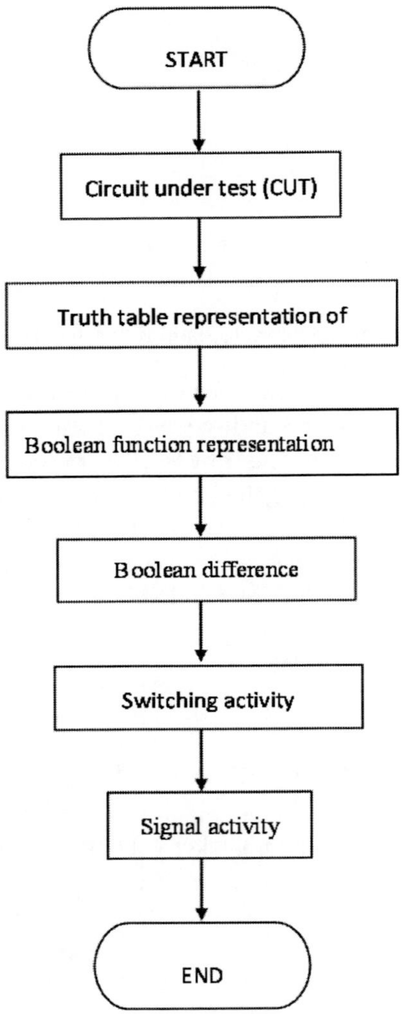

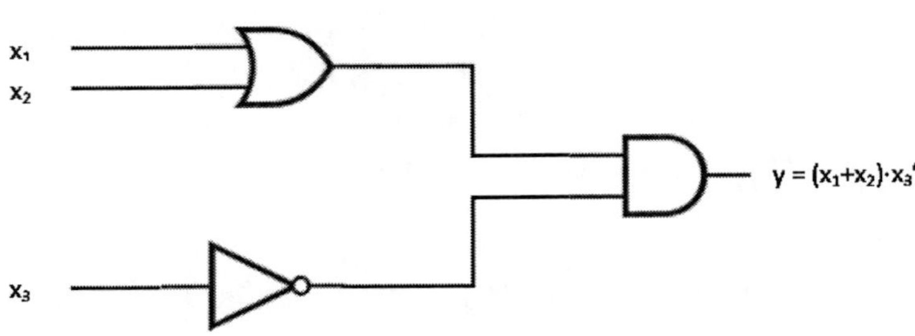

Figure 14. Logic circuit diagram for Boolean function $F = (x_1+x_2)\cdot x_3'$.

$$\begin{array}{c} \\ 000 \\ 001 \\ 010 \\ 011 \\ 100 \\ 101 \\ 110 \\ 111 \end{array}\begin{array}{c} \begin{matrix} 0 & \quad 1 \end{matrix} \\ \begin{pmatrix} 1-p & p \\ 1-p & p \\ p & 1-p \\ 1-p & p \\ p & 1-p \\ 1-p & p \\ p & 1-p \\ 1-p & p \end{pmatrix} \\ \text{PTM} \end{array} \otimes \begin{array}{c} \begin{matrix} 0 & 1 \end{matrix} \\ \begin{pmatrix} 1 & 0 \\ 1 & 0 \\ 0 & 1 \\ 1 & 0 \\ 0 & 1 \\ 1 & 0 \\ 0 & 1 \\ 1 & 0 \end{pmatrix} \\ \text{ITM} \end{array} = \begin{array}{c} \begin{matrix} 0 & \quad 1 \end{matrix} \\ \begin{pmatrix} 1-p & 0 \\ 1-p & 0 \\ 0 & 1-p \\ 1-p & 0 \\ 0 & 1-p \\ 1-p & 0 \\ 0 & 1-p \\ 1-p & 0 \end{pmatrix} \\ \text{ETM} \end{array}$$

$$A \begin{pmatrix} 0.125 \\ 0.125 \\ 0.125 \\ 0.125 \\ 0.125 \\ 0.125 \\ 0.125 \\ 0.125 \end{pmatrix} * \begin{pmatrix} 1-p & 0 \\ 1-p & 0 \\ 0 & 1-p \\ 1-p & 0 \\ 0 & 1-p \\ 1-p & 0 \\ 0 & 1-p \\ 1-p & 0 \end{pmatrix} = [0.625(1-p) \; 0.375(1-p)]$$

Figure 15. Computation of PTM, ITM and ETM for three-input majority voter logic gates. Reliability for this circuit is obtained 50%.

7. Probabilistic Techniques for Signal Activity

To analysis circuit PTM we can also use a probabilistic method to find out signal activity estimation. The average number of transitions per unit time at a circuit node is referred to as signal activity [11].

7.1. Switching Activity

A switching circuit is a realization of the mapping if n is the number of inputs of a device in nanoscale then switching probability for each cell is 2, i.e., 2^n is Cartesian product of 2 taken 'n' times. If a switching circuit $f(x_1, x_2, x_3, \ldots\ldots\ldots\ldots, x_n)$ can be realized by mapping $f/2^n \rightarrow 2$, where 2 is the set of two elements i.e. $2 = \{0, 1\}$ then f is called either a switching function or Boolean function. Switching circuits are composed of a number of sub-switching circuits [11, 44].

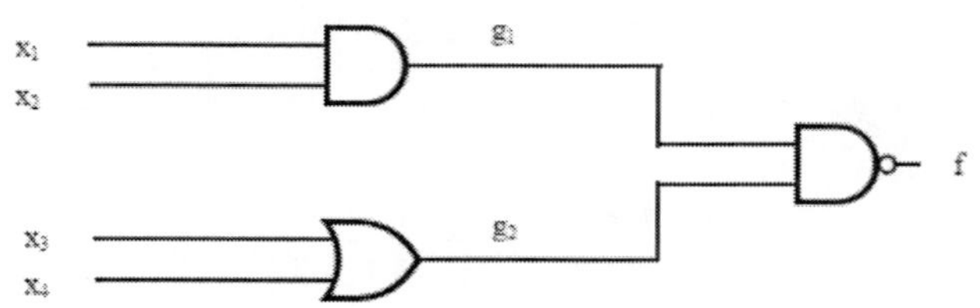

Figure 16. A switching circuit for illustration [44].

A switching circuit is illustrated in the Figure 16. It happens that the switching activity of each gate can change the output at f, where as it dependents on four inputs x_1, x_2, x_3, x_4. The equations related to Figure 16 can be written as –

$f=(g_1.g_2)'$

$g_1=x_1.x_2$

$g_2=x_3+x_4$

x_1, x_2, x_3, x_4 are called primary or input variables and g_1, g_2 are called internal or secondary variables for the circuit shown in Figure 16 whose output is f.

Suppose f is a switching function of 'n' binary variables, i.e.

$x = (x_1, x_2, x_3, \ldots\ldots\ldots, x_n)$

g_1 and g_2 are the internal variables of the circuit then the f depends on both x, g_1 and g_2. So, the functionality can be written as

$y = f(x, g(x))$ where $g = (g_1, g_2)$

This circuit is having fan-in and fan-out. g_1 and g_2 are said to be fan-in to f. To know the fault occurrence at output f, the switching activity of primary variables and secondary variables should be known. Though it is very difficult to find out the switching activity of each cell in nanoscale but the switching activity of the primary variables can be recognized. It is difficult to find the switching activity for secondary variables. For majority voters logic gates the comparison can be done between majority voter logic gate device and a logic gate device output, then the erroneous activity can easily be traced out. The majority voter logic gate output can be realized with the help of truth table and K-map representation. This Boolean expression can be represented with the help of switching activity of each cell, whereas the final output of that particular logic gate device can be represented with the help of switching activity.

These two switching activity can be compared with each other. If the polarization of each cell is known then the signal activity can be derived in nanoscale. Let x_i, where i=1, 2, ….., n represent the *'n'* inputs to a logic gate with outputs f_j, where j=1, 2, ….., m. To calculate switching activity we have derived *Boolean difference.* To understand how the error propagates through the circuit we should know the *Boolean difference* [11]. *Boolean difference* can be defined as below. Let f_j be a function that depends on input x_i. The *Boolean difference* of f_j with respect to x_i is defined as follows [11]:

$$\frac{\partial fj}{\partial xi} = f_j \mid_{xi=1} \oplus f_j \mid_{xi=0} \ldots\ldots\ldots\ldots \quad (1)$$

7.2. Signal Activity

Signal activity through basic gates can be obtained if we consider that '*p1*' and '*p2*' are the two erroneous inputs through the gate. We show the probability of erroneous output for Inverter, AND and OR gate in the Figure 17. If the two signals simultaneously switching then we do not get the reliable or deterministic output. Now, as *'p'* is the probability error rate at the output so we can say we get reliability at the output is 1 – p [11]. The signal activity of a digital logic circuit whose output is 'y' can be determined as follows:

$$A(y) = \sum_{i=1}^{n} P\left(\frac{\partial y}{\partial xi}\right) \cdot A(x_i)$$

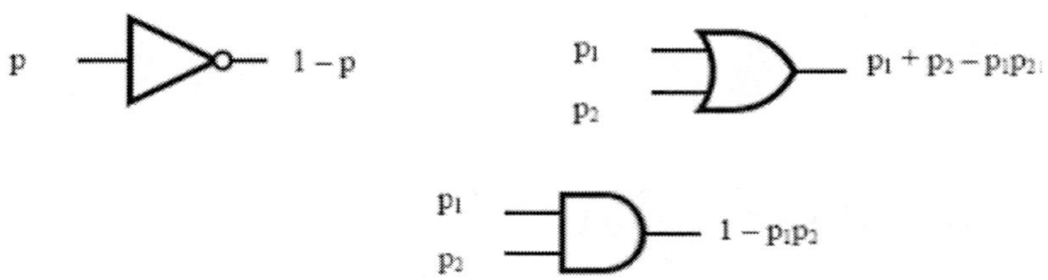

Figure 17. Propagation of probability error rate.

As the *Boolean difference* for a circuit and its MV representation are different so we can easily generate the function for MV from its truth table representation. Now we find the two different *Boolean differences* for two cases. As the *Boolean difference* differ from one to another so their signal activity are different. Hence, we can easily derive the signal activity if we know the switching activity. Simultaneous switching activity cannot be taken into account. If the two inputs switching simultaneously then PTM is in- deterministic, hence we cannot be able to find out circuit reliability. We can get Boolean expression $f_1 = x_1x_2 + x_2x_3 + x_1x_3$ for MV and $f_2 = (x_1+x_2)\cdot x_3'$ for ITM. Hence, we can find out the *Boolean difference* for these two functions. *Signal activity* can also be derived from these Boolean expressions. From these two signal activity we can easily compare the switching activity of particular nodes.

Example: in the Figure 18 we have taken a logic circuit whose output is $y = (x_1+x_2)\cdot x_3'$

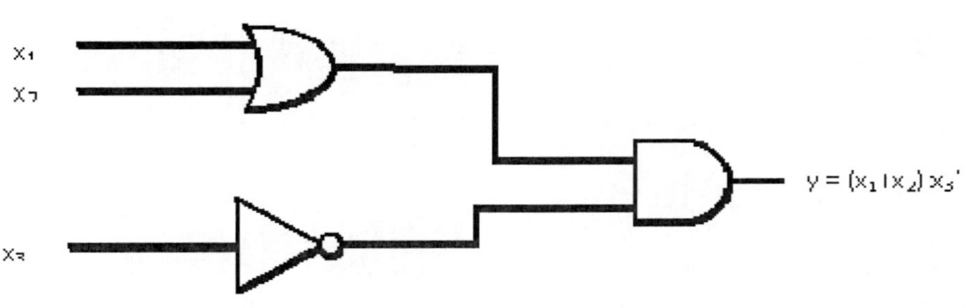

Figure 18. Logic circuit diagram for function $y = (x_1+x_2)\cdot x_3'$.

Boolean Expression for the circuit in the Figure 18 is:

$\psi = (\xi_1+\xi_2)\bullet\xi_3\neq$

$\frac{\partial y}{\partial x1} = (\xi_1+\xi_2)\bullet\xi_3\neq |x_1=1 \oplus (\xi_1+\xi_2)\bullet\xi_3\neq |x_1=0$

$= x_3\neq \oplus x_2x_3\neq = x_2\neq x_3\neq$.................................... (2)

$\frac{\partial y}{\partial x2} = (x_1+x_2)\cdot x_3\neq |x_2=1 \oplus (x_1+x_2)\cdot x_3\neq |x_2=0$

$= x_3\neq \oplus x_1x_3\neq = x_1\neq x_3\neq$.................................... (3)

$\frac{\partial y}{\partial x3} = (x_1+x_2)\cdot x_3\neq |x_3=1 \oplus (x_1+x_2)\cdot x_3\neq |x_3=0$
$= 0 \oplus x_1 + x_2$

$= x_1 + x_2$.................................... (4)

So, the signal activity for the CUT is
$A(y) = \sum_{i=1}^{n} P(\frac{\partial y}{\partial xi}) \cdot A(x_i)$ .... (5)

$= P(\frac{\partial y}{\partial x1})\cdot A(x_1) + P(\frac{\partial y}{\partial x2})\cdot A(x_2) + P(\frac{\partial y}{\partial x3})\cdot A(x_3)$

$= P(x_2\neq x_3\neq)\cdot A_1 + P(x_1\neq x_3\neq)\cdot A_2 + P(x_1+x_2)\cdot A_3$ [$A(x_i) = A_i$ where i=1,2,3,.........n]

$= [P(x_2\neq)P(x_3\neq)]\cdot A_1 + [P(x_1\neq)P(x_3\neq)]\cdot A_2 + [P(x_1) + P(x_2) – (P(x_1)P(x_2)]\cdot A_3$

$= [(1 – P(x_2)(1 – P(x_3)]\cdot A_1 + [(1 – P(x_1)(1 – P(x_3)]\cdot A_2 + [(P(x_1) + P(x_2) – (P(x_1)P(x_2)]\cdot A_3$

$= (1 – P_2)(1 – P_3)\cdot A_1 + (1 – P_1)(1 – P_3)\cdot A_2 + (P_1 + P_2 – P_1\cdot P_2)\cdot A_3$

[$P(x_i)=P_i$ where i=1,2,3,.........n]

$= (1 – P_2 – P_3 + P_2\cdot P_3)\cdot A_1 + (1 – P_1 – P_3 + P_1\cdot P_3)\cdot A_2 + (P_1 + P_2 – P_1\cdot P_2)\cdot A_3$

$= (1 – (P_2 + P_3 – P_2\cdot P_3))\cdot A_1 + (1 – (P_1 + P_3 – P_1\cdot P_3))\cdot A_2 + (P_1 + P_2 – P_1\cdot P_2)\cdot A_3$

$= (1 – (P_2 + P_3))\cdot A_1 + (1 – (P_1 + P_3))\cdot A_2 + (P_1 + P_2)\cdot A_3$.................. (6)

7.2.1. 3-Input Majority Voter Switching Circuit Analysis

The switching circuit analysis of the majority voter for the circuit shown in Figure 18 is derived as:

$y = x_1\cdot x_2 + x_2\cdot x_3 + x_1\cdot x_3$

$\frac{\partial y}{\partial x1} = (x_1{\cdot}x_2 + x_2{\cdot}x_3 + x_1{\cdot}x_3)\ |x_1{=}1 \oplus (x_1{\cdot}x_2 + x_2{\cdot}x_3 + x_1{\cdot}x_3)\ |x_1{=}0$

$= x_2 + x_2{\cdot}x_3 + x_3 \oplus x_2{\cdot}x_3$

$= x_2 + x_3 \oplus x_2{\cdot}x_3$

$= x_2{\cdot}x_3{\neq} + x_2{\neq}{\cdot}x_3$.................................... (7)

$\frac{\partial y}{\partial x2} = (x_1{\cdot}x_2 + x_2{\cdot}x_3 + x_1{\cdot}x_3)\ |x_2{=}1 \oplus (x_1{\cdot}x_2 + x_2{\cdot}x_3 + x_1{\cdot}x_3)\ |x_2{=}0$

$= x_1 + x_1{\cdot}x_3 + x_3 \oplus x_1{\cdot}x_3$

$= x_1 + x_3 \oplus x_1{\cdot}x_3$

$= x_1{\cdot}x_3{\neq} + x_1{\neq}{\cdot}x_3$........................... (8)

$\frac{\partial y}{\partial x3} = (x_1{\cdot}x_2 + x_2{\cdot}x_3 + x_1{\cdot}x_3)\ |x_1{=}1 \oplus (x_1{\cdot}x_2 + x_2{\cdot}x_3 + x_1{\cdot}x_3)\ |x_1{=}0$

$= x_1 + x_1{\cdot}x_2 + x_2 \oplus x_1{\cdot}x_2$

$= x_1 + x_2 \oplus x_1{\cdot}x_2$

$= x_1{\cdot}x_2{\neq} + x_1{\neq}{\cdot}x_2$.................................... (9)

So, the signal activity for the MV of the CUT is

$A(y) = \sum_{i=1}^{n} P(\frac{\partial y}{\partial xi})\ A(x_i)$

$= P(\frac{\partial y}{\partial x1}){\cdot}A(x_1) + P(\frac{\partial y}{\partial x2}){\cdot}A(x_2) + P(\frac{\partial y}{\partial x3}){\cdot}A(x_3)$

$= P(x_2{\cdot}x_3{\neq} + x_2{\neq}{\cdot}x_3){\cdot}A_1 + P(x_1{\cdot}x_3{\neq} + x_1{\neq}{\cdot}x_3){\cdot}A_2 + P(x_1{\cdot}x_2{\neq} + x_1{\neq}{\cdot}x_2){\cdot}A_3$

$= [P(x_2)(1 - P(x_3)) + (1 - P(x_2))P(x_3)]{\cdot}A_1 + [P(x_1)(1 - P(x_3)) + (1 - P(x_1))P(x_3)]{\cdot}A_2 + [P(x_1)(1 - P(x_2)) + (1 - P(x_1))P(x_2)]{\cdot}A_3$

$= [P_2(1 - P_3) + (1 - P_2)P_3]{\cdot}A_1 + [P_1(1 - P_3) + (1 - P_1)P_3]{\cdot}A_2 + [P_1(1 - P_2) + (1 - P_1)P_2]{\cdot}A_3$

$= (P_2 - P_2P_3 + P_3 - P_2P_3){\cdot}A_1 + (P_1 - P_1P_3 + P_3 - P_1P_3){\cdot}A_2 + (P_1 - P_1P_2 + P_2 - P_1P_2){\cdot}A_3$

$= [(P_2 + P_3 - P_2{\cdot}P_3) - P_2{\cdot}P_3]{\cdot}A_1 + [(P_1 + P_3 - P_1{\cdot}P_3) - P_1{\cdot}P_3]{\cdot}A_2 + [(P_1 + P_2 - P_1{\cdot}P_2) - P_1{\cdot}P_2]{\cdot}A_3$

$= (P_2 + P_3 - P_2{\cdot}P_3){\cdot}A_1 + (P_1 + P_3 - P_1{\cdot}P_3){\cdot}A_2 + (P_1 + P_2 - P_1{\cdot}P_2){\cdot}A_3$

$= (P_2 + P_3){\cdot}A_1 + (P_1 + P_3){\cdot}A_2 + (P_1 + P_2){\cdot}A_3$........................ (10)

The switching activity and signal activity are described above for a circuit and its majority voter circuit. In the final equation (10) probability P_i (where i=1 to n) is the switching probability of each cell, if this probability is enhanced then the circuit will be speed up, propagation delay is reduced but the power dissipation will be more. Hence to achieve high speed device we have to sacrifice more power dissipation. The switching activity A_i (where i=1 to n) expression shows the change of polarization state of the CUT in logic level. But also it is difficult to know the polarization state of each node of a QCA cell at circuit implementation level. Hence we can use the probabilistic method to compute the signal and switching activity of QCA cell at logic level.

7.3. Circuit Analysis for MQCA

Depending upon the various magnetism property magnets are categorized into various groups [6]. Among these groups many materials exist, that shows no collective magnetic interactions and are not magnetically arranged [6]. Some of the materials among these groups reveal long-range magnetic arrangement below a certain critical temperature. Iron, Chromium, Nickel etc. those are declared as Ferromagnetic materials. Very strong atomic moment shows strong interactive effect within these Ferromagnetic materials. Electronic exchange forces and its results provide this kind of interaction which tends to parallel or anti-parallel atomic alignment within the materials. The most important class of magnetic materials is the ferromagnets: iron, nickel, cobalt and manganese, or their alloys [14-18, 34, 37].

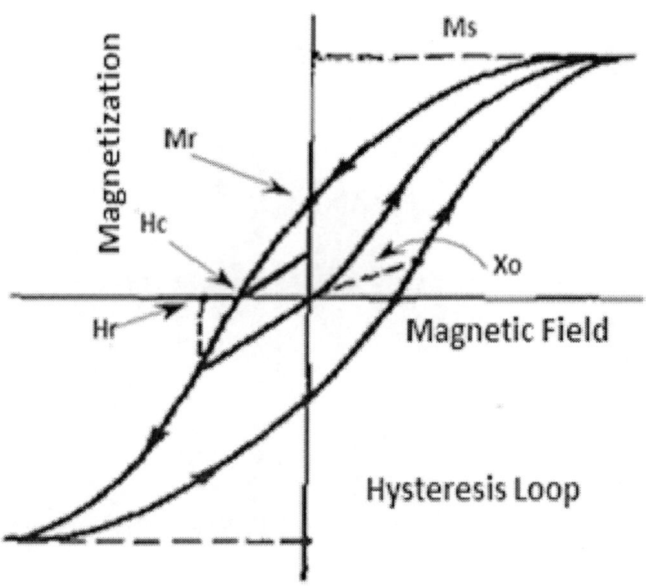

Figure 19. "Hysteresis Loop" for Ferromagnetic materials.

As referred to the Table 1, we can derive the Boolean expression for central nano magnet and the output magnet; so that from these expressions can be derived based on the switching activity hence signal activity.

7.4. Circuit Analysis for Molecular QCA

Nanotechnologies together with molecular quantum dot cellular automata are very vulnerable to high error probability. The different simultaneous testable latches such as RS-latch, D-latch, JK-latch and T-latch [38] are designed which are depends on reversible conservative logic for molecular QCA. Conservative reversible circuits are special kind of reversible circuits where there is equal number of one's in the output as well as in the inputs. In conservative logic the input parity vector is equal to the output parity vector.

In molecular quantum dot cellular automata due to the single missing or additional cell defect we use conservative reversible Fredkin gate [45] (Fredkin gate is the three-bit gate that swaps the last two bits if the first bit is 1) to analyze it. It has been seen that if there is any fault in the molecular quantum dot cellular automata implementation of Fredkin gate there will be a parity mismatch between the inputs and the outputs, otherwise input parity is equal to the corresponding output parity [37, 39, 45].

CONCLUSION

The switching activity of each node of a circuit can be determined (i.e., the polarization state of each EQCA cell) and the signal activity of whole logic circuit in nano-scale have been achieved.

Hence, power estimation of each logic circuitry can be determined with the help of switching activity. In the field of reliability the power estimation is very important aspect. We need a reliable circuit with very low power consumption [31].

FUTURE ASPECT OF RELIABILITY ESTIMATION

The QCA device is used to make gates, wires, and memories etc. To reduce the designing effort, the circuit designer can concentrate on the logical and structural design, while the physical nature of the nanoscale materials is complicated. During the early phase of digital design, the main aspect is to detect faults and defects in circuit. The relationship between system reliability and component failure rate is to form the design guideline, when we want to design a logic gate system unit on quantum-dot cellular automata in nanotechnology. The analysis is based on a decomposition of probabilistic transfer matrices which is a versatile framework for calculating the conditional probability of system failure [33, 40-44]. The molecular QCA and MQCA switching circuit can be realized using these probabilistic techniques. Hence the power dissipation and determination of noise of each node of any of the three types of QCA cell can be achieved from their switching activity.

ACKNOWLEDGMENTS

Authors are grateful to Department of Science and Technology (DST), India for sanctioning the project under "SERC Fast Track Proposal of Young Scientist" Scheme

SERC/ET-0213/2011 and UGC major project entitled "Study of quantum dot cellular automata for designing circuits and implementing them for high speed and low power fault tolerant computing" under which this chapter has been completed.

REFERENCES

[1] Timothy J. Dysart and Peter M. Kogge, "Analyzing the Inherent Reliability of Moderately Sized Magnetic and Electrostatic QCA Circuits Via Probabilistic Transfer Matrices", *IEEE Transaction on VLSI Systems*, Vol. 17, No. 4, pp. 507–516, April 2009.

[2] James R. Health, Philip J Kuekes, Gregory S Snider and R. Stanley Williams Hu and Michael Niemier, "A Defect-Tolerant Computer Architecture: Opportunities for Nanotechnology", *Science,* Vol. 280, No. 5370, pp. 1716–1721, 1998.

[3] Smita Krishnaswami, George F. Viamontes, Igor L. Markov and John P. Hayes, "Probabilistic Transfer Matrices in Symbolic Reliability Analysis of Logic Circuits", *ACM Transactions in Design Automation of Electronic Systems*, Vol. 13, No. 1, Article 8, January 2008.

[4] I. Amlani, A. O. Orlov, G. Toth, G. H. Bernstein, C. S. Lent, G. L. Snider, "Digital Logic Gate using Quantum-Dot Cellular Automata", *Science,* Vol. 284, pp. 289-291, 1999.

[5] G. L. Snider, A. O. Orlov, I. Amlani, G. H. Bernstein, C. S. Lent, J. L. Merz and W. Porod, "Quantum-Dot Cellular Automata", *Elsevier Science in Microelectronic Engineering*, Vol. 47, pp. 261–263, 1999.

[6] A. Imre, G. Csaba, L. Ji, A. Orlov, G. H. Bernstein and W. Porod, "Majority Logic Gate for Magnetic Quantum-Dot Cellular Automata", *Science*, Vol. 311, pp. 205–208, 2006.

[7] Timothy J. Dysart, "It's all about the Signal Routing: Understanding the Reliability of QCA Circuits and Systems", Dissertation for PhD, July 2009.

[8] Massimo Macucci, "Quantum Cellular Automata: Theory, Experimentation and Prospects", Imperial College Press. Copyright 2006, ISBN 1-86094-632-1.

[9] Ismo Hanninen, "Computer Arithmetic on Quantum-Dot Cellular Automata Nanotechnology", *Thesis for Doctor of Technology*, ISBN 978-952-15-2269-7, 2009.

[10] Gregory L. Snider, Alexei O. Orlov, Islamshah Amlani, Gary H. Bernstein, Craig S. Lent, James L. Merz and Wolfgang Porod, "Quantum-Dot Cellular Automata: Line and Majority Logic Gate", *Jpn. J. Appl. Phys.* Vol. 38(1999), pp. 7227-7229, Part I, No, 12B, December 1999.

[11] Kaushik Roy and Sharat Prasad, "Low-Power CMOS VLSI Circuit Design", John Wiley and Sons, Inc. May 1999, copyright ©2000, ISBN 0-471-11488-X, printed in United States of America.

[12] G. Bazan, A. O. Orlov, G. L. Snider and G. H. Bernstein, "Charge detector realization for AlGaAs/GaAs quantum-dot cellular automata", J. *Vac. Sci. Technol. B. American Vacuum Society*, Vol. 14, No. 6, pp. 4046-4050, 1996.

[13] R. P. Cowburn "Digital Nanomagnetic Logic", *Device Research Conference*, pp. 111-114, June 2003.

[14] Alexandra Imre, Gyorgy Csaba, Ling Zhou, Alexei Orlov, G.H. Bernstein, W. Porod, and Metlushko "Shape Engineering of Dipole-coupled Nanomagnets for Magnetic Logic Devices", *Device Research Conference*, Vol 1, pp. 73-74, June 2004.

[15] Imre, A., Zhou, L., Orlov, A., Csaba, G., Bernstein, G.H., Porod, W., Metlushko, V., "Application of mesoscopic magnetic rings for logic devices", 4th *IEEE Conference on Nanotechnology*, pp. 137-139, August 2004.

[16] V. Ng and L. K. Verma, "Magnetic Properties of Dumbbell Shaped Nanostructures". *IEEE Magnetic Conference*, pp. 942, May 2006.

[17] A. Imre, L. Ji, G. Csaba, A. Orlov, G. H. Bernstein, W. Porod "*Magnetic Logic based on Field-Coupled Nanomagnets*", IEEE, Vol.7, pp. 600-601, 2007.

[18] G. H. Bernstein, A. Imre, V. Metlushko, A. Orlov, L. Zhou, L. Ji, G. Csaba, W. Porod "Magnetic QCA Systems", *Microelectronics Journal*,Elsevier, Vol 36, No. 5758, pp. 619-624, January 2005.

[19] Suresh Rai, "*Majority Gate Based design for Combinational Quantum Cellular Automata Circuits*". 40th Southeastern Symposium on System Theory, IEEE, pp.222-224, 2008.

[20] Kunal Das and Debashis De, "*A Novel Approach of And-Or-Invert (AOI) Gate design for QCA*", Proc. Int. Conference on Computers and Devices for Communication, 2009.

[21] Kunal Das and Debashis De, "A study on Diverse Nanostructure for implementing Logic Gate design for QCA", *Int. Jounal of Nanoscience*, Vol. 10, No. 1, pp. 1-7, March 2011.

[22] Yuhui Lu, Mo Liu, and Craig Lent, "Molecular quantum-dot cellular automata: From molecular structure to circuit analysis", *Journal of Applied Physics*, Vol. 102, No.034311, 2007.

[23] Das, Kunal and De, Debashis, "Characterisation, applicability and defect analysis for tiles nanostructure of quantum dot cellular automata", *Molecular Simulation*, Vol. 37, No. 3, pp. 210 - 225, 2011.

[24] Kunal Das and Debashis De, "Novel Approach to design A Testable Conservative Logic Gate for QCA Implementation.", *Proc. IEEE 2nd International Advance Computing Conference*, pp. 82-87, 2010.

[25] Fijany and B. N. Toomarian,"Quantum Dots Cellular Automata: Fault Tolerant Universal Logic Gates" *J. Nanoparticle Research*, Vol. 3, pp. 27–37, 2001.

[26] M. Crocker, X. S. Hu, and M. T. Niemier. "Fault Models and Yield Analysis For QCA-Based Plas". In Proceedings of the 17th International Conference on Field Programmable Logic and Applications, pp. 435–440, 2007.

[27] M. Crocker, X. S. Hu, M. T. Niemier, Minjun Yan and Gary Bernstein, "PLAs in Quantum-Dot Cellular Automata", *IEEE Trans. Nanotech- no* l., 7(3), pp. 376-386, May 2008.

[28] J. H. Patel and L. Y. Fung, "Concurrent Error Detection in ALUs by Recomputing with Shifted Operands," *IEEE Trans. on Computers*, Vol. C-31, pp. 589-595, July 1982.

[29] T. Wei, K. Wu, R. Karri, and A. Orailoglu. "Fault tolerant quantum cellular array QCA design using triple modular redundacny with shifted operands". In *Asia and South Pacific Design Automation Conference*, January, 2005.

[30] S. Krishnaswamy, G. F. Viamontes, I. L. Markov, and J. P. Hayes. "Accurate reliability evaluation and enhancement via probabilistic transfer matrices", In Proceedings of the Design, Automation, and Test in *Europe Conference and Exhibition*, 2005.

[31] John Timler and Craig S. Lent, “Power gain and dissipation in quantum-dot cellular automata”, *Journal of Applied Physics*, Vol. 91, pp. 823-831, 2001.

[32] J. Von Neumann. Automata Studies, chapter “*Probabilistic logics and the synthesis of reliable organisms from unreliable components*”, pp. 43–98, Princeton University Press, 1956.

[33] J. Han, E. Taylor, J. Gao, and J. Fortes. “*Reliability Modeling of Nano electronic Circuits.*” Proc. of IEEE Nano, Nagoya, Japan, 2005.

[34] G. Csaba, P. Lugli, W. Porod ”Power Dissipation in Nanomagnetic Logic Devices“, 4th *IEEE Conference on Nanotechnology*, Vol. 4, pp. 346-348, 2004.

[35] Larrabee, T, “Test pattern generation using Boolean satisfiability”, *IEEE Transactions on Computer-Aided Design of Integrated Circuits and Systems*, Vol. 11, pp. 4-15, January 1992.

[36] Michael Niemier, Michael Crocker, and X. Sharon Hu, “Fabrication Variations and Defect Tolerance for Nanomagnet-based QCA”, *IEEE International Symposium on Defect and Fault Tolerance of VLSI Systems*, Vol. 8, pp. 534-542, 2008.

[37] Craig S. Lent, Beth Isaksen, Marya Lieberman, "Molecular Quantum-Dot Cellular Automata", *Journal of the American Chemical Society,* Vol. 125, pp. 1056-1063, 2003.

[38] Lei Wang, Faquir Jain and Fabrizio Lombardi, “Information-Theoretic Modeling and Analysis of Stochastic Behaviors in Quantum-Dot Cellular Automata”, pp.3-22, book edited by Alejandro Salcido, ISBN 978-953-307-172-5, InTech Publisher, April 2011.

[39] Yuhui Lu and Craig S Lent, “A metric for charcterizing bi-stability of moleqular quntum-dot cellular automata”, IOP Publishing, *Nanotechnology,* Vol. 19, No.155703, pp.1-11, 2008.

[40] L. Lewyn, Trond Ytterdal, Carsten Wulff and Kenneth Martin, “Analog Circuit Design in Nanoscale CMOS Technologies”, *Proceedings of the IEEE*, Vol. 97, No. 10, October 2009.

[41] Jianwei Dai, Lei Wang, Faqnir Jain, “A quantitative approach for analysis of defect tollerance in QCA”, 8th *IEEE Conference on Nanotechnology*, pp. 903-906, August 2008.

[42] Mo Liu, “Robustness and Power Dissipation in Quantum-Dot Cellular Automata”, A Dissertation for Ph.D, Notre Dame, Indiana, March 2006,

[43] Erin Taylor, Jie Han, Jose Fortis, “Towards Accurate and Efficient Reliability Modeling of Nano-electronic Circuits”, 6th *IEEE Conference on Nanotechnology*, pp. 395-398, June 2006.

[44] I. S. Reed, “*Boolean Difference Calculus and Fault Finding*”, JPL Technical Report 32-1526, Vol. II, pp. 65-71.

[45] X. Ma J. Huang C. Metra, F. Lombardi “Reversible Gates and Testability of One Dimensional Arrays of Molecular QCA”, *J. Electron Test*, Vol. 24, pp. 297-311, January 2008.

INDEX

A

B

C

D

E

F

G

H

N

O

P

Q

R

S

T

U

V

W

Y

Z